AEROSOL SCIENCE THEORY AND PRACTICE

With Special Applications to the Nuclear Industry

Pergamon Titles of Related Interest

ANNALS OF THE ICRP, Volume 14/2
Protection of the Public in the Event of Major Radiation Accidents: Principles of Planning

ANNALS OF THE ICRP, Volume 15/1
Principles of Monitoring for the Radiation Protection of the Population

ANNALS OF THE ICRP, Volume 15/2
Protection of the Patient in Radiation Therapy

BEI
Modern Power Station Practice, 12-vol set, 3rd edition

CHICKEN
Risk Assessment for Hazardous Installations

FULLWOOD & HALL
Probabilistic Risk Assessment in the Nuclear Power Industry

MOULD
Chernobyl: The Real Story

MURRAY
Nuclear Energy, 3rd Edition

URSU
Physics and Technology of Nuclear Materials

Pergamon Related Journals (*free specimen copy gladly sent on request*)

Accident Analysis and Prevention

Annals of Nuclear Energy

Annals of the ICRP

Energy Conversion and Management

Engineering Fracture Mechanics

Fatigue and Fracture of Engineering Materials and Structures

Health Physics: The Radiation Protection Journal

International Journal of Radiation Oncology · Biology · Physics

Plasma Physics and Controlled Fusion

Progress in Nuclear Energy

AEROSOL SCIENCE THEORY AND PRACTICE

With Special Applications to the Nuclear Industry

M. M. R. WILLIAMS
Emeritus Professor of Nuclear Engineering
University of London
and Principal Scientist
Electrowatt Engineering Services (UK) Ltd
Horsham, Sussex, UK

and

SUDARSHAN K. LOYALKA
Curators' Professor
Professor of Nuclear, Chemical, and Mechanical and Aerospace Engineering
and Director
Particulate Systems Research Center
University of Missouri-Columbia
Columbia, Missouri, USA

PERGAMON PRESS
OXFORD · NEW YORK · SEOUL · TOKYO

UK	Pergamon Press plc, Headington Hill Hall, Oxford OX3 0BW, England
USA	Pergamon Press Inc., 395 Saw Mill River Road, Elmsford, NY 10523, U.S.A
KOREA	Pergamon Press Korea, KPO Box 315, Seoul 110-603, Korea
JAPAN	Pergamon Press, 8th Floor, Matsuoka Central Building, 1-7-1 Nishi-Shinjuku, Shinjuku-ku, Tokyo 160, Japan

Copyright © 1991 M. M. R. Williams & Sudarshan K. Loyalka

All Rights Reserved. No part of this publication may be reproduced, stored in a retrieval system or transmitted in any form or by any means: electronic, electrostatic, magnetic tape, mechanical, photocopying, recording or otherwise, without permission in writing from the publisher.

First edition 1991

Library of Congress Cataloging in Publication Data
Williams, M. M. R. (Michael Maurice Rudolph)
Aerosol science: theory and practice: with special applications to the nuclear industry/M. M. R. Williams and Sudarshan K. Loyalka. –1st ed.
p.   cm.
Includes bibliographical references and indexes.
I. Nuclear reactors—Accidents. 2. Aerosols, Radioactive.
3. Radioactive pollution of the atmosphere. I. Loyalka, S. K.
II. Title.
TK9152.W55   1991   628.5'35—dc20   91-20345

British Library Cataloguing in Publication Data
Williams, M. M. R. (Michael Maurice Rudolph)
Aerosol science; theory and practice : with special applications to the nuclear industry.
I. Title II. Loyalka, Sudarshan K.
541.34515

ISBN 0 08 037209 0

Printed in Great Britain by B.P.C.C. Wheatons Ltd, Exeter

Dedication

TO

*Queen Mary College, University of London,
where I spent many happy years with
valued colleagues and friends*
　　　　　　　　　　M M R W

*My parents,
for the many joyful years that I spent in Pilani*
　　　　　　　　　S K L

Preface

The subject of aerosols goes back many years and enters into many aspects of science and technology. Optics, heat transfer, biology, meteorology, and pollution are just a few areas where the behavior of small particles suspended in a gas is of vital importance. More recently, with increasing concern about the consequences of accidents in nuclear reactors and the effects of global nuclear war (i.e., nuclear winter), a great deal of research effort has been directed towards the dispersal of radioactive aerosols both in closed containers and in the atmosphere.

The purpose of this book is twofold. First, it is intended to give a thorough treatment of the fundamentals of aerosol behavior with rigorous proofs and detailed derivations of the basic equations and removal mechanisms. Second, it is intended to provide practical examples with special attention to radioactive particles and their distribution in size following a radioactive release arising from an accident with a nuclear system.

The writing of this book has been a labor of love for both authors. Our interests in the subject go back more than 20 years and, although modern methods (especially computing) have made possible many advances, we still admire the fundamental work carried out in the 1920's and 1930's when many of the basic properties of aerosol behavior were discovered and explained. Considering the prestigious background associated with this subject, we do feel somewhat intimidated in offering our own views. Nevertheless, we feel that our approach is sufficiently general to be of interest to a wide range of readers. At the same time, we have attempted to include enough of the fundamentals to attract the attention of the specialist. There are many aspects of aerosol science described here including morphology, transport, radioactivity, coagulation, condensation and evaporation, deposition and resuspension, and practical applications. We have attempted to give them all detailed attention. One omission that is deliberate concerns the optics of aerosols. This is a vast subject and, to do it justice, would require a separate book. We have also deliberately omitted discussions of experimental measurements and available data on single and two particle equilibria and rate processes. This area is currently witnessing rapid growth and data useful for the verification of several aspects of theory are only now becoming available.

One difficulty was notation. The literature abounds with different nomenclature for the same quantity. We have tried to be reasonably consistent but where this has proven too cumbersome we have given adequate definitions within the text.

The audience to which the book is directed is wide. We see the contents being of value to graduate students, university teachers, and to researchers in aerosol science and related topics. Moreover, although the contents are aimed at the nuclear industry in particular, much of the book contains material that will be of general interest to medical physicists, workers in indoor air quality, environmentalists, and applied mathematicians.

To achieve the above objectives, we start with a a brief introduction to the applications of aerosol science and the characteristics of aerosols in Chapter 1. In Chapter 2, we devote considerable attention to single and two particle motion with respect to both translation and rotation. Chapter 3 contains extensive discussion of the aerosol general dynamical equation and the dependences of aerosol distributions on size, shape, space, composition, radioactivity, and charge are fully detailed. Important particle rate processes of coagulation, condensation, and deposition/resuspension are discussed in the chapters 4, 6 and 7, respectively. The literature in each of these areas is substantial and we have discussed both the classical and the more recent works as well as the areas where additional research efforts are needed. In Chapter 5, we provide a thorough treatment of the analytical and numerical methods used in solving the various forms of the aerosol dynamical equation. We realize fully well that the ongoing revolutions in computing hardware, software, and algorithms will make computational aerosol mechanics a much more important field of future research. We discuss the importance and applications of aerosol science to nuclear technology and, in particular, the nuclear source term in Chapter 8. Our focus in this chapter is on discussions of nuclear accidents that can potentially release large amounts of radioactivity to environment. We also discuss the progress that has been made in understanding the natural and engineered aerosol processes that limit or affect such releases.

There exist several excellent texts on aerosol science and technology. We have learned and, where appropriate, borrowed from them. Most of the present text, however, covers material that has not been previously discussed with this rigor or in this depth. We are hopeful that this book will be useful both as a text and as a reference.

Acknowledgments

One of us (MMRW) developed some of the detailed work involved in writing this book while a member of the Faculty of the Nuclear Engineering Department, University of Michigan, Ann Arbor. Thanks are due to several of those Faculty for advice and guidance. By far, the greater part of the work was carried out at Queen Mary College, London University, where MMRW was Professor of Nuclear Engineering. In that respect, thanks are extended to Dr. Stuart Simons for much good advice and scientific insight.

This book was prepared on a modern word processor. We received extensive help from several of our family members, students, friends, and associates. Ann Williams, Pranav and Prashant Loyalka, and Tad Pennington helped with the initial drafts. Cathy Erhardt, Gayla Neumeyer, Abdulraheem Kinsara, and Robert Buckley helped with the drawings. Dr. Richard C. Warder, Jr. carefully reviewed the entire document, corrected several of our oversights, and offered many useful suggestions.

The subject matter of this book draws upon the work of many distinguished scientists engineers, and mathematicians. We have attempted to cite the original work wherever possible. Shakir Hamoodi not only reviewed the text and helped with discussions and drafts on condensation and deposition but also undertook the difficult and very time consuming task of assisting with the reference materials, the citations of the references and the subject index. We are most grateful for this assistance.

Dr. Robert V. Tompson, Jr. offered, nay insisted, on the final editing as well as the preparation of the book for its camera ready form. Little had he realized that this would be an all consuming task requiring not only days, but months of dedicated work including several revisions of the entire text. We are most grateful to him for his patience, diligence, and excruciating but cheerful attention to detail regarding style, uniformity, figures, and tables. The book would have looked different without his efforts.

Finally, many of the tables and figures in the text have been adapted or reproduced from existing literature. We gratefully acknowledge the permissions received from the following for the inclusion of their copyrighted material in this work.

Academic Press.
  Figures 5.1 & 5.12
American Geophysical Union.
  Figures 2.32, 2.33, 2.38 & 2.39
American Institute of Chemical Engineers.
  Figures 2.15, 2.16, 2.17 & 2.18
American Institute of Physics.
  Figure 2.3
  Tables 2.1, 2.4, 6.1, 6.2 & 6.3
American Meteorological Society.
  Figures 4.19 & 4.20
  Tables 2.8, 2.9, 2.10 & 4.6
American Nuclear Society.
  Figures 4.21, 4.22, 4.23, 4.24, 4.25, 4.26 & 4.27
  Table 8.4
Cambridge University Press.
  Figures 2.20, 2.31 & 4.13
  Table 2.3

Electric Power Research Institute.
  Figures 8.17, 8.18, 8.21, 8.25 & 8.28
  Tables 8.11, 8.12 & 8.13
Elsevier Science Publishing Company.
  Figures 2.21, 2.22, 2.23, 4.8, 4.9, 5.13 & 5.14
  Table 5.3
Mandelbrot, B.B.
  Figures 2.27 & 2.28
Netherlands Energy Research Foundation ECN.
  Figures 2.24, 2.25 & 2.26
Royal Meteorological Society.
  Figure 4.17
Royal Society.
  Figures 3.5 & 3.6
SRI International.
  Figure 1.1
Steinkopff Verlag.
  Figures 3.7, 3.8 & 3.9

Contents

Dedication	v
Preface	vii
Acknowledgments	ix
Contents	xi
List of Tables	xv
List of Figures	xvii

CHAPTER 1 Aerosol Characterization 1
   1.1   The nature of aerosols 1
   1.2   Aerosol size and shape 2
   1.3   Distribution functions 5
          1.3.1 The Junge distribution 7
          1.3.2 The log-normal distribution 7
          1.3.3 The gamma distribution 9
   1.4   Chemical composition 9
   1.5   The nuclear aerosol 9
   References: Chapter 1 10

CHAPTER 2 The Motion of Particles in Gases 12
   2.1   Introduction 12
   2.2   Linear transport theory 13
   2.3   The drag on a body 19
   2.4   The torque on a body 19
   2.5   Slip flow 20
   2.6   The drag on an isolated sphere 21
   2.7   Stokes flow in the continuum regime 24
   2.8   The drag 28
   2.9   Drag on a sphere 29
   2.10 Drag forces on nonspherical bodies 32
   2.11 The drag force on a spheroid 34
   2.12 The drag on a closed torus 41
   2.13 The drag on two spheres in contact 42
   2.14 Upper and lower bounds on drag 44
   2.15 A general technique for axially symmetric bodies 45
   2.16 Other methods of drag calculation 49
   2.17 Irregular shapes 52
   2.18 Particles with a fractal structure 61
   2.19 Rotational motion and torque 66
   2.20 The rotating sphere 67
   2.21 Nonspherical rotating bodies 69
   2.22 Torque on an ellipsoid in free molecular flow 71
   2.23 Prolate spheroids 72
   2.24 Oblate spheroids 73
   2.25 The fluid dynamics of two interacting spheres 73
   2.26 The limiting case of touching spheres 81
   2.27 Two rotating spheres 89
   2.28 The motion of spheres moving perpendicular to their line of centers 91

2.29	Arbitrary orientation of the spheres	96
2.30	The superposition method	101
References: Chapter 2		104

CHAPTER 3 The Dynamic Equation for the Aerosol Distribution ... 109

3.1	Introduction	109
3.2	The distribution function	110
3.3	Space dependent balance equation	113
3.4	Representation in terms of radius	116
3.5	Multicompartment aerosol equations	117
3.6	The gas dynamics	121
3.7	Multispecies aerosols	125
3.8	Two species equations	128
3.9	Radioactive aerosols	130
3.10	The effect of electrical charge on the dynamic equation	132
3.11	The effect of charge on coagulation	136
3.12	The dynamic equation for a charged aerosol	142
3.13	A dynamic equation for nonspherical particles	150
References: Chapter 3		151

CHAPTER 4 Coagulation Kernels ... 154

4.1	Introduction	154
4.2	Brownian coagulation	155
	4.2.1 Diffusion theory	155
	4.2.2 Free molecular flow	159
	4.2.3 The slip flow regime	160
	4.2.4 Fuchs' method	162
4.3	Gravitational coagulation	164
4.4	Laminar shear	170
4.5	Turbulent coagulation	172
4.6	Turbulent diffusion	172
4.7	Generalized theory of aerosol coagulation	179
4.8	The balance equation	184
4.9	Diffusive and convective motion	188
4.10	Brownian and gravitational coagulation	190
4.11	Generalized turbulent coagulation	194
4.12	Turbulent coagulation by the Saffman and Turner method	198
4.13	Electrostatic forces	201
4.14	Brownian and shear coagulation	203
4.15	Brownian and gravitational coagulation	209
4.16	Acoustic coagulation	212
4.17	Trajectory analysis	217
4.18	The coagulation of nonspherical particles	231
	4.18.1 Coagulation of large spheroids and small spheres	231
	4.18.2 Coagulation of large spheres and small spheroids	233
	4.18.3 Coagulation of large and small spheroids	235
References: Chapter 4		236

CHAPTER 5 Methods of Solving the Dynamic Equation ... 241

5.1	Introduction	241
5.2	Exact solutions	241
5.3	Growth and deposition	249
5.4	Space and time dependence	256

5.5	Multicomponent aerosols	258
5.6	Radioactive aerosols	260
5.7	Self-preserving solutions	264
5.8	Brownian self-similarity	269
5.9	The moments method	271
5.10	Coagulation with deposition, condensation, and a source	273
5.11	Modified gamma distribution	275
5.12	A comparative study of the moments method	278
5.13	Shape dependence	283
5.14	Numerical methods	287
5.15	The J-space transform and cubic spline method	288
5.16	Finite element method	289
5.17	The group (sectional) method	290
5.18	Numerical solutions for condensation/evaporation without coagulation	292
5.19	The Smolarkiewicz method	294
5.20	A moving grid characteristic based finite element method	296
5.21	Numerical treatment of multicomponent aerosols	297
5.22	Spatial inhomogeneities	299
5.23	Computer programs and benchmarking	299
	References: Chapter 5	302

CHAPTER 6 Condensation and Evaporation 305

6.1	Introduction	305
6.2	Basic equations	306
6.3	Transition regime heat and mass transfer	311
6.4	Nonspherical particles and shape factors	320
6.5	Nonisothermal condensation and reaction rates	322
	References: Chapter 6	323

CHAPTER 7 Particle Deposition and Resuspension 326

7.1	Introduction	326
7.2	Gravitational settling in a well mixed volume	331
7.3	Diffusional deposition in a stagnant gas	331
7.4	Convective-diffusive deposition	332
	7.4.1 Laminar flows	332
	7.4.2 Turbulent flows	340
7.5	Convective-phoretic deposition	341
7.6	A general analysis	347
7.7	Convective-diffusiophoretic deposition	351
7.8	Convective-gravitational deposition	360
7.9	Convective-electrical deposition	361
7.10	Inertial deposition	363
7.11	Particle resuspension	364
7.12	Correlations	369
7.13	Complex flows and computer programs	369
	References: Chapter 7	370

CHAPTER 8 Nuclear Source Term 375

8.1	Nuclear source term and aerosols	375
8.2	Nuclear accidents	381
8.3	Computer programs	382
8.4	Experimental data base	386
8.5	Simulation of experiments	390

8.6	Discussions and conclusions	404
	References: Chapter 8	408
Appendix A	Common Dimensionless Groups	410
Appendix B	Commonly Encountered Constants	412
Appendix C	Typical Aerosol Properties	413
Appendix D	Differential Operators	414
Appendix E	Acronyms and Abbreviations	417
Appendix F	Journal Abbreviations	419
Author Index		421
Subject Index		425

List of Tables

Table 1.1:	Applications of aerosol science.	3
Table 2.1:	Drag on a sphere versus the inverse Knudsen number R.	23
Table 2.2:	Reduced drag, $F_D/6\pi a\mu U$, on spheroids.	40
Table 2.3:	Approach to the exact solution for flow past two spheres (values of λ).	52
Table 2.4:	Radius, a, of the equivalent sphere.	54
Table 2.5:	Various shapes investigated in the literature.	55
Table 2.6:	Values of λ_∞ and λ_o for different values of the interparticle separation, r/a.	82
Table 2.7:	Comparison of exact and asymptotic formulae for the Stokes force on either of two identical approaching spheres.	85
Table 2.8:	Equal parallel motion, $\varepsilon=0.01$.	86
Table 2.9:	Equal anti-parallel motion, $a_1/a_2=1.0$.	87
Table 2.10:	Equal anti-parallel motion, $a_1/a_2=5.0$.	87
Table 2.11:	Two equal spheres - one free. The ratio Ω_1/Ω_2 as a function of α.	90
Table 2.12:	The dynamic form factors, $\kappa_\parallel$ and $\kappa_\perp$.	102
Table 4.1:	Typical energy dissipation rates, ε_T, and corresponding values of $\tilde{G}$, the effective velocity gradient.	176
Table 4.2:	The ratio, γ_{BT}, as a function of χ^2.	179
Table 4.3:	Values of $\gamma-1$ as a function of β from Eqn. (4.229).	194
Table 4.4:	The ratios, γ_{exact} and γ_{approx}, as functions of β.	199
Table 4.5:	The gravity number, Gr, as a function of λ and a.	210
Table 4.6:	Comparison of force coefficients from Klett and Davis in the limit of zero Reynolds number with corresponding values (in parentheses) according to Stokesian hydrodynamics.	224
Table 5.1:	Particle densities, N_D, N_C, and N_A, as functions of dimensionless time, τ.	258
Table 5.2:	Particle density and variance as a function of time.	278
Table 5.3:	Comparison of computing time, number and mass concentration between different numerical schemes.	300
Table 6.1:	The dimensionless condensation rate, $\tilde{u}/\tilde{u}_{fm}$ (P_3 approximation).	317
Table 6.2:	The normalized density profile $[(n(r)-n_1)/(n_\infty-n_s)]$ for R=5.0.	317
Table 6.3:	Comparison of the theoretical dimensionless condensation rate, $\tilde{u}/\tilde{u}_{fm}$, of dioctyl phthalate (DOP) in air with the experimental results of Ray *et al.* [1988] ($\beta^2=m_{N2}/m_{DOP}=28.014/390.54=0.0717$) ($P_5$ approximation).	318
Table 7.1:	Peclet numbers and normalized deposition rates (cm^2/sec) as a function of particle diameter for different supersaturations (for a spherical particle in a mixture of air and water vapor).	357
Table 7.2:	Correlations for aerosol deposition.	369

Table 8.1:	Typical material inventories for 800 MWe pressurized water reactors (PWR) (midpoint of an equilibrium fuel cycle).	375
Table 8.2:	Important radioactive nuclides (in a 3412 MWth PWR operated for three years; as predicted by the ORIGEN code).	376
Table 8.3:	Accident sequences that could result in sizeable release of radioactive isotopes (partial list).	376
Table 8.4:	Reactor Accidents.	383
Table 8.5:	Core inventory of radionuclides at Chernobyl (based on INSAG summary as prescribed by Soviet experts).	384
Table 8.6:	Timing and dynamics of the release at Chernobyl.	384
Table 8.7:	Radionuclide composition of the discharge at Chernobyl.	385
Table 8.8:	Experimental characteristics of LWR aerosol tests performed, in progress, or planned.	388
Table 8.9:	Test conditions for CSTF tests AB1 and AB2.	393
Table 8.10:	LACE tests and objectives.	395
Table 8.11:	Participants and aerosol behavior computer codes used.	398
Table 8.12:	Participants and thermal-hydraulic computer codes used.	398
Table 8.13:	LA1 containment bypass test conditions.	400
Table A.1:	.Common dimensionless groups.	410
Table C.1:	Aerosol properties as a function of size of a unit density sphere in air at Standard Temperature and Pressure (STP).	413

List of Figures

Figure	Description	Page
Figure 1.1:	Particle size ranges for aerosols.	4
Figure 2.1:	Coordinates for torque on a body.	20
Figure 2.2:	The Knudsen layer and slip coefficient.	21
Figure 2.3:	Comparison of variational results with Millikan's formula and Sherman's law.	23
Figure 2.4:	A coordinate system for Stokes flow.	25
Figure 2.5:	An intrinsic coordinate system.	26
Figure 2.6:	A coordinate system for the motion of a sphere.	29
Figure 2.7:	Na-oxide particles at various relative humidities.	33
Figure 2.8:	A coordinate system for an axially symmetric body.	35
Figure 2.9:	A coordinate system for a spheroid.	36
Figure 2.10:	A coordinate system for an oblate spheroid moving along the direction of its major axis.	38
Figure 2.11:	A spheroid at arbitrary orientation.	39
Figure 2.12:	A closed torus moving in the direction of its major axis.	42
Figure 2.13:	A family of curves generated by $(z,\varpi)\leftrightarrow(\xi,\eta)$ transformations.	43
Figure 2.14:	A deformed spheroid, $r(\theta)=1+\varepsilon(\theta)$.	45
Figure 2.15:	Drag ratios for ellipsoids.	47
Figure 2.16:	Bodies with μ-square deformations.	47
Figure 2.17:	Drag ratios for bodies with μ-square deformations.	48
Figure 2.18:	Correlation of form factor f with change in volume.	48
Figure 2.19:	The geometry of the multiple sphere system.	50
Figure 2.20:	Drag correction factor for a 7 sphere chain at different sphere spacings.	53
Figure 2.21:	The d_s/d_n versus measured shape factor, κ_n for prisms tested by Johnson [1985] ($d_s=d_{es}$).	57
Figure 2.22:	The κ_n from Eqn. (2.204) versus measured κ_n for prisms tested by Johnson [1985].	58
Figure 2.23:	The κ_n calculated from Eqn. (2.204) versus measured κ_n for prisms (squares); sphere (plus); cylinders (diamonds); spheroids (crosses); and double conicals (triangles).	58
Figure 2.24:	Relation between the aerodynamic diameter (d_a), the number of primary particles (n), and mass (m) of branched chain-like particles.	59
Figure 2.25:	Relation between the mean primary particle diameter, $\bar{d}_1$, and f_1 obtained from exploding wire aggregates.	62
Figure 2.26:	Relation between the mean primary particle diameter, $\bar{d}_1$, and k_1 of Eqn. (2.205).	62
Figure 2.27:	Richardson's empirical data concerning the rate of increase of coastlines' lengths.	64
Figure 2.28:	Fractal umbrella trees and fractal canopies.	64
Figure 2.29:	A closed torus rotating about its major axis.	71
Figure 2.30:	Motion of two spheres along their line of centers.	74
Figure 2.31:	Graphs of f_1 plotted against ε.	81

Aerosol Science: Theory and Practice

Figure 2.32: Resistance coefficient for equal parallel motion parallel to the line of centers (equal spheres, radius a). 83
Figure 2.33: Resistance coefficient for equal anti-parallel motion parallel to the line of centers (equal spheres, radius a). 83
Figure 2.34: Two rotating spheres. 89
Figure 2.35: The torque on rotating spheres. 91
Figure 2.36: A cylindrical coordinate system for two spheres. 92
Figure 2.37: The translating and rotating spheres considered by Davis [1969]. 96
Figure 2.38: Resistance coefficient for equal anti-parallel motion perpendicular to the line of centers (equal spheres, radius a). 97
Figure 2.39: Resistance coefficient for equal parallel motion perpendicular to the line of centers (equal spheres, radius a). 97
Figure 2.40: The motion of two spheres at any arbitrary orientation. 98
Figure 2.41: A coordinate system for the motion of two spheres. 102

Figure 3.1: Aerosol flow in the various zones of a containment. 118
Figure 3.2: Aerosol inflow and outflow from zone i. 119
Figure 3.3: Aerosol flow in linearly coupled chambers. 121
Figure 3.4: Aerosol flow between two coupled chambers, i and j. 122
Figure 3.5: Variation of the surface loss constant with time. 141
Figure 3.6: Variation of the coagulation constant with time. 141
Figure 3.7: Aerosol evolution for weakly charged aerosol ($\sigma=0$). 146
Figure 3.8: Aerosol evolution for moderately charged aerosol ($\sigma=1$). 146
Figure 3.9: Aerosol evolution for strongly charged aerosol ($\sigma=2$). 147

Figure 4.1: A polar coordinate system for two particle coagulation. 157
Figure 4.2: The geometry of a two particle collision. 165
Figure 4.3: (ρ,θ) coordinates of two particle motion in a flow. 166
Figure 4.4: Stokes stream lines (normal) and particle trajectories (**bold**). 169
Figure 4.5: Two particle motion in laminar shear flow. 171
Figure 4.6: A coordinate system for laminar shear flow. 171
Figure 4.7: Stream lines of relative motion between two particles. 173
Figure 4.8: Dimensionless relative diffusion coefficient versus dimensionless particle separation for various radius ratios. 185
Figure 4.9: The enhancement factor for particles in the continuum regime (W) for the cases: 1) when only van der Waals forces operate and; 2) when both van der Waals forces and viscous forces operate. 188
Figure 4.10: Spherical polar coordinates for the solution of Eqn. (4.211). 191
Figure 4.11: A coordinate system for Brownian and shear coagulation. 203
Figure 4.12: The relationship between the coordinates, **x** and **r**. 204
Figure 4.13: The trajectories, in the plane $x_3=0$, of the centre of one sphere relative to that of another of the same size in a steady simple shearing motion. 208
Figure 4.14: Trajectories of small particles as they oscillate about a large particle. 215
Figure 4.15: Trajectories of two spheres moving towards each other. 217
Figure 4.16: A Cartesian coordinate system for two particle motion. 221
Figure 4.17: Calculated collision efficiencies for cloud droplets. 222
Figure 4.18: Two spheres moving with the velocities, $\mathbf{u}_i$ and $\mathbf{v}_i$. 223
Figure 4.19: Comparison of collision efficiencies according to the formulation of Klett and Davis [1973] in the zero-Reynolds number limit with Davis and Sartor's [1967] Stokesian values. 226

List of Figures

Figure 4.20: Collision efficiencies according to the method of Klett and Davis [1973] (solid and dotted lines) and the Stokesian treatment of Davis and Sartor [1967] (dashed lines). 226
Figure 4.21: Gravitational collision efficiency, ϵ, versus a, Stokes drag forces $\rho=\rho(a_i)$. 227
Figure 4.22: Gravitational collision efficiency, ϵ, versus a, Stokes drag forces $\rho=\rho(a_i)$. 227
Figure 4.23: Gravitational collision efficiency, ϵ, versus a, Carrier-modified Oseen drag forces $\rho=\rho(a_i)$. 228
Figure 4.24: Gravitational collision efficiency, ϵ, versus a, superposition method $\rho=\rho(a_i)$. 229
Figure 4.25: Gravitational collision efficiency, ϵ, versus a, superposition method $\rho=\rho(a_i)$. 229
Figure 4.26: A comparison of the GCEFF results with the work of atmospheric sciences (Stokesian drag forces). 230
Figure 4.27: A comparison of the GCEFF results with the work of atmospheric sciences (Carrier-modified Oseen drag forces). 230
Figure 4.28: The boundary which is the locus of the center of a sphere of radius, S, rolled over the surface of a prolate spheroid with long axis of length, A, and short axes of length, B. 232
Figure 4.29: The angle, θ, relating the orientation of a spheroid to the direction of diffusion. 234

Figure 5.1: Self-preserving particle size distribution for Brownian coagulation. 270
Figure 5.2: v for various initial values for the hydrodynamic coagulation kernel. 280
Figure 5.3: The coefficient 'ab' for the hydrodynamic coagulation kernel. 280
Figure 5.4: The similarity function at different times for the hydrodynamic coagulation kernel for constant $\beta=1$. 281
Figure 5.5: The similarity function at different times for the free molecule coagulation kernel. The normal Gamma distribution is used with $v_0=2$. 281
Figure 5.6: The similarity function at different times for the free molecule coagulation kernel. The normal Gamma distribution is used with $v_0=1$. 282
Figure 5.7: The value of α for the free molecule coagulation kernel for various initial conditions. 282
Figure 5.8: The variance, σ, as a function of the fractal index, α. 286
Figure 5.9: The inverse particle density, $1/N$, as a function of the fractal index, α. 286
Figure 5.10: The effective particle density, ϕ, as function of the fractal index, α. 287
Figure 5.11: Integration along a constant characteristic. 292
Figure 5.12: Finite elements in space and time at a typical time step. 296
Figure 5.13: The number distribution function, $3vn(v)$ for different numerical schemes after 5 seconds for Case 1. 301
Figure 5.14: The number distribution function, $3vn(v)$ for different numerical schemes after 10,000 seconds for Case 2. 301

Figure 8.1: TMI-2 nuclear plant layout. 378
Figure 8.2: Schematic of the containment design for the Grand Gulf plant. 379
Figure 8.3: A historical perspective of U.S. computational progress. 380
Figure 8.4: A schematic diagram of the CONTAIN code emphasizing the three basic phenomenological areas and their intercoupling. 386
Figure 8.5: The BMI-2104 suite of codes. 387
Figure 8.6: A schematic view of the CSTF facility. 389
Figure 8.7: A schematic view of the NSPP facility. 390

Figure 8.8: A schematic view of the DEMONA facility. ... 391
Figure 8.9: Iodine removal experiments in the CSTF. ... 391
Figure 8.10: CSTF experiments AB1 and AB2 with the corresponding HAA-3B predictions. ... 393
Figure 8.11: NSPP Test 303 with the corresponding HAARM-3 predictions. ... 394
Figure 8.12: NSPP Test 304 with the corresponding HAARM-3 predictions. ... 394
Figure 8.13: NSPP Test 305 with the corresponding HAARM-3 predictions. ... 394
Figure 8.14: NSPP Tests 401 and 209. ... 394
Figure 8.15: DEMONA experiment A7 (iron oxide aerosol). ... 396
Figure 8.16: DEMONA experiment B3. ... 396
Figure 8.17: Instrument locations for the LA1 test. ... 399
Figure 8.18: Layout of the experimental apparatus for the LA3 tests. ... 399
Figure 8.19: LA3 data and posttest code predictions. ... 401
Figure 8.20: LA3c data and posttest code predictions. ... 401
Figure 8.21: Experimental arrangement for the LA2 test. ... 402
Figure 8.22: LA2 data and posttest code predictions. ... 402
Figure 8.23: LA2 data and posttest code predictions. ... 403
Figure 8.24: LA2 data on suspended mass ratio and posttest code predictions. ... 403
Figure 8.25: A schematic diagram of the 800 m^3 containment vessel arrangement for the LA4 test. ... 405
Figure 8.26: LA4 data and posttest code predictions. ... 405
Figure 8.27: LA4 data and posttest code predictions. ... 406
Figure 8.28: Test arrangements for LA5 and LA6. ... 406
Figure 8.29: LA6 data and posttest code predictions. ... 407
Figure 8.30: LA6 data on suspended mass ratio and posttest code predictions. ... 407

Figure C.1: Aerosol properties as a function of size of a unit density sphere in air at Standard Temperature and Pressure (STP). ... 413

CHAPTER 1

Aerosol Characterization

1.1 The nature of aerosols

One cubic centimeter of atmospheric air contains approximately 2.5×10^{19} molecules. About 10^3 of these molecules may be charged (ions). The molecules of N_2, O_2, and the various trace gases have sizes (diameters) of ~3 Å. The average distance between the molecules is about ten times the molecular size. In addition to the molecules and the ions, one cubic centimeter of air also contains a substantial number of particles varying in size from a few Angstroms to several microns (μm). In relatively clean air there are about 10^3 particles with diameters 10^{-3} to 50.0 μm while in polluted air there can be 10^5 or more, including pollen, bacteria, dust, and industrial emissions. These particles, which can be both beneficial and detrimental, arise from a number of natural sources as well as from the activities of the Earth's inhabitants. The particles can have complex chemical compositions and morphologies, and may even be radioactive or toxic.

A suspension of particles in a gas is known as an aerosol. Atmospheric aerosol is of global interest and has an important impact on our lives. Aerosols are of great interest in numerous scientific and engineering applications. Perhaps no single technology can be identified that includes fundamental studies in fields as diverse as human respiration and nuclear winter and that permits the design of equipment and processes ranging from power plant emission control devices to specialized fiberoptic sensors. The various application areas include:

    a) Combustion: Fossil power plants, chemical plants, and vehicles (particularly diesel-powered ones) emit large numbers of particles. These particles are formed by fragmentation (and/or vaporization) and the subsequent homogeneous-heterogeneous nucleation and agglomeration of fuel constituents. They are characterized by the type of fuel used, the combustion process involved, and the equipment design. They may lead to or affect both local and global problems of acid rain, haze, and smog that impact the environment and contribute to respiratory diseases.

    b) Atmospheric Science: The formation of fog and clouds through nucleation, condensation and agglomeration, the scavenging of particulates by rain and snow, weather modification, ozone depletion, and solar-radiative interactions all depend strongly on particulate generation and transport processes.

    c) Nuclear Winter: It has long been realized that a nuclear war would release large numbers of particles into the atmosphere. The resulting attenuation of sunlight might be extensive enough to cause major global climatic changes. The current ability to predict climatic consequences is limited, in part, by our poor understanding of: i) the transport of dust and smoke and their feedback effects on the environment; ii) the lifetime of smoke particles in the environment, and; iii) their radiative, optical, and infrared properties.

    d) Nuclear Reactor Safety: Understandings of aerosol evolution and transport are of critical importance in estimating and controlling the radioactivity released from postulated (and real) nuclear reactor accidents. Severe core damage to reactors, core disruptive accidents in fast reactors, and the sabotage of shipping casks or of disposal sites are all potential sources of release of radioactive aerosols to the environment. The primary risk to the public, as well as to radiation workers, would be from accidents that involve large releases of radioactive aerosols to the environment.

e) Materials Manufacturing: An aerosol reactor is a gaseous system in which fine particles are formed by a chemical reaction in either a batch or flow process. Recently, such reactors have been used to produce optical fibers, catalysts, ceramics, silicon, and carbon whiskers.

f) Indoor Air Quality: There is a strong national and international concern about the quality of indoor air. Indoor air pollutants, in conjunction with the relatively large fraction of time that people spend indoors, pose significant health hazards. Radon and radon daughter products, volatile organic compounds, and several carcinogens can attach to aerosol particles. Inhalation of such particles, smoke, and of bioaerosols (viruses, bacteria, fungi) can lead to numerous diseases.

g) Control of Particulate Contamination: The manufacture of microelectronic integrated circuits is currently a major industry and that of micromachines (microrobots) is likely to become significant very soon. In the manufacture of microdevices, contamination by submicron particles must be controlled. Thus, there is a strong industrial interest in understanding the transport of such particles under a variety of process conditions.

Table 1.1, adapted from Handbook on Aerosols, provides a summary of the many applications of aerosol science and technology.

Irrespective of these diverse applications, aerosols are characterized by a few fundamental properties. Most importantly, aerosol particles have large residence times (settling speeds on the order of a fraction of a cm/sec) and, because of their small size and large number, present a large surface area for interactions with the host medium. In addition, they can have a substantial effect on the transmission of light as they tend to occur in a size range that leads to substantial interaction (scattering and absorption).

1.2 Aerosol size and shape

For particles of a given composition, size and shape determine residence time as well as other dynamical properties that bear on particle removal by filtration or collection devices. Heat and mass transport to the particles, as well as their interaction with radiation, are strongly affected by size and shape. Liquid particles are almost always spherical, but solid particles can occur in many different shapes varying from spherical to fibrous (needle-like forms). They can be chain-like agglomerates, or can be amorphous clumps. Typical size ranges of particles are shown in Fig. 1.1.

The characterization and dynamics of nonspherical particles are complicated largely because of the range of shapes that are encountered. One often considers "equivalent" spherical particles where, by proper adjustment of a fictitious particle diameter, one can mimic important properties of a nonspherical particle (*e.g.*, the settling speed). We will discuss nonspherical particles at some length in this book.

Particle size and shape can be measured by a variety of instruments. Since particle sizes cover such a wide range (0.001 to 50.0 µm), no single instrument is able to provide reliable measurement capabilities for all particles. Definitive measurements of an individual particle's size and shape can be obtained with electron or optical microscopes. Also, particles scatter light in unique ways depending on their size and shape. Thus, the laser light scattered from a suspended (trapped) particle, can often be used to measure the particle size. If the particle is nonspherical, scattered light can be used to define its shape in a somewhat approximate manner.

Table 1.1: Applications of aerosol science. Adapted from Dennis [1976]. This table excludes nuclear applications.

Public Health and Hygiene:
- Medicine — Inhalation, deposition, clearance, and retention of aerosols, therapeutic aerosol generation and characterization, occupational, and industrial medicine, physiological effects.
- Public Health — Industrial and occupational hygiene, health physics, atmospheric pollution, confined compartment atmospheres, residential dwelling and public air hygiene, characterization and engineering for control or protection of residents, occupants, or workers, clean-room technology for hospitals, treatment of biological problems.

Biological and Chemical Aerosols:
- Applications — Technical application, aerosol technology, military aerosols and agents, production, release, characterization, control, bioengineering, dissemination and transport of agricultural and forestry pesticides, herbicides.

Geophysics and Atmospheric Sciences:
- Cloud physics — Nucleation, condensation, fog, rain, scavenging by rain and snow, weather modification.
- Meteorology — Weather, haze, fogs, mist, transport, dispersion and disposal in the atmospheric reservoir.
- Chemistry — Transport in the atmosphere, formation by atmospheric processes, solar radiative interactions, photochemical smog and haze formation, deposition, rain-out, washout.
- Geophysics — Geochemistry, photochemistry, chemistry and physics of interplanetary spaces, planetary atmospheres, light transmission in the atmosphere.
- Mathematics — Slow viscous flow, sphere and cylinder problems, flow in a corner, low Reynolds number, hydrodynamics, drag.

Engineering and Technology:
- Electrical — Charging, collection, control, device development for aerosols, printing.
- Mechanical — Combustion, collection and control device development, gas dynamics in jets, gas turbines, air induction systems, clean room technology.
- Aeronautical, Astronautical — Rain, ice, clouds, air induction systems designs, aerosol jet-nozzle flow-energy interactions, jet propulsion, ion propulsion, interplanetary particle flux, protection, spacecraft-cabin atmosphere aerosol definition, lunar dust.
- Chemical — Flow of fluids, rheology, behavior of disperse systems, chemical process engineering with disperse phases, properties of aerosol systems, collection of aerosol systems, catalysis, combustion, condensation, diffusion, separation of gas-particle systems, filtering, screening, pelletizing, briquetting, sintering, fluidization and conveying of powders.
- Civil — Environmental engineering, sediment mechanics, effluent treatment.

Physics:
- Kinetic theory — Statistical mechanics, Brownian motion, coagulation.
- Physics of fluids — Gas dynamics, motion of particles, slow viscous flow.
- Electromagnetics — Light scattering, laser radar, plasma physics, optical properties.
- Electrostatics — Charging, ionization, motion, nucleation.

Chemistry:
- Heterogeneous reactions — Catalysis, combustion, flame chemistry.
- Homogeneous reactions — Generation of particulate phase, surface energy.

Physical Chemistry:
- Colloid Sci. — Study of disperse systems in fluids, agglomeration, particle beams.
- Interfacial Sci. — Adhesion, condensation, adsorption, accommodation for flux of various molecular species.
- Rheology — Aerosol modification of fluid properties, non-Newtonian flow, macromolecules.
- Photochemistry — Smog formation, radiation chemistry, heterogeneous, homogeneous.

Mathematics and Applied Mechanics:
- Fluid mechanics — Flow properties, aerosol particle interactions, energy coupling and modifications of fluid transport phenomena, turbulent flow, physico-chemical hydrodynamics.

Aerosol Science: Theory and Practice

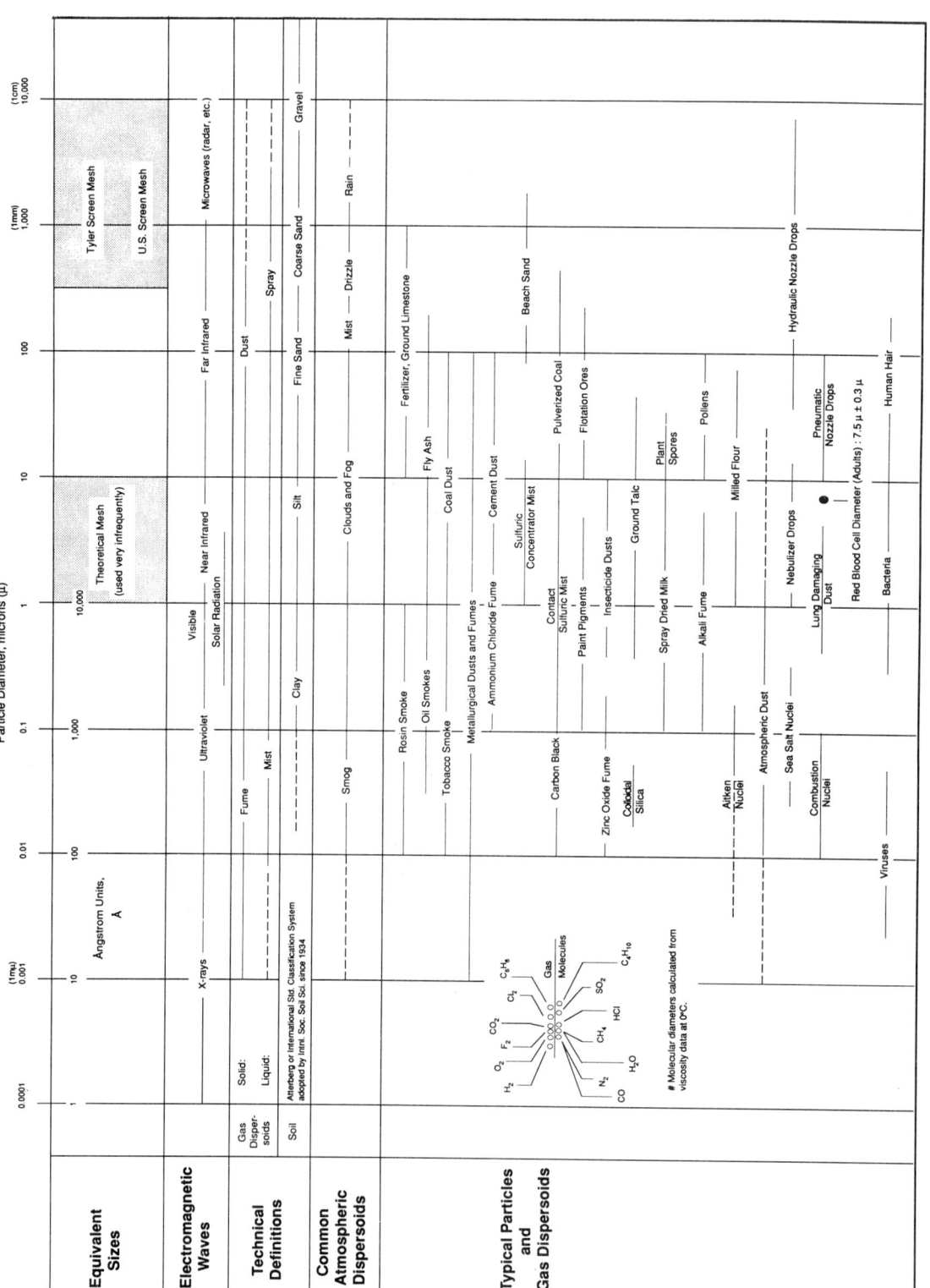

Figure 1.1: Particle size ranges for aerosols. Adapted from and reprinted courtesy of SRI International

Aerosol Characterization

Other means of particle sizing are related to the dynamical properties of particles such as the particle settling speed or the diffusion coefficient. These properties are more useful, however, in determining particle distributions rather than characterizing individual particles.

1.3 Distribution functions

Since the number of particles in an aerosol is large, and since each particle can have a multitude of physical and chemical characteristics, a description that accounts for particles individually is not sought. Rather, one focuses on the important attributes of the aerosol and considers a distribution function based on particle properties and size. If $\Delta n(d_p,\beta_i)$=the expected number of particles/cm^3 of size d_p in Δd_p of properties β_i in an interval $\Delta\beta_i$, i=1,2,3,..., then a number distribution function can be defined by:

$$n(d_p,\beta_i) = \lim_{\substack{\Delta\beta_i \to 0 \\ \Delta d_p \to 0}} \frac{\Delta n}{\Delta d_p \, \Delta\beta_i} \tag{1.1}$$

which has the interpretation that $n(d_p,\beta_i)d(d_p)d(\beta_i)$=the expected number of particles/cm^3 of size d_p in $d(d_p)$ and of properties β_i in $d(\beta_i)$, i=1,2,3,... .

In a given aerosol, this particle distribution can be measured by a combination of instruments. The number of particles in a given size interval is determined by instruments that utilize particle size dependent properties relating to diffusion (for particles <1.0 µm), light scattering (0.3–~1 µm), and motion under external forces (*e.g.*, settling, for particles >0.5 µm). The respective instruments bear names such as diffusion battery and electrical analyzer, light scattering particle counters, and inertial impactors. The particle mass in a given size interval can, for example, be determined by the use of a fractionator (classifier or impactor) and a sensitive balance. Similarly, particle constituents in a given size range can be measured by use of fractionators and neutron activation analysis, x-ray or γ-ray scattering, electron microprobe analysis, or other techniques. The particle shapes and morphologies can be characterized with electron or optical microscopes.

In general, particle surface, volume, and mass are not entirely determined by particle size (diameter) and density. If the particles are spherical and of uniform density, and if one wishes to avoid the inherent complexity, one may use the reduced distribution $n(d_p)$, where $n(d_p)d(d_p)$ is the expected number of particles/cm^3 of size d_p in $d(d_p)$ (the dependence on space or time is understood). It is often convenient to summarize the information content of these distributions by the total numbers, mean or median values, and spreads (standard deviations). Thus, if we define:

$\bar{d}_p \equiv$ The mean (average) diameter,
$N \equiv$ The total number of particles/cm^3,
$d_{p,median} \equiv$ The median diameter (N/2 particles above this size, N/2 below), and
$\sigma \equiv$ The standard deviation,

then:

$$N = \int_0^\infty n(d_p)\,d(d_p) \tag{1.2}$$

$$\bar{d}_p = \frac{1}{N}\int_0^\infty d_p\, n(d_p)\,d(d_p) \tag{1.3}$$

$$F(d_{p,median}) = \tfrac{1}{2} \tag{1.4}$$

$$\sigma^2 = \frac{1}{N}\int_0^\infty (d_p - \bar{d}_p)^2 n(d_p)\,d(d_p) \tag{1.5}$$

where $F(d_p)$ is the cumulative distribution:

$$F(d_p) = \int_0^{d_p} n(d'_p)\,d(d'_p) \tag{1.6}$$

Now, if $S(d_p)$, $V(d_p)$, and $W(d_p)$, are, respectively, the surface, volume, and mass of a particle of size, d_p, then the particle surface, volume and mass distributions can be defined as:

$$s(d_p) = S(d_p)n(d_p) \approx \pi d_p^2 n(d_p) \tag{1.7}$$

$$v(d_p) = V(d_p)n(d_p) \approx \frac{\pi}{6} d_p^3 n(d_p) \tag{1.8}$$

$$w(d_p) = W(d_p)n(d_p) \approx \frac{\pi}{6} d_p^3 \rho n(d_p) \tag{1.9}$$

and again, the corresponding mean and median diameters, based on these weighted distributions, can be defined in a similar manner. In particular, we note that d_m the <u>particle diameter of average mass</u> and d_{mm}, the <u>mean mass diameter</u> are defined by the relations:

$$\frac{\pi}{6} d_m^3 \rho N = \int_0^\infty w(d_p)\,d(d_p) \tag{1.10}$$

which leads to:

$$d_m \approx \left[\frac{1}{N}\int_0^\infty d_p^3 n(d_p)\,d(d_p)\right]^{1/3} \tag{1.11}$$

and:

$$d_{mm} = \frac{\int_0^\infty d_p\, w(d_p)\,d(d_p)}{\int_0^\infty w(d_p)\,d(d_p)} \tag{1.12}$$

which leads to:

$$d_{mm} \approx \frac{\int_0^\infty d_p^4 n(d_p)\,d(d_p)}{\int_0^\infty d_p^3 n(d_p)\,d(d_p)} \tag{1.13}$$

The cumulative mass distribution:

$$W(d_p) = \int_0^{d_p} w(d'_p)\,d(d'_p) \tag{1.14}$$

is used to define another quantity of practical interest, the <u>mass median diameter</u>:

$$W(d_{mass,median}) = \tfrac{1}{2} \tag{1.15}$$

Since the aerosol distributions of practical interest are generally skewed towards greater numbers at smaller sizes and greater mass content at larger sizes, and since the sizes span several decades, the median values are often used in preference to the mean values. Also, for some specific forms of the distribution $n(d_p)$, simple algebraic relationships can be established among the various diameters defined above.

There has always been great interest in establishing forms of the distribution function $n(d_p)$ that can be considered as somewhat universal in nature. Actually, no universal forms analogous to the Maxwell-Boltzmann distribution of gas molecules currently exist. It has been observed, however, that in many applications the aerosol distributions are close to some simple forms. Some often observed forms are discussed below.

1.3.1 The Junge distribution

In the 1940's and early 1950's, Junge showed that the observed atmospheric aerosol distribution seemed to have an interesting property in that the mass was uniformly distributed in equal geometric intervals (*i.e.*, the size interval 0.15–0.30 µm contained the same mass as the size interval 0.30–0.60 µm). Mathematically, assuming the particles to be spherical and of equal density, this implies that:

$$\frac{\frac{\pi}{6} d_p^3 \, \rho \, \Delta n(d_p)}{\Delta \log(d_p)} = \text{constant} \tag{1.16}$$

which leads to the distribution:

$$n(d_p) = A \, d_p^{-4} \tag{1.17}$$

in which A is a constant. To avoid divergence in the calculations of the mean quantities, it is necessary to specify minimum ($d_{p,min}$) and maximum ($d_{p,max}$) particle sizes. Since there may be no sharp cutoffs in the measurements, these choices are slightly arbitrary which does affect the calculated mean and median values. The Junge distribution can be more generally expressed as:

$$n(d_p) = A \, d_p^{-\alpha} \tag{1.18}$$

where the exponent α can be adjusted to account for deviations from the original Junge distribution.

The Junge distribution serves a useful purpose, but quite often urban aerosols have bimodal or other distribution forms and cannot be described by this particular form of particle distribution.

1.3.2 The log-normal distribution

The inverse power distributions do not provide good descriptions of aerosols at either extreme of particle size. Also, most test or industrial aerosols do not follow this prescription. Rather, it is often observed that the particle distribution has a certain geometrical symmetry about a mean size. To a considerable extent, this situation arises because of the lower residence times of both small (diffusion to walls or filters) and large (settling, impaction or interception by filters)

Aerosol Science: Theory and Practice

particles, as compared to particles of intermediate size (~0.3 μm). In many processes, a particle ten times the mean size is as likely to be generated as a particle one tenth the mean size. Since particles cannot have a normal distribution, the logical distribution here is one in which the logarithmic size is normally distributed. Thus, we consider the log-normal distribution:

$$\Delta \hat{n}(\ln(d_p)) = \frac{N}{\sqrt{2\pi} \ln(\sigma_g)} \exp\left[-\frac{[\ln(d_p) - \ln(d_g)]^2}{2[\ln(\sigma_g)]^2}\right] \Delta \ln(d_p) \quad (1.19)$$

in which $\Delta \hat{n}(\ln(d_p))$ is the number of particles/cm³ with diameters whose logarithms lie between $\ln(d_p)$ and $\ln(d_p) + \Delta(\ln(d_p))$. Since the number of particles between d_p and $d_p + \Delta(d_p)$ is the same as between $\ln(d_p)$ and $\ln(d_p) + \Delta(\ln(d_p))$:

$$\Delta n(d_p) = \Delta \hat{n}(\ln(d_p)) \quad (1.20)$$

and we can write:

$$\Delta n(d_p) = \frac{N}{\sqrt{2\pi} \ln(\sigma_g)} \exp\left[-\frac{[\ln(d_p) - \ln(d_g)]^2}{2[\ln(\sigma_g)]^2}\right] \frac{\Delta \ln(d_p)}{\Delta d_p} \Delta d_p \quad (1.21)$$

In the limit, this gives the following log-normal distribution:

$$n(d_p) = \frac{N}{\sqrt{2\pi} \ln(\sigma_g)} \exp\left[-\frac{[\ln(d_p) - \ln(d_g)]^2}{2[\ln(\sigma_g)]^2}\right] \frac{1}{d_p} \quad (1.22)$$

Note that:

$$\ln(d_g) = \frac{1}{N} \int_0^\infty \ln(d_p) \, n(d_p) \, d(d_p) \quad (1.23)$$

$$\ln(\sigma_g) = \frac{1}{N} \int_0^\infty [\ln(d_p) - \ln(d_g)]^2 \, n(d_p) \, d(d_p) \quad (1.24)$$

and d_g and σ_g are known as the geometric mean diameter and the geometric standard deviation, respectively. To clarify this notion, note that if:

$$n(d_p) = \sum n_i \, \delta(d_p - d_{pi}) \quad (1.25)$$

then:

$$\ln(d_g) = \frac{1}{N} \sum n_i \ln(d_{pi}) \quad (1.26)$$

and:

$$d_g = \left(d_{p1}^{n_1} \cdot d_{p2}^{n_2} \cdot d_{p3}^{n_3} \cdots\right)^{1/N} \quad (1.27)$$

Another useful property of the log-normal distribution is that the mean, the median, and the diameters of the average particles, as determined with various weighted distributions, are simply related. Such relations are known as Hatch-Choate equations and have the form:

$$d_A = d_g \exp\left[b \ln(\sigma_g^2)\right] \tag{1.28}$$

where b is a constant related to the type of conversion, A.

It is because of its simple and useful properties that the log-normal distribution is often used where its use is not quite justifiable.

1.3.3 The gamma distribution

The gamma distribution is expressed as:

$$n(d_p) = \frac{d_p^x \exp\left(-x \dfrac{d_p}{y}\right)}{\Gamma(x+1)\left(\dfrac{y}{x}\right)^{x+1}} \tag{1.29}$$

where y is the mode of the distribution (the diameter at which $n(d_p)$ is a maximum, or the most frequently occurring diameter) and x is a parameter of choice. For a given set of data, x and y can be chosen to fit the data in an appropriate least square sense. The gamma distribution is quite useful in that many aerosol distributions can be fitted adequately by using various combinations of the parameters x and y. Once obtained, the parameters x and y permit various correlations to be made and allow the assessment of specific processes.

1.4 Chemical composition

In the environment, aerosol composition is seldom simple. Usually, aerosol particles contain numerous species with varying physico-chemical properties. The compositions of the particles are determined by the processes that lead to their initial creation and their subsequent interactions with the host medium and electromagnetic radiation. Coarse particles (≥ 2 µm) are generally created by mechanical processes such as fragmentation and hence are rich in Ca, Fe, SiO_2, and other constituents of earth. Fine particles (<2 µm) are generally derived from processes such as combustion or gas to particle conversion and are rich in C, Pb, sulphates, and ammonium and nitrate ions. The trace and often toxic species such as As, Cd, Cs, Sr, Zn, and Se are also mostly concentrated in the fine particles.

The chemical composition of particles can be determined by light scattering (Raman spectroscopy), x-ray spectroscopy, Auger spectroscopy, chemical analysis, neutron activation, and electron microprobe analysis. While the first five techniques provide useful information on isotopic and species content, electron microprobe analysis can also be used to determine the internal distributions (within a particle) of multiple species. Such analyses have revealed that species can be distributed within particles in very complex ways ranging from homogeneous (evenly distributed) to heterogeneous (random, ordered-layers, or otherwise). As a result, the physico-chemical properties of an aerosol can be quite complex and, correspondingly, quite difficult to describe in simplistic terms.

1.5 The nuclear aerosol

Atmospheric nuclear tests have, in the past, released large amounts of aerosol into the environment. These aerosols contain isotopes of U, Pu, Cs, and Sr which emit α, β, and γ radiation. Atmospheric nuclear tests have been banned since 1962 such that the environmental presence of aerosols related to nuclear processes is not high. The Chernobyl accident released large amounts of both fine and coarse aerosol particles containing radioactive I, Cs, and Te (the volatile species). While most of the coarse particles settled within a 30 km zone around the plant, the fine particles travelled far and wide and were a major factor contributing to the health hazards posed by the accident.

While the Chernobyl accident is the most recent, and perhaps the most vivid accident of this type, there have also been other nuclear accidents where nuclear aerosol (by which we imply radioactive aerosol that can be traced to origins in nuclear fission related power generation processes) releases were registered.

The aerosols associated with real or hypothetical nuclear accidents are generally quite complex in terms of their shapes, sizes compositions, and properties. They are formed by vaporization, condensation, and fragmentation of material. The manners in which they evolve, in nuclear reactor vessels and containments, pose formidable challenges in the areas of analytical and experimental investigation. In the next several chapters, we will discuss the fundamentals of aerosol mechanics that are necessary for such analyses and their application to the field of nuclear reactor safety. There are several good texts and review articles that deal with aerosol mechanics and closely related topics. We have listed these in the references below.

References: Chapter 1

Davies, C.N., editor [1966] Aerosol Science (Academic, London, U.K.).

Dennis, R. [1976] Handbook on Aerosols (U.S. Department of Energy, Washington, D.C.).

Fedoseev, V.A., editor [1971-1973] Advances in Aerosol Physics, Vols. I-V (Halsted, New York, N.Y.).

Friedlander, S.K. [1977] Smoke, Dust and Haze (Wiley, New York, N.Y).

Fuchs, N.A. [1964] The Mechanics of Aerosols (Pergamon, New York, N.Y.).

Fuchs, N.A. and Sutugin, A.G. [1971] "High-Dispersed Aerosols," in Topics in Current Aerosol Research, Vol. 2, Hidy, G.M. and Brock, J.R., editors (Pergamon, Oxford, U.K.).

Happel, J. and Brenner, H. [1965] Low Reynolds Number Hydrodynamics (Prentice Hall, Englewood Cliffs, N.J.).

Hidy, G.M. [1984] Aerosols, An Industrial and Environmental Science (Academic, New York, N.Y.).

Hidy, G.M. and Brock, J.R., editors [1972] Topics in Aerosol Research, Vols. 1-3 (Pergamon, Oxford, U.K.).

Hinds, W.C. [1982] Aerosol Technology (Wiley, New York, N.Y.).

Loyalka, S.K. [1983] "Mechanics of aerosols in nuclear reactor safety: a review," *Prog. Nuc. Energy* 12, 1.

Marlow, W.H., editor [1980,1982] Aerosol Microphysics, Vols. I-II (Springer-Verlag, New York, N.Y.).

Mason, B.J. [1962] Clouds, Rain and Rainmaking (Cambridge University Press, Cambridge, U.K.).

Pruppacher, H.R. and Klett, J.D. [1978] Microphysics of Clouds and Precipitation (Reidel, New York, N.Y.).

Reist, P.C. [1984] Introduction to Aerosol Science (McMillan, New York, N.Y.).

Seinfeld, J.H. [1986] Atmospheric Chemistry and Physics of Air Pollution (Wiley, New York, N.Y.).

Shaw, D.T., editor [1978] <u>Recent Developments in Aerosol Science</u> (Wiley, New York, N.Y.).

Twomey, S. [1977] <u>Atmospheric Aerosols</u> (Elsevier, New York, N.Y.).

Voloschuk, V.M. and Sedunov, Yu.S., editors [1973] <u>Hydrodynamics and Thermodynamics of Aerosols</u> (Wiley, New York, N.Y.).

Yoshida, T., Kousaka, Y. and Okuyama, K. [1979] <u>Aerosol Science for Engineers</u> (Power Co. Ltd., Tokyo, Japan).

CHAPTER 2

The Motion of Particles in Gases

2.1 Introduction

As a particle moves through a gas it experiences drag forces which retard its motion. The magnitude of the drag depends on the shape and orientation of the particle and also on the constitution of the gas. To understand the problem it will therefore be necessary to calculate the gas properties and to understand how it interacts with the particle. We therefore develop and define in this chapter some basic facts about gases and gas-surface interactions.

The constituents of a gas are atoms or molecules in thermal motion. In order to calculate the physical properties of such a medium, e.g. pressure, temperature and density, it is necessary to know the velocity distribution function of the gas atoms. Thus we need to know $f(\mathbf{r},\mathbf{v},t)$ where $f(\mathbf{r},\mathbf{v},t)d\mathbf{r}d\mathbf{v}$ is the number of atoms with velocity between $\mathbf{v}$ and $\mathbf{v}+d\mathbf{v}$ in the volume element of space $d\mathbf{r}$ at $\mathbf{r}$ at time t. Knowing f, it is relatively straightforward to obtain the density $\rho(\mathbf{r},t)$, the temperature $T(\mathbf{r},t)$, the pressure $P(\mathbf{r},t)$, and the average flow velocity $\mathbf{u}(\mathbf{r},t)$ as follows:

$$\rho(\mathbf{r},t) = m \int d\mathbf{v}\, f(\mathbf{r},\mathbf{v},t) \tag{2.1}$$

$$\mathbf{u}(\mathbf{r},t)\rho(\mathbf{r},t) = \int d\mathbf{v}\, \mathbf{v}\, f(\mathbf{r},\mathbf{v},t) \tag{2.2}$$

$$\tfrac{3}{2} k T(\mathbf{r},t)\rho(\mathbf{r},t) = \tfrac{1}{2} m \int d\mathbf{v}\, (\mathbf{v}-\mathbf{u})^2\, f(\mathbf{r},\mathbf{v},t) \tag{2.3}$$

$$P_{ij}(\mathbf{r},t) = m \int d\mathbf{v}\, (v_i - u_i)(v_j - u_j)\, f(\mathbf{r},\mathbf{v},t) \tag{2.4}$$

or, in terms of dyadics:

$$\mathbf{P}(\mathbf{r},t) = m \int d\mathbf{v}\, (\mathbf{v}-\mathbf{u})(\mathbf{v}-\mathbf{u})\, f(\mathbf{r},\mathbf{v},t) \tag{2.5}$$

For a dilute gas, but one which covers most practical aerosol situations, $f(\mathbf{r},\mathbf{v},t)$ is given by:

$$\left(\frac{\partial}{\partial t} + \mathbf{v}_1 \cdot \nabla_r + \mathbf{a} \cdot \nabla_v\right) f(\mathbf{r},\mathbf{v},t)$$
$$= \int d\mathbf{v}_1' \int d\mathbf{v}_2 \int d\mathbf{v}_2'\, W(\mathbf{v}_1 \to \mathbf{v}_1'; \mathbf{v}_2 \to \mathbf{v}_2') \left[f(\mathbf{r},\mathbf{v}_1',t) f(\mathbf{r},\mathbf{v}_2',t) - f(\mathbf{r},\mathbf{v}_1,t) f(\mathbf{r},\mathbf{v}_2,t)\right] \tag{2.6}$$

where $\mathbf{a}$ is the acceleration on the atom due to an external force, $\mathbf{F}$. The function $W(\ldots)$ is related to the differential scattering cross section $\sigma(v,\theta)$ in the following way:

$$W(\mathbf{v}_1' \to \mathbf{v}_1; \mathbf{v}_2' \to \mathbf{v}_2) = |\mathbf{v}_1 - \mathbf{v}_2|\, \sigma(|\mathbf{v}_1 - \mathbf{v}_2|; \hat{\mathbf{v}}_1 \cdot \hat{\mathbf{v}}_2) \tag{2.7}$$

where:

The Motion of Particles in Gases

$$W(v'_1 \to v_1; v'_2 \to v_2) = W(v_1 \to v'_1; v_2 \to v'_2) \tag{2.8}$$

$\sigma(v,\theta)$ depends upon the interparticle force law and is available for a wide range of potentials (Chapman and Cowling [1960]).

In order to obtain a solution of Eqn. (2.6), it is necessary to introduce boundary conditions. To do so, we need to know how gas atoms interact with surfaces. Setting aside the physical details for a moment and considering a solid surface, we may write:

$$-(v \cdot n) f(r_s, v, t) = \int_{(v' \cdot n) > 0} dv' (v' \cdot n) \Gamma(r_s, v' \to v) f(r_s, v', t) \tag{2.9}$$

where $(v \cdot n) < 0$.

In this equation, n is a unit vector pointing out of the gas normal to the boundary point r_s. $\Gamma(r_s, v' \to v) dv$ is the probability that an atom of velocity v' striking the surface at r_s will be scattered back into the gas with a velocity between v and $v+dv$. Clearly, $\Gamma(...)$ will depend on the nature of the wall and the interaction potential between wall atoms and the gas atoms.

$\Gamma(r_s, v' \to v)$ obeys some useful relationships. For example, integrating Eqn. (2.9) over v leads to:

$$\int_{(v \cdot n) < 0} dv \, \Gamma(r_s, v' \to v) = 1 \quad ; \quad (v' \cdot n) > 0 \tag{2.10}$$

which expresses the fact that there is no net accumulation of particles at the wall.

The wall-particle scattering kernel $\Gamma(...)$ also obeys a reciprocity or detailed balance relationship:

$$|v' \cdot n| f_M(r_s, v') \Gamma(r_s, v' \to v) = |v \cdot n| f_M(r_s, v) \Gamma(r_s, v \to v') \tag{2.11}$$

where $(v' \cdot n) > 0$ and $(v \cdot n) < 0$.

$f_M(r_s, v)$ is the local Maxwell-Boltzmann distribution given by:

$$f_M(r_s, v) = n(r,t) \left(\frac{m}{2\pi k T(r,t)} \right)^{3/2} \exp\left[-\frac{m}{2 k T(r,t)} [v - u(r,t)]^2 \right] \tag{2.12}$$

where $n(r,t) = \rho(r,t)/m$.

Various models exist for the kernel, Γ, but one of the simplest is a linear combination of diffuse and specular reflection. By diffuse reflection we mean that the incident atoms are thermally accommodated by the surface and reemitted isotropically with a Maxwell-Boltzmann distribution characterized by the surface temperature, T_w. Specular reflection means that the atoms are reflected as in a mirror, the angle of incidence being equal to the angle of reflection.

2.2 Linear transport theory

Eqn. (2.6) is nonlinear and therefore difficult to solve both numerically and analytically (especially analytically). However, in many practical cases, the disturbance in the gas due to the passage of a particle or other body causes only a small deviation from the local Maxwell-

Boltzmann distribution and we are therefore justified in seeking solutions in the following forms:

$$f(\mathbf{r},\mathbf{v},t) = f_M(\mathbf{r},\mathbf{v},t)[1+h(\mathbf{r},\mathbf{v},t)] \tag{2.13}$$

where $|h|<<1$.

The local parameters in f_M (i.e., $\rho, T, \mathbf{u}$) are assumed to vary only slowly in space and time. However, h(...) can vary rapidly provided that we can neglect terms of $O(h^2)$.

Inserting Eqn. (2.13) into Eqn. (2.6) and linearizing by neglecting terms of $O(h^2)$, leads to:

$$\hat{D}(f_M) + f_M \hat{D}(h) = -f_M \hat{I}(h) \tag{2.14}$$

where:

$$\hat{D}(f_M) = \left(\frac{\partial}{\partial t} + (\mathbf{v}_1 \cdot \nabla_r) + (\mathbf{a} \cdot \nabla_v)\right) f_M \tag{2.15}$$

and:

$$\hat{I}(h) = \int d\mathbf{v}_1' \int d\mathbf{v}_2' \int d\mathbf{v}_2 \, W(\mathbf{v}_1 \to \mathbf{v}_1'; \mathbf{v}_2 \to \mathbf{v}_2') f_M(\mathbf{r},\mathbf{v}_2,t)$$
$$[h(\mathbf{r},\mathbf{v}_1',t) + h(\mathbf{r},\mathbf{v}_2',t) - h(\mathbf{r},\mathbf{v}_1,t) - h(\mathbf{r},\mathbf{v}_2,t)] \tag{2.16}$$

By means of symmetry, we can show that Eqn. (2.16) reduces to:

$$\hat{I}(h) = \tilde{V}(\mathbf{v}_1) h(\mathbf{r},\mathbf{v}_1,t) - \int d\mathbf{v}_1' \, \tilde{H}(\mathbf{v}_1' \to \mathbf{v}_1) h(\mathbf{r},\mathbf{v}_1',t) \tag{2.17}$$

where:

$$\tilde{V}(\mathbf{v}_1) = \int d\mathbf{v}_1' \int d\mathbf{v}_2' \int d\mathbf{v}_2 \, W(\mathbf{v}_1 \to \mathbf{v}_1'; \mathbf{v}_2 \to \mathbf{v}_2') f_M(\mathbf{v}_2) \tag{2.18}$$

and:

$$\tilde{H}(\mathbf{v}_1' \to \mathbf{v}_1) = 2\int d\mathbf{v}_2' \int d\mathbf{v}_2 \, f_M(\mathbf{v}_2) W(\mathbf{v}_1 \to \mathbf{v}_1'; \mathbf{v}_2 \to \mathbf{v}_2')$$
$$- f_M(\mathbf{v}_1') \int d\mathbf{v}_2' \int d\mathbf{v}_2 \, W(\mathbf{v}_1 \to \mathbf{v}_2; \mathbf{v}_1' \to \mathbf{v}_2') \tag{2.19}$$

Given W(...), $\tilde{V}$ and $\tilde{H}$ are available. We have also assumed that f_M does not depend on $\mathbf{r}$ or t.

The linear transport equation, Eqn. (2.14), has been the focus of a great deal of study for a variety of geometric and physical situations. Details may be found in Williams [1971] and Cercignani [1969, 1975]. Our interest in the equation stems from its ability to lead to the drag forces on small particles moving in a gas. We should bear in mind here that the size range of interest is very broad; going from $a=10^{-3}$ μm up to perhaps 50 or 100 μm. Now, the mean free path, λ_g, of gas atoms in air at STP is about 0.065 μm. Thus, it is clear that the Knudsen number, Kn (defined as $Kn=\lambda_g/a$) ranges from virtually zero up to around 70. It is very important to know the Knudsen number because it tells us about the flow regime of the particle. For example, for small Kn where the particle is many mean free paths in size, the granular nature of the gas becomes insignificant and it may be regarded as a continuous fluid. On the other hand, for large Kn, the particle experiences forces that depend sensitively on the gas atom distribution function. Thus, for small Kn, we may employ the hydrodynamic limit of the Boltzmann equation whereas for large Kn it is necessary to use the transport equation.

The Motion of Particles in Gases

Because the hydrodynamic limit is so much easier to work with analytically there are compelling reasons to use this approach when possible. Thus, we will show how the continuum equations can be derived from the Boltzmann equation.

Multiply Eqn. (2.6) by an arbitrary function of velocity $\phi(v_1)$ and integrate over all v_1. The result is:

$$\frac{\partial}{\partial t}(\rho\overline{\phi}) + \nabla_r \cdot \rho \overline{v_1 \phi} - \rho \overline{\nabla_v \cdot a \phi} = \rho \overline{\Delta\phi} \tag{2.21}$$

where a bar implies the following average:

$$\overline{\phi} = \frac{1}{\rho}\int dv_1 \, \phi f \tag{2.22}$$

The term on the right hand side can be reduced by use of symmetry to the form (Williams [1971]):

$$\rho \overline{\Delta\phi} = \tfrac{1}{4} \int dv_1 \int dv_1' \int dv_2 \int dv_2' \, W(v_1 \to v_1'; v_2 \to v_2')$$
$$[f(v_1')f(v_2') - f(v_1)f(v_2)][\phi(v_1) + \phi(v_2) - \phi(v_1') - \phi(v_2')] \tag{2.23}$$

We note from this that if $\phi(v)=1$, v, or v^2, that $\overline{\Delta\phi}=0$. This follows from the conservation of mass, momentum, and energy. More generally, we can take ϕ to be the mass, m, the relative momentum, $m(v-u)$, or the thermal energy, $m(v-u)^2/2$. Eqn. (2.21) now leads to the continuity equation:

$$\frac{\partial \rho}{\partial t} + \nabla \cdot (\rho u) = 0 \tag{2.24}$$

the momentum equation (using the continuity equation):

$$\rho\left(\frac{\partial u_i}{\partial t} + \sum_{j=1}^{3} u_j \frac{\partial u_i}{\partial x_j}\right) = -\sum_{j=1}^{3} \frac{\partial P_{ij}}{\partial x_j} + \rho F_i \tag{2.25}$$

which in terms of vectors and dyadics reads:

$$\rho\left(\frac{\partial u}{\partial t} + (u \cdot \nabla u)\right) = -(\nabla \cdot P) + \rho F \tag{2.26}$$

and the energy equation:

$$\frac{3k}{2m}\rho\left(\frac{\partial T}{\partial t} + (u \cdot \nabla T)\right) + (\nabla \cdot q) = -(P \colon D) \tag{2.27}$$

where the heat flux vector:

$$q = \tfrac{1}{2} m \rho \overline{(v-u)(v-u)^2} \tag{2.28}$$

and:
$$D_{ij} = \tfrac{1}{2}\left(\frac{\partial u_i}{\partial x_j} + \frac{\partial u_j}{\partial x_i}\right)$$

or:
$$\mathbf{D} = \tfrac{1}{2}\left(\nabla \mathbf{u} + (\nabla \mathbf{u})^+\right) \tag{2.29}$$

Eqns. (2.24)-(2.27) are not a closed set since we have more unknowns than equations. Before proceeding it is therefore necessary to obtain further relationships for $\mathbf{P}$ and $\mathbf{q}$. In other words, we need the distribution function $f(\mathbf{r},\mathbf{v},t)$. As an exercise, let us assume that the gas is everywhere in local equilibrium with a distribution:

$$f(\mathbf{r},\mathbf{v},t) = n(\mathbf{r},t)\left(\frac{m}{2\pi k T(\mathbf{r},t)}\right)^{3/2} \exp\left\{-\frac{m}{2kT(\mathbf{r},t)}[\mathbf{v}-\mathbf{u}(\mathbf{r},t)]^2\right\} \tag{2.30}$$

In that case we readily find $P_{ij} = nkT\delta_{ij}$ and $\mathbf{q} = 0$. Thus, there is no net heat flux and the pressure is given by the normal pressure p=nkT. The equations therefore become:

$$\rho\left(\frac{\partial \mathbf{u}}{\partial t} + (\mathbf{u} \bullet \nabla \mathbf{u})\right) + \nabla p = \rho \mathbf{F} \tag{2.31}$$

and:
$$\left(\frac{\partial}{\partial t} + (\mathbf{u} \bullet \nabla)\right)T + \frac{2}{3m} T \nabla \bullet \mathbf{u} = 0 \tag{2.32}$$

These equations are recognizable as the hydrodynamic equations of inviscid flow, *i.e.* the Euler equations. They are useful in regions where the effect of viscosity and heat conduction can be neglected. For our purposes, however, we need to include viscosity and heat flow.

It is possible from the linearized Boltzmann equation to obtain relationships for the pressure and heat flux vector in terms of the fundamental force laws between atoms or molecules. Such matters are discussed in Chapman and Cowling [1960] and will not be elaborated on here. Expressions for $\mathbf{P}$ and $\mathbf{q}$ are available as:

$$P_{ij} = p\delta_{ij} - 2\mu\left[D_{ij} - \tfrac{1}{3}\delta_{ij}(\nabla \bullet \mathbf{u})\right]$$

or:
$$\mathbf{P} = p\mathbf{I} - 2\mu\left\{\tfrac{1}{2}\left[\nabla \mathbf{u} + (\nabla \mathbf{u})^+\right] - \tfrac{1}{3}\mathbf{I}(\nabla \bullet \mathbf{u})\right\} \tag{2.33}$$

and:
$$\mathbf{q} = -\lambda \nabla T \tag{2.34}$$

where μ is the viscosity and λ is the coefficient of thermal conductivity and both are given in terms of certain averages over the differential scattering kernel $\sigma(v,\theta)$.

Utilizing Eqn. (2.33) we find that the momentum equation, Eqn. (2.26), can be written:

$$\rho\frac{\partial \mathbf{u}}{\partial t} + \rho \mathbf{u} \bullet \nabla \mathbf{u} = -\nabla p + \mu \nabla^2 \mathbf{u} + \tfrac{1}{3}\mu \nabla(\nabla \bullet \mathbf{u}) + \rho \mathbf{F} \tag{2.35}$$

The Motion of Particles in Gases

where we note the relationships:

$$\nabla \cdot (\nabla u)^+ \equiv \nabla(\nabla \cdot u)$$
$$\nabla \cdot (\nabla u) \equiv \nabla^2 u$$
$$\nabla \cdot (I(\nabla \cdot u)) \equiv \nabla(\nabla \cdot u)$$

If the fluid is incompressible, which is generally true for the low speed flow of gases, then we have $(\nabla \cdot u) = 0$. Thus, the equation governing the fluid is:

$$\rho \frac{\partial u}{\partial t} + \rho u \cdot \nabla u = -\nabla p + \mu \nabla^2 u + \rho F \tag{2.36}$$

which is one form of the Navier-Stokes equations. When $\mu = 0$ we regain the Euler equations.

Eqn. (2.36) may be written in various coordinate systems and, as an example, we show it in Cartesian coordinates (x,y,z) where:

$$u = i u_x + j u_y + k u_z$$

$$F = i F_x + j F_y + k F_z$$

$$\frac{\partial u_x}{\partial t} + u_x \frac{\partial u_x}{\partial x} + u_y \frac{\partial u_x}{\partial y} + u_z \frac{\partial u_x}{\partial z} = -\frac{1}{\rho}\frac{\partial p}{\partial x} + \frac{\mu}{\rho}\left(\frac{\partial^2 u_x}{\partial x^2} + \frac{\partial^2 u_x}{\partial y^2} + \frac{\partial^2 u_x}{\partial z^2}\right) + F_x \tag{2.37a}$$

$$\frac{\partial u_y}{\partial t} + u_x \frac{\partial u_y}{\partial x} + u_y \frac{\partial u_y}{\partial y} + u_z \frac{\partial u_y}{\partial z} = -\frac{1}{\rho}\frac{\partial p}{\partial y} + \frac{\mu}{\rho}\left(\frac{\partial^2 u_y}{\partial x^2} + \frac{\partial^2 u_y}{\partial y^2} + \frac{\partial^2 u_y}{\partial z^2}\right) + F_y \tag{2.37b}$$

and:

$$\frac{\partial u_z}{\partial t} + u_x \frac{\partial u_z}{\partial x} + u_y \frac{\partial u_z}{\partial y} + u_z \frac{\partial u_z}{\partial z} = -\frac{1}{\rho}\frac{\partial p}{\partial z} + \frac{\mu}{\rho}\left(\frac{\partial^2 u_z}{\partial x^2} + \frac{\partial^2 u_z}{\partial y^2} + \frac{\partial^2 u_z}{\partial z^2}\right) + F_z \tag{2.37c}$$

In addition, the continuity equation is written:

$$\frac{\partial u_x}{\partial x} + \frac{\partial u_y}{\partial y} + \frac{\partial u_z}{\partial z} = 0 \tag{2.38}$$

Similarly, the components of the pressure tensor are:

$$P_{xx} = p - 2\mu \frac{\partial u_x}{\partial x} \tag{2.39}$$

$$P_{yy} = p - 2\mu \frac{\partial u_y}{\partial y} \tag{2.40}$$

$$P_{zz} = p - 2\mu \frac{\partial u_z}{\partial z} \tag{2.41}$$

$$P_{xy} = P_{yx} = \mu \left(\frac{\partial u_x}{\partial y} + \frac{\partial u_y}{\partial x} \right) \tag{2.42}$$

$$P_{yz} = P_{zy} = \mu \left(\frac{\partial u_y}{\partial z} + \frac{\partial u_z}{\partial y} \right) \tag{2.43}$$

and:

$$P_{zx} = P_{xz} = \mu \left(\frac{\partial u_z}{\partial x} + \frac{\partial u_x}{\partial z} \right) \tag{2.44}$$

The boundary conditions on Eqn. (2.36) are that the relative tangential velocity at the surface is zero, *i.e.* no slip, and the normal velocity is equal to that of the boundary. Thus, for a stationary body, the overall vector boundary condition is:

$$v = 0 \tag{2.45}$$

Eqn. (2.36) is a nonlinear equation and in general admits only a very few exact analytical solutions. However, for low speed flows where inertial effects are small and there are no rapid changes with time, it seems physically reasonable that the left hand side of Eqn. (2.36) will be small compared with the right hand side. In order to quantify this, let us introduce two dimensionless numbers, $Re_S = aU/\nu$ and $Re_T = a^2\omega/\nu$, where $\nu = \mu/\rho$ is the kinematic viscosity, a is the typical dimension of a body, U is the asymptotic flow velocity, and ω is a typical frequency of oscillation for some time dependent disturbance. Re_S and Re_T are Reynolds numbers for space and time dependent variations, respectively. If we scale space, time, velocity, and pressure as follows:

$$\tilde{t} = \omega t \quad , \quad \tilde{r} = \frac{r}{a} \quad , \quad \tilde{u} = \frac{u}{U} \quad , \quad \tilde{p} = \frac{p\,a}{\mu U}$$

we find that Eqn. (2.36) may be written:

$$Re_T \frac{\partial \tilde{u}}{\partial t} + Re_S \, \tilde{u} \bullet \tilde{\nabla} \tilde{u} = -\tilde{\nabla} \tilde{p} + \tilde{\nabla}^2 \tilde{u} \tag{2.46}$$

where the force term has been absorbed into $\tilde{p}$.

Three situations arise: (1) if Re_T is sufficiently small we can neglect the time derivatives; (2) if Re_S is sufficiently small we can neglect the nonlinear inertial terms, and; (3) if both Re_S and Re_T are small the entire left hand side may be neglected.

For most situations that arise in aerosol problems, case (3) is valid and we have the equation:

$$\mu \nabla^2 u = \nabla p \tag{2.47}$$

to describe the fluid. Eqn. (2.47) is known as the slow viscous flow equation and the physical process it describes is called creeping motion or Stokes flow. As long as the Reynolds number is small this equation will describe the flow over bodies and surfaces very accurately. Eqn. (2.47) has the marked advantage of being linear.

The Motion of Particles in Gases

If there are rapid time variations in the fluid (*e.g.*, harmonic oscillations) then it may not be possible to neglect the time derivative term. In that case we must solve:

$$\rho \frac{\partial \mathbf{u}}{\partial t} = \mu \nabla^2 \mathbf{u} - \nabla p \tag{2.48}$$

which is still linear.

An improvement to the slow viscous flow equations was made by Oseen [1910, 1927] who approximated the inertial term by:

$$\mathbf{u} \cdot \nabla \mathbf{u} \cong \mathbf{U} \cdot \nabla \mathbf{u} \tag{2.49}$$

where $\mathbf{U}$ is the uniform stream velocity prevailing at some distance from the body under investigation. In this way, the Stokes equations become:

$$\mu \nabla^2 \mathbf{u} - \nabla p = \rho \, \mathbf{U} \cdot \nabla \mathbf{u} \tag{2.50}$$

This equation, which is also linear, extends the range of calculated properties (*e.g.*, drag) to higher values of Reynolds number. It has been studied extensively and its theory placed on a sound footing by Proudman and Pearson [1957]. A useful discussion on this matter is given by Happel and Brenner [1973] and we also make use of the theory in Chapter 4.

2.3 The drag on a body

Before further discussion of either the Boltzmann equation or the equations of hydrodynamics, it will be useful to define the drag force on a body. Clearly, as a body moves through a fluid or as the fluid moves over a fixed body, it will exert a frictional force or a drag. It is vital to calculate this drag for various shapes of the body since it will tell us, via the equation of motion, how rapidly the body moves under various external forces. The classic case would be the gravitational settling of a sphere in air. In order to quantify the resistive force we must calculate the net effect of friction over the surface of the body.

Consider a body immersed in a fluid and let dS be an element of surface normal to the body and pointing into the fluid. The resultant force due to the local stresses exerted by the fluid on the body is:

$$\mathbf{F} = \int_S \mathbf{P} \cdot d\mathbf{S} \tag{2.51}$$

where $\mathbf{P}$ is the pressure tensor given either by Eqn. (2.5) or by Eqn. (2.33). In the case of the Boltzmann equation it will be necessary to know the complete velocity distribution function of the particles and in the case of hydrodynamic theory the somewhat less demanding solution of the continuum equations is required.

2.4 The torque on a body

Due to nonuniform flow or because of an asymmetry in its shape, a body moving in a fluid can experience rotation. The forces leading to this rotation generate a torque. In order to calculate the net effect, it is necessary to sum the turning moments over the surface of the body.

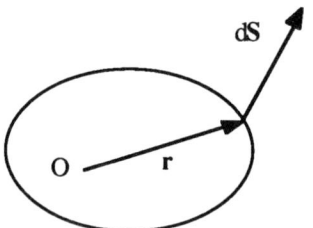

Figure 2.1: Coordinates for torque on a body

If in Fig. 2.1, O is an origin and **r** is a vector from that origin to the point on the surface where the normal area pointing out of the body is dS, then, the elementary torque at that point is $\mathbf{r} \times \mathbf{P} \cdot d\mathbf{S}$ and hence the total torque is:

$$\mathbf{T} = \int_S \mathbf{r} \times \mathbf{P} \cdot d\mathbf{S}$$

Examples of drag and torque are given below.

2.5 Slip flow

The boundary condition imposed on the velocity in the equations of hydrodynamics is that the tangential velocity at the surface is zero. That is, at the surface the fluid is firmly attached to the surface. Solution of the Boltzmann equation shows, however, that this is not the case (Williams [1971]). For example, in the case of a gas (x>0) moving parallel to a fixed wall (at x=0), it is found that at the wall surface the gas velocity is not zero, *i.e.* there is a slippage of the gas over the surface. This effect arises from the atomic nature of the gas and cannot be predicted from continuum theory.

In the very special case of the gas moving over a plane wall it turns out that an exact solution can be obtained for certain special models of scattering. The average flow velocity of the gas can be shown to assume the form:

$$u(x) = K_0 x + \tfrac{1}{2} B_0 + u_{tr}(x) \tag{2.51}$$

As Fig. 2.2 shows, there are two well defined regions as we move away from the wall. Near the wall, to be precise within a few mean free paths of the wall, the average flow velocity $u(x)$ varies rapidly from a finite value $u(0)$ at the surface to an asymptotic value $K_0 x + B_0/2$. The term $u_{tr}(x)$ is a rapidly decaying function of x which is negligible at a few mean paths from the surface. The asymptotic form:

$$u_{asy}(x) = K_0 x + \tfrac{1}{2} B_0 \tag{2.52}$$

satisfies the hydrodynamic equation, provided that the following boundary condition is imposed:

$$\zeta_v \frac{d}{dx} u_{asy}(x) \bigg|_{x=0} = u_{asy}(0) \tag{2.53}$$

The Motion of Particles in Gases

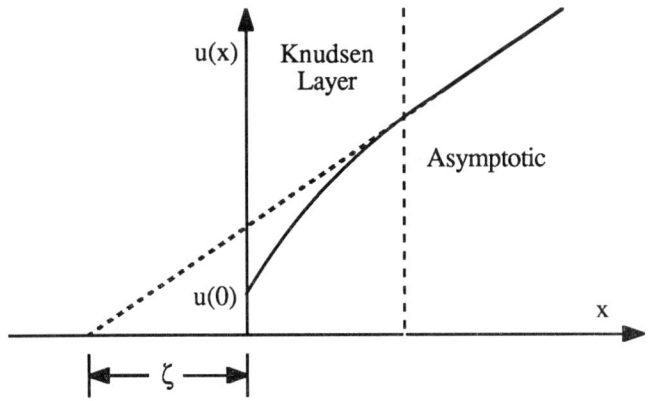

Figure 2.2: The Knudsen layer and slip coefficient.

where $\zeta_v = B_0/2K_0$. The value of ζ_v depends to some extent on the gas-atom scattering kernel and the wall-gas interaction but is generally on the order of a mean free path. It is clear, therefore, that if the scale of distance at which we are working is comparable to or less than a mean free path then this slip correction will be important. If, on the other hand, the body is many mean free paths in extent, the effect is negligible and the condition $u_{asy}(0)=0$ is adequate.

Although the above argument was based on a plane boundary, it may be shown that all solutions of the Boltzmann equation for flow over surfaces consist of an asymptotic solution, which satisfies the hydrodynamic equations, plus a rapidly varying term which decays to a negligible value at several mean free paths from the surface, but which is appreciable near the surface. Thus small spheres whose radii are comparable to a mean free path will certainly be influenced by this slip effect.

Use of the slip condition, as shown in its most simple form in Eqn. (2.53), enables the equations of hydrodynamics to be used in a range of Knudsen numbers much larger than might be expected. Indeed, in many aerosol problems, the slip correction covers the complete range of particle sizes of practical interest.

When the theory of slip is extended to bodies of more complex shape, the boundary condition has to be generalized. Thus, we write:

$$u_s - C_m \frac{\lambda_g}{\mu} P_{ns} = 0 \tag{2.54}$$

where u_s is the tangential velocity and P_{ns} is a component of the stress tensor with **n** being the direction normal to the surface and **s** the direction tangential to it. C_m is a slip correction factor and is around unity but depends weakly on the gas-surface interaction. If the body is moving, then the right hand side of Eqn. (2.54) would be equal to the tangential component of this velocity.

2.6 The drag on an isolated sphere

The most fundamental problem in aerosol dynamics concerns the manner in which a sphere moves in an infinite fluid. Historically, this problem was first tackled by Stokes [1845, 1851] using hydrodynamic theory, when he derived his famous formula for the frictional drag:

$$F_d = 6\pi a \mu U \tag{2.55}$$

This formula is limited to spheres that are large compared to a mean free path and therefore is invalid for small particles. Stokes' formula was modified by various workers for larger Knudsen numbers by the introduction of an empirical factor C such that:

$$F_d = \frac{6\pi a \mu U}{C} \tag{2.56}$$

where:

$$C = 1 + Kn\left[A_1 + A_2 \exp\left(-\frac{A_3}{Kn}\right)\right] \tag{2.57}$$

This factor is called variously the Stokes-Cunningham factor, the Cunningham factor, or the Knudsen-Weber factor. Kn is the Knudsen number, λ_g/a, where λ_g is the mean free path and a is the radius of the sphere.

A_i are constants the values of which vary from author to author but are given by Friedlander [1977] as $A_1=1.257$, $A_2=0.400$, $A_3=1.1$. This modification enables the Stokes drag to be used up to about Kn=10 although, with some further adjustment of the A_i, Eqn. (2.56) can be made to span the complete range of Kn.

Very small particles, where Kn>>1, can be treated as large molecules and a particle velocity distribution function can be found using kinetic theory from which, by direct averaging, the drag is obtained. For a simple model of diffuse reflection it may be shown that (Williams [1973]):

$$F_D = \frac{4\pi a^2}{3} \rho_g \left(\frac{8kT}{\pi m}\right)^{1/2} \left\{1 + \frac{\pi}{8}\right\} U \tag{2.58}$$

where ρ_g is the gas density and m is the mass of a gas atom. Thus, Eqns. (2.55), (2.56), and (2.58) span the complete range of Knudsen numbers although there is a gap in the range 10<Kn<∞.

It is useful to consider the limiting value of F_D as Kn→∞ in Eqn. (2.57). Then we find:

$$F_D \rightarrow \frac{6\pi a^2 \mu U}{\lambda_g (A_1 + A_2)} \tag{2.58a}$$

Noting that $A_1+A_2=(1.657)$, and from kinetic theory (Kennard [1938]) that $\mu/\lambda_g=\rho_g\bar{v}/3$ we find that Eqn. (2.58a) differs from the exact value only by a factor of 0.65. Thus, as we mentioned above, some adjustment of the A_i could lead to a formula which is accurate over the whole range of Knudsen numbers.

Useful as these formulae have been, it is clearly desirable to have a result which is self-consistent and predicts the drag over the complete range of Knudsen numbers. This calculation was first carried out by Cercignani and Pagani [1968] and Cercignani et al. [1968] using the integral form of the Boltzmann transport equation. In order to simplify the equation, these authors employed the BGK (Bhatnagar, Gross, and Krook) model for the gas atom scattering and a perfect accommodation model for the gas-wall interaction. To obtain the drag, a variational method is used the results of which are presented in the form:

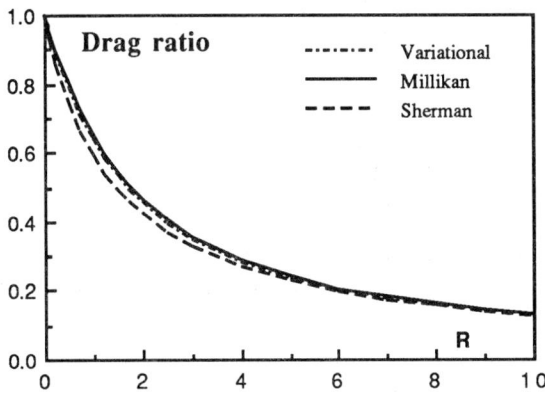

Figure 2.3: Comparison of variational results with Millikan's formula and Sherman's law. Adapted from Cercignani *et al.* [1968] with permission of the American Institute of Physics.

Table 2.1: Drag on a sphere versus the inverse Knudsen number R. Adapted from Cercignani *et al.* [1968] with permission of the American Institute of Physics.

R	Variational results for D	Millikan's formula for D	R	Variational results for D	Millikan formula for D
0.05	0.9778	0.9784	2.0	0.4546	0.4645
0.075	0.9651	0.9677	2.5	0.3951	0.4029
0.1	0.9529	0.9571	3.0	0.3488	0.3551
0.25	0.8864	0.8959	4.0	0.2818	0.2863
0.5	0.7900	0.8036	5.0	0.2360	0.2396
0.75	0.7088	0.7236	6.0	0.2029	0.2058
1.0	0.6404	0.6549	7.0	0.1779	0.1804
1.25	0.5824	0.5961	8.0	0.1583	0.1606
1.5	0.5332	0.5456	9.0	0.1426	0.1447
1.75	0.4910	0.5021	10.0	0.1297	0.1317

$$D = \frac{F_D(Kn)}{F_D(\infty)} = \frac{1}{8+\pi}\left(8 - \frac{3 J_{st}}{2\sqrt{\pi} R^2 U^2}\right) \tag{2.59}$$

where $U = U_\infty / \bar{c}$, $R = a/\theta\bar{c}\,(= Kn^{-1})$, $\theta = \mu/p$, $\lambda_g = \theta\bar{c}$, and $\bar{c} = (2kT/m)^{1/2}$.

U_∞ is the velocity of the fluid. J_{st} is the stationary value of a complicated functional which has been numerically calculated by Cerergnani *et al.* [1968]. The results are shown in Fig. 2.3 and in Table 2.1.

It is clear that for large and small values of Knudsen number the variational result is exact. The authors state that errors due to the BGK model and the approximate nature of the trial function are likely to lead to errors in D not exceeding 1%.

In the figure the curve marked 'Millikan' refers to the use of a modified Stokes law including the factor, C, while the curve marked 'Sherman' refers to a simple interpolation formula of the form:

$$D = [1 + (0.685) R]^{-1} \tag{2.60}$$

Sherman's formula can deviate from the exact results by as much as 10% but may be useful in certain approximate calculations. This has been verified by Lea and Loyalka [1980] and Law and Loyalka [1986] who solved numerically the integral, BGK based, transport equation.

2.7 Stokes flow in the continuum regime

In most problems in aerosol dynamics we shall find that the continuum equations of hydrodynamics provide an adequate description of particle motion if they are supplemented by the slip boundary conditions. There are exceptions, but they will be dealt with as appropriate. Thus, in this section, we examine the solution of the slow viscous flow equations for a number of practical situations.

In general, three dimensional fluid flow problems are very difficult mathematically. There exists, however, a certain class of problems in which axial symmetry permits significant simplification. Such axi-symmetric problems are characterized by a scalar function called a stream function, ψ. Consider the coordinate system shown in Fig. 2.4.

The curve AB is the surface of a body of revolution about the z-axis. **n** and **t** are unit vectors normal and tangential to the curve, respectively. Symmetry is assumed about the z-axis and we consider (z,ϖ) to be the independent variables. The vector $\mathbf{i}_\phi$ is pointing out of the plane of the paper so that the system is described by the set of unit vectors $(\mathbf{n},\mathbf{t},\mathbf{i}_\phi)$.
The Stokes stream function ψ is related to the velocity of the fluid by (Happel and Brenner [1973]):

$$\mathbf{u} = \frac{1}{\varpi} \mathbf{i}_\phi \times \nabla \psi \tag{2.61}$$

Thus, in the cylindrical coordinates (z,ϖ):

$$u_\varpi = \frac{1}{\varpi}\frac{\partial \psi}{\partial z} \quad \text{and} \quad u_z = -\frac{1}{\varpi}\frac{\partial \psi}{\partial \varpi} \tag{2.62}$$

The Motion of Particles in Gases

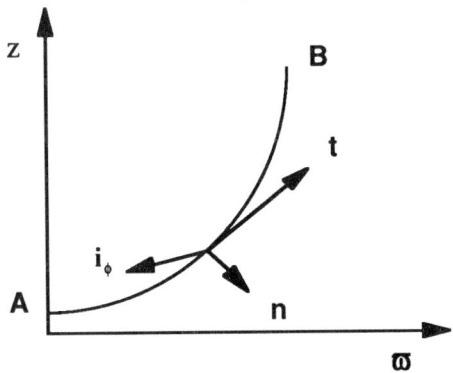

Figure 2.4: A coordinate system for Stokes flow.

More generally, if q_1, q_2, and ϕ constitute a right-handed system of orthogonal curvilinear coordinates, then:

$$u_1 = -\frac{h_2}{\varpi}\frac{\partial \psi}{\partial q_2} \quad \text{and} \quad u_2 = \frac{h_1}{\varpi}\frac{\partial \psi}{\partial q_1} \tag{2.63}$$

where $\varpi = \varpi(q_1, q_2)$ and h_1 and h_2 are the metric coefficients. An equation for the stream function ψ can be obtained as follows. The vorticity vector, ζ, is:

$$\zeta = \nabla \times \mathbf{u} = \mathbf{i}_\phi \left(\frac{\partial u_\varpi}{\partial z} - \frac{\partial u_z}{\partial \varpi}\right) = \mathbf{i}_\phi \left\{\frac{\partial}{\partial z}\left(\frac{1}{\varpi}\frac{\partial \psi}{\partial z}\right) + \frac{\partial}{\partial \varpi}\left(\frac{1}{\varpi}\frac{\partial \psi}{\partial \varpi}\right)\right\} = \mathbf{i}_\phi \frac{1}{\varpi}E^2\psi \tag{2.64}$$

where the operator E^2 is given by:

$$E^2 = \varpi \frac{\partial}{\partial \varpi}\left(\frac{1}{\varpi}\frac{\partial}{\partial \varpi}\right) + \frac{\partial^2}{\partial z^2} \tag{2.65}$$

It is readily verified that:

$$\nabla \times (\nabla \times \zeta) = -\mathbf{i}_\phi \frac{1}{\varpi}E^4\psi \tag{2.66}$$

Let us now return to Eqn. (2.48):

$$\frac{\partial \mathbf{u}}{\partial t} = \nu \nabla^2 \mathbf{u} - \frac{1}{\rho}\nabla p \tag{2.67}$$

and note that:

$$\nabla^2 \mathbf{u} = \nabla(\nabla \cdot \mathbf{u}) - \nabla \times (\nabla \times \mathbf{u}) \tag{2.68}$$

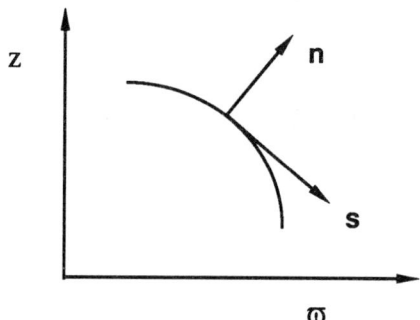

Figure 2.5: An intrinsic coordinate system.

Taking the curl throughout and using $(\nabla \cdot \mathbf{u})=0$, we get:

$$\frac{\partial \zeta}{\partial t} = -\nu \nabla \times (\nabla \times \zeta) \qquad (2.69)$$

where $(\nabla \times \nabla p)=0$. Then from Eqns. (2.64) and (2.66) we find:

$$\frac{\partial}{\partial t}(E^2 \psi) = \nu (E^4 \psi) \qquad (2.70)$$

which is a linear equation for the scalar ψ. In most cases we may neglect the time dependent term such that Eqn. (2.70) reduces to:

$$E^4 \psi = 0 \qquad (2.71)$$

The boundary conditions on Eqn. (2.71) can be obtained by reference to what Happel and Brenner [1973] call 'intrinsic coordinates'. Fig. 2.5 illustrates these coordinates where $\mathbf{n}$ is the unit normal to the surface and $\mathbf{s}$ is the unit vector in the tangent direction.

In terms of these coordinates, the magnitude of the normal velocity, u_n, and the magnitude of the tangential velocity, u_s, are given by:

$$u_n = -\frac{1}{\varpi}\frac{\partial \psi}{\partial s} \qquad (2.72)$$

and:

$$u_s = \frac{1}{\varpi}\frac{\partial \psi}{\partial n} \qquad (2.73)$$

The problem that will most commonly be encountered is one in which an axi-symmetric body is moving in the z-direction with a velocity $\mathbf{u}=\mathbf{i}_z U$. Thus, with respect to an observer at rest, the impenetrability of the surface is described by:

$$(\mathbf{u} - \mathbf{U}) \cdot \mathbf{n} = 0 \qquad (2.74)$$

The Motion of Particles in Gases

In the case of no tangential relative velocity:

$$(\mathbf{u} - \mathbf{U}) \cdot \mathbf{s} = 0 \tag{2.75}$$

Eqn. (2.74) may be written, using Eqn. (2.72), and the fact that:

$$\mathbf{n} \cdot \mathbf{i}_z = \frac{\partial \varpi}{\partial s}$$

as:

$$-\frac{1}{\varpi} \frac{\partial \psi}{\partial s} - \frac{\partial \varpi}{\partial s} U = 0$$

or:

$$\frac{\partial}{\partial s}\left(\psi + \tfrac{1}{2} \varpi^2 U\right) = 0$$

Integrating this around the body, we find:

$$\psi + \tfrac{1}{2} \varpi^2 U = 0 \tag{2.76}$$

where we have defined $\psi=0$ at $\varpi=0$, i.e., along the axis of the body.

Similarly, Eqn. (2.75) may be written with the help of Eqn. (2.73) and:

$$\mathbf{s} \cdot \mathbf{i}_z = -\frac{\partial \varpi}{\partial n} \tag{2.77}$$

as:

$$\frac{\partial}{\partial n}\left(\psi + \tfrac{1}{2} \varpi^2 U\right) = 0 \tag{2.78}$$

If the surface coincides with a member of the curvilinear coordinate system, say q_N, then Eqn. (2.78) becomes:

$$\frac{\partial}{\partial q_N}\left(\psi + \tfrac{1}{2} \varpi^2 U\right) = 0 \tag{2.79}$$

Finally, it is also necessary that:

$$\lim_{r \to \infty}\left(\frac{\psi}{r^2}\right) = 0$$

where $r^2 = \varpi^2 + z^2$ and the velocity is zero at infinity.

If the tangential boundary condition on $\mathbf{u}$ is of the slip type, then Eqn. (2.75) becomes:

$$(\mathbf{u} - \mathbf{U}) \cdot \mathbf{s} = C_m \frac{\lambda_g}{\mu} P_{ns} \tag{2.80}$$

where:

$$P_{ns} = \mu\left(\frac{\partial u_n}{\partial s} + \frac{\partial u_s}{\partial n}\right) \tag{2.81}$$

whence:

$$\frac{\partial}{\partial n}\left(\psi + \tfrac{1}{2}\varpi^2 U\right) = C_m \lambda_g \left(-\frac{\partial}{\partial s}\frac{1}{\varpi}\frac{\partial \psi}{\partial s} + \frac{\partial}{\partial n}\frac{1}{\varpi}\frac{\partial \psi}{\partial n}\right) \qquad (2.82)$$

2.8 The drag

We also wish to know how to calculate the drag directly from ψ. Let us recall Eqn. (2.34) for the stress dyadic:

$$\mathbf{P} = p\mathbf{I} - \mu\left[\nabla\mathbf{u} + (\nabla\mathbf{u})^*\right] \qquad (2.83)$$

Then, in terms of the coordinate system $(\mathbf{n},\mathbf{s},\mathbf{i}_\phi)$ and with:

$$\mathbf{I} = \mathbf{nn} + \mathbf{ss} + \mathbf{i}_\phi\mathbf{i}_\phi \quad \text{and} \quad \mathbf{u} = \mathbf{n}\,u_n + \mathbf{s}\,u_s$$

we find:

$$\mathbf{P}_n = \mathbf{P}\cdot\mathbf{n} = \mathbf{n}\left(p - 2\mu\frac{\partial u_n}{\partial n}\right) - \mathbf{s}\mu\left(\frac{\partial u_n}{\partial s} + \frac{\partial u_s}{\partial n}\right) \qquad (2.84)$$

Rearranging and noting that:

$$\boldsymbol{\zeta} = \nabla\times\mathbf{u} = \mathbf{i}_\phi\left(\frac{\partial u_s}{\partial n} - \frac{\partial u_n}{\partial s}\right) \qquad (2.85)$$

we can write $\mathbf{P}_n$ as:

$$\mathbf{P}_n = p\mathbf{n} - 2\mu\,\nabla u_n - \mathbf{s}\mu\left(\mathbf{i}_\phi\cdot\boldsymbol{\zeta}\right) = p\mathbf{n} + \nabla\left(\frac{1}{\varpi}\frac{\partial\psi}{\partial s}\right) - \mathbf{s}\mu\frac{1}{\varpi}E^2\psi \qquad (2.86)$$

Since the symmetry of the flow leads only to a force in the axial direction we find:

$$F_z = \int_S \mathbf{P}_z\cdot d\mathbf{S} = \int_S P_{nz}\,dS \qquad (2.87)$$

where $dS = 2\pi\varpi\,ds$.

It is shown by Happel and Brenner [1973], using Eqn. (2.86), that the drag reduces to the calculation of:

$$F_z = \mu\pi\int_S \varpi^3 \frac{\partial}{\partial n}\left(\frac{E^2\psi}{\varpi^2}\right)dS \qquad (2.88)$$

Sometimes it is convenient to divide the drag into two components, one due to the normal stress and one due to the tangential stress. Thus, we write Eqn. (2.86) as $\mathbf{P}_n = \mathbf{P}_N + \mathbf{P}_T$ where $\mathbf{P}_N = \mathbf{n}p$ and $\mathbf{P}_T$ is the remaining part.

The Motion of Particles in Gases

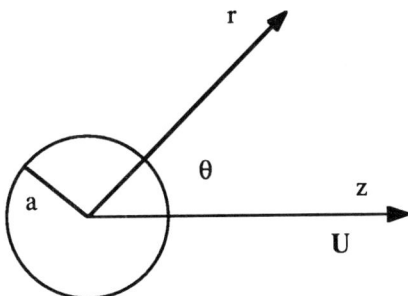

Figure 2.6: A coordinate system for the motion of a sphere.

In this case the drag is:

$$F_z = p \int_S (\mathbf{n} \cdot \mathbf{i}_z) \, dS + \int_S (\mathbf{P}_T \cdot \mathbf{i}_z) \, dS \tag{2.88a}$$

The term involving $\mathbf{P}_N$ is called the form or profile drag while that involving $\mathbf{P}_T$ is called the skin or viscous drag.

This formula is completely general and can be used for bounded fluids. On the other hand, it is usually very tedious to evaluate the integral. For this reason an alternative representation of the drag is often used. This representation is restricted to an unbounded fluid but in practice this is not often a hindrance. The result, due to Payne and Pell [1960], can be written:

$$F_z = 8\pi\mu \lim_{r \to \infty} \left(\frac{r\psi}{\varpi^2} \right) \tag{2.89}$$

This expression arises if one assumes that the stream function at a sufficiently large distance from the sphere is the same as that produced by a point force of magnitude equal to the drag on the sphere. If the fluid is not at rest at infinity but can be described by a stream function ψ_∞, then Eqn. (2.89) is modified to:

$$F_z = 8\pi\mu \lim_{r \to \infty} \left(\frac{r(\psi - \psi_\infty)}{\varpi^2} \right) \tag{2.90}$$

2.9 Drag on a sphere

To illustrate some of the general ideas outlined above, let us consider the case of a sphere of radius a moving in a stationary fluid with a steady velocity U in the z-direction. Fig. 2.6 shows the situation.

In this case, $\varpi = r \sin(\theta)$, $h_1 = 1$, $h_2 = 1/r$, $q_1 = r$, $q_2 = \theta$.

Thus:

$$u_r = -\frac{1}{r^2 \sin(\theta)} \frac{\partial \psi}{\partial \theta} \tag{2.91}$$

and:
$$u_\theta = \frac{1}{r\sin(\theta)} \frac{\partial \psi}{\partial r} \tag{2.92}$$

and we have to satisfy:
$$\left[\frac{\partial^2}{\partial r^2} + \frac{\sin(\theta)}{r^2}\frac{\partial}{\partial \theta}\left(\frac{1}{\sin(\theta)}\frac{\partial}{\partial \theta}\right)\right]^2 \psi = 0 \tag{2.93}$$

subject to:
$$\psi(a,\theta) + \tfrac{1}{2} U a^2 \sin^2(\theta) = 0 \tag{2.94}$$

$$\left.\frac{\partial \psi}{\partial r}\right|_{r=a} + U a \sin^2(\theta) = 0 \tag{2.95}$$

and:
$$\lim_{r\to\infty}\left(\frac{\psi}{r^2}\right) = 0 \tag{2.96}$$

Let us seek a solution in the form:
$$\psi(r,\theta) = \tfrac{1}{2} U f(r) \sin^2(\theta) \tag{2.97}$$

which, when substituted into Eqn. (2.93), leads to:
$$\left(\frac{d^2}{dr^2} - \frac{2}{r^2}\right)^2 f(r) = 0 \tag{2.98}$$

The general solution of Eqn. (2.98) is:
$$f(r) = \frac{A}{r} + Br + Cr^2 + Dr^4 \tag{2.99}$$

Clearly, from Eqn. (2.96), D=0 and C=0. Also, from Eqns. (2.94) and (2.95), we obtain $A=a^3/2$ and $B=-3a/2$. Thus:
$$\psi(r,\theta) = \tfrac{1}{4} U a^2 \sin^2(\theta)\left(\frac{a}{r} - \frac{3r}{a}\right) \tag{2.100}$$

The corresponding fluid speeds are:
$$u_r = -\tfrac{1}{2} U \cos(\theta)\left(\frac{a}{r}\right)^2\left(\frac{a}{r} - \frac{3r}{a}\right) \tag{2.101}$$

and:
$$u_\theta = -\tfrac{1}{4} U \sin(\theta)\left(\frac{a}{r}\right)\left(\left(\frac{a}{r}\right)^2 + 3\right) \tag{2.102}$$

The Motion of Particles in Gases

The drag on the sphere can be obtained directly from Eqn. (2.89):

$$F_z = 8\pi\mu \lim_{r\to\infty}\left[\tfrac{1}{4}U a^2 \sin^2(\theta)\left(\frac{a}{r}-\frac{3r}{a}\right)\right] = -8\pi\mu\frac{3a}{4}U = -6\pi\mu a U \tag{2.103}$$

which is the classical Stokes value.

The other expression for F_z is Eqn. (2.88) from which we see that $dS=rd\theta$, $dn=dr$, and integration is over θ from 0 to π. Thus:

$$F_z = \mu\pi \int_0^\pi r^3 \sin^3(\theta)\frac{\partial}{\partial r}\left(\tfrac{3}{2}U\frac{a}{r^3}\right)r\,d\theta = -6\pi\mu a U \tag{2.104}$$

which is, of course, identical to the result obtained by the method of Payne and Pell [1960]. We may also write the drag as the sum of the form drag and the skin drag ($F_z = F_N + F_T$) where $F_N = -2\pi\mu a U$ and $F_T = -4\pi\mu a U$. These results can be obtained by integrating the normal and tangential stresses P_{rr} and $P_{r\theta}$ over the surface of the sphere.

It is possible to modify the boundary condition to incorporate the slip condition discussed above and given by Eqn. (2.54). In our coordinate system, we have $u_s = u_\theta$ and:

$$P_{ns} = P_{r\theta} = \mu\left[\frac{1}{r}\frac{\partial u_r}{\partial \theta} + r\frac{\partial}{\partial r}\left(\frac{u_\theta}{r}\right)\right] \tag{2.105}$$

Hence, the boundary condition now becomes (bearing in mind that the sphere is moving with speed, U):

$$u_\theta - C_m \lambda_g \left[\frac{1}{r}\frac{\partial u_r}{\partial \theta} + r\frac{\partial}{\partial r}\left(\frac{u_\theta}{r}\right)\right] = -U\sin(\theta) \tag{2.106}$$

But, since $u_r = U\cos(\theta)$ on the sphere, this reduces to:

$$u_\theta + \frac{1}{r}C_m \lambda_g U\sin(\theta) - C_m \lambda_g a\frac{\partial}{\partial r}\left(\frac{u_\theta}{r}\right) = -U\sin(\theta) \tag{2.107}$$

which in terms of the Stokes' stream function becomes:

$$\frac{\partial\psi}{\partial r} + C_m \lambda_g U\sin^2(\theta) - C_m \lambda_g a^2 \frac{\partial}{\partial r}\left(\frac{1}{r^2}\frac{\partial\psi}{\partial r}\right) = -aU\sin^2(\theta) \tag{2.108}$$

Using Eqns. (2.97) and (2.99), we find that the modified value of B is:

$$B = -\frac{3a(a + 2C_m \lambda_g)}{2(a + 3C_m \lambda_g)}$$

and hence, that the drag is:

$$F_z = -6\pi\mu a U\left(\frac{1 + 2C_m \lambda_g / a}{1 + 3C_m \lambda_g / a}\right) \tag{2.109}$$

Because we only expect this slip correction to be valid for $C_m\lambda_g/a \ll 1$, Eqn. (2.109) can also be written:

$$F_z = -\frac{6\pi\mu a U}{1 + C_m Kn}$$

where Kn is the Knudsen number.

While for the sphere kinetic theory calculations give accurate values for the drag over the whole range of Knudsen numbers, such calculations would be very difficult for other shapes. On the other hand, the equations of slow viscous flow can be solved analytically for quite complex shapes (*e.g.*, ellipsoids, toroids, dumbells, *etc.*). Using numerical methods, the drag on very irregularly shaped bodies may also be obtained (Segal [1979]). If the slip boundary conditions are incorporated, then the drag on such nonspherical bodies may be extended into the kinetic theory regime. Examples of this procedure will be given in Section 2.20 and following sections.

Finally, we note that an alternative procedure for calculating the drag on a sphere (or indeed any shape) is to consider the body to be fixed and allow the fluid to move over it with a steady velocity. The modifications necessary in that case are to the boundary conditions. For example, at infinity, the fluid will not be stationary but moving with a steady velocity, U, or to put it another way, the Stokes stream function at large distances from the sphere will be:

$$\psi_\infty = \tfrac{1}{2} U r^2 \sin^2(\theta)$$

Also, the boundary conditions at the surface of the sphere will be $u_r=0$ and $u_\theta=0$ for the no-slip condition which, in terms of the stream function, become:

$$\psi(a,\theta) = 0$$

and:

$$\left.\frac{\partial \psi}{\partial r}\right|_{r=a} = 0$$

2.10 Drag forces on nonspherical bodies

In practical situations, virtually all aerosol particles are nonspherical. The only case where particles remain spherical throughout their lifetime is when they are liquid drops and, even then, there may be some distortion due to external forces. Most aerosols, even if they are created as spheres, will coagulate and the product will tend to be irregular aggregations which can vary from 'fluffy' balls to long dendritic forms. Fig. 2.7 shows some typical particle shapes. If there is considerable moisture in the gas then aerosols do tend to develop a close-packed form resembling spheres but, in general, this is not the case and it is necessary to develop a procedure for calculating the drag on and effective sizes of such agglomerates.

Before discussing methods for dealing with the highly irregular shapes discussed above, it is useful to examine how the drag behaves for some simple shapes such as spheroids, dumbells, and toroids. Such shapes will give some indication of the effect of geometry and will also enable the accuracy of the 'equivalent sphere approximation' to be assessed. The equivalent sphere approximation simply assumes that the nonspherical particle has the same drag as a sphere of the same volume.

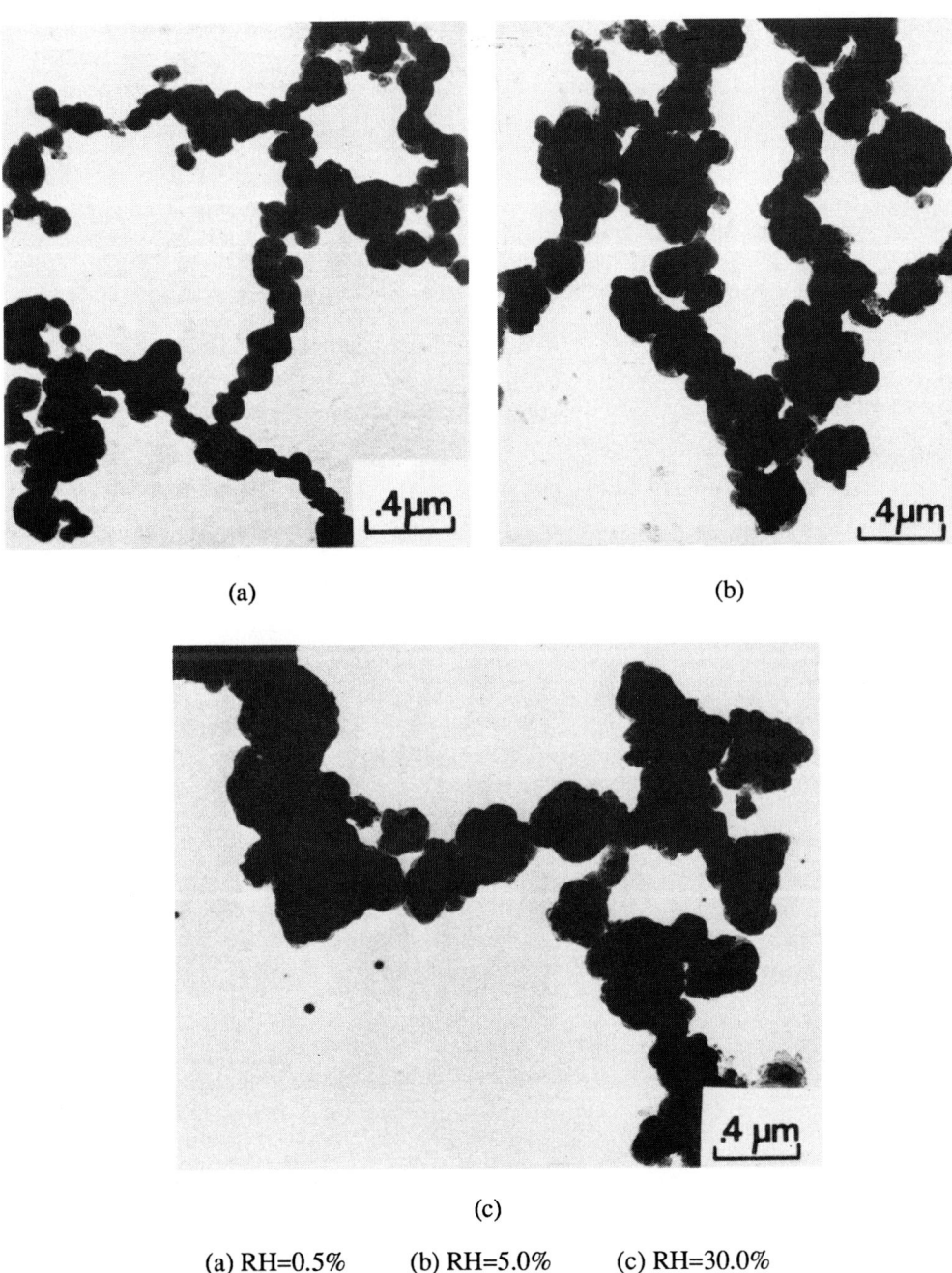

(a) RH=0.5%   (b) RH=5.0%   (c) RH=30.0%

Figure 2.7: Na-oxide particles at various relative humidities. Adapted from van de Vate *et al.* [1980].

2.11 The drag force on a spheroid

We have seen in Section 2.7 that, for axially symmetric bodies, the Stokes stream function satisfies the following partial differential equation:

$$E^4\psi = 0 \qquad (2.110)$$

The coordinate system for the body is shown in Fig. 2.8.

n and **s** are the unit normal and tangent, respectively, and $\mathbf{i}_z$ and $\mathbf{i}_\varpi$ are unit vectors in the z and ϖ directions, respectively. For axial symmetry, it is of considerable assistance to introduce a conformal transformation of the type (Happel and Brenner [1973]):

$$z + i\varpi = f(\xi + i\eta) \qquad (2.111)$$

where $f(\xi+i\eta)$ is defined so that the coordinate surfaces correspond to the boundaries of the geometry of interest. It is readily shown that in the new coordinate system:

$$E^2\psi = \varpi h^2 \left[\frac{\partial}{\partial \xi}\left(\varpi \frac{\partial \psi}{\partial \xi}\right) + \frac{\partial}{\partial \eta}\left(\varpi \frac{\partial \psi}{\partial \eta}\right) \right] \qquad (2.112)$$

where $\varpi = \varpi(\xi,\eta)$ and h is the metric coefficient. In the present case, we choose ξ as the coordinate normal to the surface and η as the one tangent to it.

In the case of an oblate spheroid the appropriate transformation is:

$$f(\xi + i\eta) = c\sinh(\xi + i\eta) \qquad (2.113)$$

where $0 \le \xi < \infty$ and $0 \le \eta \le \pi$. This leads to the relations:

$$z = c\sinh(\xi)\cosh(\eta) \qquad (2.114)$$

and:

$$\varpi = c\cosh(\xi)\sinh(\eta) \qquad (2.115)$$

Eliminating η we obtain:

$$\frac{z^2}{c^2\sinh^2(\xi)} + \frac{\varpi^2}{c^2\cosh^2(\xi)} = 1 \qquad (2.116)$$

From this we can readily establish the fact that the coordinate surfaces, ξ=constant, are a family of confocal oblate spheroids.

Now, the boundary conditions on the surface of the spheroid are:

$$u_\xi = U\mathbf{n}\cdot\mathbf{i}_z \qquad (2.117)$$

and:

$$u_\eta = U\mathbf{s}\cdot\mathbf{i}_z \qquad (2.118)$$

The Motion of Particles in Gases

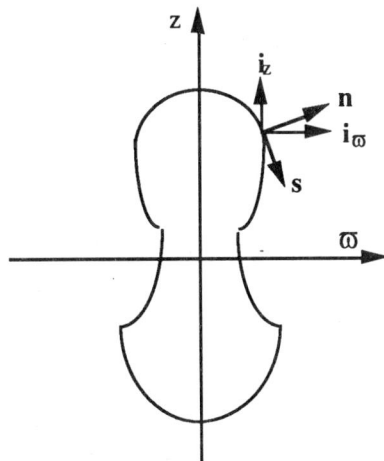

Figure 2.8: A coordinate system for an axially symmetric body.

where u_ξ and u_η are the fluid normal and tangential velocities and U is the translational velocity of the sphere in the z-direction. If slip boundary conditions are to be used, then the condition for the tangential velocity becomes:

$$u_\xi - \frac{C_m \lambda_g}{\mu} P_{\xi\eta} = U \mathbf{s} \cdot \mathbf{i}_z \tag{2.119}$$

where:

$$P_{\xi\eta} = \mu \left[\frac{\partial}{\partial \xi}(h u_\eta) + \frac{\partial}{\partial \eta}(h u_\xi) \right] \tag{2.120}$$

We shall not, however, employ this enhanced boundary condition here.

It is not difficult to show that:

$$\mathbf{n} \cdot \mathbf{i}_z = h \frac{\partial \varpi}{\partial \eta} \tag{2.121}$$

and:

$$\mathbf{s} \cdot \mathbf{i}_z = -h \frac{\partial \varpi}{\partial \xi} \tag{2.122}$$

and, moreover, from Eqns. (2.72) and (2.73):

$$u_\xi = -\frac{h}{\varpi} \frac{\partial \psi}{\partial \eta} \tag{2.123}$$

and:

$$u_\eta = \frac{h}{\varpi} \frac{\partial \psi}{\partial \xi} \tag{2.124}$$

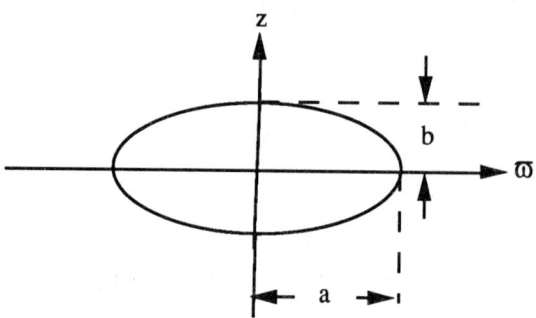

Figure 2.9: A coordinate system for a spheroid.

In order to proceed it is useful to define two new variables, $\lambda=\sinh(\xi)$ and $\zeta=\cosh(\eta)$. Then it follows that:

$$\varpi = c\left[\left(\lambda^2+1\right)\left(1-\zeta^2\right)\right]^{1/2} \tag{2.125}$$

$$z = c\lambda\zeta \tag{2.126}$$

$$h = \frac{1}{c\left(\lambda^2+\zeta^2\right)^{1/2}} \tag{2.127}$$

and:

$$c^2 = a^2 - b^2 \tag{2.128}$$

where, from Fig. 2.9, a and b are the major and minor radii.

In terms of λ and ζ, the boundary conditions become:

$$\frac{\partial \psi}{\partial \zeta} = c^2 \, U\left(\lambda_0^2+1\right)\zeta \tag{2.129}$$

and:

$$\left.\frac{\partial \psi}{\partial \lambda}\right|_{\lambda=\lambda_0} = -U\,c^2\left(1-\zeta^2\right)\lambda_0 \tag{2.130}$$

where $\lambda_0 = [(a/b)^2-1]^{-1/2}$ defines the surface of the spheroid.

The equation for ψ takes the form $E^4\psi=0$ where, in the new system of variables:

$$E^2 = \frac{1}{c^2\left(\lambda^2+\zeta^2\right)}\left[\left(\lambda^2+1\right)\frac{\partial^2}{\partial\lambda^2}+\left(1-\zeta^2\right)\frac{\partial^2}{\partial\zeta^2}\right] \tag{2.131}$$

Happel and Brenner [1973, page 145] show that the solution to the equation for ψ can be written as:

$$\psi(\lambda,\zeta) = \tfrac{1}{2}\left(1-\zeta^2\right)\left\{-C_1\,\lambda + C_2\left[\lambda-\left(\lambda^2+1\right)\cot^{-1}(\lambda)\right]\right\} \tag{2.132}$$

The Motion of Particles in Gases

where C_1 and C_2 are to be determined from the boundary conditions. The result is:

$$C_1 = \frac{2c^2 U}{\lambda_0 + (1-\lambda_0^2)\cot^{-1}(\lambda_0)} \qquad (2.133)$$

and:

$$C_2 = \frac{c^2 U (1-\lambda_0^2)}{\lambda_0 + (1-\lambda_0^2)\cot^{-1}(\lambda_0)} \qquad (2.134)$$

To obtain the drag, we use the result of Payne and Pell [1960] given in Eqn. (2.89):

$$F_z = \lim_{\lambda \to \infty} \left(\frac{\psi}{c\lambda(1-\zeta^2)} \right) \qquad (2.135)$$

$$= -\frac{4\pi\mu}{c} C_1 \qquad (2.136)$$

We have, therefore, for the drag on the oblate spheroid:

$$F_\|(O) = -\frac{8\pi\mu c U}{\lambda_0 + (1-\lambda_0^2)\cot^{-1}(\lambda_0)} \qquad (2.137)$$

$$= -\frac{8\pi\mu b \kappa^3 U}{\kappa + (\kappa^2 - 1)\tan^{-1}(\kappa)} \qquad (2.138)$$

where:

$$\kappa^2 = \frac{a^2}{b^2} - 1$$

The drag on a prolate spheroid can be obtained directly from the result for the oblate case if we note that coordinates appropriate to the prolate case are defined by:

$$f(\xi + i\eta) = c \cosh(\xi + i\eta) \qquad (2.139)$$

Then, with $\tau = \cosh(\xi)$ and $\zeta = \cosh(\eta)$, we note that there is a one-to-one correspondence if we replace λ by $i\tau$ and c by $-ic$. Since the surface of the prolate spheroid is defined by $\tau_0 =$ constant, it is now only necessary to set $\lambda_0 = i\tau_0$ where:

$$\tau_0 = \left(1 - \frac{b^2}{a^2}\right)^{-1/2} \qquad (2.140)$$

to obtain:

$$F_\|(P) = -\frac{8\pi\mu c U}{(\tau_0^2 + 1)\coth^{-1}(\tau_0) - \tau_0} \qquad (2.141)$$

$$= -\frac{8\pi\mu b \kappa^3 U}{(2\kappa^2 + 1)\ln\left(\kappa + \frac{a}{b}\right) - \frac{a\kappa}{b}} \qquad (2.142)$$

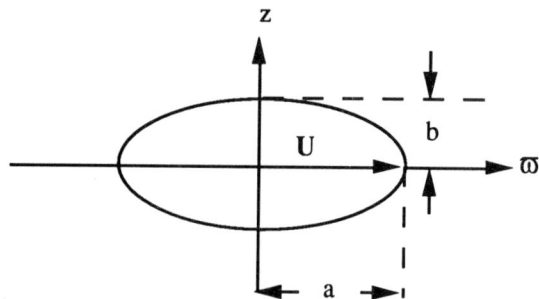

Figure 2.10: A coordinate system for an oblate spheroid moving along the direction of its major axis.

Two interesting limiting cases can be obtained from Eqns. (2.137) and (2.141). In the oblate case if we allow a>>b it reduces to a disc of radius, a, then we find:

$$F_{\|}(\text{disk}) = -16\mu a U = -6\pi\mu a U \left(\frac{8}{3\pi}\right) \tag{2.143}$$

which is a factor of (0.85) less than a sphere of the same radius.

In the case of the prolate spheroid, if we allow b<<a, it reduces to a thin rod. After a careful limiting procedure we find:

$$F_{\|}(\text{rod}) = -\frac{4\pi\mu a U}{\ln\left(\frac{a}{b}\right)} \tag{2.144}$$

The resistance is therefore relatively insensitive to the ratio (a/b).

The results for the oblate and prolate spheroids are relatively easy to obtain because the problems are axially symmetric. However, we may also ask for the drag on an oblate spheroid when its major axis is in the direction of motion.

As Fig. 2.10 shows, the direction of motion is now in the ϖ direction. This requires a more detailed treatment of the flow and has been solved by Oberbeek [1876] who obtains the following expressions for the drag (denoted by the subscript $\perp$). For the oblate spheroid:

$$F_{\perp}(O) = -\frac{16\pi\mu b \kappa^3 U}{(3\kappa^2 + 1)\tan^{-1}(\kappa) - \kappa} \tag{2.145}$$

In the case of the prolate spheroid moving in the direction ϖ, the result is:

$$F_{\perp}(P) = -\frac{16\pi\mu b \kappa^3 U}{(2\kappa^2 - 1)\ln\left(\kappa + \frac{a}{b}\right) + \frac{a\kappa}{b}} \tag{2.146}$$

The Motion of Particles in Gases

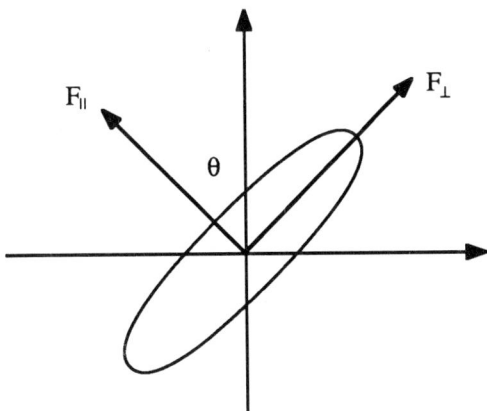

Figure 2.11: A spheroid at arbitrary orientation.

In practice, the spheroid may be at some arbitrary orientation in space as shown in Fig. 2.11. However, because the equations of fluid flow describing the drag forces are linear, we can use superposition to obtain the net drag force by resolving the values obtained in the two separate cases: $F_\parallel$ and $F_\perp$. Thus, the compound drag is:

$$F_D = F_\parallel \cos(\theta) + F_\perp \sin(\theta) \tag{2.147}$$

Bearing in mind that the value of U in the expression for $F_\parallel$ is now $U\cos(\theta)$ and in the expression for $F_\perp$ is $U\sin(\theta)$, we obtain:

$$F_D(O) = -\left[\frac{8\pi\mu b\kappa^3\, U\cos^2(\theta)}{\kappa + (\kappa^2-1)\tan^{-1}(\kappa)} + \frac{16\pi\mu b\kappa^3\, U\sin^2(\theta)}{(3\kappa^2-1)\tan^{-1}(\kappa) - \kappa}\right] \tag{2.148}$$

Similarly, for the prolate case:

$$F_D(P) = -\left[\frac{8\pi\mu b\kappa^3\, U\cos^2(\theta)}{(2\kappa^2+1)\ln\left(\kappa+\frac{a}{b}\right) - \frac{a\kappa}{b}} + \frac{16\pi\mu b\kappa^3\, U\sin^2(\theta)}{(2\kappa^2-1)\ln\left(\kappa+\frac{a}{b}\right) + \frac{a\kappa}{b}}\right] \tag{2.149}$$

The limiting case of a disk moving at angle, θ, is obtained by allowing $\kappa \to \infty$ in Eqn. (2.148), whence:

$$F_D(\text{disk}) = -\tfrac{16}{3}\mu a\kappa U\left[3\cos^2(\theta) + 2\sin^2(\theta)\right] \tag{2.150}$$

Clearly, $F_D(0°)/F_D(90°) = 1.5$ and, as we expect, the drag is much less when the disk moves edgeways.

Table 2.2: Reduced drag, $F_D/6\pi a\mu U$, on spheroids.

κ	Oblate Spheroid				Prolate Spheroid			
	$\theta=0$	$\theta=\pi/2$	average	equiv. sphere	$\theta=0$	$\theta=\pi/2$	average	equiv. sphere
0.0	1.0	1.0	1.0	1.0	1.0	1.0	1.0	1.0
0.1	0.999	0.998	0.998	0.998	0.996	0.997	0.997	0.997
0.2	0.996	0.992	0.994	0.993	0.984	0.988	0.987	0.987
0.5	0.979	0.957	0.965	0.963	0.916	0.936	0.929	0.928
1.0	0.943	0.880	0.901	0.891	0.767	0.821	0.803	0.794
2.0	0.896	0.770	0.812	0.765	0.560	0.654	0.623	0.585
5.0	0.861	0.658	0.726	0.581	0.354	0.471	0.432	0.338
10.0	0.852	0.613	0.693	0.463	0.264	0.381	0.342	0.215
20.0	0.849	0.590	0.676	0.368	0.208	0.318	0.282	0.136
50.0	0.849	0.575	0.667	0.271	0.162	0.261	0.228	0.0737
100.0	0.849	0.571	0.663	0.215	0.139	0.230	0.200	0.0464
∞	0.849	0.566	0.660	0.0	0.0	0.0	0.0	0.0

If the spheroids are assumed to be rotating due to Brownian motion, then an average over the orientations can be taken leading to the replacement of $\cos^2(\theta)$ by 1/3 and $\sin^2(\theta)$ by 2/3.

We illustrate these results numerically in Table 2.2 for $\theta=0$ and $\pi/2$ and also for the average case. The values shown in Table 2.2 are those of the reduced drag, $F_D/6\pi a\mu U$, i.e. deviations from the Stokes result for a sphere of radius a.

We also show the result for the equivalent sphere approximation in which the Stokes drag on a sphere is employed with the effective radius, R, being determined from $4\pi R^3/3 = V_{ps}$ or V_{os} where $V_{ps}=4\pi ab^2/3$ and $V_{os}=4\pi a^2b/3$.

As might be expected in the case of the oblate spheroid, the drag is less when it moves in the direction of smallest cross sectional area. Similarly, in the case of a prolate spheroid, the drag is smallest when it moves in the direction of the 'needle' axis. As far as the equivalent sphere is concerned, we note that for κ<2 (i.e. (a/b)<√5) its accuracy is better than 6%. It is, therefore, a convenient approximation for slightly deformed spheres.

It is appropriate to note at this point that the mobility, B, of a particle is defined as the velocity per unit force. Thus, we may write U=BF. Then, from the Einstein relation, D=kTB, we can write D=kTU/F. It follows therefore, from Eqn. (2.147), that the diffusion coefficient for an arbitrarily orientated spheroid is:

$$\frac{1}{D} = \frac{\cos(\theta)}{D_{\parallel}} + \frac{\sin(\theta)}{D_{\perp}}$$

(2.151)

where $D_{\parallel}=kTU/F_{\parallel}$ and $D_{\perp}=kTU/F_{\perp}$. Using Eqns. (2.137), (2.141), (2.145), and (2.146) leads to the desired result.

The Motion of Particles in Gases

2.12 The drag on a closed torus

Another nonspherical shape for which the drag may be obtained exactly is the closed torus. This problem was first solved by Dorrepaal et al. [1976a] for the case where the toroid moves in the direction of its major axis. Fig. 2.12 shows the situation.

In order to convert the equation, $E^4\psi=0$, to a suitable system, the following tangent sphere coordinates are introduced:

$$z = \frac{\xi}{\xi^2+\eta^2} \quad \text{and} \quad \varpi = \frac{\eta}{\xi^2+\eta^2}$$

These coordinates define closed toroids of circular cross-section of radius $a=1/2\eta_0$, i.e. η=constant defines a family of closed toroids. In terms of these new coordinates we find:

$$E^2 = \tfrac{1}{4}\eta\left(\xi^2+\eta^2\right)\left[\frac{\partial}{\partial\xi}\left(\frac{\left(\xi^2+\eta^2\right)}{\eta}\frac{\partial}{\partial\xi}\right) + \frac{\partial}{\partial\eta}\left(\frac{\left(\xi^2+\eta^2\right)}{\eta}\frac{\partial}{\partial\eta}\right)\right] \quad (2.152)$$

The boundary conditions on the surface of the toroid are $u_n = U\,\mathbf{n}\cdot\mathbf{i}_z$ or, in terms of ψ:

$$\psi + \tfrac{1}{2}\frac{U\eta_0^2}{\left(\xi^2+\eta_0^2\right)^2} = 0 \quad (2.153)$$

The transverse condition is $u_s = U\,\mathbf{s}\cdot\mathbf{i}_z$ or, if we include slip:

$$u_s - \frac{C_m\lambda_g}{\mu}P_{ns} = U\,\mathbf{s}\cdot\mathbf{i}_z \quad (2.154)$$

where:

$$P_{ns} = -\mu\left[\frac{\partial}{\partial\eta}\left[\left(\xi^2+\eta^2\right)u_s\right] + \frac{\partial}{\partial\xi}\left[\left(\xi^2+\eta^2\right)u_n\right]\right]$$

with:

$$u_s = -\frac{\left(\xi^2+\eta^2\right)^2}{\eta}\frac{\partial\psi}{\partial\eta}$$

and:

$$u_n = \frac{\left(\xi^2+\eta^2\right)^2}{\eta}\frac{\partial\psi}{\partial\xi}$$

The slip condition therefore takes the form:

$$\frac{\left(\xi^2+\eta^2\right)^2}{\eta}\frac{\partial\psi}{\partial\eta} + C_m\lambda_g\left[\frac{\partial}{\partial\eta}\left(\frac{\left(\xi^2+\eta^2\right)^3}{\eta}\frac{\partial\psi}{\partial\eta}\right) - \frac{1}{\eta}\frac{\partial}{\partial\xi}\left(\frac{\left(\xi^2+\eta^2\right)^3}{\eta}\frac{\partial\psi}{\partial\xi}\right)\right] = U\frac{\left(\eta^2-\xi^2\right)}{\left(\xi^2+\eta^2\right)} \quad (2.155)$$

Now, Dorrepaal et al. have shown that the general solution for ψ is:

$$\psi(\xi,\eta) = \frac{\eta}{\left(\xi^2+\eta^2\right)^{3/2}}\int_0^\infty ds\,\{A(s)\,I_1(s\eta) + B(s)\,\eta\,I_0(s\eta)\}\cos(s\xi) \quad (2.156)$$

Aerosol Science: Theory and Practice

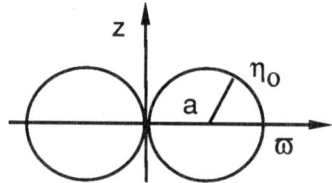

Figure 2.12: A closed torus moving in the direction of its major axis.

where I_0 and I_1 are modified Bessel functions and A and B are unknowns to be determined from the boundary conditions. In the case of no slip, Dorrepaal *et al.* obtained an exact value for the drag:

$$F_z = -8\mu a U \int_0^\infty \frac{1-2[I_1(x) K_1(x) + I_2(x) K_0(x)]}{I_1^2(x) - I_0(x) I_2(x)} dx = -(35.26) \mu a U \tag{2.157}$$

Using a variational method, Williams [1987a] obtained an approximate result for the drag, including slip, in the form:

$$F_z = -(35.26) \mu a U \left\{ \frac{1+(2.6306)\lambda}{1+(5.0418)\lambda} \right\} \tag{2.158}$$

where $\lambda = C_m \lambda_g / 2a$. This slip correction enables the result to be used over a much wider range of physical conditions. It is useful to compare Eqn. (2.158) with the corresponding value for a sphere as given by Eqn. (2.109) and is instructive to calculate the drag on a sphere with the same volume as the toroid. Since the toroid volume is $2\pi^2 a^3$, we find the effective sphere radius is:

$$R = \left(\frac{3\pi}{2}\right)^{1/3} a \tag{2.159}$$

from which, using Eqn. (2.109), we obtain:

$$F_{eff} = -(31.60) \mu a U \left\{ \frac{1+(1.193)\lambda}{1+(1.789)\lambda} \right\} \tag{2.160}$$

While the leading term is in error by about 10%, the slip corrections are seen to differ significantly from the exact result. Results for an open torus with no slip are given by Majumdar and O'Neill [1977].

2.13 The drag on two spheres in contact

The problem of drag on two spheres in contact has been solved for no slip by Cooley and O'Neill [1969]. In this case both spheres are moving along their line of centers. We shall discuss the procedure for calculating the drag.

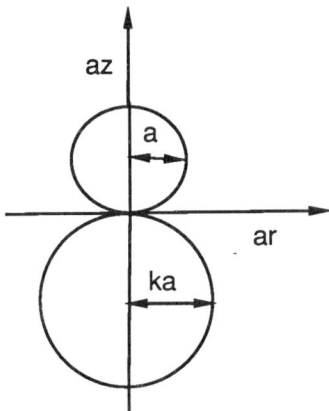

Figure 2.13: A family of curves generated by (z,ϖ)↔(ξ,η) transformations.

The problem is axially symmetric and so we have to find a solution of $E^4\psi=0$ subject to the following boundary conditions:

$$u_n = U\,\mathbf{n}\cdot\mathbf{i}_z$$

and:

$$u_s - \frac{C_m \lambda_g}{\mu} P_{ns} = U\,\mathbf{s}\cdot\mathbf{i}_z$$

The following transformations are convenient:

$$z = \frac{2\xi}{\xi^2 + \eta^2}$$

and:

$$\varpi = \frac{2\eta}{\xi^2 + \eta^2}$$

because they generate a family of curves as shown in Fig. 2.13.

In Fig. 2.13, $a_1=a$ and $a_2=ka$. However, for simplicity, we only consider the case of $a_1=a_2$ deferring the general case until Section 2.25. We note that ξ=constant defines circles of radii $a=1/\xi_0$.

The appropriate form for E^2 is then identical to Eqn. (2.152) and, as shown by Cooley and O'Neill, the general solution is:

$$\psi(\xi,\eta) = \frac{\eta}{(\xi^2+\eta^2)^{3/2}} \int_0^\infty ds\,\bigl[B(s)\cosh(s\xi) + C(s)\,\xi\,\sinh(s\xi)\bigr] J_1(s\eta) \tag{2.161}$$

where B and C are determined by the boundary conditions. For the case of no slip, Cooley and O'Neill obtain for the drag:

$$F_z = -6\pi\mu a f U \tag{2.162}$$

where:

$$f = \tfrac{2}{3} \int_0^\infty \left\{ 1 - \frac{2\sinh^2(s) - 2s^2}{\sinh(2s) + 2s} \right\} ds = 1.2902819 \tag{2.163}$$

With slip, Williams [1987b] has used a variational method to find an approximate result as follows:

$$F_z = -6\pi\mu a U (1.2804147) \left\{ \frac{1 + (6.950)\lambda + (8.595)\lambda^2}{1 + (8.128)\lambda + (14.95)\lambda^2} \right\} \tag{2.164}$$

For the equivalent sphere approximation, in which $R = 2^{1/3}a$, we find:

$$F_{eff} = -6\pi\mu a U (1.259921) \left\{ \frac{1 + (1.587)\lambda}{1 + (2.381)\lambda} \right\} \tag{2.165}$$

The error in the leading term is 14%, the term to $O[\lambda]$ is 4%, and the term to $O[1/\lambda]$ is 4%.

2.14 Upper and lower bounds on drag

In a study of Newtonian viscous fluids in quasi-static flow, Hill and Power [1956] developed some extremum principles. These authors were able to obtain upper and lower bounds on the energy dissipation of a body translating and rotating in a viscous fluid. Consequently upper and lower bounds for the drag were available. One useful general conclusion of this work is that the drag on a body of arbitrary shape can be bracketed from below by the drag on any inscribed body and above by any circumscribing body. Thus, no matter what the complexity of the body, it should always be possible to find simple inscribed and circumscribing shapes: the simplest situation being spheres. As an example, let us consider the case of the closed toroid mentioned above.

For the inscribed body we see that a torus of radius a contains a disc of radius, 2a. The drag on such a disc is, as we have seen, given by:

$$F_{disk} = 32\mu a U = (10.186)\pi\mu a U \tag{2.166}$$

The body can be circumscribed by a sphere of radius 2a, the drag of which is:

$$F_{sphere} = 12\pi\mu a U \tag{2.167}$$

A somewhat closer circumscription would be obtained with an oblate spheroid whose generating ellipse has a major axis of length 4a and a radius of curvature at the ends of the major axis equal to 2a. This requires the eccentricity to be $1/\sqrt{2}$ and the minor axis to be of length $2\sqrt{2}a$. The drag on such an ellipsoid is:

$$F_{ellipse} = 8\sqrt{2}\,\pi\mu a U = (11.314)\pi\mu a U \tag{2.168}$$

Now, the exact result as shown in Eqn. (2.157) is:

$$F_{toroid} = (11.224)\pi\mu a U \tag{2.169}$$

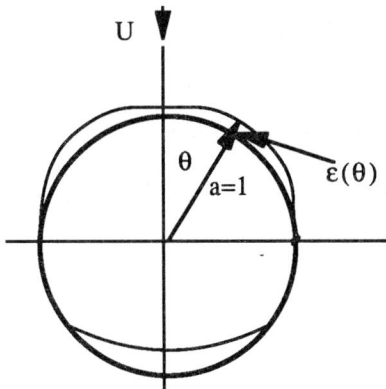

Figure 2.14: A deformed spheroid, $r(\theta)=1+\varepsilon(\theta)$.

Thus, we see that:

$$F_{disk}(9.4) < F_{toroid} < F_{ellipse}(0.8) < F_{sphere}(7) \qquad (2.170)$$

where the % error is given in brackets. We see that Hill and Power's method is indeed useful but that in this case the lower bound is less accurate.

2.15 A general technique for axially symmetric bodies

In an attempt to obtain some general trends for the drag of objects of variable shape, O'Brien [1968] carried out a semianalytical study based on the work of Savic [1953]. The coordinate system is not specialized to any specific configuration as with the ellipsoid or torus but rather the (r,θ) system is retained and the surface of the body is defined by a relationship of the form:

$$r(\theta) = 1 + \varepsilon(\theta) \qquad (2.171)$$

where r is normalized to the equatorial radius, a (see Fig. 2.14).

Thus, the equation to be solved, $E^4\psi=0$, is in (r,θ) or rather (r,μ) coordinates, where $\mu=\cos(\theta)$:

$$\left(\frac{\partial^2}{\partial r^2} + \frac{1-\mu^2}{r^2}\frac{\partial^2}{\partial \mu^2}\right)^2 \psi(r,\mu) = 0 \qquad (2.172)$$

In this case the body is assumed to be stationary and the fluid flows over it so that the boundary conditions are:

$$\psi \to \frac{r^2}{2}(1-\mu^2) \quad \text{as} \quad r \to \infty \qquad (2.173)$$

$$u_r = \frac{1}{r^2}\frac{\partial \psi}{\partial \mu} = 0 \qquad (2.174)$$

and:

$$u_\theta = \frac{1}{r(1-\mu^2)^{1/2}} \frac{\partial \psi}{\partial r} = 0 \qquad (2.175)$$

on the surface of the body and assuming no slip.

A general solution to Eqn. (2.172) can be written (Savic [1953], Sampson [1891]):

$$\psi(r,\mu) = \sum_{n=2}^{\infty} C_n^{(-1/2)}(\mu) \left[A_n \, r^n + C_n \, r^{n+2} + B_n \, r^{-n+1} + D_n \, r^{-n+3} \right] \qquad (2.176)$$

where $C_n^{(-1/2)}(\mu)$ are Gegenbauer polynomials. There is also the restriction that coefficients for powers of r greater than two are zero. Thus, a simpler form for Eqn. (2.176) is:

$$\psi(r,\mu) = \tfrac{1}{2} U \left(1-\mu^2\right) a^2 \, r^2 + \sum_{n=2}^{\infty} C_n^{(-1/2)}(\mu) \left[B_n \, r^{-n+1} + D_n \, r^{-n+3} \right] \qquad (2.177)$$

where we have set $A_2 = Ua^2$ to ensure the correct asymptotic behavior.

The coefficients of ψ are computed by matching boundary conditions at a finite number of points on the body contour using Eqn. (2.171). Of course, only a finite number of terms are taken in the summation according to the accuracy desired. The greater the distortion from a sphere, the larger the number of terms required for a given accuracy.

In order to calculate the coefficients B_n and D_n it is necessary to solve a set of algebraic equations. Then, ψ can be reconstructed and hence the fluid velocity. The drag is obtained directly from Eqn. (2.90) which leads to:

$$F = 4\pi\mu a U D_2 \qquad (2.178)$$

where a is the equatorial radius and we have noted that:

$$C_2^{(-1/2)}(\mu) = \tfrac{1}{2}(1-\mu^2)$$

Thus, knowledge of D_2 gives the drag.

O'Brien has written a computer program which solves the problem for arbitrary $\varepsilon(\theta)$. If m collocation points are taken then there are (2m+2) equations to solve. In most cases O'Brien takes m=10 and hence solves 22 simultaneous equations. The analytical form used for $r(\theta)$ is:

$$r(\theta) = 1 + \sum_{k=1}^{K} \varepsilon_k \, \mu^k \qquad (2.179)$$

The accuracy of the drag on oblate and prolate spheroids is shown in Fig. 2.15 where the circles are the numerical approximations and the full lines the exact results from Eqns. (2.138) and (2.142). It is obvious that more collocation points are required for highly eccentric oblate spheroids. In this case eccentricity is defined by:

$$e = \left(1 - \frac{b^2}{a^2}\right)^{1/2} \qquad (2.180)$$

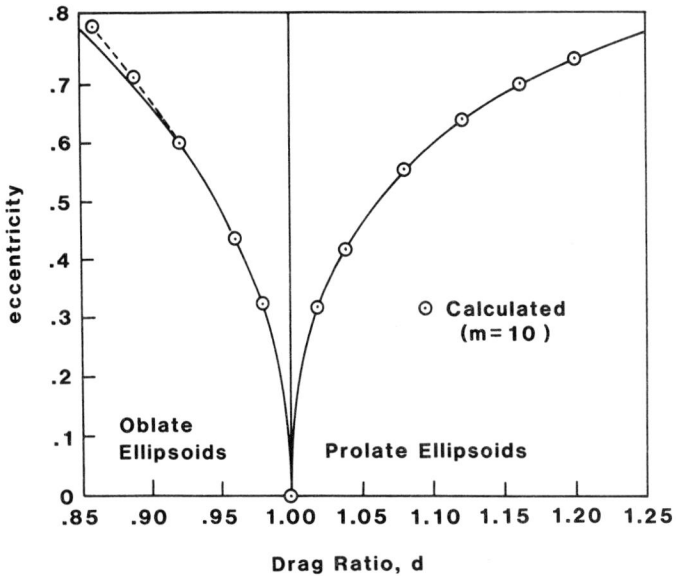

Figure 2.15: Drag ratios for ellipsoids. Adapted from O'Brien [1968] with permission of the American Institute of Chemical Engineers.

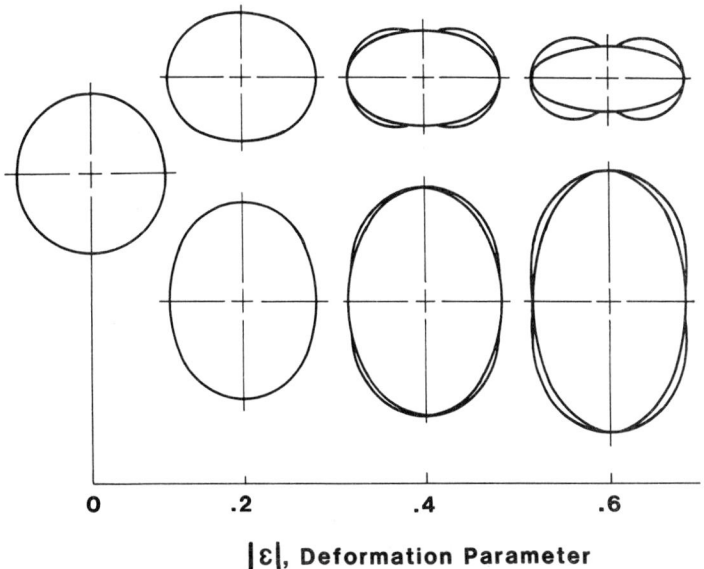

Figure 2.16: Bodies with µ-square deformations. Adapted from O'Brien [1968] with permission of the American Institute of Chemical Engineers.

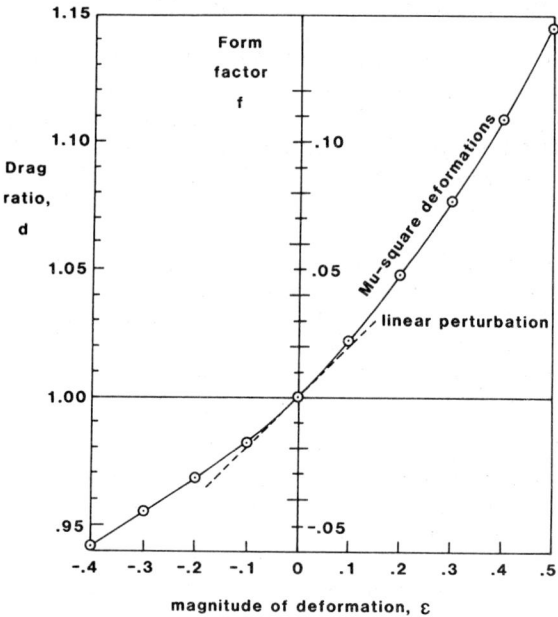

Figure 2.17: Drag ratios for bodies with μ-square deformations. Adapted from O'Brien [1968] with permission of the American Institute of Chemical Engineers.

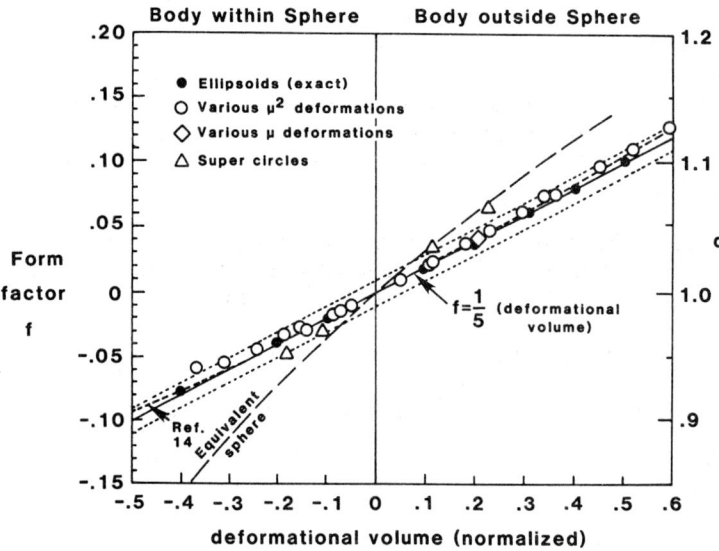

Figure 2.18: Correlation of form factor f with change in volume. Adapted from O'Brien [1968] with permission of the American Institute of Chemical Engineers. Ref. 14 refers to Heiss and Coull [1952].

The Motion of Particles in Gases

The drag ratio, d, is defined as the drag divided by $6\pi a\mu U$.

This method is, of course, most useful for bodies where no analytical result is available. Fig. 2.16 shows a selection of shapes generated by the equation $r=1+\varepsilon\mu^2$, with their inscribed prolate and oblate ellipsoids. The results for the drag ratio, d, and the form factor, $f=d-1$, are shown in Fig. 2.17 as a function of ε. The dashed line is the linear perturbation result of Brenner [1964]. Some further results are given for shapes in the form $r=1+\varepsilon_1\mu+\varepsilon_2\mu^2$ with positive and negative values of ε_1 and ε_2.

O'Brien discusses the application of the above technique to arbitrary shapes which can be entered numerically. He finds that there is some difficulty in convergence for highly distorted bodies and suggests that improvements will result if more points and more precision are adopted. Since his work is now more than 20 years old, it would seem likely that modern computers should be able to overcome these difficulties. There are also alternative methods of solution available based on the finite element method (Segal [1979]).

O'Brien's work continues with some discussion of the drag ratio. For example, he uses the theorem of Hill and Power [1956] of inscribed and circumscribed spheres and also the equivalent sphere approximation. Fig. 2.18 shows the results for spheroids. Clearly, both the equivalent sphere results and the exact results fall between the bounds, but the equivalent sphere is not very accurate. This is consistent with the results given in Table 2.2. Various correlations between drag ratio and what O'Brien calls the deformational volume are given.

One major difficulty with the above theories concerns particle orientation. Our theory assumes that the particle falls with its axis vertical. Oblate spheroids of uniform density will tend to do this naturally, but a prolate spheroid will tend to rotate to a position of maximum stability. Although this tilt will not be appreciable it does lead to a correction that would have to be accounted for via Eqn. (2.149).

2.16 Other methods of drag calculation

An ingenious method for dealing with particles that are composed of finite linear arrays of axisymmetric bodies has been developed by Gluckman et al. [1971, 1972]. We shall describe this theory with an illustration of flow past arrays of spheres which conform to natural coordinate systems.

The procedure begins by using a system of coordinates which is appropriate for spheres, such as the one used by O'Brien in the preceding section. Indeed, Eqn. (2.177) is the basis of the calculation. Consider a linear array of spheres as shown in Fig. 2.19.

The equation for ψ is linear, which means that at the point $(r,\mu=\cos(\theta))$ it is possible to write the stream function as a superposition of stream functions generated by each sphere. Thus, if ψ_j is the stream function at (r,μ) due to sphere j, the net effect will be:

$$\psi = \sum_{-(N-1)/2}^{(N-1)/2} \psi_j \qquad (2.181)$$

for N odd, with the origin taken at the center of sphere zero.

For an even number of spheres, the origin is taken on one of the spheres closest to the center of the chain:

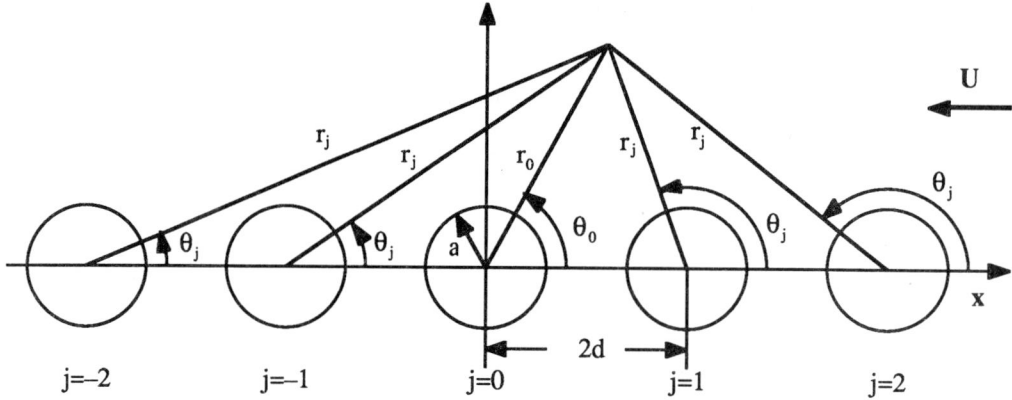

Figure 2.19: The geometry of the multiple sphere system.

$$\psi = \sum_{-(N-2)/2}^{N} \psi_j \tag{2.182}$$

ψ_j is of the form of Eqn. (2.177) as follows:

$$\psi_j(r,\mu) = \tfrac{1}{2} U r_0^2 \sin^2(\theta_0) + \sum_{0}^{\infty} \left[B_{nj} r_j^{-n+1} + D_{nj} r_j^{-n+3} \right] C_n^{(-1/2)}(\mu_j) \tag{2.183}$$

where:

$$r_j = \left[(x - 2jd)^2 + y^2 \right]^{1/2} \tag{2.184}$$

and:

$$\mu_j = \frac{(x - 2jd)}{r_j} \tag{2.185}$$

It is now necessary to choose the constants B_{nj} and D_{nj} to satisfy the boundary conditions. Since the force on each sphere is given by:

$$F_j = \mu \pi \int_0^\pi r_j^3 \sin^3(\theta_j) \frac{\partial}{\partial r_j} \left[\frac{E^2 \psi}{r_j^2 \sin^2(\theta_j)} \right] r_j \, d\theta_j \tag{2.186}$$

we can use the orthogonality properties of the Gegenbauer polynomials to obtain:

$$F_j = 4\pi\mu D_{2j} \tag{2.187}$$

The terms in Eqn. (2.181) are called multipoles by Gluckman et al. [1971, 1972] and each multipole has associated with it two constants, B_{nj} and D_{nj}, which are related to the intensity of

The Motion of Particles in Gases

the multipole. For flow past spherical objects the no-slip condition results in two equations for each discrete point on the arc of the body, *i.e.* at $(\theta_j, r_j = a)$. At these points:

$$\frac{\partial \psi}{\partial \theta_j} = 0 \qquad (2.188)$$

and:

$$\frac{\partial \psi}{\partial r_j} = 0 \qquad (2.189)$$

The two arbitrary constants can therefore be used to satisfy the boundary conditions at one point on the sphere j. To satisfy the boundary conditions over the complete surface, corresponding to an infinite number of points, it will be necessary to retain an infinite number of multipoles. In practice only a finite number can be retained and the viability of the model depends on the rate of convergence for a desired accuracy. It is to be expected, however, that the method becomes more accurate as $d/a \rightarrow \infty$ since then the problem would describe an isolated sphere. On the other hand, touching spheres would represent a severe test of the method.

As a practical matter, we note that if there are N spheres in the chain and the boundary conditions are satisfied at M points on the arc of each sphere, then there will be 2NM simultaneous algebraic equations to be solved. Thus, differentiating Eqn. (2.181), using Eqn. (2.183), and applying Eqns. (2.188) and (2.189) leads to:

$$-U \cos(\theta_{jm}) + \sum_{q=1}^{N} \sum_{n=2}^{M+1} \left[B'_{nqm} B_{nq} + D'_{nqm} D_{nq} \right] = 0 \qquad (2.190)$$

and:

$$U \sin(\theta_{jm}) + \sum_{q=1}^{N} \sum_{n=2}^{M+1} \left[B''_{nqm} B_{nq} + D''_{nqm} D_{nq} \right] = 0 \qquad (2.191)$$

where the B'_{nqm}, B''_{nqm}, D'_{nqm}, and D''_{nqm} are known functions of r_{qm} and μ_{qm}. μ is the collocation point and $1 \leq \mu \leq M$. Eqns. (2.190) and (2.191) must be solved numerically.

As an example of the accuracy of this work, Gluckman et al. have applied it to the case of two spheres with flow along their line of centers. This problem was chosen because it is one of the few multibody problems for which an exact result is known. Such a solution was provided by Stimson and Jeffrey [1926] as we shall illustrate in Section 2.25. In order to represent the drag on the spheres the force on each one is written $F = 6\pi\mu aU\lambda$. Clearly, for widely separated spheres, $d >> a$, $\lambda \rightarrow 1$ and we have the classical Stokes value. As the spheres approach one another there will be an interaction effect and λ will be different from unity; in fact λ is less than unity since the effect of two spheres moving together is to reduce the drag. We have the curious fact, therefore, that two identical spheres moving downwards under gravity, fall faster than each individually. However, this is incidental to the problem in which the most important piece of information required is to see how many terms are needed in the multipole expansion to lead to satisfactory accuracy.

Omitting discussion on some purely practical matters concerned with the computation (given in the original papers, Gluckman et al. [1971, 1972]) we show the results in Table 2.3.

Table 2.3: Approach to the exact solution for flow past two spheres (values of λ). Adapted from Gluckman *et al.* [1971] with permission of the Cambridge University Press.

Number of points, M	Spacing, d/a					
	1	2	3	4	8	16
1	0.66152	0.75065	0.80851	0.84604	0.91484	0.95530
3	0.64411	0.74244	0.80477	0.84414	0.91454	0.95525
5	0.64487	0.74266	0.80472	0.84412	0.91454	0.95525
7	0.64514	0.74266	0.80472	0.84412		
9	0.64515					
11	0.64515					

In Table 2.3 the convergence of the method is investigated over the range $1 \leq d/a \leq \infty$. Convergence is seen to be good as the last number given is exact except in the case d=a when it is (0.64514). For example, in the most restrictive case of touching spheres, we see that four significant figure accuracy is achieved with only nine equally spaced points. As a increases, the convergence improves dramatically. Even one point leads to an error of only 2.5% for d=a.

These encouraging results suggest that the drag for multiple arrays will be given very accurately. As an example of the technique, we show in Fig. 2.20 how the drag varies from sphere to sphere in a chain of seven spheres. The associated λ_j values are given at different spacings. It is interesting to note that the drag increases as we move to the outer spheres. Such results give some indications about the nature of settling of long dendritic aerosol particles which, if not rigidly connected, would tend to fall in a bent shape.

Gluckman *et al.* [1971, 1972] have extended this superposition technique to arrays of spheroids. In order to do this, the coordinate system has to be changed from that of a sphere to that of a natural confocal system for the spheroid (oblate or prolate). Then such elementary solutions are superposed as for the spherical case. Numerical results are given in the original paper and extensions to arbitrary shapes can be found in Gluckman *et al.* [1972].

Finally, we note that some recent work on dendritic chains attached to a collector have been reported by Ramarao and Tien [1988]. These authors used slender body theory and report values of drag on various configurations.

2.17 Irregular shapes

In virtually all practical situations aerosol particles are not only nonspherical but are highly irregular and have no particularly classifiable shape. It is useful, however, to look at irregular shapes that are composed of regularly shaped sub-units. This procedure was described in the previous sections. Some experimental work along these lines was carried out by Kunkel [1948] with glass beads in oil. While this is not an aerosol it does exhibit similar shape effects. Table 2.4 shows the shapes considered. It was noticed that the linear chains always rotated until their long axis was horizontal. Plate-like objects would fall with their flat surfaces in a horizontal position although any asymmetry would cause a sideways drift as the particle fell.

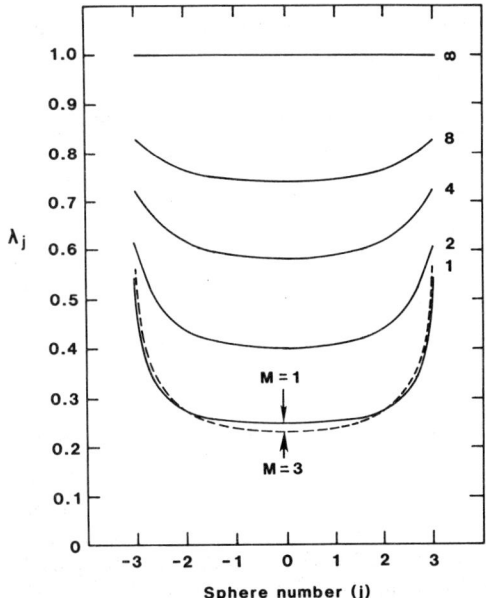

Figure 2.20: Drag correction factor for a 7 sphere chain at different sphere spacings. Adapted from Gluckman *et al.* [1971] with permission of the Cambridge University Press.

Table 2.4 shows the radius, a, of the sphere of equal volume and the radius, a_s, of the sphere, assuming Stokes' law, that has the same fall velocity. It is evident from these results that $a_s < a$, indicating that nonspherical shapes always have greater drag than the equivalent sphere. We also note that the discrepancy is larger the greater the distortion from spherical symmetry. These results are not unexpected but it is useful to see them quantified. Kunkel also looks at other flat plate shapes (circles, squares, and rectangles) and obtains similar conclusions.

Since it is very difficult, if not impossible, to calculate the drag analytically for irregularly shaped particles, it is convenient to introduce some semiempirical description that can be linked to experiment but which is also useful for incorporation into theories. In view of the fact that most theoretical formulations of aerosol dynamics are based on spherical particles, it is conventional to define effective particle diameters. Since several such definitions exist we shall discuss each one and see how useful it may be. We shall concentrate on the continuum regime using slip corrections where necessary.

We recall that the classical result of Stokes' law for the drag on a sphere of diameter, d, is (see Eqn. (2.56)):

$$F = -\frac{3\pi\mu d U}{C(d)}$$

We have written the result in terms of diameter because most work on the drag of irregularly shaped particles employs the concept of effective diameters rather than effective radii. For nonspherical particles it is convenient to retain the Stokes' law form and to write the drag as:

Table 2.4: Radius, a, of the equivalent sphere. Adapted from Kunkel [1948] with permission of the American Institute of Physics.

	Arrangement	Equal volume radius, a(mm)	Equal fall velocity radius, a_s(mm)	Difference in Percent
linear	O	1.18	1.20	2
	OO	1.51	1.40	7
	OOO	1.72	1.50	13
	OOOO	1.95	1.56	20
	OOOOOOOO	2.40	1.64	32
plane	(3 bead)	1.72	1.53	11
	(7 bead)	2.28	1.83	20
space	(6 bead cluster)	2.12	1.85	13

$$F = -\frac{3\pi\mu d_e \kappa U}{C(d_e)} \tag{2.192}$$

where d_e is the equivalent diameter, *i.e.* the diameter of the sphere having the same volume as the body. The parameter κ is called the dynamic shape factor. As we have seen above, κ can only be calculated theoretically for certain regular shapes. The shapes that have been investigated are summarized in Table 2.5.

To the summary in Table 2.5 must be added the semi-analytical/numerical results of O'Brien [1968] and of Gluckman *et al.* [1971, 1972] discussed above, as well as the modified results which include slip (Williams [1987a,b]).

We should also note some inconsistencies in the use of Eqn. (2.192) arising from slip corrections. It is conventional to use the Cunningham correction factor with the effective diameter to correct for slip but to use the uncorrected value of κ. As we have seen it is sometimes possible to obtain a value of κ which itself contains slip corrections, in which case it would no longer be necessary to use C as well.

For regular shapes, Kops [1976] has given an excellent summary of the various forms taken by κ for cylinders and ellipsoids in various orientations. Also Dahneke [1973a,b,c] has given a prescription for modifying the constants appearing in C to account for shape over the whole range of Knudsen numbers.

The case of highly irregular particles cannot be dealt with theoretically and so recourse must be made to experiment. In Eqn. (2.192) we see that the drag is characterized by two parameters, κ and d_e. However, from an experimental point of view, one parameter describing the drag would be more convenient. In this respect, two different but related effective diameters have been introduced: the Stokes diameter, d_s, and the aerodynamic diameter, d_a.

Table 2.5: Various shapes investigated in the literature.

Spheroids and needles	Gans [1911]; Payne and Pell [1960]
Lenses and hemispheres	Payne and Pell [1960]
Hollow spherical caps	Collins [1963]; Dorrepaal et al. [1976b]
Toroids	Payne and Pell [1960]
Bispherical units	Stimson and Jeffrey [1926]; Cooley and O'Neill [1969]; O'Neill [1969]; O'Neill and Majumdar [1970a,b]

The Stokes diameter, d_s, of a particle is defined as the diameter of a sphere with the same density and settling rate as the particle in question. On the other hand, the aerodynamic diameter, d_a, is the diameter of a sphere of unit density with the same settling rate as the particle in question. The aerodynamic diameter is more useful than the Stokes diameter because it does not require knowledge of the particle density which is often difficult to obtain. For this reason, it is generally preferred by experimentalists.

Relations between d_a, κ, and d_e may be obtained by considering the terminal velocity of the particle. If U is the terminal velocity, then:

$$\frac{3\pi \mu d_e \kappa U}{C(d_e)} = \frac{\pi}{6} d_e^3 \rho_p g \tag{2.193}$$

where we neglect the density of the gas. Thus:

$$U = \frac{d_e^2 \rho g C(d_e)}{18 \mu \kappa} \tag{2.194}$$

For a reference sphere having the same settling velocity but unit density, ρ_0, we have:

$$U = \frac{d_a^2 \rho_0 g C(d_a)}{18 \mu} \tag{2.195}$$

whence:

$$d_a = \frac{1}{\sqrt{\kappa}} \left(\frac{\rho}{\rho_0}\right)^{1/2} \left\{\frac{C(d_e)}{C(d_a)}\right\}^{1/2} d_e \tag{2.196}$$

The relationship between d_s and d_a is clearly:

$$d_a^2 = \left(\frac{\rho}{\rho_0}\right) \left\{\frac{C(d_s)}{C(d_a)}\right\} d_s^2 \tag{2.197}$$

Further classifications of irregularly shaped bodies have been introduced by Leith [1987]. Leith notes that the total drag on a body is the sum of the form drag (due to the normal stress) and the skin drag (due to the tangential stress; see Eqn. (2.88a)). Thus, the standard Stokes formula is actually:

$$F_z = -3\pi \mu d U \left(\tfrac{1}{3} + \tfrac{2}{3}\right) \tag{2.198}$$

where the 1/3 corresponds to form drag and the 2/3 to skin drag. It is then argued that form drag, because it arises from integration of the fluid pressure over the object surface, should be associated with the area of the object projected normal to its direction of motion. By analogy, therefore, form drag for a nonspherical particle might be associated with a sphere with same projected area as the nonspherical object. The diameter of such a sphere is d_n.

Friction or skin drag is due to integration over the whole surface of the body and therefore the friction drag of a nonspherical object should be expressed through a sphere with the same total area as the object. The diameter of such a sphere is d_{es}.

Such definitions raise questions about how to calculate the effective surface area of an irregular body, especially those with cusps and perforations where there are some hydrodynamically inactive zones.

Bearing in mind the two effective diameters, Leith suggests that the drag be written:

$$F_z = -3\pi\mu U \left(\tfrac{1}{3} d_n + \tfrac{2}{3} d_{es}\right) \tag{2.199}$$

This may be rewritten as:

$$F_z = -3\pi\mu d_n U \kappa_n \tag{2.200}$$

where the dynamic shape factor:

$$\kappa_n = \tfrac{1}{3} + \tfrac{2}{3}\frac{d_{es}}{d_n} \tag{2.201}$$

Alternatively, if the dynamic shape factor is defined in the traditional way in terms of the equivalent volume diameter:

$$F_z = -3\pi\mu d_e U \kappa \tag{2.202}$$

where:

$$\kappa = \tfrac{1}{3}\frac{d_n}{d_e} + \tfrac{2}{3}\frac{d_{es}}{d_e} \tag{2.203}$$

In order to evaluate κ_n, Leith has used some measured results on prisms obtained by Johnson [1985].

Fig. 2.21 shows how κ_n varies with d_{es}/d_n from the Johnson results and compares it with the simple hypothesis of Eqn. (2.199). There are deviations which increase as the object becomes less spherical. Using a least squares fit to the Johnson data, Leith arrives at:

$$\kappa_n = (0.357) + (0.684)\frac{d_{es}}{d_n} + (0.00154)(\text{length ratio}) + (0.0104)(\text{axis ratio}) \tag{2.204}$$

where:

$$\text{length ratio} = \frac{(\text{length of the axis parallel to the direction of motion})^2}{\text{projected area normal to the direction of motion}} \tag{2.205}$$

$$\text{axis ratio} = \frac{\text{longest axis in the projected area normal to the direction of motion}}{\text{shortest axis in the projected area normal to the direction of motion}} \tag{2.206}$$

The Motion of Particles in Gases

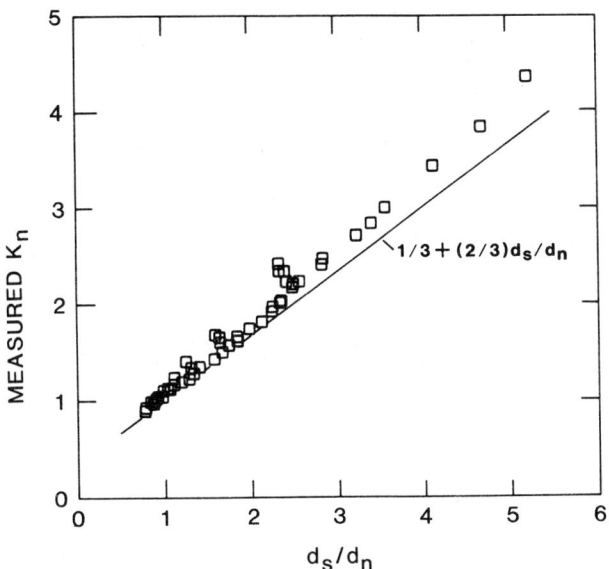

Figure 2.21: The d_s/d_n versus measured shape factor, κ_n for prisms tested by Johnson [1985] ($d_s=d_{es}$). Adapted from Leith [1987] with permission of the Elsevier Science Publishing Company.

As a check on this semi-empirical result, the value for a sphere is $\kappa_n=(1.053)$ and is 5.3% in error. Fig. 2.22 shows κ_n as measured for Johnson prisms versus Eqn. (2.204) and a satisfactory correlation is observed. To see how the same formula can be used for other shapes we show Fig. 2.23. This compares the measured κ_n values for prisms, spheres, cylinders, spheroids, and double cones with Eqn. (2.204). Again we note the satisfactory agreement. Thus, while the raw theory expressed by Eqn. (2.201) shows some deviations, the modified expression, with the two additional terms involving axis ratio and length ratio, leads to excellent agreement. While these results are still related to nonspherical but regular shapes, Leith argues that they are not completely empirical and highlights parameters that should affect drag for less regular objects, *e.g.* the effective surface concept.

One of the most systematic and accurate investigations carried out on irregularly shaped particles in recent times is due to Kops [1976] and deals with the physical characterization of branched chain-like aggregates. Such aggregates are found as a result of the combustion of hydrocarbon fuels or by the condensation of metallic vapors. They are composed of a large number of primary particles. In the experiments carried out, the aerosol particles are produced by means of the exploding wire technique. Fig. 2.7 shows typical aerosols generated by this method. Branched, chain-like, dendritic forms are well illustrated and the nonsphericity is evident. The important aspect of this study as far as drag is concerned is the determination of a relationship between the aerodynamic diameter and the number of primary particles, n, in the aggregates. Results show that for n up to a certain critical value, the aerodynamic diameter is found to be proportional to $n^{1/6}$ which corresponds to more or less linear chains. For n exceeding the critical value, the aerodynamic diameter is proportional to $n^{1/3}$ which corresponds to irregularly shaped three dimensional networks. Some typical results for the

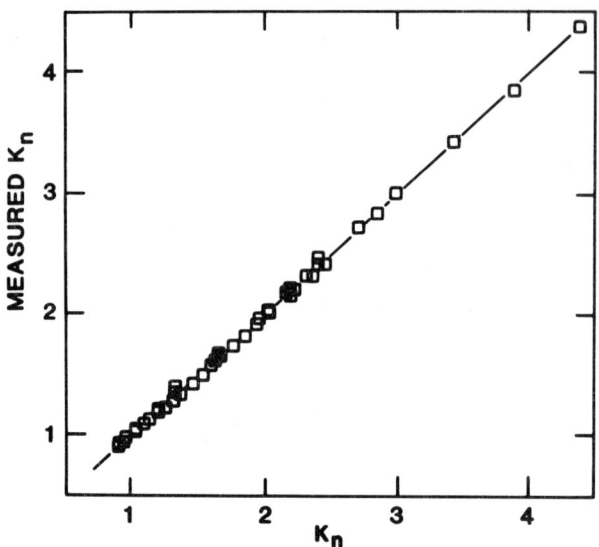

Figure 2.22: The κ_n from Eqn. (2.204) versus measured κ_n for prisms tested by Johnson [1985]. Adapted from Leith [1987] with permission of the Elsevier Science Publishing Company.

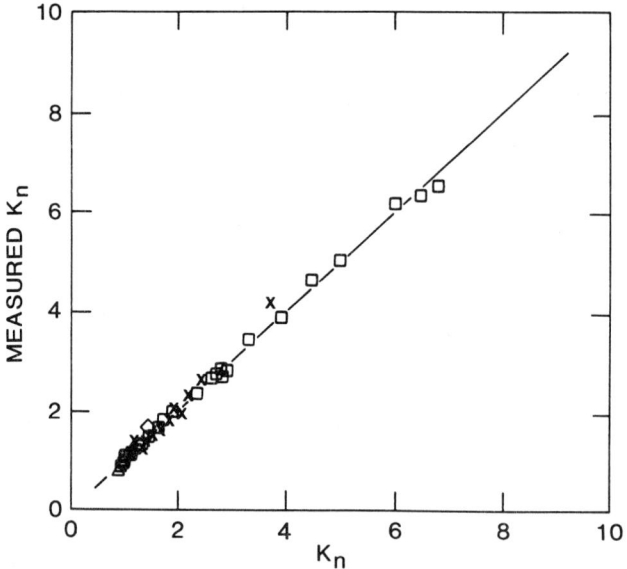

Figure 2.23: The κ_n calculated from Eqn. (2.204) versus measured κ_n for prisms (squares); sphere (plus); cylinders (diamonds); spheroids (crosses); and double conicals (triangles). Adapted from Leith [1987] with permission of the Elsevier Science Publishing Company.

The Motion of Particles in Gases

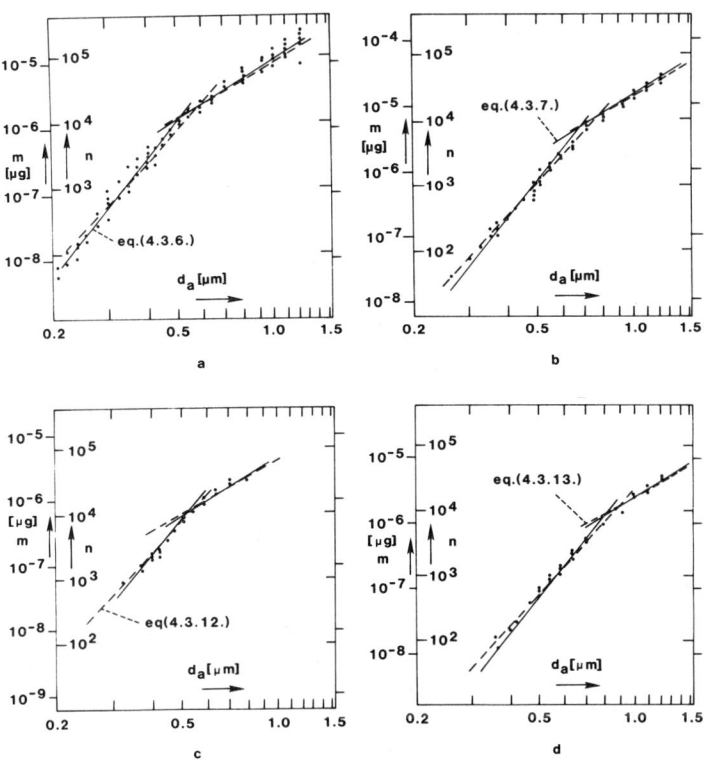

a: Fe oxide, d_{1g}=0.028 μm, σ_{1g}=1.67. b: Fe oxide, d_{1g}=0.042 μm, σ_{1g}=1.68.
c: Cu oxide, d_{1g}=0.020 μm, σ_{1g}=1.55. d: gold, d_{1g}=0.018 μm, σ_{1g}=1.60.

Figure 2.24: Relation between the aerodynamic diameter (d_a), the number of primary particles (n), and mass (m) of branched chain-like particles. Equation numbers refer to the original work. Adapted from Kops [1976] with permission of the Netherlands Energy Research Foundation ECN.

relationship between the aerodynamic diameter and the mass of the chain (and consequently the value of n) are shown in Fig. 2.24.

On the assumption that the particle size distribution can be described by a log-normal distribution (see Chapter 1), Kops shows that the two linear portions can be written:

$$d_a = k_1 n^{1/6} \left(\frac{\rho}{\rho_0}\right)^{1/2} d_{1g} \exp\left[2 \ln^2(\sigma_{1g})\right]$$

(2.205)

and:

$$d_a = f_1 n^{1/3} \left(\frac{\rho}{\rho_0}\right)^{1/2} d_{1g} \exp\left[\tfrac{3}{2} \ln^2(\sigma_{1g})\right]$$

(2.206)

where k_1 and f_1 are constants independent of d_a or n. d_{1g} is the geometric mean diameter of a primary particle and σ_{1g} is the geometric standard deviation of the primary particle size distribution.

The log-normal distribution is:

$$f(x) = \frac{1}{x\sqrt{2\pi}\ln(\sigma_g)} \exp\left\{-\frac{\ln^2(x/x_g)}{2\ln^2(\sigma_g)}\right\} \tag{2.207}$$

where $0 \leq x \leq \infty$. The rth moment of $f(x)$ is:

$$\overline{x^r} = x_g^r \exp\left[\tfrac{1}{2} r^2 \ln^2(\sigma_g)\right] \tag{2.208}$$

Thus, for example:

$$\overline{x} = x_g \exp\left[\tfrac{1}{2} \ln^2(\sigma_g)\right] \tag{2.209}$$

and:

$$\overline{x^3} = x_g^3 \exp\left[\tfrac{9}{2} \ln^2(\sigma_g)\right] \tag{2.210}$$

Eqns. (2.205) and (2.206) can be deduced from Eqn. (2.196) in the following way. If we introduce a new dynamic shape factor:

$$\tilde{\kappa} = \kappa C(d_a)/C(d_e) \tag{2.211}$$

then:

$$d_a = \frac{1}{\sqrt{\tilde{\kappa}}} \left(\frac{\rho}{\rho_0}\right)^{1/2} d_e \tag{2.212}$$

But the volume, v_e, of the equivalent sphere is given by:

$$v_e = \frac{\pi}{6} d_e^3 = n \frac{\pi}{6} \overline{d_1^3} \tag{2.213}$$

where $\pi \overline{d_1^3}/6$ is the mean volume of a primary particle. Thus, from Eqn. (2.210):

$$d_e = n^{1/3} \left(\overline{d_1^3}\right)^{1/3} = n^{1/3} d_{1g} \exp\left[\tfrac{3}{2} \ln^2(\sigma_{1g})\right] \tag{2.214}$$

Accordingly, we may write:

$$d_a = \frac{1}{\sqrt{\tilde{\kappa}}} \left(\frac{\rho}{\rho_0}\right)^{1/2} n^{1/3} d_{1g} \exp\left[\tfrac{3}{2} \ln^2(\sigma_{1g})\right] \tag{2.215}$$

This equation is equivalent to Eqn. (2.206) if we write:

$$\tilde{\kappa} = \frac{1}{f_1^2} = \text{constant} \tag{2.216}$$

The Motion of Particles in Gases

Such a result has been verified by Stöber et al. [1970] and Mercer et al. [1972] for cluster aggregates composed of up to 23 monodisperse polystyrene spheres.

In order to establish Eqn. (2.205) it is necessary to introduce the sphericity, ϕ, where:

$$\phi = \pi d_e^2 / A$$

and A is the surface area of the aggregate. Now for an aggregate of n primary particles:

$$A = n \pi \overline{d_1^2} = n \pi d_{1g}^2 \exp\left[2 \ln^2(\sigma_{1g})\right] \qquad (2.217)$$

from Eqn. (2.208). Thus, we find:

$$\phi = \frac{1}{n^{1/3}} \exp\left[\ln^2(\sigma_{1g})\right] \qquad (2.218)$$

and hence:

$$d_a = \frac{1}{(\tilde{\kappa}\phi)^{1/2}} \left(\frac{\rho}{\rho_0}\right)^{1/2} n^{1/6} d_{1g} \exp\left[2 \ln^2(\sigma_{1g})\right] \qquad (2.219)$$

This equation corresponds to Eqn. (2.205) if we identify:

$$\tilde{\kappa}\phi = \frac{1}{k_1^2} = \text{constant} \qquad (2.220)$$

It has been shown by Stober that, for linear aggregates of not more than eight monodisperse polystyrene spheres, this relation is indeed true. Figs. 2.25 and 2.26 show how f_1 and k_1 vary with the mean primary particle diameter, $\overline{d}_1$, for various materials.

One curious but interesting feature of aerosol behavior is the very abrupt transition from linear chains to three dimensional networks. Kops argues that this could be due to the effect of Brownian rotation on the coagulation process. We envisage two rotating linear chains coagulating and forming a larger chain. However, Brownian rotation will decrease with increasing chain length and this will lead to a greater chance of forming three dimensional networks. Thus, at least a qualitative change in structure is expected due to the coagulation process. Kops also advanced some quantitative arguments based on the speed of Brownian rotation but indicates that, while the proposed mechanism is plausible, his calculations provide no strong argument for the abrupt transition.

2.18 Particles with a fractal structure

The dimensions of points, lines, surfaces, and solids can be ordered because points can divide lines, lines can divide surfaces, and surfaces can divide solids into separate parts. Such ordering defines the so-called topological dimensions of an object. In the above cases, the dimensions are 0, 1, 2, 3, respectively. There is, however, another aspect of dimensionality which until relatively recently has been overlooked, namely, that the size of an irregular object

Aerosol Science: Theory and Practice

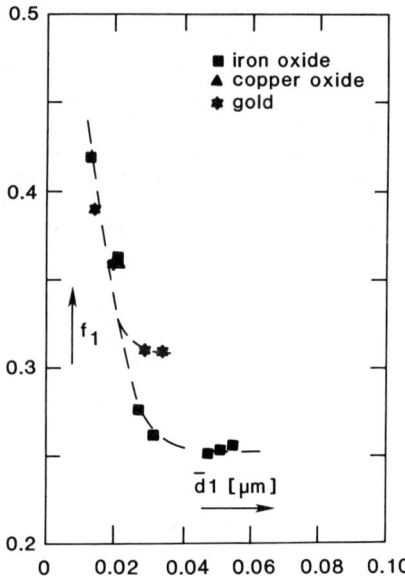

Figure 2.25: Relation between the mean primary particle diameter, $\bar{d}_1$, and f_1 obtained from exploding wire aggregates. Adapted from Kops [1976] with permission of the Netherlands Energy Research Foundation ECN.

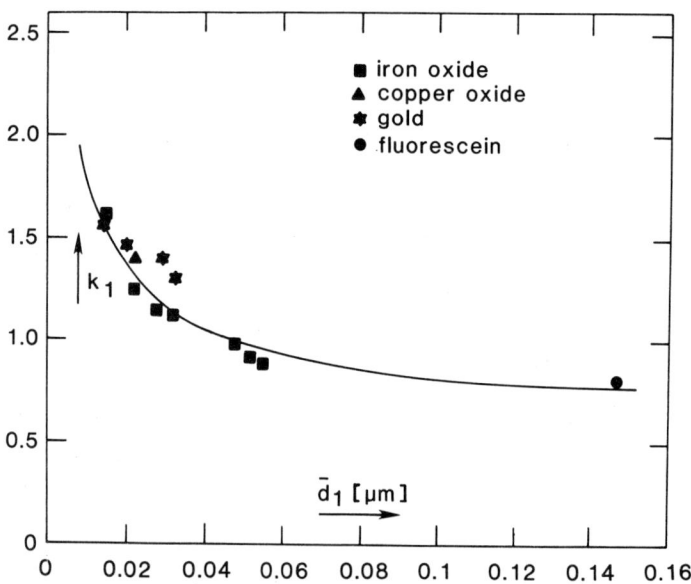

Figure 2.26: Relation between the mean primary particle diameter, $\bar{d}_1$, and k_1 of Eqn. (2.205). Adapted from Kops [1976] with permission of the Netherlands Energy Research Foundation ECN.

or, for example, its perimeter depends on the length scale used to measure it. A classic example of this length scale can be understood if one were asked to measure the length of the coastline of the British Isles. This could be done by pacing it out following all the twists and turns in various inlets and coves to obtain a certain length, ℓ_1. On the other hand, the measurement could be done on a much finer scale using a rod one cm long. In that case a much finer structure of the coastline would be observed leading to a perimeter, ℓ_2, where ℓ_2 is certainly greater than ℓ_1. As the scale of measurement is refined, ultimately leading to atomic dimensions, the perimeter continues to increase. It seems, therefore, that the definition of size of an irregular object is scale dependent. Such problems do not arise with mathematically smooth curves such as spheres and parallelepipeds but such bodies are ideals that do not occur in nature.

These arguments would suggest that it is not possible to obtain any fundamental parameters to define an irregular body. However, Richardson [1960] suggested that the length of the perimeter of an irregular object, $L(\varepsilon)$, measured with a scale length, ε, could be written as:

$$L(\varepsilon) \approx F\varepsilon^{1-D} \tag{2.221}$$

where D is independent of ε. Richardson argued that to approximate a coastline by a broken line, one needs roughly $F\varepsilon^{-D}$ intervals of length ε, thereby leading to the length $L(\varepsilon)$ above. To Richardson, D had no special meaning other than that of a useful classification parameter of a particular coastline. It was Mandelbrot [1977] who proposed that in spite of the fact that D was not an integer it should be interpreted as a dimension: a fractional or 'fractal' dimension. Such ideas, at the mathematical level, had been previously discussed by Hausdorf [1919].

Eqn. (2.221) has been verified experimentally by Richardson and the results are discussed by Mandelbrot [1977]. Fig. 2.27 shows the curve:

$$\log_{10}(L) = \log_{10}(F) + (1-D)\log_{10}(\varepsilon) \tag{2.222}$$

for various coastlines and for a circle.

It is clear that, for the circle, as $\varepsilon \to 0$ its length rapidly converges to a fixed value and the zero slope leads to the value, D=1, which corresponds to a line. On the other hand, for the irregularly shaped bodies, the size continues to increase as $\varepsilon \to 0$ but, most significantly, the slope 1–D is constant. It varies for different coastlines, i.e. different fractal structures, but is a well defined parameter for a particular structure. For example, the west coast of Britain has a fractal dimensionality of approximately 1.15.

There are many other shapes in nature that are irregular and can be interpreted as fractals. For example, snowflakes, clouds, water and jets, leaves, root formations, and the bronchial tract of the lung. Fractal structure is not confined to closed shapes but also describes branching processes such as lightning and tree shapes. Fig. 2.28 shows such an example.

A fractal nature can also be associated with irregularly shaped aerosol particles in the sense that if a particle is made up of a collection of primary particles, its effective 'size', R, can be written:

$$R(k) = Ak^\alpha \tag{2.223}$$

where k is the number of primary particles and A and α are parameters associated with the fractal structure. If the composite particle were to form a close packed structure, we would

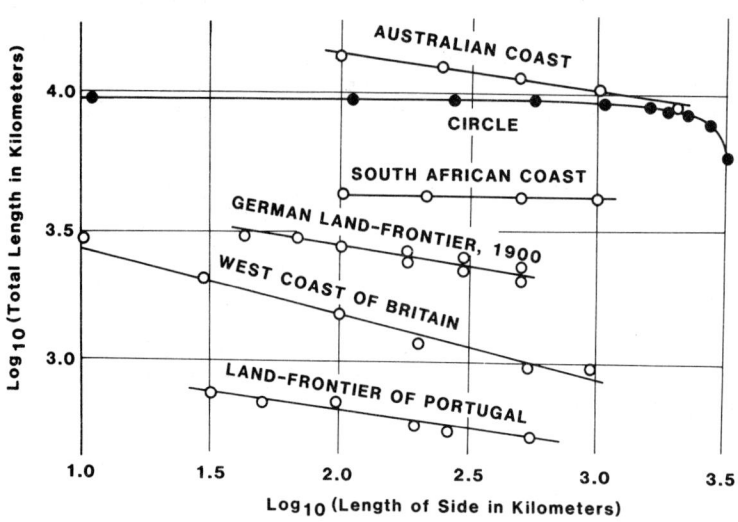

Figure 2.27: Richardson's empirical data concerning the rate of increase of coastlines' lengths. Adapted from Mandelbrot [1977] with permission of B.B. Mandelbrot.

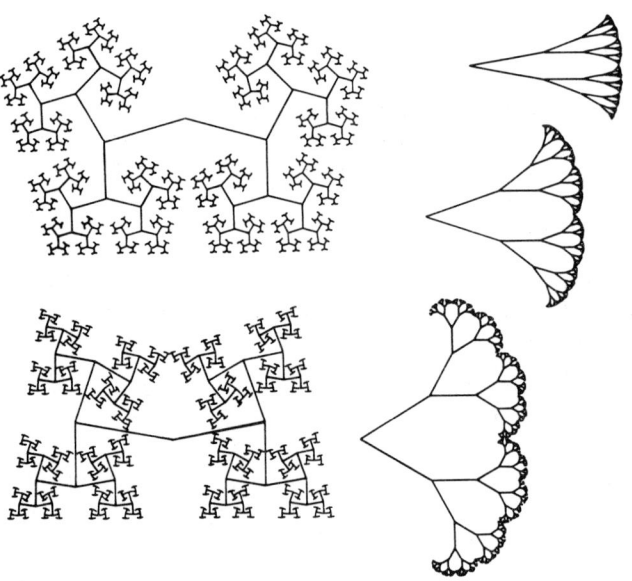

Figure 2.28: Fractal umbrella trees and fractal canopies. Adapted from Mandelbrot [1977] with permission of B.B. Mandelbrot.

expect, for sufficiently large k, that a roughly spherical object would be found with $\alpha \approx 1/3$. $\alpha=1/3$ corresponds to a volume that is proportional to the number of primary particles.

Recently, Mountain and Mulholland [1984] have carried out some tests on the shape of aggregates formed during the process of Brownian coagulation. In this case, it was assumed that the particles were very small compared with a mean free path and so they are in the free molecular regime.

The simulation is carried out using the following equation of motion, or Langevin equation:

$$m_0 \frac{dv}{dt} = -m_0 \beta v + F$$

where v is the particle velocity, m_0 the mass of a sub-unit, $m_0\beta$ a friction coefficient based on free molecular flow and F is a random force due to molecular bombardment. Thus, we have the classical Brownian situation.

The simulations involved 500 particles. Also Mountain and Mulholland assumed that the friction coefficient $m_0\beta$ increases linearly with the number of sub-units, k. This is an approximation that has not been justified but is, nevertheless, a reasonable first approximation. Essentially, it means that as clusters collide and stick m_0 changes to the new mass and therefore so does $m_0\beta$. Thus, the larger clusters will have greater friction coefficients and hence smaller speeds. The simulation solves the Langevin equation at each time step and checks to see whether a new cluster has been formed. When this occurs, the radius of gyration of the cluster is determined and the process continued. The outcome of the simulation is a relationship for the radius of gyration, $R_g(k)$, as a function of k. It was found that:

$$R_g(k) = A k^\alpha \tag{2.224}$$

where A=(0.34) and α=(0.56). The dimensionality of the clusters is therefore $D=1/\alpha=(1.78)$ rather than 3 for close packed units. That we have a fractal structure is established by virtue of the fact that α remains constant as the cluster grows. This implies that the growth follows the self-similar replication properties associated with fractal geometries. On the other hand, from a purely pragmatic point of view, it could be argued that Eqn. (2.224) is nothing more than an 'experimentally' or numerically determined recipe. There is no fundamental theory of fractals involved, only the observation that some form of self-similarity is present. This is useful, but in no way assists in formulating any new equation of aerosol dynamics or leads to better methods for calculating drag on such particles. What tends to be done in practice is to use the fractal relationship between effective size and number of sub-units to modify the existing formulae. For example, we could argue that the volume of a sphere containing k sub-units is proportional to k. Thus, Eqn. (2.224) could be rewritten as:

$$R = R_0 \left(\frac{v}{v_0}\right)^\alpha \tag{2.225}$$

where R is the radius of the equivalent sphere describing the cluster. This has indeed been done in aerosol physics (Simons [1987]) and in the scattering of light by fractal particles (Berry and Percival [1986]). The results of the computations using $D_\alpha=(1.78)$ instead of 3 leads to significant differences and shows the importance of the dimensionality of the object. However, it is not clear that a single parameter, D_α, will cover all aspects of the problem. For example, in Brownian coagulation the size of the particle enters in both the magnitude of the drag and in the effective cross section for interaction. Now, while a radius of gyration may be a good measure of the effective cross section of a particle, it may not describe the drag very well, which depends on other features of the shape such as its axial ratio and other asymmetries. Thus, the use of fractal recipes is certainly a step in the right direction, but they must not be used uncritically.

2.19 Rotational motion and torque

In order to study rotational motion it is useful to return to the vector form of slow viscous flow as represented by Eqn. (2.47). Then, if we write:

$$\mathbf{u} = \mathbf{i}_\varpi u_\varpi + \mathbf{i}_\phi u_\phi + \mathbf{i}_z u_z$$

(cylindrical coordinates) and insert this into Eqn. (2.47), we find for the components, u_ϖ, u_ϕ, and u_z:

$$\nabla^2 u_\varpi - \frac{u_\varpi}{\varpi^2} - \frac{2}{\varpi^2}\frac{\partial u_\phi}{\partial \phi} = \frac{1}{\mu}\frac{\partial p}{\partial \varpi} \tag{2.226}$$

$$\nabla^2 u_\phi - \frac{u_\phi}{\varpi^2} + \frac{2}{\varpi^2}\frac{\partial u_\varpi}{\partial \phi} = \frac{1}{\mu}\frac{1}{\varpi}\frac{\partial p}{\partial \phi} \tag{2.227}$$

and:

$$\nabla^2 u_z = \frac{1}{\mu}\frac{\partial p}{\partial z} \tag{2.228}$$

where:

$$\nabla^2 = \frac{\partial^2}{\partial z^2} + \frac{\partial^2}{\partial \varpi^2} + \frac{1}{\varpi}\frac{\partial}{\partial \varpi} + \frac{1}{\varpi^2}\frac{\partial^2}{\partial \phi^2}$$

If p, u_ϖ, and u_z do not depend on ϕ, then it is readily shown that Eqns. (2.226) and (2.228) describe the axial symmetry case discussed earlier. On the other hand, the equation for u_ϕ describes rotational motion about the z-axis and will therefore be useful for calculating the torque on axially symmetric bodies. This particular form of the viscous flow equations was first discussed in detail by Jeffery [1915a,b].

The equation for u_ϕ where $\mathbf{u}$ is independent of ϕ, becomes:

$$\left(\frac{\partial^2}{\partial z^2} + \frac{\partial^2}{\partial \varpi^2} + \frac{1}{\varpi}\frac{\partial}{\partial \varpi} - \frac{1}{\varpi^2}\right)u_\phi = 0 \tag{2.229}$$

The boundary condition associated with the problem is:

$$u_\phi = \Omega \varpi \tag{2.230}$$

where Ω is the angular speed (magnitude of the angular velocity) of the sphere. If there is slip at the surface, then the condition is modified to:

$$u_\phi - \frac{C_m \lambda_g}{\mu} P_{n\phi} = \Omega \varpi \tag{2.231}$$

The Motion of Particles in Gases

As we have shown earlier, the torque, **T**, can be written:

$$\mathbf{T} = \int_S \mathbf{r} \times \mathbf{P} \cdot d\mathbf{S} = \int_S \mathbf{r} \times \mathbf{P}_n \, dS$$

which can be transformed to:

$$\mathbf{T} = 2\pi\mu \int_c \varpi^3 \frac{\partial}{\partial n}\left(\frac{u_\phi}{\varpi}\right) ds \qquad (2.232)$$

where s is measured along the arc of the contour, c.

An alternative expression for the torque, valid if the fluid is unbounded, was derived by Kanwal [1961]. The result, which is analogous to that of Payne and Pell [1960] for the drag (see Eqn. (2.89)), can be written:

$$\mathbf{T} = 8\pi\mu \lim_{r \to \infty}\left(r^3 \frac{u_\phi}{\varpi}\right) \qquad (2.233)$$

We will apply these results to some geometries of interest.

2.20 The rotating sphere

Let us consider a sphere of radius a rotating about an axis. Eqn. (2.229) can be written in spherical coordinates as:

$$\left(\frac{\partial^2}{\partial r^2} + \frac{2}{r}\frac{\partial}{\partial r} + \frac{1}{r^2}\frac{\partial^2}{\partial \theta^2} + \frac{\cot(\theta)}{r^2}\frac{\partial}{\partial \theta} - \frac{1}{r^2 \sin^2(\theta)}\right) u_\phi = 0 \qquad (2.234)$$

The boundary condition for no slip becomes:

$$u_\phi(a,\theta) = a\Omega_0 \sin(\theta) \qquad (2.235)$$

and we seek a solution in the form:

$$u_\phi(r,\theta) = \eta(r) \sin(\theta) \qquad (2.236)$$

whence:

$$r^2 \eta'' + 2r\eta' - 2\eta = 0 \qquad (2.237)$$

subject to $\eta(a) = a\Omega_0$.

The general solution of the equation for $\eta(r)$ is:

$$\eta(r) = Ar + \frac{B}{r^2} \qquad (2.238)$$

But, since $u_\phi \to 0$ as $r \to \infty$, $A=0$. From the boundary condition we have that, $B=a^3\Omega_0$, and so:

$$u_\phi(r,\theta) = \frac{a^3 \Omega_0}{r^2} \sin(\theta)$$
(2.239)

Using Eqn. (2.232) with $\varpi = r \sin(\theta)$ and $ds = a\, d\theta$ we find for the torque:

$$T = 8\pi\mu a^3 \Omega_0$$
(2.240)

An identical results ensues from Eqn. (2.233).

It is interesting to extend this result to the case of slip boundary conditions. Then, we find that Eqn. (2.231) is written:

$$u_\phi - \frac{C_m \lambda_g}{\mu} P_{n\phi} = a\Omega_0 \sin(\theta)$$
(2.241)

In this case, since $u_r = 0$, we can write:

$$P_{n\phi} = \mu r \frac{\partial}{\partial r}\left(\frac{u_\phi}{r}\right)$$
(2.242)

Using the transformations of Eqn. (2.236) we find that the boundary condition on $\eta(r)$ becomes:

$$\eta(a) - a\Omega_0 = C_m \lambda_g a \frac{\partial}{\partial r}\left(\frac{\eta}{r}\right)\bigg|_{r=a}$$
(2.243)

Thus, after some algebra:

$$u_\phi(r,\theta) = \frac{a^4 \Omega_0}{a + 3 C_m \lambda_g} \frac{\sin(\theta)}{r^2}$$
(2.244)

whence:

$$T = \frac{8\pi\mu a^3 \Omega_0}{1 + 3 C_m \lambda_g / a}$$
(2.245)

The effect of slip, therefore, is to reduce the torque. Of course, as the ratio λ_g/a becomes very large, the result would fail since we then enter the region of free molecular flow. In that case Halbritter [1974] has used kinetic theory to obtain:

$$T_K = \frac{2\alpha\pi}{3} \rho \bar{v} a^4 \Omega_0$$
(2.246)

where $\bar{v} = (8kT/\pi m)^{1/2}$ is the mean speed of the gas molecules and α is the accommodation coefficient. Halbritter also gives results for rotating oblate and prolate spheroids in the free molecular regime. We shall discuss his technique below.

It is interesting to note that if λ_g/a is very large, then Eqn. (2.245) reduces to:

$$T = \frac{8\pi}{3} \frac{\mu}{C_m \lambda_g} a^4 \Omega_0$$
(2.247)

The Motion of Particles in Gases

But, from the kinetic theory of gases:

$$\frac{\mu}{\lambda_g} = \tfrac{1}{3}\rho\bar{v} \tag{2.248}$$

whence:

$$T = \frac{4}{3\alpha C_m} T_K \tag{2.248}$$

which is very close to the exact result. This is a case where the slip correction is valid over a wide range of Knudsen numbers on the assumption that Eqn. (2.245) is a reasonable interpolation between the two limiting cases.

2.21 Nonspherical rotating bodies

There are a variety of shapes for which the equation for u_ϕ can be transformed into a suitable coordinate system such that exact solutions are obtainable. For example, a prolate spheroid results from the transformation (Jeffery [1915b]):

$$z + i\varpi = c\cosh(\xi + i\eta)$$

Then a solution of the form:

$$u_\phi(\xi,\eta) = \sum_{n=1}^{\infty} B_n P_n^{(1)}(\cos(\eta)) Q_n^{(1)}(\cosh(\xi))$$

arises, where $P_n^{(1)}$ and $Q_n^{(1)}$ are Legendre functions. When this solution is made to satisfy the no-slip boundary conditions the coefficients B_n are found, from which:

$$T = \tfrac{16}{3}\pi\mu c^3 \Omega_0 \left[\tfrac{1}{2}\ln\left(\frac{a_0+c}{a_0-c}\right) - \frac{a_0 c}{b_0^2}\right]^{-1} \tag{2.249}$$

where $c^2 = a_0^2 - b_0^2$, $a_0 = c\cosh(\xi_0)$, and $b_0 = c\sinh(\xi_0)$. Thus:

$$\xi_0 = \tfrac{1}{2}\log\left(\frac{a_0+b_0}{a_0-b_0}\right)$$

Using the slip boundary conditions it is found that, to a reasonable first approximation:

$$T(\lambda_g/a_0) = \frac{\tfrac{16}{3}\pi\mu c^3 \Omega_0}{\dfrac{2c^3 C_m \lambda_g}{a_0 b_0^3} + \dfrac{a_0 c}{b_0^2} - \tfrac{1}{2}\ln\left(\dfrac{a_0+c}{a_0-c}\right)} \tag{2.250}$$

For large Knudsen numbers, we find:

$$T(\lambda_g/a_0) \to \tfrac{8}{3}\pi\Omega_0 \frac{a_0 b_0^3}{C_m \lambda_g} \tag{2.251}$$

which, as we shall note below, is not the same as the free molecular value.

The oblate spheroid requires the transformation:

$$z + i\varpi = c \sinh(\xi + i\eta)$$

and hence all prolate results are valid provided we replace c by $-ic$ and interchange a_0 and b_0. Thus, the torque on an oblate spheroid spinning about its axis of symmetry is (Jeffery [1915b]):

$$T = \tfrac{16}{3} \pi \mu c^3 \Omega_0 \left[\frac{b_0 c}{a_0^2} - \cot^{-1}\left(\frac{b_0}{c}\right) \right]^{-1} \tag{2.252}$$

where $c^2 = a_0^2 - b_0^2$.

The limiting case of a disk arises when $b_0 \to 0$. Then we find:

$$T_{disk} = \tfrac{32}{3} \mu a_0^3 \Omega_0 \tag{2.253}$$

If we apply slip corrections we find, to a first approximation:

$$T(\lambda_g / a_0) = \frac{\tfrac{16}{3} \pi \mu c^3 \Omega_0}{\dfrac{2c^3 C_m \lambda_g}{a_0^3 b_0} + \dfrac{c b_0}{a_0^2} - \cot^{-1}\left(\dfrac{b_0}{c}\right)} \tag{2.254}$$

which, for $\lambda_g/b_0 \to \infty$, becomes:

$$T(\lambda_g / b_0) = \tfrac{8}{3} \pi \mu \Omega_0 \frac{a_0^3 b_0}{C_m \lambda_g} \tag{2.255}$$

This is not the correct limit, which according to Halbritter [1974] is:

$$T_k = \tfrac{1}{4} \rho \bar{v} a_0^4 \Omega_0 \tag{2.255a}$$

Another shape that has received some attention is that of a closed torus rotating about its major axis as shown in Fig. 2.29.

This problem was solved by Dorrepaal et al. [1976a] for no slip and by Williams [1987a] with slip. Results for the torque are given in the form:

$$T(\lambda) = \frac{T(0)}{1 + (3.3055)\lambda} \tag{2.256}$$

where:

$$T(0) = (50.9968) \pi \mu a^3 \Omega_0 \tag{2.257}$$

and $\lambda = C_m \lambda_g / 2a$.

The Motion of Particles in Gases

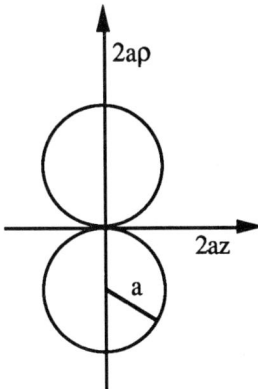

Figure 2.29: A closed torus rotating about its major axis.

Jeffery [1915b] has also considered the case of two nonconcentric spheres rotating about their line of centers, but we will defer discussion of that case until later.

2.22 Torque on an ellipsoid in free molecular flow

If the size of a body is very small compared with a mean free path of gas atoms, then its motion can be studied very precisely. This is because the presence of the body does not affect the velocity distribution of impinging atoms which is given by the gas conditions. We describe how to calculate the torque and the drag for such bodies and illustrate the method by some results obtained by Halbritter [1974].

In principle, the limit of Kn→∞ can be dealt with by solving the collision free form of the Boltzmann equation, Eqn. (2.6). However, in practice it is more convenient to start from first principles. Let us assume, therefore, that the prevailing gas atom velocity distribution function is $f^-(v)$ and that this is incident on the body. Due to the boundary condition, the distribution which leaves the body will be changed to $f^+(v)$. Thus the elementary force dF on a surface element dS of the body will be:

$$dF = -dS \left[m \int_{(v' \cdot n) < 0} v'(v' \cdot n) f^-(v') dv' + m \int_{(v' \cdot n) > 0} v'(v' \cdot n) f^+(v') dv' \right] \quad (2.258)$$

where n is the outward pointing unit normal at dS. On the left hand side of the equation is the rate of change of momentum on dS. f^+ and f^- are related by the boundary condition.

If the body is rotating with an angular velocity, Ω, with respect to the laboratory system, then at every point r on the surface of the body, the velocity v' is connected with the velocity in the laboratory system v by $v' = v - \Omega \times r$. For a body rotating with an angular velocity $-\Omega \times r$, the Maxwellian distribution will be:

$$f^-(v') = n_0 \left(\frac{m}{2\pi k T} \right)^{3/2} \exp\left\{ -\frac{m}{2kT} (v' + \Omega \times r)^2 \right\} \quad (2.259)$$

where n_0 is the number density of the gas and T is its physical temperature.

We use a simple boundary condition of perfect accommodation although the more general case of specular and diffuse reflection is considered by Halbritter. Thus we can write:

$$f^+(v) = \frac{1}{2\pi}\left(\frac{m}{kT}\right)^2 \exp\left\{-\frac{mv^2}{2kT}\right\} \int_{(v'\cdot n)<0} (-v'\cdot n) f^-(v') dv' \qquad (2.260)$$

where $(v\cdot n)>0$.

Inserting this expression into Eqn. (2.258) leads to:

$$dF = -dS\left[m \int_{(v'\cdot n)<0} v'(v'\cdot n) f^-(v') dv' + n\frac{\pi}{4}\bar{v} m \int_{(v'\cdot n)<0} (-v'\cdot n) f^-(v') dv'\right] \qquad (2.261)$$

where $\bar{v} = (8kT/\pi m)^{1/2}$.

If the condition $\Omega L << \bar{v}$ is satisfied, where L is a typical dimension of the body, then we may write:

$$f^-(v') \approx n_0 \left(\frac{m}{2\pi kT}\right)^{3/2} \exp\left\{-\frac{mv'^2}{2kT}\right\}\left[1 - \frac{m}{kT}v'\cdot(\Omega\times r)\right] \qquad (2.262)$$

For a rotating Brownian particle where $\Omega L/\bar{v} \approx (m/M)^{1/2}$, M being the mass of the body, this is clearly a good approximation. After inserting Eqn. (2.262) into Eqn. (2.261) and integrating over v' we find:

$$dF = -dS\,p\,n - dS\,\rho_g\,\bar{v}\left[\tfrac{1}{4}(\Omega\times r) + (\tfrac{1}{4}+\tfrac{1}{8}\pi)n\,n\cdot(\Omega\times r)\right] \qquad (2.263)$$

where the gas density $\rho_g = mn_0$ and the pressure $p = kT\rho_g/m$.

The torque on element dS is $dT = r\times dF$. Integrating over the surface we find the total torque:

$$T = -\tfrac{1}{4}\rho\bar{v}\left[\int r\times(\Omega\times r)\,dS + (1+\tfrac{1}{2}\pi)\int (r\times n)\,\Omega\cdot(r\times n)\,dS\right] \qquad (2.264)$$

This expression is valid for a body of arbitrary shape.

For a sphere and a disk the approximation of Eqn. (2.262) is not required and it is only necessary to change the velocity variable to $v = v' + \Omega\times r$ and note that $n\times(\Omega\times r) = 0$. We then find Eqns. (2.246) and (2.255a), respectively. After some lengthy integrations, Halbritter finds the following results for prolate and oblate spheroids.

2.23 Prolate spheroids

The spheroid is characterized by the unit vector, **u**. The major and minor semi-axes are a and b (a>b) and the eccentricity $e = (1-b^2/a^2)^{1/2}$. Then it is found that:

$$T = -\gamma_\|\,u\,u\cdot\Omega - \gamma_\perp(\Omega - u\,u\cdot\Omega) \qquad (2.265)$$

where $\gamma_\|$ and $\gamma_\perp$ are functions of the eccentricity.

2.24 Oblate spheroids

An expression is obtained similar to Eqn. (2.265) but with different values for $\gamma_\parallel$ and $\gamma_\perp$. One interesting limiting case is for the disk when the angular velocity lies in the plane of the disk. Then we find:

$$\mathbf{T} = -\tfrac{1}{4}\left(1+\tfrac{1}{4}\pi\right)\rho \bar{\nu} a^4 \Omega \tag{2.266}$$

which is about twice the value for the case when the disk rotates about its central axis.

Full details of $\gamma_\parallel$, $\gamma_\perp$, and other matters may be found in the original reference (Halbritter [1974]).

2.25 The fluid dynamics of two interacting spheres

The basis of aerosol coagulation involves the collision of two particles. The simplest case is that of two spheres and it is clear that this problem must be fully understood before problems of greater complexity can be considered.

While there were a number of very early attempts to tackle the problem of the motion of two spheres in a fluid, the problem for one important special case was not solved exactly until 1926 by Stimson and Jeffery [1926]. This particular problem was that of two spheres of different radii moving with the same velocity along their line of centers. Some years earlier Jeffery [1915b] had also solved the problem of two spheres rotating about their line of centers with constant but differing angular velocities.

In this section we will discuss Jeffery's method for the two spheres but generalize it somewhat so that the spheres have different velocities.

As the two spheres are moving along their line of centers, we have an axially symmetric problem as shown in Fig. 2.30. We have seen that for axially symmetric bodies, the Stokes stream function obeys the equation:

$$E^4 \psi = 0 \tag{2.267}$$

Since we assume that the fluid is stationary at infinity, the boundary conditions on each sphere are:

$$\psi + \tfrac{1}{2}\varpi^2 V_1 = 0 \quad \text{on 1} \tag{2.268}$$

$$\psi + \tfrac{1}{2}\varpi^2 V_2 = 0 \quad \text{on 2} \tag{2.269}$$

$$\frac{\partial}{\partial n}\left(\psi + \tfrac{1}{2}\varpi^2 V_1\right) = 0 \quad \text{on 1} \tag{2.270}$$

$$\frac{\partial}{\partial n}\left(\psi + \tfrac{1}{2}\varpi^2 V_2\right) = 0 \quad \text{on 2} \tag{2.271}$$

and $\psi \to 0$ at infinity.

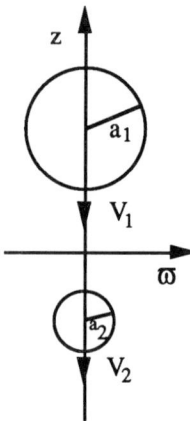

Figure 2.30: Motion of two spheres along their line of centers.

The key to this problem is the choice of an appropriate coordinate system. The bipolar transformation is suitable and is defined by:

$$\xi + i\eta = \ln\left(\frac{\varpi + i(z+c)}{\varpi + i(z-c)}\right) \qquad (2.272)$$

which leads to:

$$\varpi = \frac{c \sin(\eta)}{\cosh(\xi) - \cos(\eta)} \qquad (2.273)$$

and:

$$z = \frac{c \sinh(\xi)}{\cosh(\xi) - \cos(\eta)} \qquad (2.274)$$

The surfaces obtained by rotating the curves ξ=constant about the z-axis are a family of spheres having z=0 (or ξ=0) for a common radial plane. Two spheres external to each other are defined by $\xi=\alpha$, $=\beta$ ($\alpha>0$, $\beta<0$). α, β, and the constant, c, may be chosen so that these spheres have any radii and any center distance greater than the sum of their radii. If the radii are a_1 and a_2 and have their centers at distances, d_1 and d_2, on opposite sides of the origin, then:

$$a_1 = c \, \text{cosech}(\alpha) \qquad (2.275)$$

$$a_2 = -c \, \text{cosech}(\beta) \qquad (2.276)$$

$$d_1 = c \, \coth(\alpha) \qquad (2.277)$$

$$d_2 = -c \, \coth(\beta) \qquad (2.278)$$

$$c^2 = d_1^2 - a_1^2 = d_2^2 - a_2^2 \qquad (2.279)$$

The Motion of Particles in Gases

Solving these equations leads to:

$$\alpha = \ln\left[\frac{d_1}{a_1} + \left(\frac{d_1^2}{a_1^2} - 1\right)^{1/2}\right] \geq 0 \qquad (2.280)$$

and:

$$\beta = \ln\left[\frac{d_2}{a_2} - \left(\frac{d_2^2}{a_2^2} - 1\right)^{1/2}\right] \leq 0 \qquad (2.281)$$

It is also useful to have α and β in terms of the interparticle distance, $r=d_1+d_2$. In that case:

$$\alpha = \ln\left[\frac{\frac{r^2}{a_1^2} - \frac{a_2^2}{a_1^2} + 1}{\frac{2r}{a_1}} + \left\{\left(\frac{\frac{r^2}{a_1^2} - \frac{a_2^2}{a_1^2} + 1}{\frac{2r}{a_1}}\right)^2 - 1\right\}^{1/2}\right] \qquad (2.281)$$

and:

$$\beta = \ln\left[\frac{\frac{r^2}{a_1^2} + \frac{a_2^2}{a_1^2} - 1}{\frac{2r}{a_1}} - \left\{\left(\frac{\frac{r^2}{a_1^2} + \frac{a_2^2}{a_1^2} - 1}{\frac{2r}{a_1}}\right)^2 - 1\right\}^{1/2}\right] \qquad (2.282)$$

We see, therefore, that these are natural coordinates for systems of two spheres. Before transforming to these new coordinates, let us note that the ordinary theory of linear partial differential equations with constant coefficients (Sneddon [1957]) shows that a solution of Eqn. (2.267) is:

$$\psi = \psi_1 + z\psi_2 \qquad (2.283)$$

where ψ_1 and ψ_2 are any solutions of $E^2\psi=0$. Transforming to the new coordinates, we find:

$$E^2\psi = \frac{\cosh(\xi) - \mu}{c^2}\left\{\frac{\partial}{\partial \xi}\left([\cosh(\xi) - \mu]\frac{\partial \psi}{\partial \xi}\right) + (1-\mu^2)\frac{\partial}{\partial \mu}\left([\cosh(\xi) - \mu]\frac{\partial \psi}{\partial \mu}\right)\right\} \qquad (2.284)$$

where $\mu=\cos(\eta)$.

The general solution of $E^2\psi=0$ is readily shown to be:

$$\psi = (\cosh(\xi) - \mu)^{-1/2} \sum_{n=1}^{\infty} \{a_n \cosh((n+\tfrac{1}{2})\xi) + b_n \sinh((n+\tfrac{1}{2})\xi)\} V_n(\mu) \qquad (2.285)$$

where:

$$V_n(\mu) = P_{n-1}(\mu) - P_{n+1}(\mu) \propto P_n^{(1)}(\mu) \sin(\eta) \qquad (2.286)$$

Thus, the general solution of Eqn. (2.267) is, using Eqn. (2.283):

$$\psi = (\cosh(\xi) - \mu)^{-3/2} \sum_{n=1}^{\infty} V_n(\mu) \{[\cosh(\xi) - \mu][a_n \cosh((n+\tfrac{1}{2})\xi) + b_n \sinh((n+\tfrac{1}{2})\xi)]$$
$$+ a \sinh(\xi)[c_n \cosh((n+\tfrac{1}{2})\xi) + d_n \sinh((n+\tfrac{1}{2})\xi)]\}$$
(2.287)

After some rearrangement and use of the relation:

$$\mu V_n = \frac{n-1}{2n-1} V_{n-1} + \frac{n+2}{2n+3} V_{n+1}$$

we have:

$$\psi = [\cosh(\xi) - \mu]^{-3/2} \sum_{n=1}^{\infty} U_n(\xi) V_n(\mu)$$
(2.288)

where:

$$U_n(\xi) = A_n \cosh((n-\tfrac{1}{2})\xi) + B_n \sinh((n-\tfrac{1}{2})\xi)$$
$$+ C_n \cosh((n+\tfrac{3}{2})\xi) + D_n \sinh((n+\tfrac{3}{2})\xi)$$
(2.289)

Writing:

$$\chi = \sum_{n=1}^{\infty} U_n(\xi) V_n(\mu)$$
(2.290)

we find that the boundary conditions become:

$$\chi(\alpha,\mu) = -\tfrac{1}{2} V_1 \frac{c^2(1-\mu^2)}{(\cosh(\alpha) - \mu)^{1/2}}$$

and:

$$\chi(\beta,\mu) = -\tfrac{1}{2} V_2 \frac{c^2(1-\mu^2)}{(\cosh(\beta) - \mu)^{1/2}}$$

since $\partial/\partial n = \partial/\partial \xi$, the normal boundary conditions transform to:

$$\left.\frac{\partial \psi}{\partial \xi}\right|_\alpha = \tfrac{1}{4} V_1 \frac{c^2(1-\mu^2)\sinh(\alpha)}{(\cosh(\alpha) - \mu)^{3/2}}$$

and:

$$\left.\frac{\partial \psi}{\partial \xi}\right|_\beta = \tfrac{1}{4} V_2 \frac{c^2(1-\mu^2)\sinh(\beta)}{(\cosh(\beta) - \mu)^{3/2}}$$

In view of the fact that:

$$V_n = \frac{2n+1}{n(n+1)} \sin(\eta) P_n^{(1)}(\mu)$$

we can write:

$$\chi = \sum_{n=1}^{\infty} \frac{2n+1}{n(n+1)} U_n(\xi) \sin(\eta) P_n^{(1)}(\mu)$$
(2.291)

Thus, at $\xi = a$:

$$\sum_{n=1}^{\infty} \frac{2n+1}{n(n+1)} U_n(\alpha) \sin(\eta) P_n^{(1)}(\mu) = -\tfrac{1}{2} V_1 \frac{c^2(1-\mu^2)}{(\cosh(\alpha)-\mu)^{1/2}}$$

Multiplying by $P_m^{(1)}(\mu)$ and integrating over $\eta(0,\pi)$, we find:

$$\sum_{n=1}^{\infty} \frac{2n+1}{n(n+1)} U_n(\alpha) \int_0^\pi d\eta \, \sin(\eta) P_m^{(1)}(\mu) P_n^{(1)}(\mu) = -\tfrac{1}{2} V_1 c^2 \int_0^\pi \frac{d\eta \, \sin^2(\eta) P_m^{(1)}(\mu)}{(\cosh(\alpha)-\mu)^{1/2}}$$

But, the integrals may be carried out on both sides of the equation and lead to:

$$U_m(\alpha) = -\frac{V_1 c^2}{\sqrt{2}} \frac{m(m+1)}{2m+1} \left\{ \frac{\exp[-(m-\tfrac{1}{2})\alpha]}{2m-1} - \frac{\exp[-(m+\tfrac{3}{2})\alpha]}{2m+3} \right\}$$

Similarly:

$$U_m(\beta) = -\frac{V_2 c^2}{\sqrt{2}} \frac{m(m+1)}{2m+1} \left\{ \frac{\exp[(m-\tfrac{1}{2})\beta]}{2m-1} - \frac{\exp[(m+\tfrac{3}{2})\beta]}{2m+3} \right\}$$

Employing the other two boundary conditions, we find integrals of the type encountered previously. These can be evaluated with the net result that:

$$U'_m(\alpha) = \frac{V_1 c^2}{\sqrt{2}} \sinh(\alpha) \frac{m(m+1)}{2m+1} \exp[-(m+\tfrac{1}{2})\alpha]$$

and:

$$U'_m(\beta) = \frac{V_2 c^2}{\sqrt{2}} \sinh(\beta) \frac{m(m+1)}{2m+1} \exp[(m+\tfrac{1}{2})\beta]$$

where the prime denotes differentation with respect to ξ. The four boundary conditions lead to four equations for the unknowns, A_n, B_n, C_n, and D_n.

With the Stokes stream function known, we return to Eqn. (2.88) to calculate the drag forces on each sphere. This is a lengthy and tedious procedure and requires evaluation of:

$$F_{1,2} = \mu \pi \int_0^\pi \omega^3 \frac{\partial}{\partial \xi}\left(\frac{E^2 \psi}{\omega^2}\right) d\eta \qquad \text{on } \xi = \alpha, \beta \tag{2.292}$$

The viscosity here should not be confused with the variable, $\mu = \cos(\eta)$. The net result is that:

$$F_1 = -\frac{2\sqrt{2}\,\mu\pi}{c} \sum_{n=1}^{\infty} (2n+1)[A_n + B_n + C_n + D_n] \tag{2.293}$$

and:

$$F_2 = -\frac{2\sqrt{2}\,\mu\pi}{c} \sum_{n=1}^{\infty} (2n+1)[A_n - B_n + C_n - D_n] \tag{2.294}$$

Aerosol Science: Theory and Practice

The total drag on both spheres can be calculated using Eqn. (2.90). This is much easier to employ than the integral form but only gives the total effect F_1+F_2 and not the more important individual values. We observe that:

$$F_1 + F_2 = -\frac{4\sqrt{2}\,\mu\pi}{c} \sum_{n=1}^{\infty} (2n+1)(A_n + C_n) \tag{2.295}$$

If the results for A_n, B_n, C_n, and D_n are collected together as required in Eqns. (2.293) and (2.294) we find that the forces may be written as follows:

$$F_1 = -\kappa_1 V_1 + \lambda_1 V_2 \tag{2.296}$$

$$F_2 = -\kappa_2 V_2 + \lambda_2 V_1 \tag{2.297}$$

where:

$$\kappa_1 = \frac{\mu\pi\sqrt{2}}{c} \sum_{n=1}^{\infty} (2n+1)(A'_n - A_n + B'_n - B_n + C'_n - C_n + D'_n - D_n) \tag{2.298}$$

$$\kappa_2 = \frac{\mu\pi\sqrt{2}}{c} \sum_{n=1}^{\infty} (2n+1)(-A'_n - A_n + B'_n + B_n - C'_n - C_n + D'_n + D_n) \tag{2.299}$$

$$\lambda_1 = \frac{\mu\pi\sqrt{2}}{c} \sum_{n=1}^{\infty} (2n+1)(A'_n + A_n + B'_n + B_n + C'_n + C_n + D'_n + D_n) \tag{2.300}$$

$$\lambda_2 = \frac{\mu\pi\sqrt{2}}{c} \sum_{n=1}^{\infty} (2n+1)(A_n - A'_n - B_n + B'_n + C_n - C'_n - D_n + D'_n) \tag{2.301}$$

$$\begin{aligned}
A_n \Delta = (2n+3) K \{ &4 \exp[(n+\tfrac{1}{2})(\beta-\alpha)] \sinh((n+\tfrac{1}{2})(\alpha-\beta)) \\
&+(2n+1)^2 \exp(\alpha-\beta) \sinh(\alpha-\beta) \\
&+2(2n-1) \sinh((n+\tfrac{1}{2})(\alpha-\beta)) \cosh((n+\tfrac{1}{2})(\alpha+\beta)) \\
&-2(2n+1) \sinh((n+\tfrac{3}{2})(\alpha-\beta)) \cosh((n-\tfrac{1}{2})(\alpha+\beta)) \\
&-(4n^2-1) \sinh(\alpha-\beta) \cosh(\alpha+\beta) \}
\end{aligned}$$

$$\begin{aligned}
B_n \Delta = -(2n+3) K \{ &2(2n-1) \sinh((n+\tfrac{1}{2})(\alpha-\beta)) \sinh((n+\tfrac{1}{2})(\alpha+\beta)) \\
&-2(2n+1) \sinh((n+\tfrac{3}{2})(\alpha-\beta)) \sinh((n-\tfrac{1}{2})(\alpha+\beta)) \\
&+(4n^2-1) \sinh(\alpha-\beta) \sinh(\alpha+\beta) \}
\end{aligned}$$

The Motion of Particles in Gases

$$C_n \Delta = -(2n-1) K \left\{ 4 \exp[(n+\tfrac{1}{2})(\beta-\alpha)] \sinh((n+\tfrac{1}{2})(\alpha-\beta)) \right.$$
$$-(2n+1)^2 \exp(\beta-\alpha) \sinh(\alpha-\beta)$$
$$+2(2n+1) \sinh((n-\tfrac{1}{2})(\alpha-\beta)) \cosh((n+\tfrac{3}{2})(\alpha+\beta))$$
$$-2(2n+3) \sinh((n+\tfrac{1}{2})(\alpha-\beta)) \cosh((n+\tfrac{1}{2})(\alpha+\beta))$$
$$\left. +(2n+3)(2n+1) \sinh(\alpha-\beta) \cosh(\alpha+\beta) \right\}$$

$$D_n \Delta = (2n-1) K \left\{ 2(2n+1) \sinh((n-\tfrac{1}{2})(\alpha-\beta)) \sinh((n+\tfrac{3}{2})(\alpha+\beta)) \right.$$
$$-2(2n+3) \sinh((n+\tfrac{1}{2})(\alpha-\beta)) \sinh((n+\tfrac{1}{2})(\alpha+\beta))$$
$$\left. +(2n+1)(2n+3) \sinh(\alpha-\beta) \sinh(\alpha+\beta) \right\}$$

$$A'_n \Delta = (2n+3) K \left\{ 2(2n-1) \sinh((n+\tfrac{1}{2})(\alpha-\beta)) \sinh((n+\tfrac{1}{2})(\alpha+\beta)) \right.$$
$$-2(2n+1) \sinh((n+\tfrac{3}{2})(\alpha-\beta)) \sinh((n-\tfrac{1}{2})(\alpha+\beta))$$
$$\left. -(4n^2-1) \sinh(\alpha-\beta) \sinh(\alpha+\beta) \right\}$$

$$B'_n \Delta = -(2n+3) K \left\{ -4 \exp[(n+\tfrac{1}{2})(\beta-\alpha)] \sinh((n+\tfrac{1}{2})(\alpha-\beta)) \right.$$
$$-(2n+1)^2 \exp(\alpha-\beta) \sinh(\alpha-\beta)$$
$$+2(2n-1) \sinh((n+\tfrac{1}{2})(\alpha-\beta)) \cosh((n+\tfrac{1}{2})(\alpha+\beta))$$
$$-2(2n+1) \sinh((n+\tfrac{3}{2})(\alpha-\beta)) \cosh((n-\tfrac{1}{2})(\alpha+\beta))$$
$$\left. +(4n^2-1) \sinh(\alpha-\beta) \cosh(\alpha+\beta) \right\}$$

$$C'_n \Delta = -(2n-1) K \left\{ 2(2n+1) \sinh((n-\tfrac{1}{2})(\alpha-\beta)) \sinh((n+\tfrac{3}{2})(\alpha+\beta)) \right.$$
$$-2(2n+3) \sinh((n+\tfrac{1}{2})(\alpha-\beta)) \sinh((n+\tfrac{1}{2})(\alpha+\beta))$$
$$\left. -(2n+1)(2n+3) \sinh(\alpha-\beta) \sinh(\alpha+\beta) \right\}$$

$$D'_n \Delta = (2n-1) K \left\{ -4 \exp[(n+\tfrac{1}{2})(\beta-\alpha)] \sinh((n+\tfrac{1}{2})(\alpha-\beta)) \right.$$
$$+(2n+1)^2 \exp(\beta-\alpha) \sinh(\alpha-\beta)$$
$$+2(2n+1) \sinh((n-\tfrac{1}{2})(\alpha-\beta)) \cosh((n+\tfrac{3}{2})(\alpha+\beta))$$
$$-2(2n+3) \sinh((n+\tfrac{1}{2})(\alpha-\beta)) \cosh((n+\tfrac{1}{2})(\alpha+\beta))$$
$$\left. -(2n+3)(2n+1) \sinh(\alpha-\beta) \cosh(\alpha+\beta) \right\}$$

$$\Delta = 4 \sinh^2((n+\tfrac{1}{2})(\alpha-\beta)) - (2n+1)^2 \sinh^2(\alpha-\beta)$$

and:

$$K = \frac{c^2 n(n+1)}{(4n^2-1)(2n+3)\sqrt{2}}$$

These formulae are taken from the paper by Spielman [1970] and eliminate some errors which appear in the paper by Maude [1961].

The special case of two spheres moving in the same direction with equal velocities, U, has been studied numerically by Cooley and O'Neill [1969]. In this case, Eqns. (2.296) and (2.297) become:

$$F_1 = -(\kappa_1 - \lambda_1) U$$

and:

$$F_2 = -(\kappa_2 - \lambda_2) U$$

To simplify the notation, Cooley and O'Neill set $a_1=a$ and $a_2=ka$. (see Fig. 2.13). In that case, the forces can be written:

$$F_1 = -6\pi\mu a U f_1\left(\tfrac{1}{k}, \tfrac{\varepsilon}{k}\right)$$

$$F_2 = -6\pi\mu a U f_2\left(\tfrac{1}{k}, \tfrac{\varepsilon}{k}\right)$$

where ε is the normalized interparticle separation (*i.e.*, the gap between the surfaces). In terms of the center to center distance, r, $\varepsilon=(r/a)-1-k$. The spheres touch when $r=a(1+k)$.

The following relations hold for the functions f_1 and f_2:

$$f_1\left(\tfrac{1}{k}, \tfrac{\varepsilon}{k}\right) = \tfrac{1}{k} f_2(k,\varepsilon)$$

$$f_2\left(\tfrac{1}{k}, \tfrac{\varepsilon}{k}\right) = \tfrac{1}{k} f_1(k,\varepsilon)$$

Consequently when f_1 and f_2 are known for $0<k\leq 1$, they are known for all k. When $k>1$ we replace ε by ε/k.

Extensive tables of f_1 are given by Cooley and O'Neill [1969] but these are summarized in Fig. 2.31 which shows f_1 versus interparticle separation, ε, for various ratio of radii, k. Cooley and O'Neill note the interesting behavior of the forces acting on the larger and smaller spheres as contact is approached. For example, if the ratio of the larger sphere's radius to the smaller sphere's radius is λ (*i.e.*, either k or 1/k), then the results shown in Fig. 2.31 indicate that the force on the smaller sphere decreases monotonically with ε from its maximum value at infinite separation to its minimum value at contact. However, the force on the larger sphere decreases monotonically with ε to its minimum value at contact for $1<\lambda\leq\lambda^*$, where $\lambda^*\approx 10/7$. For $\lambda>\lambda^*$, the force on the larger sphere passes through a minimum at some finite distance from contact before rising again to a larger value at contact.

A useful review of the topics discussed in this and earlier sections can be found in O'Neill [1981].

The Motion of Particles in Gases

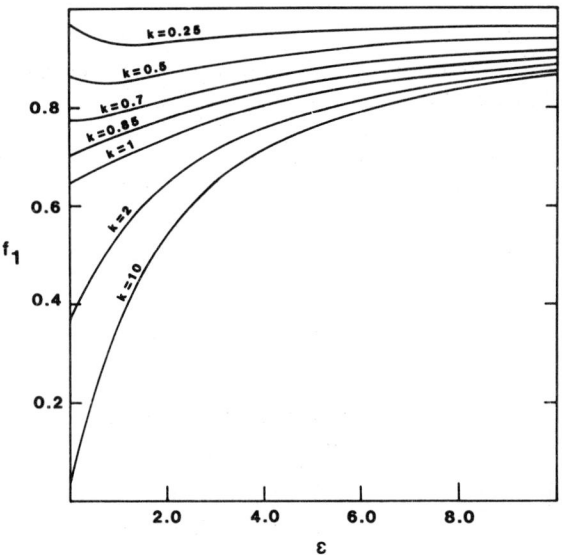

Figure 2.31: Graphs of f_1 plotted against ε. Adapted from Cooley and O'Neill [1969] with permission of the Cambridge University Press.

2.26 The limiting case of touching spheres

While the results of the previous section can be used for any value of the separation distance, the infinite sums become very slowly converging as the spheres approach each other closely. For this reason, Cooley and O'Neill [1969] have calculated from first principles the drag on two unequal touching spheres moving along their line of centers. This procedure involves the coordinate transformation:

$$\varpi = \frac{2\eta}{\xi^2 + \eta^2} \quad \text{and} \quad z = \frac{2\xi}{\xi^2 + \eta^2}$$

Cooley and O'Neill find it useful to measure distance in units of the radius of the sphere in $z>0$. This sphere is of radius a. The sphere in $z<0$ is of radius ka. The two spheres are therefore defined by $\xi=1$ and $\xi=-\alpha$, respectively. Following the procedure described above and details in Cooley and O'Neill, the general solution for the Stokes stream function is:

$$\psi = \frac{\eta}{\left(\xi^2 + \eta^2\right)^{3/2}} \int_0^\infty \left\{ (A + \xi C)\sinh(s\xi) + (B + \xi D)\cosh(s\xi) \right\} J_1(s\eta)\, ds \tag{2.302}$$

where J_1 is a Bessel function and A, B, C, and D are unknown functions of s. Applying the four boundary conditions:

Table 2.6: Values of λ_s and λ_o for different values of the interparticle separation, r/a.

r/a	λ_s	λ_o	r/a	λ_s	λ_o
2.0	0.6451	∞	3.0	0.6983	2.039
2.01	0.6457	53.48	3.5	0.7214	1.754
2.05	0.6480	12.71	4.0	0.7423	1.597
2.10	0.6509	7.413	5.0	0.7772	1.424
2.25	0.6593	4.036	7.0	0.8265	1.270
2.50	0.6729	2.772	12.0	0.8895	
2.75	0.6859	2.298	∞	1.0	1.0

$$\psi = -\frac{2\eta^2 U}{\left(\xi^2 + \eta^2\right)^2}$$

and:

$$\frac{\partial \psi}{\partial \xi} = \frac{8\xi\eta^2 U}{\left(\xi^2 + \eta^2\right)^3} \tag{2.303}$$

at $\xi=1$ and $-\alpha$, leads to equations for A, B, C, and D. Moreover, use of Eqn. (2.88) gives, for the forces on the spheres, an expression of the following form:

$$F = -6\pi\mu a U f \tag{2.304}$$

where U is the common velocity of the spheres and:

$$f = \tfrac{1}{6} \int_0^\infty s(B \pm A)\, ds \tag{2.305}$$

with the (+) sign referring to the upper sphere and the (–) sign to the lower one. In the case of equal size spheres a very simple result emerges for f:

$$f = \tfrac{1}{3} \int_0^\infty \left\{ 1 - \frac{2\sinh^2(s) - 2s^2}{\sinh(2s) + 2s} \right\} ds = 0.645141 \tag{2.306}$$

Thus, the force on each sphere is significantly less than the isolated value. Cooley and O'Neill give extensive tables of the factors, f_1 and f_2, for each sphere for various values of the ratio, k. As we have discussed in the last section they also give results for spheres separated by a center to center distance of r. In order to illustrate these results in more detail we consider spheres of equal size moving with equal velocities, U. Then, the force acting on one sphere may be written:

$$F_s = -6\pi\mu a U \lambda_s \tag{2.307a}$$

where:

The Motion of Particles in Gases

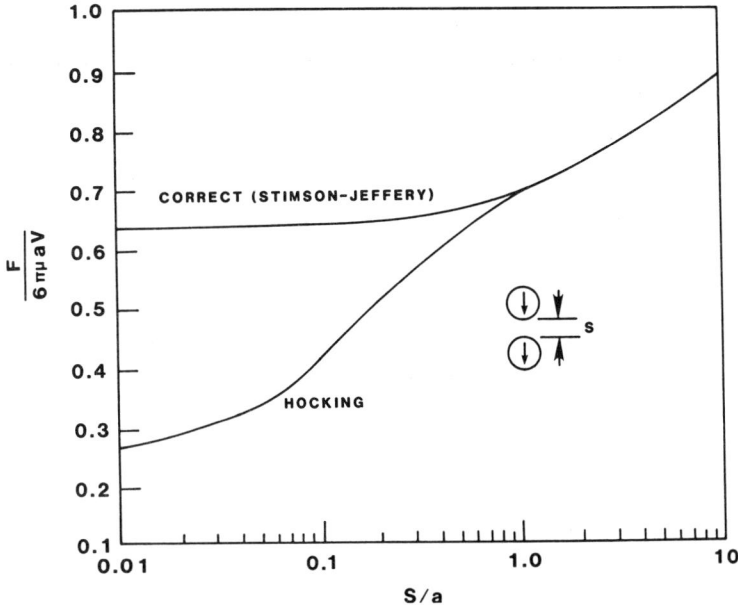

Figure 2.32: Resistance coefficient for equal parallel motion parallel to the line of centers (equal spheres, radius a). Adapted from Davis [1966] with permission of the American Geophysical Union.

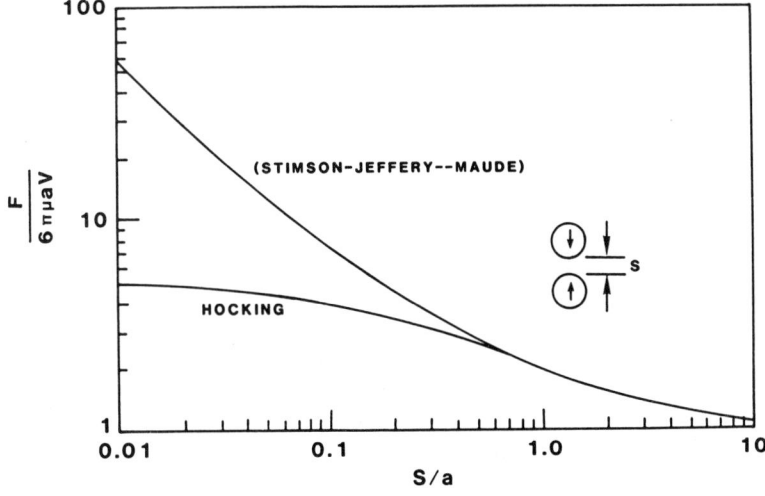

Figure 2.33: Resistance coefficient for equal anti-parallel motion parallel to the line of centers (equal spheres, radius a). Adapted from Davis [1966] with permission of the American Geophysical Union.

$$\lambda_s = \tfrac{4}{3}\sinh(\alpha)\sum_{n=1}^{\infty}\frac{n(n+1)}{(2n-1)(2n+3)}\left[1-\frac{4\sinh^2\left(\left(n+\tfrac{1}{2}\right)\alpha\right)-(2n+1)\sinh^2(\alpha)}{2\sinh((2n+1)\alpha)+(2n+1)\sinh(2\alpha)}\right]$$
(2.308a)

where $\alpha = \cosh^{-1}(r/2a)$.

We also consider the same spheres moving with equal but opposite velocities towards each other. In that case we write:

$$F_o = -6\pi\mu a U \lambda_o$$
(2.307b)

Table 2.6 shows the values of λ_s and λ_o for different values of the interparticle separation r/a. Figs. 2.32 and 2.33 illustrate the results graphically as a function of (r/a)–2.

We note that for particles moving in the same direction the force decreases as they move closer together. On the other hand, particles moving toward each other experience an increasing force as they approach (Maude [1961]). Hocking's [1959] approximate results are shown in Figs. 2.32 and 2.33 for comparison. It is noted that in all of these calculations the unsteady term in the Stokes equation is assumed to be negligible (Cooley and O'Neill [1969]).

The case of spheres moving together is an important one since it corresponds to the motion that arises in coagulation. It is of some concern, therefore, that the force is seen to become infinite when the spheres touch.

It is very important to calculate the manner in which the force approaches infinity as the spheres approach each other. To study this we note that λ_o can be written explicitly as:

$$\lambda_o = \tfrac{4}{3}\sinh(\alpha)\sum_{n=1}^{\infty}\frac{n(n+1)}{(2n-1)(2n+3)}\left[\frac{4\cosh^2\left(\left(n+\tfrac{1}{2}\right)\alpha\right)+(2n+1)^2\sinh^2(\alpha)}{2\sinh((2n+1)\alpha)-(2n+1)\sinh(2\alpha)}-1\right]$$
(2.308b)

where $\alpha = \cosh^{-1}(r/2a)$.

The above series is mathematically convergent for all ratios $1<r/2a<\infty$ but numerical convergence is poor near r=2a or as $\alpha \to 0$. However, a method based upon 'inner and outer expansions' has been developed by Cox and Brenner [1967] and used by Hansford [1970]. It leads to the following result:

$$\lambda_o \approx \frac{1}{2\alpha^2} + \tfrac{9}{10}\ln\left(\frac{1}{\alpha}\right) + (1.303913...) + O(1)$$
(2.309)

The accuracy of this result is shown in Table 2.7. The exact and asymptotic results are in good agreement even for $\alpha=1$.

Since $\alpha^2 \approx (r/a -2)$ for small α, we see that when the spheres are very close:

$$\lambda_o \approx \frac{a}{2(r-2a)}$$
(2.310)

In practice, this behavior means that the particles will never touch. This may be seen if we assume that one of the particles is stationary and the other is moving under the action of a

Table 2.7: Comparison of exact and asymptotic formulae for the Stokes force on either of two identical approaching spheres. Adapted from Hansford [1970].

a	r/2a	exact λ_o	Eqn. (2.309)
0.05	1.001	204.001	204.000
0.25	1.031	10.571	10.552
0.50	1.128	3.987	3.928
0.75	1.295	2.562	2.451
1.00	1.543	1.974	1.804

constant force, say gravity. Then, if h is the separation distance between the sphere surfaces, we can write $U = \dot{h}$ and solve the simple equation:

$$\frac{dh}{dt} = -K h$$

where $K = m g / 3\pi a^2 \mu$.
Clearly, the time to move from an initial separation, h_0, to a reduced separation, h, is:

$$t = \frac{1}{K} \ln\left(\frac{h_0}{h}\right) \tag{2.311}$$

As h→0, the time is infinite. Three procedures have been used to overcome this embarrassing and clearly false behavior. (1) In calculating collision behavior a collision is defined to have taken place when the spheres are a certain fraction of their size apart. This somewhat *ad hoc* assumption is useful because the final results for trajectories are not too sensitive to the separation distance chosen (Hocking [1959]). (2) It is assumed that some surface forces of the van der Waals - London type exist which are attractive and which overcome the repulsive fluid force. (3) Kinetic theory considerations become important when the gap between surfaces is comparable to a mean free path and therefore slip boundary conditions should be employed.

Procedure (1) is a practical solution. Procedure (2) is probably true but is sometimes difficult to incorporate into trajectory calculations. Procedure (3) is the most satisfactory method and has been carried out by Davis [1972] and by Hocking [1973], although a combination of procedures (2) and (3) would presumably be the most desirable.

Davis introduced slip boundary conditions into the Stimson-Jeffery analysis discussed above. The expression for the transverse component now becomes, from Eqn. (2.82):

$$\frac{\partial}{\partial n}\left(\psi + \tfrac{1}{2}\varpi^2 U\right) = \pm \frac{C_m \lambda_g}{\mu} P_{ns} \tag{2.312}$$

where:

Table 2.8: Equal parallel motion, ε=0.01. Adapted from Davis [1972] with permission of the American Meteorological Society.

Kn	F/F$_s$	F$_1$/F$_{1s}$	F$_2$/F$_{2s}$
0.01	0.992	0.998	0.992
0.02	0.985	0.996	0.984
0.05	0.964	0.991	0.966

$$P_{ns} = P_{\xi\eta} = \frac{\mu}{c^3}(\cosh(\xi)-\mu)\{(\cosh(\xi)-\mu)[\sin(\eta)\,\psi_{\mu\mu} - \text{cosec}(\eta)\,\psi_{\xi\xi}]$$
$$-3[\sin(\eta)\,\psi_{\mu} + \sin(\xi)\,\text{cosec}(\eta)\,\psi_{\xi}]\}$$

The upper sign is to be used for $\xi=\alpha$ and the lower one for $\xi=-\beta$. To be more precise about the historical origin of this term we note that Bart [1968] first introduced it in connection with the motion of viscous fluid spheres settling in the presence of a plane surface. It was not, however, used in the kinetic theory context.

When these boundary conditions are used to determine the unknowns, A_n, B_n, C_n, and D_n in Eqn. (2.288), they lead to a set of recursive algebraic equations which must be solved numerically. Details of the calculations are not given by Davis but his results are of considerable interest. Two cases are examined: parallel motion in which the spheres move in the same direction with the same velocity and anti-parallel motion in which they move towards each other with the same velocity.

Results are given in terms of the effective Knudsen number, Kn=$C_m\lambda_g/a_2$, as a ratio of the force including slip, F, to that excluding slip (i.e. the Stimson-Jeffery calculation), F_s. Also, for numerical reasons, it was not possible to calculate the forces for spheres whose separation was less than ε, where:

$$\varepsilon = (r - a_1 - a_2)/a_2$$

Table 2.8 shows the results for parallel motion.

It is clear from these results that the effect is small for Kn<(0.05). Unfortunately, Davis does not give values in the interesting region of large Kn although the gap width is very small; h=a_2/100.

In the case of anti-parallel motion along the line of centers, Tables 2.9 and 2.10 show the results.

For the case of equal size spheres shown in Table 2.9, we note that deviations from the case of no-slip become marked at close separations, e.g. when the Knudsen number is (0.002) there is a decrease in the repulsive force between the spheres by a factor of four. Similar results are noted in Table 2.10 for the case of unequal spheres. The effect does not seem to discriminate

Table 2.9: Equal anti-parallel motion, $a_1/a_2=1.0$. Adapted from Davis [1972] with permission of the American Meteorological Society.

Kn	$\varepsilon=0.10$		$\varepsilon=0.01$		$\varepsilon=0.001$	
	F/F_s	f	F/F_s	f	F/F_s	f
0.002	0.985	0.980	0.754	0.833	0.260	0.333
0.005	0.944	0.952	0.586	0.667	0.152	0.167
0.010	0.888	0.909	0.451	0.500	0.097	0.091
0.020	0.804	0.833	0.329	0.333	0.061	0.048

Table 2.10: Equal anti-parallel motion, $a_1/a_2=5.0$. Adapted from Davis [1972] with permission of the American Meteorological Society.

Kn	$\varepsilon=0.10$			$\varepsilon=0.01$			$\varepsilon=0.001$		
	F_1/F_{1s}	F_2/F_{2s}	f	F_1/F_{1s}	F_2/F_{2s}	f	F_1/F_{1s}	F_2/F_{2s}	f
0.002	0.973	0.965	0.980	0.755	0.747	0.833	0.301	0.298	0.333
0.005	0.937	0.919	0.952	0.587	0.574	0.667	0.175	0.172	0.167
0.010	0.887	0.856	0.909	0.452	0.435	0.500	0.111	0.109	0.091
0.020	0.814	0.763	0.833	0.332	0.310	0.333	0.070	0.068	0.048

between larger and smaller spheres, the forces on both of which are affected by much the same factor.

Davis has developed an extremely simple yet accurate interpolation formula for the ratio F/F_s in the form:

$$f = \left(1 + \frac{Kn}{\varepsilon}\right)^{-1} \tag{2.313}$$

Values for f are also given in Tables 2.9 and 2.10 and are seen to compare very favorably with the exact results. We also note that at very close separations the force, with this correction factor, becomes:

$$F = -6\pi\mu a U \frac{a}{2(r-2a)} \frac{\varepsilon}{Kn} = -3\pi\mu a U / Kn \tag{2.314}$$

This eliminates the $1/\varepsilon$ behavior of the pure Stokesian force and thereby allows collisions to occur in a finite time. It is also of interest to note that the result is similar to that for a single sphere with a Cunningham correction factor as shown in Eqn. (2.56).

A different approach to the small gap problem has been taken by Hocking [1973]. Here, use is made of lubrication theory which enables certain terms in the Navier-Stokes equations to be eliminated thereby leading to an exact solution for the drag. Using slip boundary conditions Hocking shows that the forces on the two spheres, one of which is at rest and the other moving towards it with a velocity, U, can be written as:

$$F_2 = -6\pi a_2 \mu U f_N^{(2)}$$
(2.315)

and:

$$F_1 = -6\pi a_1 \mu U f_N^{(1)}$$
(2.316)

where:

$$f_N^{(2)} = \frac{2 a_1^2 a_2}{(a_1 + a_2)^2 h \beta^2} \left[(1+\beta)\ln(1+\beta) - \beta\right]$$
(2.317)

and:

$$f_N^{(1)} = -\frac{a_2}{a_1} f_N^{(2)}$$
(2.318)

with $h = r - a_1 - a_2$ and $\beta = 6\lambda_g / h$. Defining $Kn = \lambda_g / a$ and looking at small separations where $\beta \gg 1$, we find, for spheres of equal size:

$$F = 2\pi a \mu U \frac{1}{Kn} \ln\left(\frac{6 a Kn}{h}\right)$$
(2.319)

This behavior is not in agreement with the work of Davis, but does appear to be more soundly based. It does lead to a finite collision time and so, again, we see that kinetic theory considerations can remove the unphysical limit of the Stokes no-slip theory.

Further considerations on this problem involving the rate of approach of a sphere to a plane surface have been given by Brenner [1961], Cox and Brenner [1967], Mackay et al. [1963], and Hocking [1973].

Another case which has been dealt with fairly accurately in the slip flow regime is that of two spheres of equal size joined rigidly together (Williams [1987b]). The solution follows closely that of Cooley and O'Neill described by Eqn. (2.302) except that now one of the boundary conditions becomes:

$$\frac{(\xi^2 + \eta^2)}{4\eta} \frac{\partial \psi}{\partial \xi} + \frac{C_m \lambda_g}{\mu} P_{\eta\xi} = \frac{2 U \eta \xi}{(\xi^2 + \eta^2)^2} \quad \text{at } \xi = 1$$

where:

$$P_{\eta\xi} = \mu \left\{ \frac{\partial}{\partial \xi}\left(\frac{(\xi^2 + \eta^2)^3}{8\eta} \frac{\partial \psi}{\partial \xi}\right) - \frac{\partial}{\partial \eta}\left(\frac{(\xi^2 + \eta^2)^3}{8\eta} \frac{\partial \psi}{\partial \eta}\right) \right\}$$

The resulting equation for the stream function cannot be solved explicitly but a variational principle was used to calculate the drag on the bispherical unit. The result is:

$$F = -6\pi a \mu (1.2804147) \left\{ \frac{1 + (6.950)\lambda + (8.595)\lambda^2}{1 + (8.128)\lambda + (14.95)\lambda^2} \right\}$$

where $\lambda = C_m \lambda_g / a$.

The Motion of Particles in Gases

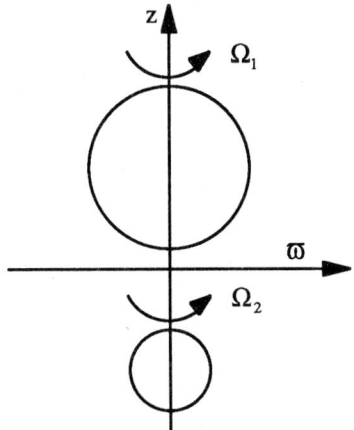

Figure 2.34: Two rotating spheres.

Clearly, the case for $\lambda=0$ corresponds to the result of Cooley and O'Neill for the double unit. However, the term in curly brackets describes the effect of slip. In the limit of $\lambda \to \infty$, we find $F=-6\pi a\mu U(0.7363)$, whence:

$$\frac{F(\lambda=\infty)}{F(\lambda=0)} = 0.5750$$

which is an exact result because the limiting cases can be solved without approximation. For a sphere, the ratio is 2/3. (We also note a typographical error in the original reference regarding this result). It should be noted here that, while the limit $\lambda \to \infty$ may be useful for air bubbles in a fluid, it is not necessarily good for aerosol particles in the free molecular regime (as we pointed out earlier in Section 2.22).

2.27 Two rotating spheres

A further problem regarding two spheres, solved by Jeffery [1915a], is when they rotate about their line of centers. Here one desires to find the torque required to maintain them in motion with angular speeds Ω_1 and Ω_2. Fig. 2.34 shows the situation.

Eqn. (2.229) describes such motion and, when it is transformed to bipolar coordinates by means of Eqn. (2.272), leads to the solution (Jeffery [1912]):

$$u_\phi = (\cosh(\xi)-\mu)^{1/2} \sum_{n=0}^{\infty} \{A_n \cosh((n+\tfrac{1}{2})\xi) + B_n \sinh((n+\tfrac{1}{2})\xi)\} P_n^{(1)}(\mu) \qquad (2.320)$$

The boundary condition for no-slip is:

$$u_\phi = \Omega_1 \varpi \quad \text{on } \xi = \alpha$$
$$u_\phi = -\Omega_2 \varpi \quad \text{on } \xi = -\beta$$

Table 2.11: Two equal spheres - one free. The ratio Ω_1/Ω_2 as a function of α. Adapted from Jeffery [1915b].

α	2a/r	Ω_1/Ω_2	α	2a/r	Ω_1/Ω_2
0.2	0.9803	0.1278	1.4	0.4649	0.0126
0.4	0.9250	0.1023	1.6	0.3880	0.0073
0.6	0.8435	0.0759	1.8	0.3218	0.0042
0.8	0.7477	0.0524	2.0	0.2658	0.0023
1.0	0.6481	0.0340	2.5	0.1631	0.0005
1.2	0.5523	0.0211	3.0	0.0993	0.0001

Using some integral relations involving the Legendre polynomials, we find A_n and B_n and hence the torque from Eqn. (2.232). Details may be found in the two papers by Jeffery [1912, 1915b] but the result is:

$$T_1 = 8\pi \mu c^3 \left[\Omega_1 \sum_{m=0}^{\infty} \text{cosech}^3((m+1)\alpha + m\beta) - \Omega_2 \sum_{m=0}^{\infty} \text{cosech}^3((m+1)(\alpha+\beta)) \right] \quad (2.321)$$

If one sphere is inside the other, then β is negative and the calculation shows that:

$$T_1 = 8\pi \mu c^3 (\Omega_1 - \Omega_2) \sum_{m=0}^{\infty} \text{cosech}^3[(m+1)\alpha - m\beta] \quad (2.322)$$

Suppose that one sphere, β, is made to rotate with a given angular speed, Ω_2. Then there will be a certain value of Ω_1 for which T_1 vanishes. This is the steady angular speed with which sphere α would rotate if allowed to move freely. In the case when one sphere encloses the other this occurs when $\Omega_1 = \Omega_2$. That is, if one sphere is allowed to move freely it will acquire the speed of the other sphere and both will move as a rigid body. When the spheres are separated and external, the other sphere will move with the angular speed:

$$\Omega_1 = \Omega_2 \frac{\sum_{m=0}^{\infty} \text{cosech}^3[(m+1)(\alpha+\beta)]}{\sum_{m=0}^{\infty} \text{cosech}^3[(m+1)\alpha + m\beta]} \quad (2.323)$$

To illustrate these results numerically, we consider spheres of equal size, when $\text{sech}(\alpha) = 2a/r$. In Table 2.11 we show the ratio Ω_1/Ω_2 as a function of α.

From these results we infer that there is very little interaction between the spheres. Even when the gap to interparticle ratio is as small as (0.02), one sphere communicates only about 1/8 of its spin to the other.

Jeffery also gives results for a sphere spinning in the neighborhood of a plane which corresponds to the special case of $\beta = 0$.

The Motion of Particles in Gases

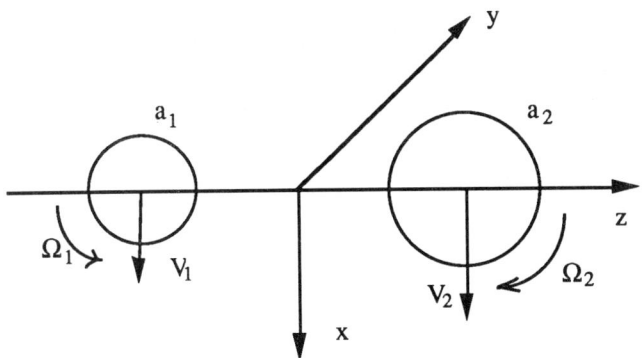

Figure 2.35: The torque on rotating spheres.

As far as we are aware, no slip corrections have been made to this problem although the modification is straightforward. Extensions have been made by Dean and O'Neill [1963] who considered a sphere rotating at a distance, d, from a rigid plane with the axis of rotation parallel to the plane. Because the equations of viscous flow are linear, it is possible, by combining Jeffery's result with that of Dean and O'Neill, to find the torque on a sphere rotating with its axis at an arbitrary angle to the plane. Some numerical results are given by Dean and O'Neill but the calculations are far more complex than for the case considered by Jeffery.

2.28 The motion of spheres moving perpendicular to their line of centers

One of the more difficult problems of slow viscous flow involving two spheres concerns the understanding of their motion when they move perpendicular to their line of centers. While several approximate techniques have been used to obtain the general trend of motion and drag forces (Hocking [1959] and Happel and Brenner [1973]), a definitive treatment was first given by Dean and O'Neill [1963] and O'Neill [1964]. This work formed the basis for the later investigations of Wakiya [1967], Davis [1969], and O'Neill and Majumdar [1970a,b].

We shall review below the procedure for dealing with this problem. The basic equations for study are:

$$\mu \nabla^2 \mathbf{u} = \nabla p \qquad (2.324)$$

and:

$$\nabla \cdot \mathbf{u} = 0 \qquad (2.325)$$

We also note that as the spheres move a torque is generated by the relative motion and so they tend to rotate. This rotation has to be included in the fluid dynamics. The situation is shown pictorially in Fig. 2.35.

The spheres move in the same direction with speeds U_1 and U_2 and have angular speeds of Ω_1 and Ω_2, respectively. It is convenient to introduce a cylindrical system of coordinates as shown in Fig. 2.36.

In such a system, Eqns. (2.324) and (2.325) take the form given by Eqns. (2.226) and (2.228). In addition, the continuity equation becomes:

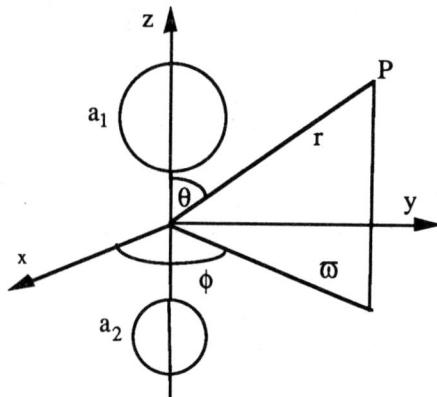

Figure 2.36: A cylindrical coordinate system for two spheres.

$$\left(\frac{\partial}{\partial \varpi}+\frac{1}{\varpi}\right)u_\varpi+\frac{1}{\varpi}\frac{\partial u_\phi}{\partial \phi}+\frac{\partial u_z}{\partial z}=0 \tag{2.326}$$

If there is symmetry about the plane, $\phi=0$, solutions can be found in the form:

$$p=\frac{\mu}{c}\sum_{m=0}^{\infty} Q_m \cos(m\phi) \tag{2.327}$$

$$u_\varpi = \tfrac{1}{2}\sum_{m=0}^{\infty}\left(\frac{\varpi}{c}Q_m+q_m+s_m\right)\cos(m\phi) \tag{2.328}$$

and:

$$u_\phi = \tfrac{1}{2}\sum_{m=0}^{\infty}(q_m-s_m)\sin(m\phi) \tag{2.329}$$

$$u_z = \tfrac{1}{2}\sum_{m=0}^{\infty}\left(\frac{z}{c}Q_m+2w_m\right)\cos(m\phi) \tag{2.330}$$

where c is a constant length to be defined later and Q_m, q_m, s_m, and w_m are functions only of ϖ and z satisfying:

$$L_m^2 Q_m = L_m^2 w_m = L_{m-1}^2 s_m = L_{m+1}^2 q_m \tag{2.331}$$

with L_m^2 the operator:

$$L_m^2 = \frac{\partial^2}{\partial \varpi^2}+\frac{1}{\varpi}\frac{\partial}{\partial \varpi}-\frac{m^2}{\varpi^2}+\frac{\partial^2}{\partial z^2}$$

The equation of continuity becomes:

The Motion of Particles in Gases

$$\left(3+\varpi\frac{\partial}{\partial\varpi}+z\frac{\partial}{\partial z}\right)Q_m + c\left[\left(\frac{\partial}{\partial\varpi}+\frac{m+1}{\varpi}\right)q_m + \left(\frac{\partial}{\partial\varpi}-\frac{m-1}{\varpi}\right)s_m + 2\frac{\partial w_m}{\partial z}\right] = 0 \quad (2.332)$$

The symmetry of the problem, i.e. translation and rotation perpendicular to the z-axis (line of centers), requires that m=1. The solution will then describe translational motion parallel to the x-axis and rotation around an axis parallel to the y-axis.

The equations for Q_1, q_1, w_1, and s_1 are now expressed in terms of the bispherical coordinates introduced by Jeffery [1922]. Then, it is readily shown that:

$$w_1 = (\cosh(\xi)-\mu)^{1/2} \sum_{n=1}^{\infty} \Lambda_n(A,B,\xi) P_n^{(1)}(\mu) \quad (2.333)$$

$$Q_1 = (\cosh(\xi)-\mu)^{1/2} \sum_{n=1}^{\infty} \Lambda_n(C,D,\xi) P_n^{(1)}(\mu) \quad (2.334)$$

$$q_1 = (\cosh(\xi)-\mu)^{1/2} \sum_{n=1}^{\infty} \Lambda_n(G,H,\xi) P_n^{(2)}(\mu) \quad (2.335)$$

and:

$$s_1 = (\cosh(\xi)-\mu)^{1/2} \sum_{n=1}^{\infty} \Lambda_n(E,F,\xi) P_n^{(2)}(\mu) \quad (2.336)$$

where $P_n^{(m)}(\mu)$ are associated Legendre polynomials and:

$$\Lambda_n(A,B,\xi) = A_n \exp[(n+\tfrac{1}{2})\xi] + B_n \exp[-(n+\tfrac{1}{2})\xi] \quad (2.337)$$

The constants A_n, B_n, C_n, etc. are determined from the boundary conditions.

Substitution of these equations into the continuity equation also leads to two further relations between A_n, B_n, C_n, etc.

Using the appropriate expressions for the drag forces and the torque, it may be shown that:

$$F_1 = -6\pi\mu a_1 \frac{2\sqrt{2}}{3} \sinh(\alpha) \sum_{n=0}^{\infty} E_n \quad (2.338)$$

$$F_2 = -6\pi\mu a_2 \frac{2\sqrt{2}}{3} \sinh(\beta) \sum_{n=0}^{\infty} F_n \quad (2.339)$$

$$T_1 = -8\pi\mu a_1^2 \frac{1}{\sqrt{2}} \sinh^2(\alpha) \sum_{n=0}^{\infty} (2n+1-\coth(\alpha)) E_n \quad (2.340)$$

and:

$$T_2 = -8\pi\mu a_2^2 \frac{1}{\sqrt{2}} \sinh^2(\beta) \sum_{n=0}^{\infty} (2n+1-\coth(\beta)) F_n \quad (2.341)$$

The forces are in the x-direction and the torques are in the y-direction.

To obtain a closed set of equations for the coefficients A_n, B_n, C_n, etc., it is necessary to employ the boundary conditions. We assume that sphere 1 moves in the x-direction with a speed, U_1 and rotates about the y-axis with angular speed, Ω_1. Similar conditions apply to sphere 2. Then, on $\xi=\alpha$:

$$u_\varpi = \Omega_1(z-d_1)\cos(\phi) + U_1\cos(\phi)$$
$$u_\phi = -\Omega_1(z-d_1)\sin(\phi) - U_1\sin(\phi)$$
$$u_z = -\Omega_1\varpi\cos(\phi)$$

while for $\xi=-\beta$:

$$u_\varpi = \Omega_2(z+d_2)\cos(\phi) + U_2\cos(\phi)$$
$$u_\phi = -\Omega_2(z+d_2)\sin(\phi) - U_2\sin(\phi)$$
$$u_z = -\Omega_2\varpi\cos(\phi)$$

Comparing with Eqns. (2.327)-(2.330) leads to:

$$Q_1 = \frac{2\Omega_1 c\varpi}{z} - \frac{2cw_1}{z}$$

$$q_1 = \frac{\Omega_1\varpi^2}{z} + \frac{w_1\varpi}{z}$$

and:

$$s_1 = 2\Omega_1 z - 2\Omega_1 d_1 + 2U_1 + \frac{\Omega_1\varpi^2}{z} + \frac{w_1\varpi}{z}$$

for $\xi=\alpha$, with a similar result for $\xi=\beta$.

Inserting Eqns. (2.333)-(2.336) and using some well known properties of Legendre polynomials leads to a set of six linear algebraic equations for the A_n, B_n, C_n, etc. Unfortunately, these equations are recursive and therefore have to be solved numerically. Such calculations have been carried out by Davis [1969] and by O'Neill and Majumdar [1970a,b]. We shall discuss their results below since they have important consequences for coagulation processes and aerosol behavior in general.

Davis [1969] reports his results in the form:

$$F_1 = 6\pi\mu a_1\left[C_1 U_1 + C_5 U_2 + C_9 a_2 \Omega_1 + C_{13} a_2 \Omega_2\right] \tag{2.342}$$

$$F_2 = 6\pi\mu a_2\left[C_2 U_1 + C_6 U_2 + C_{10} a_2 \Omega_1 + C_{14} a_2 \Omega_2\right] \tag{2.343}$$

$$T_1 = 8\pi\mu a_1^2\left[C_3 U_1 + C_7 U_2 + C_{11} a_2 \Omega_1 + C_{15} a_2 \Omega_2\right] \tag{2.344}$$

and:

$$T_2 = 8\pi\mu a_2^2\left[C_4 U_1 + C_8 U_2 + C_{12} a_2 \Omega_1 + C_{16} a_2 \Omega_2\right] \tag{2.345}$$

The Motion of Particles in Gases

where the C_i are functions of mass ratio a_1/a_2 and interparticle distance $s=r-a_1-a_2$.

On the other hand, O'Neill and Majumder [1970] write:

$$F_1 = -6\pi \mu\, a_1 \left[f_{21}(k,\varepsilon) U_1 + f_{22}(\tfrac{1}{k},\tfrac{\varepsilon}{k}) U_2 + f_{11}(k,\varepsilon) a_1 \Omega_1 - f_{12}(\tfrac{1}{k},\tfrac{\varepsilon}{k}) a_1 \Omega_2 \right] \tag{2.346}$$

$$F_2 = -6\pi \mu\, a_2 \left[f_{22}(k,\varepsilon) U_1 + f_{21}(\tfrac{1}{k},\tfrac{\varepsilon}{k}) U_2 + f_{12}(k,\varepsilon) a_2 \Omega_1 - f_{11}(\tfrac{1}{k},\tfrac{\varepsilon}{k}) a_2 \Omega_2 \right] \tag{2.347}$$

$$T_1 = -8\pi \mu\, a_1^2 \left[g_{21}(k,\varepsilon) U_1 + g_{22}(\tfrac{1}{k},\tfrac{\varepsilon}{k}) U_2 + g_{11}(k,\varepsilon) a_1 \Omega_1 - g_{12}(\tfrac{1}{k},\tfrac{\varepsilon}{k}) a_1 \Omega_2 \right] \tag{2.348}$$

and:

$$T_2 = -8\pi \mu\, a_2^2 \left[-g_{22}(k,\varepsilon) U_1 - g_{21}(\tfrac{1}{k},\tfrac{\varepsilon}{k}) U_2 - g_{12}(k,\varepsilon) a_2 \Omega_1 + g_{11}(\tfrac{1}{k},\tfrac{\varepsilon}{k}) a_2 \Omega_2 \right] \tag{2.349}$$

In these formulae, $\varepsilon=(r/a_1)-1-k$ where $k=a_2/a_1$. Extensive tables of the f_{ij} and g_{ij} are given. However, other than the case for which $a_1=a_2$, the numerical results are not directly comparable except in some special cases. Where comparison of the numerical results is possible they do agree. For purposes of illustration, we use Davis' results although, as the results of O'Neill and Majumdar show, several of Davis' C_i are related.

The forces are reported in the literature in two different ways. Some work ignores the rotation of the particles and the forces are quoted in the form:

$$F_1 = 6\pi\mu a_1 \left[C_1 U_1 + C_5 U_2 \right] \tag{2.350}$$

and:

$$F_2 = 6\pi\mu a_2 \left[C_2 U_1 + C_6 U_2 \right] \tag{2.351}$$

In practice, however, the particles are free to rotate (for experimental confirmation see Jayaweera et al. [1964]). Some earlier but less accurate work by Eveson et al. [1959] should also be noted. This means that there is no net applied torque. Rather, the translational motion of the particles leads to a torque which causes rotation. The corresponding angular velocities are obtained by setting $T_1=T_2=0$ in Eqns. (2.344) and (2.345). Thus, we may solve for Ω_1 and Ω_2 in terms of U_1 and U_2 and eliminate the Ω_i from Eqns. (2.342) and (2.343). The net effect is that:

$$a_2 \Omega_1 = C_{19} U_1 + C_{20} U_2$$

and:

$$a_2 \Omega_2 = C_{17} U_1 + C_{18} U_2$$

where:

$$C_{17} = \frac{C_{11} C_4 - C_3 C_{12}}{C_{12} C_{15} - C_{11} C_{16}}$$

$$C_{18} = \frac{C_{11} C_8 - C_7 C_{12}}{C_{12} C_{15} - C_{11} C_{16}}$$

$$C_{19} = \frac{C_{15} C_4 - C_3 C_{16}}{C_{11} C_{16} - C_{15} C_{12}}$$

$$C_{20} = \frac{C_{15} C_8 - C_7 C_{16}}{C_{11} C_{16} - C_{15} C_{12}}$$

Aerosol Science: Theory and Practice

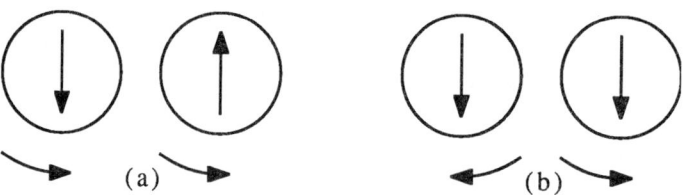

Figure 2.37: The translating and rotating spheres considered by Davis [1969].

$$F_1 = 6\pi\mu a_1\left[\left(C_1 + C_9 C_{19} + C_{13} C_{17}\right) U_1 + \left(C_5 + C_9 C_{20} + C_{13} C_{18}\right) U_2\right] \quad (2.352)$$

and:

$$F_2 = 6\pi\mu a_2\left[\left(C_2 + C_{10} C_{19} + C_{14} C_{17}\right) U_1 + \left(C_6 + C_{10} C_{20} + C_{14} C_{18}\right) U_2\right] \quad (2.353)$$

The rotations therefore affect the drag forces. In order to illustrate the magnitude of these corrections, Davis [1966] has given graphs of F_1 and F_2 for equal spheres as a function of interparticle separation for the two cases. To be more specific, he has considered two dynamical situations as shown in Figs. 2.37(a) and 2.37(b).

Thus, parallel and anti-parallel motion are studied. We also note that for anti-parallel motion the spheres rotate in the same direction whereas in parallel motion they rotate in opposite directions. Figs. 2.38 and 2.39 show the actual values of the F_i wherein we observe that for anti-parallel motion the effect of rotation becomes significant as the spheres get closer. The difference for touching spheres amounts to about –18%. Thus, neglect of rotation leads to an overestimate of the drag. It should be noted, however, that at a separation of 3a, *i.e.* a gap distance of one radius, the difference is negligible.

In the case of parallel motion, the error caused by the neglect of rotation is much smaller, amounting to only about –1.5%. Once again the drag is overestimated by neglect of rotation. We note that O'Neill and Majumdar [1970a,b] have given expressions for f_{ij} and g_{ij} when ε is very small.

2.29 Arbitrary orientation of the spheres

We now have relations between the drag forces and velocities for spheres of arbitrary size ratio moving along their line of centers and perpendicular to their line of centers. From these two cases it is possible to obtain a general expression for two spheres at any arbitrary orientation. Fig. 2.40 shows two spheres which, for the sake of example, we will assume are moving downward under the force of gravity.

Note that the force $m_i g$ can be resolved into components along the line of centers and perpendicular to the line of centers. F_{ir} are the forces along the line of centers and U_{ir} the corresponding velocities. $F_{i\theta}$ are the forces perpendicular to the line of centers with $U_{i\theta}$ the corresponding velocities. But F_{ir} and U_{ir} correspond to motion along the line of centers for which we have the Stimson-Jeffery solution of Eqns. (2.296) and (2.297), *i.e.*:

$$F_{1r} = -\kappa_1 U_{1r} + \lambda_1 U_{2r} \quad (2.354)$$

The Motion of Particles in Gases

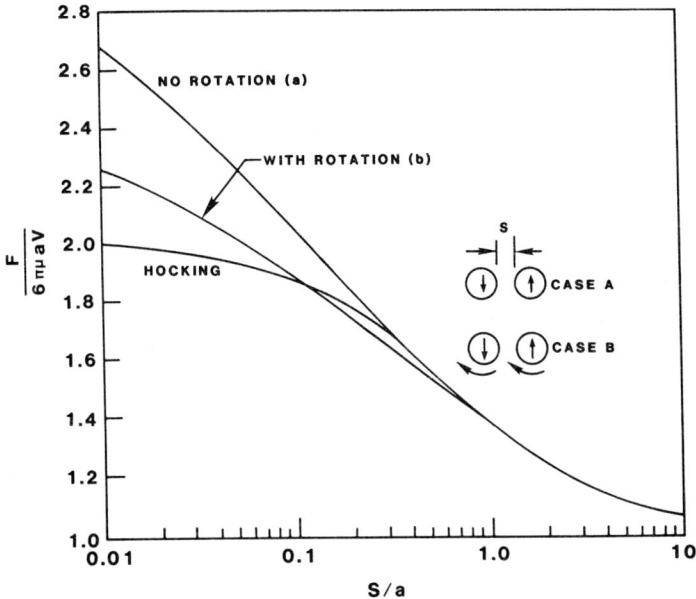

Figure 2.38: Resistance coefficient for equal anti-parallel motion perpendicular to the line of centers (equal spheres, radius a). Adapted from Davis [1966] with permission of the American Geophysical Union.

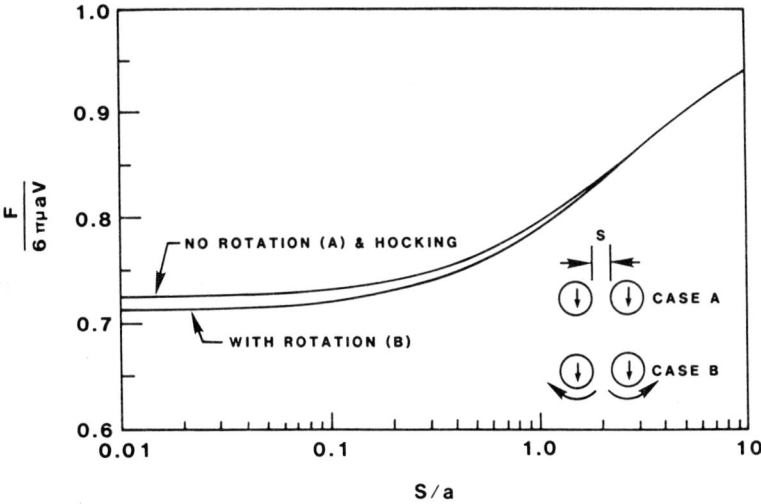

Figure 2.39: Resistance coefficient for equal parallel motion perpendicular to the line of centers (equal spheres, radius a). Adapted from Davis [1966] with permission of the American Geophysical Union.

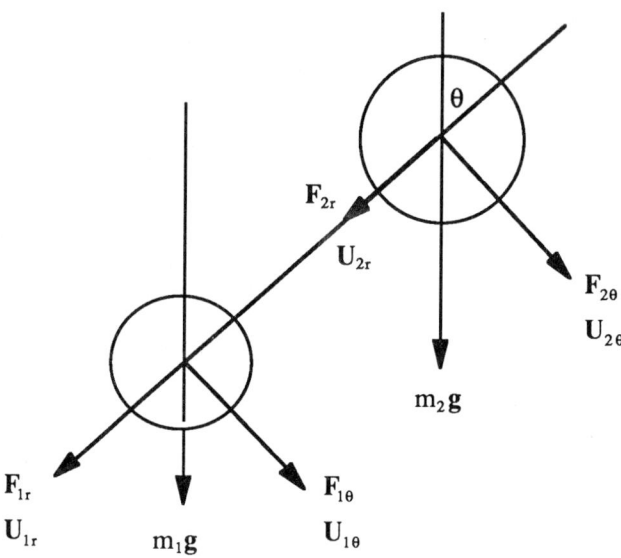

Figure 2.40: The motion of two spheres at any arbitrary orientation.

and:
$$F_{2r} = -\kappa_2 U_{2r} + \lambda_2 U_{1r} \tag{2.355}$$

Similarly, $F_{i\theta}$ and $U_{i\theta}$ correspond to force and motion perpendicular to the line of centers for which we also have exact solutions in Eqns. (2.352) and (2.353):

$$F_{1\theta} = \zeta_1 U_{1\theta} + \theta_1 U_{2\theta} \tag{2.356}$$

and:
$$F_{2\theta} = \zeta_2 U_{2\theta} + \theta_2 U_{1\theta} \tag{2.357}$$

(Because F_{ir} and $F_{i\theta}$ are drag related they are negative).

These four relations may be combined in a convenient way by writing:

$$F_1 = -k_{11} \cdot U_1 - k_{12} \cdot U_2 \tag{2.358}$$

and:
$$F_2 = -k_{21} \cdot U_1 - k_{22} \cdot U_2 \tag{2.359}$$

where k_{ij} are the resistance tensors. Such tensors are conveniently written as:

$$k_{ij} = 3\pi \mu (a_i + a_j) \left\{ S_{ij}(r) \frac{rr}{r^2} + T_{ij}(r) \left[I - \frac{rr}{r^2} \right] \right\} \tag{2.360}$$

where I is the unit dyadic and r is the interparticle distance between the sphere centers. Clearly, S_{ij} and T_{ij} are directly related to κ_i, λ_i, ζ_i, and θ_i above in the following way:

The Motion of Particles in Gases

$$\kappa_1 = 6\pi \mu \, a_1 \, S_{11} \qquad\qquad \zeta_1 = -6\pi \mu \, a_1 \, T_{11}$$

$$\lambda_1 = -3\pi \mu \, (a_1 + a_2) \, S_{12} \qquad\qquad \theta_1 = -3\pi \mu \, (a_1 + a_2) \, T_{12}$$

$$\kappa_2 = 6\pi \mu \, a_2 \, S_{22} \qquad\qquad \zeta_2 = -6\pi \mu \, a_2 \, T_{22}$$

$$\lambda_2 = -3\pi \mu \, (a_1 + a_2) \, S_{21} \qquad\qquad \theta_2 = -3\pi \mu \, (a_1 + a_2) \, T_{21}$$

S_{ij} and T_{ij} depend only on r and the ratio of the radii, a_2/a_1.

Eqns. (2.358) and (2.359) are in a particularly useful form since they can be expressed easily in any system of coordinates (*e.g.* Cartesian (x,y,z)) when necessary.

In the case considered above, the forces acting against the drag forces are those due to gravity, with:

$$\mathbf{F}_{ir} = -m_i \, g \, \cos(\theta)$$

and:

$$\mathbf{F}_{i\theta} = -m_i \, g \, \sin(\theta)$$

However, Eqns. (2.358) and (2.359) are valid for any arbitrary situation.

Batchelor [1976, 1982] and Jeffery and Onishi [1984] have found it convenient to rewrite Eqns. (2.358) and (2.359) in terms of mobility tensors, $\mathbf{b}_{ij}$, in the following way:

$$\mathbf{U}_1 = \mathbf{b}_{11} \cdot \mathbf{F}_1 + \mathbf{b}_{12} \cdot \mathbf{F}_2 \tag{2.361}$$

$$\mathbf{U}_2 = \mathbf{b}_{21} \cdot \mathbf{F}_1 + \mathbf{b}_{22} \cdot \mathbf{F}_2 \tag{2.362}$$

The sign of Batchelors forces are opposite to those of Eqns. (2.358) and (2.359) because he considers them to be applied external forces whereas our forces are assumed to be drag forces. Batchelor writes:

$$\mathbf{b}_{ij} = \frac{1}{3\pi \mu \, (a_i + a_j)} \left\{ A_{ij}(r) \frac{\mathbf{rr}}{r^2} + B_{ij}(r) \left[\mathbf{I} - \frac{\mathbf{rr}}{r^2} \right] \right\} \tag{2.363}$$

where the A_{ij} and B_{ij} are readily related to the resistance matrix elements, S_{ij} and T_{ij}.

Batchelor notes some useful relationships between A_{ij} and B_{ij}:

$$A_{11}(\rho,\lambda) = A_{22}(\rho,1/\lambda), \qquad B_{11}(\rho,\lambda) = B_{22}(\rho,1/\lambda)$$
$$A_{12}(\rho,\lambda) = A_{21}(\rho,1/\lambda), \qquad B_{12}(\rho,\lambda) = B_{21}(\rho,1/\lambda) \tag{2.364}$$

Moreover, the reciprocal theorem of Lorentz [1906] leads to:

$$A_{12}(\rho,\lambda) = A_{21}(\rho,\lambda), \qquad B_{12}(\rho,\lambda) = B_{21}(\rho,\lambda)$$
$$A_{12}(\rho,1/\lambda) = A_{21}(\rho,1/\lambda), \qquad B_{12}(\rho,1/\lambda) = B_{21}(\rho,1/\lambda) \tag{2.365}$$

where $\rho=2r/(a_1+a_2)$ and $\lambda=a_2/a_1$. Similar relations hold for the S_{ij} and T_{ij}.

Some useful expansions of A_{ij} and B_{ij} are given by Batchelor for large separations:

$$A_{11} = 1 - \frac{60 \lambda^3}{(1+\lambda)^4 \rho^4} + \frac{32 \lambda^3 (15 - 4\lambda^2)}{(1+\lambda)^6 \rho^6} - \frac{192 \lambda^3 (5 - 22\lambda^2 + 3\lambda^4)}{(1+\lambda)^8 \rho^8} + O(\rho^{-10})$$

$$B_{11} = 1 - \frac{68 \lambda^5}{(1+\lambda)^6 \rho^6} - \frac{32 \lambda^3 (10 - 9\lambda^2 + 9\lambda^4)}{(1+\lambda)^8 \rho^8} + O(\rho^{-10})$$

$$A_{12} = \frac{3}{2\rho} - \frac{2(1+\lambda^2)}{(1+\lambda)^2 \rho^3} + \frac{1200 \lambda^3}{(1+\lambda)^6 \rho^7} + O(\rho^{-9})$$

and:

$$B_{12} = \frac{3}{4\rho} + \frac{1+\lambda^2}{(1+\lambda)^2 \rho^3} + O(\rho^{-9})$$

From these expansions it is interesting to observe the deviations from Stokes' law at wide separations. Thus, for spheres moving along their line of centers we find for the force on sphere 1:

$$F_1 = 6\pi\mu\, a_1 \left\{ U_1 \left[1 + \tfrac{9}{4}\frac{a_1 a_2}{r^2} + \tfrac{3}{4} r^{-4}\left(-2 a_1^3 a_2 + \tfrac{27}{4} a_1^2 a_2^2 + 3 a_1 a_2^3\right) \right] \right.$$

$$\left. + U_2 \left[-\tfrac{3}{2}\frac{a_2}{r} + \tfrac{1}{2} r^{-3}\left(a_1^2 a_2 - \tfrac{27}{4} a_1 a_2^2 + a_2^3\right) - \tfrac{9}{4} r^{-5}\left(a_1^3 a_2^2 + \tfrac{27}{8} a_1^2 a_2^3 + a_1 a_2^4\right) \right] \right\}$$

(2.366)

For sphere 2, it is necessary to interchange the subscripts 1 and 2.

For spheres moving perpendicular to their line of centers we obtain:

$$F_1 = 6\pi\mu\, a_1 \left\{ U_1 \left[1 + \tfrac{9}{16}\frac{a_1 a_2}{r^2} + \tfrac{3}{8} r^{-4}\left(a_1^3 a_2 + \tfrac{27}{32} a_1^2 a_2^2 + 3 a_1 a_2^3\right) \right] \right.$$

$$\left. - U_2 \left[\tfrac{3}{4}\frac{a_2}{r} + \tfrac{1}{4} r^{-3}\left(a_1^2 a_2 + \tfrac{27}{16} a_1 a_2^2 + a_2^3\right) + \tfrac{63}{64} r^{-5}\left(a_1^3 a_2^2 + \tfrac{27}{112} a_1^2 a_2^3 + a_1 a_2^4\right) \right] \right\}$$

(2.367)

again, interchanging the subscripts 1 and 2 for sphere 2.

There are four special cases where it is useful to have readily available numerical results. These cases are as follows:

(i) Equal spheres moving with equal velocities along their line of centers:

$$F_\parallel^s = 6\pi\mu\, a\, \kappa_\parallel^s$$

(ii) Equal spheres moving with equal and opposite velocities towards each other along their line of centers:

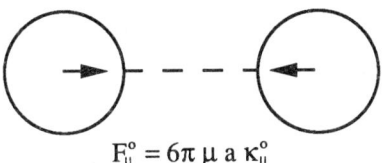

$$F_{||}^o = 6\pi \mu a \kappa_{||}^o$$

(iii) Equal spheres moving with equal velocities perpendicular to their line of centers:

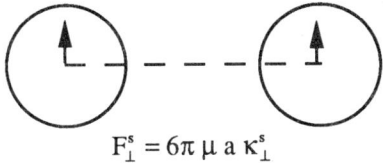

$$F_{\perp}^s = 6\pi \mu a \kappa_{\perp}^s$$

(iv) Equal spheres moving with equal and opposite velocities perpendicular to their line of centers:

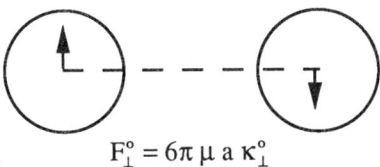

$$F_{\perp}^o = 6\pi \mu a \kappa_{\perp}^o$$

The dynamic form factors $\kappa_{||}$ and $\kappa_{\perp}$ are listed in Table 2.12 as a function of the interparticle distance $\rho = r/a$. The curious aspect is the slight minimum in $\kappa_{\perp}^s$ around $\rho = (2.01)$.

2.30 The superposition method

An early and somewhat primitive procedure for dealing with the interaction of two spheres moving in a fluid is called the superposition method. While this method is not used for precision calculations, it has a certain physical appeal that is worth noting and, moreover, it could be of value in dealing with the interaction of nonspherical bodies. We shall therefore give a brief description.

Consider two spheres moving in a fluid in the coordinates of Fig. 2.41. To a first approximation we may write the forces on the two spheres as:

$$F_1 = -6\pi \mu a_1 \left[V_1 - U_2(d, \theta + \pi) \right] \tag{2.368}$$

and:

$$F_2 = -6\pi \mu a_2 \left[V_2 - U_1(d, \theta) \right] \tag{2.369}$$

where $U_1(d,\theta)$ is the fluid velocity at the center of sphere 2 due to Stokes flow around sphere 1. $U_2(d,\theta+\pi)$ is the fluid velocity at the center of sphere 1 due to the Stokes flow around sphere 2.

Now, in terms of the coordinates, x and y, we may write:

Table 2.12: The dynamic form factors, $\kappa_\parallel$ and $\kappa_\perp$. Adapted from Batchelor [1976] who in turn obtained them from Cooley and O'Neill [1969], O'Neill and Majumdar [1970a,b], and Nir and Acrivos [1973].

ρ	$\kappa_\parallel^s$	$\kappa_\parallel^o$	$\kappa_\perp^s$	$\kappa_\perp^o$
2.0	0.6452	∞	0.7241	2.494
2.001	–	–	0.7141	2.141
2.005	–	–	0.7129	2.069
2.01	0.6457	53.48	0.7127	2.026
2.05	0.6480	12.71	–	–
2.10	0.6509	7.413	0.7185	1.797
2.25	0.6593	4.036	–	–
2.5	0.6729	2.772	–	–
2.75	0.6859	2.298	–	–
3.0	0.6983	1.039	0.7894	1.370
3.5	0.7214	1.754	–	–
4.0	0.7423	1.597	–	–
5.0	0.7772	1.424	–	–
7.0	0.8264	1.270	0.9020	1.122
∞	1.0	1.0	1.0	1.0

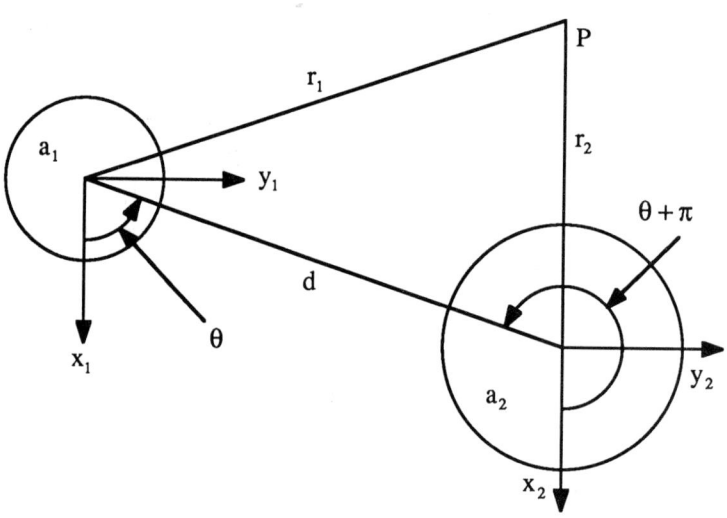

Figure 2.41: A coordinate system for the motion of two spheres.

The Motion of Particles in Gases

$$V_1 = i\, V_{1x} + j\, V_{1y}$$
$$U_2 = i_{r2}\, U_{r2} + i_{\theta 2}\, U_{\theta 2}$$
$$V_2 = i\, V_{2x} + j\, V_{2y}$$
$$U_1 = i_{r1}\, U_{r1} + i_{\theta 1}\, U_{\theta 1}$$

But also, $i_{r2} = -i_{r1}$ and $i_{\theta 2} = -i_{\theta 1}$:

$$i_{r1} = i\cos(\theta) + j\sin(\theta)$$
$$i_{\theta 1} = -i\sin(\theta) + j\cos(\theta)$$

Therefore:

$$V_1 - U_2 = \left(V_{1x} + U_{r2}\cos(\theta) - U_{\theta 2}\sin(\theta)\right) i$$
$$+ \left(V_{1y} + U_{r2}\sin(\theta) + U_{\theta 2}\cos(\theta)\right) j$$

and:

$$-\frac{F_1}{6\pi\mu\, a_1\, V_{1x}} = \left[1 + \frac{U_{r2}}{V_{1x}}\cos(\theta) - \frac{U_{\theta 2}}{V_{1x}}\sin(\theta)\right] i + \left[\frac{V_{1y}}{V_{1x}} + \frac{U_{r2}}{V_{1x}}\sin(\theta) + \frac{U_{\theta 2}}{V_{1x}}\cos(\theta)\right] j$$

$$\equiv \Lambda_1\, i + \left[\frac{V_{1y}}{V_{1x}} + \Gamma_1\right] j \qquad (2.370)$$

A similar calculation leads to:

$$-\frac{F_2}{6\pi\mu\, a_2\, V_{2x}} = \Lambda_2\, i + \left[\frac{V_{2y}}{V_{2x}} + \Gamma_2\right] j \qquad (2.371)$$

where:

$$\Lambda_1 = 1 + \frac{U_{r2}}{V_{1x}}\cos(\theta) - \frac{U_{\theta 2}}{V_{1x}}\sin(\theta)$$

$$\Lambda_2 = 1 - \frac{U_{r1}}{V_{2x}}\cos(\theta) + \frac{U_{\theta 1}}{V_{2x}}\sin(\theta)$$

$$\Gamma_1 = \frac{U_{r2}}{V_{1x}}\sin(\theta) + \frac{U_{\theta 2}}{V_{1x}}\cos(\theta)$$

and:

$$\Gamma_2 = \frac{V_{2y}}{V_{2x}} - \frac{U_{r1}}{V_{2x}}\sin(\theta) - \frac{U_{\theta 1}}{V_{2x}}\cos(\theta)$$

But Stokes flow predicts:

$$U_r = \tfrac{1}{2} U_\infty\, a^2 \cos(\theta)\left(\frac{3}{a\, r} - \frac{a}{r^3}\right)$$

103

and:

$$U_\theta = -\tfrac{1}{4} U_\infty a^2 \sin(\theta)\left(\frac{3}{ar} + \frac{a}{r^3}\right)$$

Using these results in Eqns. (2.370) and (2.371) leads to:

$$\Lambda_1 = 1 + \frac{V_{2x}}{V_{1x}}\left[\tfrac{1}{2}\left(\frac{a_2^3}{d^3} - \frac{3a_2}{d}\right)\cos^2(\theta) - \tfrac{1}{4}\left(\frac{3a_2}{d} + \frac{a_2^3}{d^3}\right)\sin^2(\theta)\right] \tag{2.372}$$

$$\Lambda_2 = 1 - \frac{V_{1x}}{V_{2x}}\left[\tfrac{1}{2}\left(\frac{3a_1}{d} - \frac{a_1^3}{d^3}\right)\cos^2(\theta) + \tfrac{1}{4}\left(\frac{3a_1}{d} + \frac{a_1^3}{d^3}\right)\sin^2(\theta)\right] \tag{2.373}$$

$$\Gamma_1 = -\tfrac{3}{4}\frac{V_{2x}}{V_{1x}}\sin(\theta)\cos(\theta)\left(\frac{a_2}{d} - \frac{a_2^3}{d^3}\right) \tag{2.374}$$

and:

$$\Gamma_2 = -\tfrac{3}{4}\frac{V_{1x}}{V_{2x}}\sin(\theta)\cos(\theta)\left(\frac{a_1}{d} - \frac{a_1^3}{d^3}\right) \tag{2.375}$$

Eqns. (2.370) and (2.372) are therefore the results for arbitrary interparticle distance, d, and orientation, θ. To compare with known exact results we set $a_1 = a_2$ and $\theta = 0, \pi/2$. When $\theta = 0$, we have spheres moving along their line of centers, and for touching spheres we find:

$$\kappa_\parallel^s = 0.3135\ (0.6452)$$

$$\kappa_\parallel^o = 1.6875\ (\infty)$$

$$\kappa_\perp^s = 0.594\ (0.724)$$

$$\kappa_\perp^o = 1.406\ (2.494)$$

The exact values follow in brackets. It is clear therefore that the superposition method gives only a general trend. However, for large separations it is not an unreasonable approximation. Further details may be found in Pruppacher and Klett [1978].

References: Chapter 2

Bart, E. [1968] "The slow unsteady settling of a fluid sphere toward a flat fluid interface," *Chem. Eng. Sci.* **23**, 193.

Batchelor, G.K. [1976] "Brownian diffusion of particles with hydrodynamic interaction," *J. Fluid Mech.* **74**, 1.

Batchelor, G.K. [1982] "Sedimentation in a dilute polydisperse system of interacting spheres, Part 1. General Theory," *J. Fluid Mech.* **119**, 379.

Berry, M.V. and Percival, I.C. [1986] "Optics of fractal clusters such as smoke," *Opt. Acta* **33**, 577.

Brenner, H. [1961] "The slow motion of a sphere through a viscous fluid towards a plane surface," *Chem. Eng. Sci.* **16**, 242.

Brenner, H. [1964] "The Stokes resistance of a slightly deformed sphere," *Chem. Eng. Sci.* **19**, 519.

Cercignani, C. [1969] <u>Mathematical Methods in Kinetic Theory</u> (Plenum Press, New York, N.Y.).

Cercignani, C. [1975] <u>Theory and Application of the Boltzmann Equation</u> (Scottish Academic Press, Edinburgh, U.K.).

Cercignani, C. and Pagani, C.D. [1968] "Flow of a rarefied gas past an axisymmetric body. I. General Remarks," *Phys. Fluids* **11**, 1395.

Cercignani, C., Pagani, C.D. and Bassanini, P. [1968] "Flow of a rarefied gas past an axisymmetric body II. Case of a sphere," *Phys. Fluids* **11**, 1399.

Chapman, S. and Cowling, T.G. [1970] <u>The Mathematical Theory of Non-Uniform Gases</u> (Cambridge University Press, Cambridge, U.K.).

Collins, W.D. [1963] "A note on the axisymmetric Stokes flow of viscous fluid past a spherical cap," *Mathematika* **10**, 72.

Cooley, M.D.A. and O'Neill, M.E. [1969] "On the slow motion of two spheres in contact along their line of centres through a viscous fluid," *Proc. Camb. Philos. Soc.* **66**, 407.

Cox, R.G. and Brenner, H. [1967] "The slow motion of a sphere through a viscous fluid towards a plane surface-II. Small gap widths, including inertial effects," *Chem. Eng. Sci.* **22**, 1753.

Dahneke, B.E. [1973a] "Slip correction factors for nonspherical bodies I. Introduction and continuum flow," *J. Aerosol Sci.* **4**, 139.

Dahneke, B.E. [1973b] "Slip correction factors for nonspherical bodies II. Free molecule flow," *J. Aerosol Sci.* **4**, 147.

Dahneke, B.E. [1973c] "Slip correction factors for nonspherical bodies III. The form of the general law," *J. Aerosol Sci.* **4**, 163.

Davis, M.H. [1966] "Collisions of very small cloud drops," *J. Geophys. Res.* **71**, 3101.

Davis, M.H. [1969] "The slow translation and rotation of two unequal spheres in a viscous fluid," *Chem. Eng. Sci.* **24**, 1769.

Davis, M.H. [1972] "Collisions of small cloud droplets: Gas kinetic effects," *J. Atmos. Sci.* **29**, 911.

Dean, W.R. and O'Neill, M.E. [1963] "A slow motion of viscous liquid caused by the rotation of a solid sphere," *Mathematika* **10**, 13.

Dorrepaal, J.M., Majumdar, S.R., O'Neill, M.E. and Ranger, K.B. [1976a] "A closed torus in Stokes flow," *Q. J. Mech. and Appl. Math.* **29**, 381.

Dorrepaal, J.M., O'Neill, M.E. and Ranger, K.B. [1976b] "Axisymmetric Stokes flow past a spherical cap," *J. Fluid Mech.* **75**, 273.

Eveson G.F., Hall, E.W. and Ward, S.G. [1959] "Interaction between two equal-sized equal settling spheres moving through a viscous liquid," *Br. J. Appl. Phys.* **10**, 43.

Friedlander, S.K. [1977] <u>Smoke, Dust and Haze</u> (Wiley, New York, N.Y.).

Gans, R. [1911] "Wie fallen stabe und scheiben in einer reibenden flussigkeit," *Sitzungsber. Math.-Phys. Kl. Bayer. Akad. Wiss. München* **41**, 191 (in German).

Gluckman, M.J., Pfeffer, R. and Weinbaum, S. [1971] "A new technique for treating multiparticle slow viscous flow: axisymmetric flow past spheres and spheroids," *J. Fluid Mech.* **50**, 705.

Gluckman, M.J., Weinbaum, S. and Pfeffer, R. [1972] "Axisymmetric slow viscous flow past an arbitrary convex body of revolution," *J. Fluid Mech.* **55**, 677.

Halbritter, J. [1974] "Torque on a rotating ellipsoid in a rarefied gas," *Z. Naturforsch.* **29a**, 1717.

Hansford, R.E. [1970] "On converging solid spheres in a highly viscous fluid," *Mathematika* **17**, 250.

Happel, J. and Brenner, H. [1973] <u>Low Reynolds Number Hydrodynamics</u> (Martinus Nijhoff, Dordrecht).

Hausdorff, F. [1919] "Dimension und ausseres mass," *Math. Ann.* **79**, 157 (in German).

Heiss, J.F. and Coull, J. [1952] "The effect of orientation and shape on the settling velocity of non-isometric particles in a viscous medium," *Chem. Eng. Prog.* **48**, 133.

Hill, R. and Power, G. [1956] "Extremum principles for slow viscous flow and the approximate calculation of drag," *Q. J. Mech. and Appl. Math.* **9**, 313.

Hocking, L.M. [1959] "The collision efficiency of small drops," *Q. J. Roy. Meteor. Soc.* **85**, 44.

Hocking, L.M. [1973] "The effect of slip on the motion of a sphere close to a wall and of two adjacent spheres," *J. Eng. Math.* **7**, 207.

Jayaweera, K.O.L.F., Mason, B.J. and Slack, G.W. [1964] "The behaviour of clusters of spheres falling in a viscous fluid, 1. Experiment," *J. Fluid Mech.* **20**, 121.

Jeffery, G.B. [1912] "On a form of the solution of Laplace's equation suitable for problems relating to two spheres," *Proc. Roy. Soc. (London)* **A87**, 109.

Jeffery, G.B. [1915a] "The equations of motion of a viscous fluid," *Philos. Mag.* **29**, 445.

Jeffery, G.B. [1915b] "On the steady rotation of a solid of revolution in a viscous fluid," *Proc. London Math. Soc.* **14**, 327.

Jeffery, G.B. [1922] "The motion of ellipsoidal particles immersed in a viscous fluid," *Proc. Roy. Soc. (London)* **A102**, 161.

Jeffrey, D.J. and Onishi, Y. [1984] "Calculation of the resistance and mobility functions for two unequal rigid spheres in low-Reynolds-number flow," *J. Fluid Mech.* **139**, 261.

Johnson, D.L. [1985] Drag Forces on Orthotropic Aerosol Particles (University of North Carolina, Chapel Hill) Ph.D. Thesis.

Kanwal, R.P. [1961] "Slow steady rotation of axially symmetric bodies in a viscous fluid," *J. Fluid Mech.* **10**, 17.

Kennard, E.H. [1938] Kinetic Theory of Gases (McGraw Hill, New York, N.Y.).

Kops, J.A.M.M. [1976] "The Aerodynamic Diameter and Specific Surface Area of Branched Chain-like Aggregates" (Netherlands Energy Research Foundation) Report # ECN-5.

Kunkel, W.B. [1948] "Magnitude and character of errors produced by shape factors in Stokes' law estimates of particle radius," *J. Appl. Phys.* **19**, 1056.

Law, W.S. and Loyalka, S.K. [1986] "Motion of a sphere in a rarefied gas II: role of temperature variation in the Knudsen layer," *Phys. Fluids* **29**, 3886.

Lea, K.C. and Loyalka, S.K. [1982] "Motion of a sphere in a rarefied gas," *Phys. Fluids* **25**, 1550.

Leith, D. [1987] "Drag on nonspherical objects," *Aerosol Sci. and Tech.* **6**, 153.

Lorentz, H.A. [1906] *Abh. Theoret. Phys.* **1**, 23.

Mackay, G.D.M., Suzuki, M. and Mason, S.G. [1963] "Approach of a solid sphere to a rigid plane interface. Part 2," *J. Colloid Sci.* **18**, 103.

Majumdar, S.R. and O'Neill, M.E. [1977] "On axisymmetric Stokes flow past a torus," *Z.A.M.P.* **28**, 541.

Mandelbrot, B.B. [1977] The Fractal Geometry of Nature (W.H. Freeman, New York, N.Y.).

Maude, A.D. [1961] "End effects in a falling-sphere viscometer," *Br. J. Appl. Phys.* **12**, 293.

Mercer, T.T., Morrow, P.E., and Stöber, W., editors [1972] Assessment of Airborne Particles (Charles C. Thomas, Springfield, Illinois).

Mountain, R.D. and Mulholland, G.W. [1984] "Stochastic Dynamics Simulation of Particle Aggregation," in Kinetics of Aggregation and Gelation, Family, F. and Landau, D.P., editors (Elsevier, New York, N.Y.) pp. 83-86.

Nir, A. and Acrivos, A. [1973] "On the creeping motion of two arbitrary-sized touching spheres in a linear shear field," *J. Fluid Mech.* **59**, 209.

O'Brien, V. [1968] "Form factors for deformed spheroids in Stokes flow," *A.I.Ch.E. J.* **14**, 870.

O'Neill, M.E. [1964] "A slow motion of viscous liquid caused by a slowly moving solid sphere," *Mathematika* **11**, 67.

O'Neill, M.E. [1969] "On asymmetrical slow viscous flows caused by the motion of two equal spheres almost in contact," *Proc. Camb. Philos. Soc.* **65**, 543.

O'Neill, M.E. [1981] "Small particles in viscous media," *Sci. Prog. (Oxford)* **67**, 149.

O'Neill, M.E. and Majumdar, S.R. [1970a] "Asymmetrical slow viscous fluid motions caused by the translation or rotation of two spheres. Part I: the determination of exact solutions for any values of the ratio of radii and separation parameters," *Z.A.M.P.* **20**, 164.

O'Neill, M.E. and Majumdar, S.R. [1970b] "Asymmetrical slow viscous fluid motions caused by the translation or rotation of two spheres. Part II: asymptotic forms of the solutions when the minimum clearance between the spheres approaches zero," *Z.A.M.P.* **20**, 180.

Oberbeck, A. [1876] "Concerning stationary motion of fluids with internal friction" (Crelle's Journal) *Reine Angew. Math.* **81**, 62 (in German).

Oseen, C.W. [1910] "Ueber die Stokes'sche formel und uber eine verwardte aufgabe in der hydrodynamik,"*Ark. Matemat.* **6(29)**.

Oseen, C.W. [1927] Hydrodynamik (Akademische Verlag, Leipzig, Germany).

Payne, L.E. and Pell, W.H. [1960] "The Stokes flow problem for a class of axially symmetric bodies," *J. Fluid Mech.* **7**, 529.

Proudman, I. and Pearson, J.R.A. [1957] "Expansions at small Reynolds numbers for the flow past a sphere and a circular cylinder," *J. Fluid Mech.* **2**, 237.

Pruppacher, H.R. and Klett, J.D. [1978] Microphysics of Clouds and Precipitation (Reidel, New York, N.Y.).

Ramarao, B.V. and Tien Chi [1988] "Calculations of drag forces acting on particle dendrites," *Aerosol Sci. and Tech.* **8**, 81.

Richardson, L.F. [1960] Statistics of Deadly Quarrels (Boxwood press, Pittsburgh) Chapter XII: "Contiguity and Deadly Quarrels: The Local Pacifying Influence."

Sampson, R.A. [1891] "On Stokes current function," *Philos. Trans. Roy. Soc. London Ser. A* **A182**, 449.

Savic, P. [1953] "Circulation and Distortion of Liquid Drops Falling Through a Viscous Medium" (National Research Council of Canada) Report # MT-22.

Segal, A. [1979] "On the numerical solution of the Stokes equations using the finite element method," *Comput. Meth. Appl. Mech. and Eng.* **19**, 165.

Simons, S. [1987] "The effect of coagulation on steady state Brownian diffusion for particles with a fractal structure," *J. Phys. D Appl. Phys.* **20**, 1197.

Sneddon, I.N. [1957] Elements of Partial Differential Equations (McGraw Hill, New York, N.Y.).

Spielman, L.A. [1970] "Viscous interactions in Brownian coagulation," *J. Colloid Interface Sci.* **33**, 562.

Stimson, M. and Jeffery, G.B. [1926] "The motion of two spheres in a viscous fluid," *Proc. Roy. Soc. (London)* **A111**, 110.

Stöber, W., Flachsbart, H. and Hochrainer, D. [1970] "The aerodynamic diameter of latex aggregates and asbestos fibers," *Staub J.* **30(7)**, 1.

Stokes, G.G. [1845] "On the theories of the internal friction of fluids in motion and of the equilibrium and motion of elastic solids," *Proc. Camb. Philos. Soc.* **1**, 16.

Stokes, G.G. [1851] "On the effect of the internal friction of fluids on the motion and of pendulums," *Camb. Philos. Soc. Trans.* **1**, [8].

van de Vate, J.F., van Leeuwan, W.F., Plomp, A. and Smit, H.C.D. [1980] "Morphology and Aerodynamics of Sodium Oxide Aerosol at Low Relative Humidities," in *Proceedings of the CSNI Specialists Meeting on Nuclear Aerosols in Reactor Safety* (Oak Ridge National Laboratory, Oak Ridge, Tennessee) Report # NUREG/CR-1724, ORNL/NUREG/TM-404, CSNI-45.

Wakiya, S. [1967] "Slow motions of a viscous fluid around two spheres," *J. Phys. Soc. Japan* **22**, 1101.

Williams, M.M.R. [1971] *Mathematical Methods in Particle Transport Theory* (Butterworths, London, U.K.).

Williams, M.M.R. [1973] "On the motion of small spheres in gases. III. Drag and heat transfer," *J. Phys. D Appl. Phys.* **6**, 744.

Williams, M.M.R. [1987a] "A closed torus in Stokes flow with slip boundary condition," *Q. J. Mech. and Appl. Math.* **40**, 235.

Williams, M.M.R. [1987b] "The drag on two spheres in contact in the slip flow regime," *Z.A.M.P.* **38**, 92.

CHAPTER 3

The Dynamic Equation for the Aerosol Distribution

3.1 Introduction

The pioneering work on the coalescence of particles to form successively larger particles was carried out by Smoluchowski [1916, 1917]. His particular interests were directed towards hydrosols rather than aerosols but the principle remains the same. Before proceeding with a detailed modern treatment of coagulating particles as a whole, it is worthwhile outlining Smoluchowski's approach because of the insight that it contains.

If we consider a single test sphere of radius, a_1, within an ensemble of such particles whose density is N per unit volume, then the number striking the test particle per unit time will be $4\pi D_1 a_1 N$ where D_1 is the diffusion coefficient. If we consider the total number of collisions due to all the particles acting as centers of removal, the net rate of removal will be:

$$\frac{dN}{dt} = -4\pi D_1 a_1 \left(\frac{N^2}{2}\right) \tag{3.1}$$

where the factor of 1/2 is due to the fact that we must avoid counting the collision twice.

Since all particles are diffusing, we must write instead of D_1, the sum (D_1+D_2) and a_1 must be replaced by a_1+a_2. Thus, a more realistic rate of change of particle concentration will be:

$$\frac{dN}{dt} = -2\pi (D_1 + D_2)(a_1 + a_2) N^2 \tag{3.2}$$

The diffusion process discussed above refers to Brownian motion and therefore according to Einstein:

$$D = \frac{kT}{6\pi\mu a} \tag{3.3}$$

where k is Boltzmann's constant, T the temperature of the gas, and μ its viscosity. Thus:

$$\frac{dN}{dt} = -\frac{kT}{3\mu}\left(\frac{1}{a_1} + \frac{1}{a_2}\right)(a_1 + a_2) N^2 \tag{3.4}$$

In order to proceed, some assumptions must be made about the way in which the radii, a_i, change with time. Clearly, some average value must be used and we suggest setting $a=a(t)=a_0 f(t)$ where a_0 is the initial average radius. It is not necessary to know $f(t)$ because only the ratio a_1/a_2 occurs. Thus, Eqn. (3.4) becomes:

$$\frac{dN}{dt} = -\frac{4kT}{3\mu} N^2 \tag{3.5}$$

The solution of this differential equation subject to the initial condition $N(0)=N_0$ is:

$$N(t) = \frac{N_0}{1+KN_0 t/2} \tag{3.6}$$

where $K=8kT/3\mu$. The average volume can therefore be calculated from:

$$\bar{v} = \frac{N_0 \bar{V}_0}{N(t)} \tag{3.7}$$

where $\bar{v}_0 = 4\pi a_0^3 / 3$. Thus:

$$\bar{v} = \bar{v}_0 \left(1 + \tfrac{1}{2} K N_0 t\right) \tag{3.8}$$

i.e., a linear increase with time.

3.2 The distribution function

The type of analysis described above was employed to study coagulation problems during the 1930's with some improvements to account for deviations from Stokes' law for very small particles. It is clearly unsatisfactory because it gives only a measure of the total number of particles present and some indication of the average volume. To extend the method to more complex coagulation mechanisms, and to losses due to settling and deposition, seems unproductive although it has been done. It is necessary, therefore, to seek a more comprehensive description of the aerosol distribution.

To this end we introduce the particle distribution function $n(v,r,t)$ where $n(v,r,t)dvdr$ is the number of aerosol particles with volumes in the range v to $v+dv$ in the volume element of physical space dr at r, at time t. Before discussing this general function we consider a simpler situation in which there is no spatial dependence and no loss of particles. Thus, the only process occurring is that of coagulation, i.e. as time proceeds an initial distribution of very small particles becomes fewer in number but larger in size.

The process may be illustrated by noting first that coagulation takes place in discrete steps. Indeed, the volume distribution function is really a histogram rather than a smooth curve and we should speak of $n_k(t)$, rather than $n(v,t)$, where $n_k(t)$ is the number of particles with k sub-units at time t. Here we assume that the sub-units are primary particles of a given size. A balance equation may be constructed for $n_k(t)$ as follows:

$$\frac{dn_k}{dt} = \tfrac{1}{2} \sum_{i=1}^{k-1} b_{i,k-i} n_i n_{k-i} - n_k \sum_{i=1}^{\infty} b_{i,k} n_i \tag{3.9}$$

where we have suppressed the argument t. The parameter b_{ij} is the number of collisions per unit time between particles containing i sub-units and j sub-units. The first term on the right hand side of Eqn. (3.9) is the rate at which k-type particles are formed as a result of the collision of i-type and (k–i)-type particles. The factor of 1/2 is again present to prevent collisions being counted twice. The second term describes the rate at which particles of type k are being removed due to collisions which convert them to some other size. The initial condition would be $n_i(0)=n_{0i}$, i.e. a given number of particles with i primary sub-units is present at the beginning of the coagulation process. Given b_{ij}, the above equation may be solved recursively to obtain $n_k(t)$.

The Dynamic Equation for the Aerosol Distribution

In general it is more convenient to write the discrete distribution function n_k as a quasi-continuous function of volume or some other convenient size descriptor. For example, in his classic work on this subject Muller [1928a,b] let:

$$n_i(t) = n(v_i, t)\, \Delta v_i$$

be the number of aerosol particles per unit volume at time t whose volumes lie between v_i and $v_i + \Delta v_i$. Thus, as $\Delta v_i \to 0$, $n(v_i,t)$ becomes the continuous function $n(v,t)$. At the same time, the coagulation rate $b_{i,j}$ becomes $K(v_i, v_j)$ where v_i and v_j are the volumes of the colliding particles.

Thus, Eqn. (3.9) can now be written as:

$$\frac{\partial n(v_i, t)}{\partial t} \Delta v_i = \tfrac{1}{2} \sum_{j=1}^{i-1} K(v_j, v_i - v_j)\, n(v_j, t)\, n(v_i - v_j, t)\, \Delta v_j\, \Delta(v_i - v_j)$$

$$- n(v_i, t)\, \Delta v_i \sum_{j=1}^{\infty} K(v_i, v_j)\, n(v_j, t)\, \Delta v_j$$

Dividing by Δv_i and taking the limit as $\Delta v_j \to 0$ but with Δv_i fixed, leads to:

$$\frac{\partial n(v,t)}{\partial t} = \tfrac{1}{2} \int_0^v du\, K(u, v-u)\, n(u,t)\, n(v-u,t)$$

$$- n(v,t) \int_0^\infty du\, K(u,v)\, n(u,t) \qquad (3.10)$$

where $v_i = v$ and $v_j = u$. While the first formulation of the aerosol balance equation was given by Muller [1928a,b], it was pointed out by Zebel [1958a] that Muller had neglected a term inside one of the integrals and his equation is therefore incomplete. The corrected integro-differential equation for heterogeneous aerosols first appeared in the Handbook of Aerosols [1950] with authorship credited to D.E. Goldman who, however, never published it in a journal. We should also mention the apparently independent work of Schumann [1940] who derived Eqn. (3.10) and solved it exactly for some special cases.

Interpreting Eqn. (3.10) physically, we can say that the first term on the right hand side expresses the fact that particles of volume, v, are created when particles of volume, u, collide with particles of volume, v−u. The second term states that a particle of volume, v, disappears after colliding with a particle of volume, u.

In some applications it may be more convenient to write the first integral on the right hand side of Eqn. (3.10) in another way. Thus, if we note that:

$$\tfrac{1}{2}\int_0^v du\, K(u,v-u)\, n(u)\, n(v-u) = \tfrac{1}{2}\int_0^{v/2} du\, K(u,v-u)\, n(u)\, n(v-u)$$

$$+ \tfrac{1}{2}\int_{v/2}^v du\, K(u,v-u)\, n(u)\, n(v-u)$$

and then in the first integral on the right hand side set $u = v/2 - w$ and in the second integral set $u = w + v/2$, we find that the right hand side becomes:

$$\tfrac{1}{2}\int_0^{v/2} dw\, K(v/2-w, v/2+w)\, n(v/2+w)\, n(v/2-w)$$

$$+\tfrac{1}{2}\int_0^{v/2} dw\, K(v/2+w, v/2-w)\, n(v/2+w)\, n(v/2-w)$$

From the symmetry of $K(u,v)$ this becomes:

$$\tfrac{1}{2}\int_0^{v/2} dw\, K(v/2+w, v/2-w)\, n(v/2+w)\, n(v/2-w)$$

Now setting $w = v/2 - u$, we find:

$$\int_0^{v/2} du\, K(u, v-u)\, n(u)\, n(v-u)$$

Thus the factor of 1/2 disappears but the upper limit of the integral becomes $v/2$.

Yet another way of writing the right hand side of Eqn. (3.10) which is sometimes more instructive is as:

$$\left(\frac{\partial n}{\partial t}\right)_{\text{coag}} = \tfrac{1}{2}\int_0^\infty du \int_0^\infty dw\, K(u,w)\, \delta(v-u-w)\, n(u,t)\, n(w,t)$$

$$- \int_0^\infty dw\, \delta(w-v)\, n(w,t) \int_0^\infty du\, K(u,v)\, n(u,t)$$

The delta function expresses conservation of volume in a collision and, bearing in mind the limits on v, is equivalent to Eqn. (3.10).

Eqn. (3.10) has some interesting integral properties which can be demonstrated by multiplying it by $\psi(v)$ and integrating over $v(0,\infty)$. Thus:

$$\frac{d}{dt}\int_0^\infty dv\, \psi(v)\, n(v,t) = \tfrac{1}{2}\int_0^\infty du \int_0^\infty dv\, [\psi(v+u) - \psi(v) - \psi(u)]\, K(u,v)\, n(u,t)\, n(v,t) \tag{3.11}$$

Eqn. (3.11) is found by judicious changes of the orders of integration and use of the symmetry condition $K(u,v) = K(v,u)$.

Two very useful relations arise from Eqn. (3.11). First, if we set $\psi = 1$, then we obtain:

$$\frac{dN(t)}{dt} = -\tfrac{1}{2}\int_0^\infty du \int_0^\infty dv\, K(u,v)\, n(u,t)\, n(v,t) \tag{3.12}$$

and secondly, if $\psi = v$, then we find:

$$\frac{d\phi(t)}{dt} = 0 \tag{3.13}$$

where the total number of particles per unit volume of space is given by:

The Dynamic Equation for the Aerosol Distribution

$$N(t) = \int_0^\infty dv\, n(v,t) \tag{3.14}$$

and the total volume of particulate per unit volume of space is given by:

$$\phi(t) = \int_0^\infty dv\, v\, n(v,t) \tag{3.15}$$

We also note that the average volume $\bar{v}(t)$ is given by $\phi(t)/N(t)$.

The fact that $d\phi/dt=0$, implies that ϕ=constant irrespective of the nature of the coagulation. This is clearly physically true because, for pure coagulation, the amount of material remains constant. It is simply being converted from a large number of small particles into a smaller number of large ones. Of course, if there is a loss mechanism such as settling or deposition this result is no longer true.

The result given by Eqn. (3.12) is also very useful and instructive. For example, if the coagulation kernel is independent or insensitive to particle volume, then we may write:

$$\frac{dN}{dt} = -\tfrac{1}{2} K N^2 \tag{3.16}$$

which corresponds precisely with Smoluchowski's original formulation with $K=8kT/3\mu$ for Brownian coagulation. We shall see in later sections that Eqn. (3.11) is particularly valuable for obtaining approximate solutions to complex coagulation processes.

3.3 Space dependent balance equation

We consider now the balance equation for $n(v,r,t)$ in which mechanisms for loss and gain of particles are present. Thus, we have (Friedlander [1977]):

$$\frac{\partial n(v,r,t)}{\partial t} + \nabla \cdot [U(v,r,t)\, n(v,r,t)] - \nabla \cdot [D(v,r,t)\, \nabla n(v,r,t)]$$

$$+ \frac{\partial}{\partial v}[I(v,r,t)\, n(v,r,t)] = S(v,r,t) + \left(\frac{\partial n}{\partial t}\right)_{coag} \tag{3.17}$$

where $U(v,r,t)$ is the vector sum of the fluid and particle velocities, $D(v,r,t)$ is the Brownian diffusion coefficient, $I(v,r,t)$ is the rate of growth of an individual particle due to evaporation or condensation, and $S(v,r,t)$ is an independent source term. The quantity $(\partial n/\partial t)_{coag}$ denotes the coagulation terms as displayed on the right hand side of Eqn. (3.10).

Physically, therefore, we note that the second term on the left hand side of Eqn. (3.17) describes losses due to convective motion of the particle, the third term is due to diffusion and the fourth to vapor condensation or evaporation. On the right hand side, there are gains due to the source and the coagulation processes. In general, all of the parameters U, D, I, etc. are functions of position and time. Eqn. (3.17) is intractable as it stands and therefore a number of assumptions are made in order to simplify it. One of these is spatial homogenization. That is, the aerosol is assumed to be in a well stirred atmosphere so that its density is spatially constant except in the neighborhood of surfaces. We shall describe such a homogenization procedure below but first we note that in a turbulent fluid, such as that expected in a reactor containment building, it is possible to replace the turbulent component of U by an eddy viscosity term (Friedlander [1977]) such that:

$$U n(v,r,t) \Rightarrow \varepsilon(r) \nabla n(v,r,t) + (\overline{U} + V_p) n(v,r,t) \qquad (3.18)$$

In this equation, ε is the eddy viscosity, V_p is the velocity of the particles due to external forces (such as gravity or electric fields), and $\overline{U}$ is the mean gas velocity. The terms involving spatial derivatives in Eqn. (3.17) can therefore be replaced by:

$$\nabla \cdot (D + \varepsilon) \nabla n(v,r,t) + \nabla \cdot \left[(\overline{U} + V_p) n(v,r,t)\right] \qquad (3.19)$$

Let us now average Eqn. (3.17) over the volume of the vessel, V. In doing this, we define the average particle distribution function as:

$$n(v,t) = \frac{1}{V} \int_V dr\, n(v,r,t) \qquad (3.20)$$

Then, averaging the gradient term and using Gauss' theorem, we find:

$$\frac{1}{V} \int_V dr\, \nabla \cdot \left[(\overline{U} + V_p) n(v,r,t)\right] = \frac{1}{V} \int_A dA \cdot (\overline{U} + V_p) n(v,r_s,t) \qquad (3.21)$$

where A is the surface of V and $n(v,r_s,t)$ is the value of n at the wall, r_s. Eqn. (3.21) is further simplified if we set $n(v,r_s,t) \approx n(v,t)$ such that the right hand side of Eqn. (3.21) becomes:

$$R_s = \frac{1}{V} A \cdot (\overline{U} + V_p) n(v,t) \qquad (3.22)$$

As an example, let the mean gas velocity, U, be zero at the wall and let V_p be the settling velocity, kV_s, due to gravity (k being a unit vector in the direction of gravity). We find:

$$R_s = \frac{A_f V_s(v)}{V} n(v,t) \qquad (3.23)$$

where A_f is the area of the floor and $V_s(v)$ is the Stokes' velocity. For a vessel of height, H, we may write R_s as:

$$R_s = \frac{V_s(v)}{H} n(v,t) \qquad (3.24)$$

In the case of diffusion, the averaging leads to:

$$\frac{1}{V} \int_V dr\, \nabla \cdot (D + \varepsilon) \nabla n(v,r,t) = \frac{1}{V} \int_A dA \cdot \nabla n(v,r_s,t) (D + \varepsilon(r_s))$$

But at the wall, $\varepsilon=0$ and deposition is due to diffusion only. We therefore have to approximate the term:

The Dynamic Equation for the Aerosol Distribution

$$\frac{D}{V} \int_A dA \cdot \nabla n(v, r_s, t)$$

To do this we assume that there exists a thin boundary layer of thickness, δ_D, near the wall and that normal to the wall we can set:

$$\mathbf{n} \cdot \nabla n(v, r_s, t) \approx \frac{n(v, t)}{\delta_D}$$

with **n** being a unit vector normal to the wall.

Thus, the loss term due to wall deposition by diffusion is:

$$R_d = \frac{D(v) A_D}{V \delta_D} n(v, t) \tag{3.25}$$

where A_D is the area of surfaces exposed to diffusion.

It has been shown by a systematic study (van de Vate [1980]) that the diffusional boundary layer thickness, δ_D, is given by:

$$\delta_D = (4.6) D(v)^{0.265}$$

where δ_D is in cm and D in cm^2 s^{-1}. Thus, an estimate of R_d is:

$$R_d = \frac{A_d}{(4.6) V} D(v)^{0.735} n(v, t) \tag{3.27}$$

The rate of growth, I, can be due to several different mechanisms. Condensation, which may be viewed as accretion of vapor molecules through Brownian diffusion, takes the form:

$$I(v, t) = a(t) v^{1/3} \tag{3.28}$$

where a(t) involves various physical parameters and the difference in vapor pressures of the diffusing species in the bulk gas and at the particle surface. If the particle is very small its radius of curvature will influence I but we neglect that effect here (Twomey [1977]). For chemical reactions, which occur on the surface of a particle, I is proportional to area, viz: $v^{2/3}$, while for volume reactions I is proportional to v. A general form which covers all cases may be written:

$$I(v, t) = I_\alpha(t) v^\alpha \tag{3.29}$$

or a sum of such terms.

Finally, we note that the well-mixed hypothesis assumes the volume average of the nonlinear coagulation term to be equal to the product of the averages:

$$\frac{1}{V} \int_V dr\, n(u, r, t) n(v, r, t) \approx \frac{1}{V} \int_V dr\, n(u, r, t) \frac{1}{V} \int_V dr\, n(v, r, t) = n(u, t) n(v, t) \tag{3.30}$$

It is difficult to assess the error involved here, but it is well to keep it in mind.

Collecting up terms, we see that the balance equation for the particle volume distribution function takes the form:

$$\frac{\partial n(v,t)}{\partial t} + R(v,t)\,n(v,t) + \frac{\partial}{\partial v}(I(v,t)\,n(v,t)) = \tfrac{1}{2}\int_0^v du\, K(u,v-u)\,n(u,t)\,n(v-u,t)$$
$$-n(v,t)\int_0^\infty du\, K(u,v)\,n(u,t) + S(v,t) \quad (3.31)$$

where $R(v,t)$ is of the form, $av^m + bv^n$.

Eqn. (3.31) is basic to many studies of aerosol behavior in reactor environments.

3.4 Representation in terms of radius

Sometimes it is convenient to describe the particle distribution function $n(v,t)$ in terms of the radius of the equivalent spherical particle, where $v = 4\pi r^3/3$. Let us therefore define:

$$\tilde{n}(r,t)\,dr = n(v,t)\,dv$$

or:

$$\tilde{n}(r,t) = 4\pi r^2\, n(v,t)$$

Now with $u = 4\pi s^3/3$, we find by direct transformation that $\tilde{n}(r,t)$ satisfies:

$$\frac{\partial \tilde{n}(r,t)}{\partial t} + \tilde{R}(r,t)\,\tilde{n}(r,t) + \frac{\partial}{\partial r}\left[\tilde{I}(r,t)\,\tilde{n}(r,t)\right]$$
$$= 2\pi r^2 \int_0^r ds\, \tilde{K}\!\left(s,(r^3-s^3)^{1/3}\right) \tilde{n}(s,t)\,n\!\left(4\pi(r^3-s^3)/3,t\right) - \tilde{n}(r,t)\int_0^\infty ds\, \tilde{K}(s,r)\,\tilde{n}(s,t)$$

where:

$$\tilde{R}(r,t) = R\!\left(\frac{4\pi r^3}{3},t\right)$$

$$\tilde{I}(r,t) = \frac{1}{4\pi r^2}\, I\!\left(\frac{4\pi r^3}{3},t\right)$$

and:

$$\tilde{K}(s,r) = K\!\left(\frac{4\pi s^3}{3},\frac{4\pi r^3}{3}\right)$$

There only remains the term $n(4\pi(r^3-s^3)/3,t)$ to deal with. This is done by setting $w = 4\pi q^3/3$ and $q^3 = r^3 - s^3$. Then:

$$n(w,t)\,dw = \tilde{n}(q,t)\,dq$$

whence:

The Dynamic Equation for the Aerosol Distribution

$$n(w,t) = \tilde{n}(q,t)\frac{dq}{dw} = \frac{\tilde{n}\left((r^3-s^3)^{1/3},t\right)}{4\pi(r^3-s^3)^{2/3}}$$

and the equation, in terms of radius, becomes:

$$\frac{\partial \tilde{n}(r,t)}{\partial t} + \tilde{R}(r,t) + \frac{\partial}{\partial r}\left[\tilde{I}(r,t)\,\tilde{n}(r,t)\right]$$
$$= \tfrac{1}{2}\int_0^r ds\, \tilde{K}\!\left(s,(r^3-s^3)^{1/3}\right)\tilde{n}(s,t)\,\tilde{n}\!\left((r^3-s^3)^{1/3},t\right)\frac{r^2}{(r^3-s^3)^{2/3}} - \tilde{n}(r,t)\int_0^\infty ds\, \tilde{K}(s,t)\,\tilde{n}(s,t) \tag{3.32}$$

Some workers use an equation in which the upper limit on the integral is $r/2^{1/3}$ rather than r (Zebel [1958a]) and eliminate the factor of 1/2 preceding the integral.

3.5 Multicompartment aerosol equations

The well-mixed hypothesis is considered to be conservative in terms of mass leaked to the environment, *i.e.*, through any breaches in the outer containment, and it has the advantage of allowing calculations to be carried out without a detailed knowledge of the internal structure of the containment vessel. It is clear, however, that certain situations may arise (*e.g.*, in long term sodium fires) in which convective loops are created where the containment atmosphere will be mixed in a periodic manner. Such matters have been considered by Jordan *et al.* [1980] in relation to the 'chimney effect' caused by a pool of burning sodium. They argue that the aerosol particles created immediately above the flame zone of the fire will rise vertically in a 'chimney' and not 'see' the containment walls until they reach the return loop. During this transport period, which is characterized by high aerosol concentration, rapid particle growth occurs and indicates that particles will be significantly larger when they eventually meet the container walls than might be expected from the well-mixed hypothesis. The degree of deviation will depend on a characteristic mixing time and intuitively we might expect the well-mixed model to break down when the characteristic coagulation time is much smaller than the characteristic mixing time. If, therefore, particles remain in the chimney region for several coagulation times, some degree of failure of the homogeneous model can be expected.

This particular problem has been modelled by Jordan *et al.* by splitting the containment vessel into a number of zones and allowing particle flow from one zone to another according to the expected gas flow in the system. Fig. 3.1 shows the general arrangement which contains four zones although two are identical.

The volumetric flow from zone i to zone j is denoted by $Q^{i \to j}$. In the case described above it is clear that in a steady state and for incompressible flow:

$$Q^{1\to 2} = Q^{3\to 1}$$
$$Q^{2\to 3} = Q^{3\to 1} + Q^{3\to 2}$$

More accurate relations can be obtained by considering the gas dynamics more carefully as we will show below.

The zonal concept may be expressed quantitatively by averaging the space and time dependent Eqn. (3.17) over a zone of volume, V_k, within the total volume, V. Then, assuming that transfer between zones i and j is governed by a leakage probability per unit time, $\alpha_{i \to j}$, we can obtain N coupled aerosol equations, where N is the number of zones.

Aerosol Science: Theory and Practice

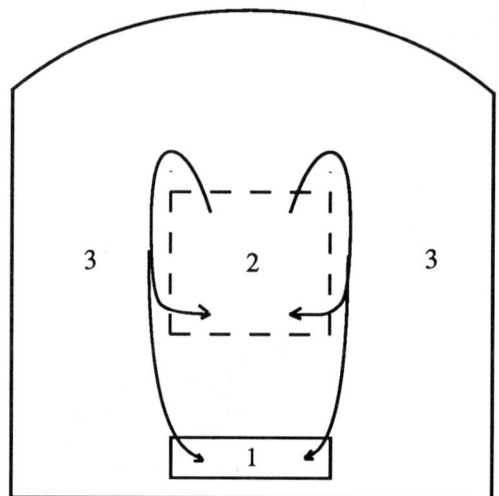

1) Na source zone, 2) chimney zone, 3) external zone

Figure 3.1: Aerosol flow in the various zones of a containment.

In order to establish a useful theory of coupled zones or chambers, we consider Eqn. (3.17) and integrate it over the volume, V_i, of a single zone i which is coupled to other zones through various openings or flow regions.

Fig. 3.2 illustrates zone i with gas moving from it into zones i+1 and i+2 and with gas being received from zones j and j+1.

Integrating Eqn. (3.17) we find, after using Gauss' theorem, that:

$$\frac{\partial}{\partial t}[V_i\, n_i(v,t)] + \int_{A_i} d\mathbf{A} \cdot \nabla[n(v,\mathbf{r}_s,t) D(v,\mathbf{r}_s)] + \int_{A_i} d\mathbf{A} \cdot \mathbf{U}(v,\mathbf{r}_s,t)\, n(v,\mathbf{r}_s,t) = Q_i \qquad (3.33)$$

where A_i is the surface area of V_i and the average density of particles in zone i is:

$$n_i(v,t) = \frac{1}{V_i} \int_{V_i} d\mathbf{r}\, n(v,\mathbf{r},t)$$

Q_i is the average of the right hand side.

Let us assume that A_i consists of j parts through which gas can flow either into or out of V_i. Let the areas of such parts be A_{ij} where j=1,2,...,N.

The remaining part of A_i, which we call A_{id}, is the area of solid surfaces. It is now assumed that, while passing through a flow route, the convective term, nU, dominates the diffusive term, $D\nabla n$. On the other hand, on the solid part of A_i, both diffusion and convective flow play a role in depositing material on the surface.

With this division of A_i in mind, we can write Eqn. (3.33) as:

The Dynamic Equation for the Aerosol Distribution

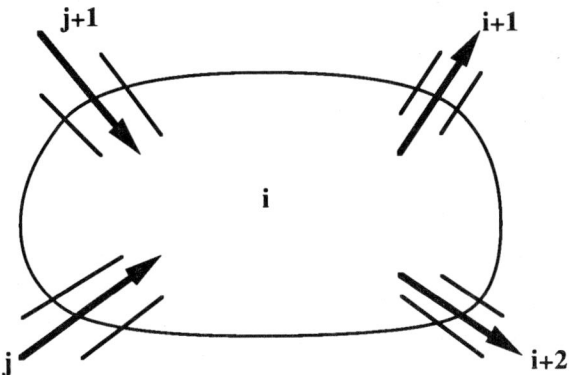

Figure 3.2: Aerosol inflow and outflow from zone i.

$$\frac{\partial}{\partial t}[V_i\, n_i(v,t)] + \int_{A_{id}} dA \bullet \{\mathbf{V}[n(v,\mathbf{r}_s,t)\, D(v,\mathbf{r}_s)] + \mathbf{U}(v,\mathbf{r}_s,t)\, n(v,\mathbf{r}_s,t)\}$$

$$+ \sum_{j=1}^{M_i} \int_{A_{ij}} dA \bullet \mathbf{U}(v,\mathbf{r}_s,t)\, n(v,\mathbf{r}_s,t) = Q_i \qquad (3.34)$$

i=1,2,...,N, where N=number of zones. M_i is the number of entry and exit points in zone i.

We now assume that, on A_{id} and A_{ij}, the values of n, ∇n, D, and U remain sensibly constant, so that we may write:

$$\frac{\partial}{\partial t}[V_i\, n_i(v,t)] + \{\mathbf{A}_{id} \bullet \mathbf{V}[n(v,\mathbf{r}_{si},t)\, D(v,\mathbf{r}_{si})] + \mathbf{A}_{is} \bullet \mathbf{U}(v,\mathbf{r}_{si},t)\}$$

$$+ \sum_{j=1}^{M_i} \mathbf{A}_{ij} \bullet \mathbf{U}(v,\mathbf{r}_{si},t)\, n(v,\mathbf{r}_{si},t) = Q_i \qquad (3.35)$$

It is useful at this point to separate the flow terms into those entering i and those leaving it. Thus, if there are $M_i^{(o)}$ exit points:

$$\sum_{j=1}^{M_i} \mathbf{A}_{ij} \bullet \mathbf{U}\, n(v,\mathbf{r}_{sj},t) = \sum_{j=1}^{M_i^{(o)}} \mathbf{A}_{ij} \bullet \mathbf{U}\, n(v,\mathbf{r}_{sj},t) + \sum_{j=M_i^{(o)}+1}^{M_i} \mathbf{A}_{ij} \bullet \mathbf{U}\, n(v,\mathbf{r}_{sj},t) \qquad (3.36)$$

But for outflow, the density $n(v,\mathbf{r}_{sj},t)$ will be characterized by the average density in i, thus we may write approximately:

$$\sum_{j=1}^{M_i^{(o)}} \mathbf{A}_{ij} \bullet \mathbf{U}\, n(v,\mathbf{r}_{sj},t) \approx -n_i(v,t) \sum_{j=1}^{M_i^{(o)}} \mathbf{A}_{ij}\, U_{i\to j}(v,t) \qquad (3.37)$$

On the other hand, for inflow, the density $n(v,\mathbf{r}_{sj},t)$ will be characterized by the average density in the outside zone j, thus:

$$\sum_{j=M_i^{(o)}+1}^{M_i} A_{ij} \cdot U\, n(v, r_{sj}, t) \approx \sum_{j=M_i^{(o)}+1}^{M_i} A_{ij}\, U_{i \to j}(v, t)\, n_j(v, t) \tag{3.38}$$

We also write for the surface term:

$$\int_{A_{id}} dA \cdot U\, n(v, r_s, t) \approx A_{id}\, U_{id}\, n_i(v, t) \tag{3.39}$$

where U_{id} is the deposition velocity.

As far as the diffusive term is concerned we use the well-mixed hypothesis in each zone, and in a thin boundary layer of thickness, δ_i, near the surface, we write:

$$A_{id} \cdot \nabla n \approx \frac{A_{id}}{\delta_i} n_i(v, t) \tag{3.40}$$

Collecting all these terms together, we find:

$$\frac{\partial}{\partial t}[V_i\, n_i(v, t)] + \frac{A_{is}\, D_i(v)}{\delta_i\, V_i} V_i\, n_i(v, t) + \frac{A_{id}}{V_i} U_{id}(v, t)\, V_i\, n_i(v, t)$$

$$+ \sum_{j=M_i^{(o)}+1}^{M_i} \frac{A_{ij}}{V_j} U_{j \to i}(v, t)\, V_j\, n_j(v, t) - V_i\, n_i(v, t) \sum_{j=1}^{M_i^{(o)}} \frac{A_{ij}}{V_i} U_{i \to j}(v, t) = Q_i \tag{3.41}$$

The averaging of Q_i is straightforward and takes the form:

$$Q_i = -\frac{\partial}{\partial v}[I_i(v, t)\, V_i\, n_i(v, t)] + \tfrac{1}{2} V_i \int_0^v du\, K_i(u, v-u)\, n_i(u, t)\, n_i(v-u, t)$$

$$- V_i\, n_i(v, t) \int_0^\infty du\, K_i(u, v)\, n_i(u, t) + V_i\, S_i(v, t) \tag{3.42}$$

Now we set:

$$R_{id}(v, t) = \frac{A_{id}\, D_i(v)}{\delta_i\, V_i} \tag{3.44}$$

$$R_{is}(v, t) = \frac{A_{id}\, U_{id}(v, t)}{V_i} \tag{3.45}$$

and:

$$\alpha_{j \to i}(v, t) = \frac{A_{ij}}{V_j} U_{j \to i}(v, t) \tag{3.46}$$

and obtain:

The Dynamic Equation for the Aerosol Distribution

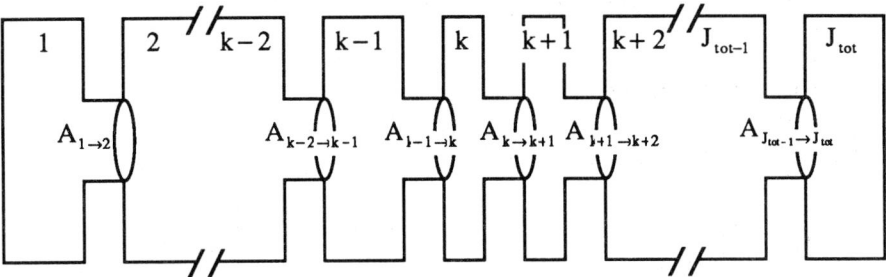

Figure 3.3: Aerosol flow in linearly coupled chambers.

$$\frac{\partial}{\partial t}[V_i \, n_i(v,t)] + [R_{id}(v,t) + R_{is}(v,t)] V_i \, n_i(v,t) + \frac{\partial}{\partial v}[I_i(v,t) V_i \, n_i(v,t)]$$

$$= \sum_{j=M_i^{(o)}+1}^{M_i} \alpha_{j \to i}(v,t) V_j \, n_j(v,t) - V_i \, n_i(v,t) \sum_{j=1}^{M_i^{(o)}} \alpha_{i \to j}(v,t)$$

$$+ \tfrac{1}{2} V_i \int_0^v du \, K_i(u, v-u) \, n_i(u,t) \, n_i(v-u,t) - V_i \, n_i(v,t) \int_0^\infty du \, K_i(u,v) \, n_i(u,t) + S_i(v,t)$$

(3.47)

These equations form a coupled set and are substantially more difficult to deal with than that for a single zone.

3.6 The gas dynamics

The problem discussed in the previous section has provided a means of dealing with coupled zones or chambers if the coupling coefficients $\alpha_{i \to j}$ can be obtained. This may be done by means of an analysis of the gas dynamics. As a simple example of how to carry out such a calculation we consider a series of linearly coupled chambers as shown in Fig. 3.3 (Simpson et al. [1989]).

If r_k is the gas density in chamber k and $\dot{W}_{k,j}$ is the mass flow rate of gas from chamber k to chamber j, then from continuity:

$$V_k \frac{d\rho_k}{dt} = \dot{W}_{k-1,k} - \dot{W}_{k,k+1} \quad ; \quad k = 1,2,3,\ldots,J \tag{3.47}$$

where J is the number of chambers. We assert that $\dot{W}_{0,1} = \dot{W}_{J,J+1} = 0$.

In terms of $U_{i \to j}$ above we note that $\dot{W}_{i,j} = \rho_i \, A_{ij} \, U_{i \to j}$. We may also write an energy balance. In doing so, however, we assume that the control surface is adiabatic and that no external work is done. Treating the fluid like a perfect gas, we may write for chamber i, $P_i = \rho_i R T_i$, where R is the gas constant, P_i the gas pressure, and T_i the gas temperature. Then:

$$C_v \, V_i \frac{d}{dt}(\rho_i \, T_i) = \dot{W}_{i-1,i}\left(h + \tfrac{1}{2} v^2\right)_{i-1,i} - \dot{W}_{i,i+1}\left(h + \tfrac{1}{2} v^2\right)_{i,i+1} \tag{3.48}$$

Aerosol Science: Theory and Practice

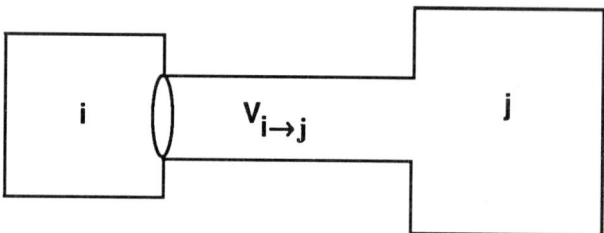

Figure 3.4: Aerosol flow between two coupled chambers, i and j.

where C_v is the isochoric specific heat, h the specific enthalpy, and v the gas speed.

$\dot{W}_{i,j}$ can be obtained from an adiabatic analysis. The calculations can be rather tedious but because of its importance in aerosol transport, we give below a detailed treatment for two coupled chambers. For this analysis we are indebted to Dr. R.C. Raichura of Queen Mary College, London University.

Consider the two coupled chambers as shown in Fig. 3.4. In chamber 1, we have temperature, $T_1(t)$, pressure, $P_1(t)$, and density, $\rho_1(t)$. The volume is constant and has value, V_1, with an area, A_1, normal to the orifice. This orifice is a narrow opening leading to the second chamber in which the temperature, pressure, density, volume and area are T_2, P_2, ρ_2, V_2, and A_2, respectively. The velocity of the flow through the orifice is u(t). The velocities in chambers 1 and 2 are considered negligible because the area of each chamber is much greater than the area of the connecting orifice. We write the equation of continuity for the system:

$$0 = \frac{dm_{cv}}{dt} + \dot{m}_{out} - \dot{m}_{in} \tag{3.49}$$

where m_{cv} is the mass of gas in the control volume and $\dot{m}$ is a mass flow rate. For chamber 1, we have:

$$\dot{m}_{in} = 0 \quad , \quad m_{cv} = \rho_1 V \quad \text{and} \quad \dot{m}_{out} = \dot{m}(t)$$

and hence:

$$V_1 \frac{d\rho_1}{dt} = -\dot{m} \tag{3.49a}$$

Similarly, for chamber 2:

$$V_2 \frac{d\rho_2}{dt} = \dot{m} \tag{3.49b}$$

The energy equation takes the form:

$$0 = \frac{dE_{cv}}{dt} + \int_{A_{out}} \left(h + \tfrac{1}{2} u^2\right) d\dot{m}_{out} - \int_{A_{in}} \left(h + \tfrac{1}{2} u^2\right) d\dot{m}_{in} \tag{3.50}$$

where we have assumed no external heating or work done on the gas. E_{cv} is the energy in the control volume and h(t) is the enthalpy. u=u(t) is the velocity at the edge of the orifice, A_{out} and A_{in} are the areas at the orifice. For chamber 1, the integral over A_{in} is zero and we find:

The Dynamic Equation for the Aerosol Distribution

$$\int_{A_{out}} \left(h + \tfrac{1}{2} u^2\right) d\dot{m}_{out} = \dot{m}\left(h + \tfrac{1}{2} u^2\right)$$

The energy of the gas in the control volume is:

$$E_{cv} = (T_1 - T_R)\rho_1 V_1 C_v$$

where T_R is a convenient reference temperature. Thus:

$$\frac{dE_{cv}}{dt} = C_v V_1 \frac{d}{dt}(\rho_1 T_1)$$

or:

$$C_v V_1 \frac{d}{dt}(\rho_1 T_1) = -\dot{m}\left(h + \tfrac{1}{2} u^2\right) \qquad (3.51)$$

Similarly, for chamber 2, the integral over A_{out} is zero and we find:

$$C_v V_2 \frac{d}{dt}(\rho_2 T_2) = \dot{m}\left(h + \tfrac{1}{2} u^2\right) \qquad (3.52)$$

We also have the perfect gas equation: $P = \rho RT$.

It is now necessary to obtain the equation for the mass flow rate. To do this we assume that the flow is steady, adiabatic, and frictionless. Then from standard gas dynamic theory (Shapiro [1953]):

$$\dot{m} = q A \rho_1 \left\{\frac{2\gamma}{\gamma-1} R T_1\right\}^{1/2} \qquad (3.53)$$

where:

$$q = \left(\frac{2}{\gamma+1}\right)^{\frac{1}{\gamma-1}} \left(\frac{\gamma-1}{\gamma+1}\right)^{1/2} = q_f(\text{constant}) \quad ; \text{ if } \frac{P_2}{P_1} \leq x_{cr}$$

$$= \left(\frac{P_2}{P_1}\right)^{\frac{1}{\gamma}} \left\{1 - \left(\frac{P_2}{P_1}\right)^{\frac{\gamma-1}{\gamma}}\right\}^{1/2} \quad ; \text{ if } \frac{P_2}{P_1} \geq x_{cr} \qquad (3.54)$$

and the critical pressure ratio is:

$$x_{cr} = \left(\frac{2}{\gamma+1}\right)^{\frac{\gamma}{\gamma-1}}$$

Also in deriving Eqn. (3.54) we have assumed isentropic flow such that:

$$\frac{P_2}{P_1} = \left(\frac{\rho_2}{\rho_1}\right)^{\gamma} = \left(\frac{T_2}{T_1}\right)^{\frac{\gamma}{\gamma-1}} \qquad (3.55)$$

We note at this point that, if friction cannot be neglected, then Eqns. (3.53) and (3.54) are still valid if γ is replaced by a polytropic index of expansion. There are also corrections that can be made to account for the type of orifice (*e.g.*, smooth nozzle or standard orifice).

Proceeding with our calculation, we combine Eqns. (3.49a) and (3.49b) and integrate to obtain:

$$V_1 \rho_1(t) + V_2 \rho_2(t) = V_1 \rho_1(0) + V_2 \rho_2(0) = M$$

where M is a constant. Thus:

$$\rho_2 = \rho_{2,0} + (\rho_{1,0} - \rho_1) R_V = f(\rho_1) \qquad (3.56)$$

where $R_V = V_1/V_2$ is the volume ratio. Then, from Eqn. (3.55):

$$P_2 = P_1 \left[\frac{f(\rho_1)}{\rho_1} \right]^\gamma \equiv P_1\, g(\rho_1) \qquad (3.57)$$

Similarly:

$$T_2 = T_1 \left[\frac{f(\rho_1)}{\rho_1} \right]^{\gamma-1} \equiv T_1\, h(\rho_1) \qquad (3.58)$$

is obtained using $P_1 = \rho_1 R T_1$.

Eqns. (3.56), (3.57) and (3.58) give ρ_2, P_2, and T_2 in terms of P_1 and ρ_1. However, we have not yet used the energy equations, Eqns. (3.51) and (3.52). Adding them and integrating leads to:

$$C_v V_1 \rho_1(t) T_1(t) + C_v V_2 \rho_2(t) T_2(t) = C_v V_1 \rho_1(0) T_1(0) + C_v V_2 \rho_2(0) T_2(0) = E$$

where E is a constant.

Since P=ρRT, this becomes $P_1 V_1 + P_2 V_2 = E (\gamma - 1)$. Using Eqn. (3.57) for P_2, we find:

$$P_1 = \frac{(\gamma - 1) E}{V_1 + g(\rho_1) V_2} \equiv p(\rho_1) \qquad (3.59)$$

Returning now to Eqn. (3.49a):

$$V_1 \frac{d\rho_1}{dt} = -\dot{m} \qquad (3.60)$$

we make the assumption that Eqns. (3.53), (3.54) and (3.55) are valid, such that the properties P, ρ, T, *etc.* are instantaneous values. Thus:

$$\dot{m} = q\, A\, \rho_1 \left\{ \frac{2\gamma}{\gamma - 1} R\, T_1 \right\}^{1/2} \qquad (3.61)$$

where $q = q(P_2/P_1)$ and $P_2/P_1 = \gamma(\rho_1)$, so that:

The Dynamic Equation for the Aerosol Distribution

$$q = \left(\frac{2}{\gamma+1}\right)^{\frac{1}{\gamma-1}} \left(\frac{\gamma-1}{\gamma+1}\right)^{\frac{1}{2}} \quad ; \quad \text{if } g(\rho_1) \leq x_{cr}$$

$$= g(\rho_1)^{\frac{1}{\gamma}} \left\{1 - g(\rho_1)^{\frac{\gamma-1}{\gamma}}\right\}^{1/2} \quad ; \quad \text{if } g(\rho_1) \geq x_{cr}$$

In any case, $q = q(\rho_1)$.

Substituting in Eqn. (3.60), we find:

$$V_1 \frac{d\rho_1}{dt} = -q(\rho_1) A \rho_1 \left\{\frac{2\gamma}{\gamma-1} \frac{p(\rho_1)}{\rho_1}\right\}^{1/2} \equiv -\phi(\rho_1) \tag{3.62}$$

Integration of this equation leads to a relationship $t=t(\rho_1)$ or $\rho_1=\rho_1(t)$. From Eqn. (3.59) we find $P_1(t)$ and hence $T_1(t) = P_1/\rho R$. It is then straightforward to obtain ρ_2, T_2, and P_2 from Eqns. (3.56), (3.57), and (3.58). The extension of this procedure to a series of such chambers is obvious. The mass flow rate, $\dot{m}$, can be obtained easily from Eqn. (3.62); it is simply $\phi(\rho_1)$ and hence:

$$\alpha_{1\to 2} = \frac{\dot{m}}{\rho_1 V_1}$$

Further work concerning discharge of gases through orifices and pipes may be found in Kestin and Glass [1951], Horlock and Woods [1965], El-Hakeem and Ryley [1972], Schmidt [1965], and Sakurai and Takayama [1983].

3.7 Multispecies aerosols

Thus far, we have considered the aerosol as a single species (*e.g.* water or sodium). A much more practical situation, however, is when the aerosol consists of a mixture of different species (*e.g.*, cesium, iodine, tellurium, and water). In such a case, at any given time after the initial injection of material, a single aerosol particle will consist of a certain fraction of each species present initially. Thus, we must be able to calculate the following extended aerosol distribution function:

$$n(v_1, v_2, \ldots v_N; m_1, m_2, \ldots, m_N, t) \, dv_1 \, dv_2 \, \ldots \, dv_N \, dm_1 \, dm_2 \, \ldots \, dm_N \tag{3.63}$$

This is the number of aerosol particles which contain a volume, v_p to v_p+dv_p, and mass, m_p to m_p+dm_p, of species p $(1 \leq p \leq N)$. The total volume of the aerosol particle will be:

$$v = \sum_{p=1}^{N} v_p \tag{3.64}$$

and the total mass:

$$m = \sum_{p=1}^{N} m_p \tag{3.65}$$

The conventional distribution function, discussed above, for the total volume regardless of its constituents, is:

$$N(v,t) = \int dv_1 \int dv_2 \ldots \int dv_N \int dm_1 \int dm_2 \ldots \int dm_N$$
$$n(v_1,v_2,\ldots,v_N;m_1,m_2,\ldots,m_N,t) \delta\left(v - \sum_{p=1}^{N} v_p\right) \quad (3.66)$$

We can also define other useful reduced distribution functions. For example, if $M_p(v,t)$ is the total volume of species p per unit volume of space which resides in particles whose own volume lies in the range, v to v+dv, then we have:

$$M_p(v,t) = \int dv_1 \int dv_2 \ldots \int dv_p \, v_p \ldots \int dv_N \int dm_1 \int dm_2 \ldots \int dm_N$$
$$n(v_1,v_2,\ldots,v_p,\ldots,v_N;m_1,m_2,\ldots,m_N,t) \delta\left(v - \sum_{p=1}^{N} v_p\right) \quad (3.67)$$

Similarly for the mass fraction, $M_p(m,t)$. Such quantities are useful for they give a measure of the concentration of a particular species in a given particle of volume, v, or mass, m.

It remains to develop a dynamic equation for the generalized distribution function. We do this in the well-mixed hypothesis, although the extension to the more general case is available. We write the equation down and then explain the origin of each term. First, however, we introduce a shorthand notation, writing:

$$n(v_1,v_2,\ldots,v_N;m_1,m_2,\ldots,m_N,t) = n(v,m,t) \quad (3.68)$$

and:

$$\int dv_1 \int dv_2 \ldots \int dv_N = \int dv$$

Now we have:

$$\frac{\partial}{\partial t} n(v,m,t) + R(v,m,t) n(v,m,t) + \sum_{p=1}^{N} \frac{\partial}{\partial v_p} [I_p(v,m,t) n(v,m,t)]$$

$$= \tfrac{1}{2} \int_0^\infty dv \int_0^\infty dw \int_0^\infty dq \int_0^\infty ds \, n(u,q,t) n(w,s,t) K(u,q|w,s) \prod_{p=1}^{N} \delta(v_p - u_p - w_p) \delta(m_p - q_p - s_p)$$

$$- n(v,m,t) \int_0^\infty du \int_0^\infty dq \, K(u,q|v,m) n(u,q,t) + S(v,m,t) \quad (3.69)$$

where:

$$q = \sum_{p=1}^{N} q_p$$

and:

$$s = \sum_{p=1}^{N} s_p$$

The terms of Eqn. (3.69) have the following meanings. The second term on the left hand side describes the losses due to settling and deposition. These can be due to gravitational settling

The Dynamic Equation for the Aerosol Distribution

and various diffusive mechanisms. In general, such processes are functions of the total volume, v, of the particle, which determines the drag, and the total mass, m, which determines the force in the case of gravitational settling. Examples may be seen in Eqns. (3.24) and (3.25). The third term on the left hand side describes condensation with $I_p(v,m,t)$ being the condensation rate for species p. This term affects the total volumes and masses of species p by either the accretion or evaporation of material which is residing in the atmosphere in the form of vapor. Since there can be several types of vapor species, the net effect will be the sum over all species present. The nature of $I_p(v,m,t)$ will depend on the type of growth law involved. It is most likely that the growth law will depend only on the total volume of the particle. However, it could also be envisaged that the vapor-solid interaction might depend on the material near the surface in which case I_p would be sensitive to the individual values of v_p. A not unreasonable assumption would be to assume that I_p depends on some power of v_p, e.g. $I_p = s_p(t) v_p^{\gamma}$.

The first term on the right hand side describes the rate at which coagulation creates particles of individual volumes **v** and masses **m**. K(u,q|w,s) is the coagulation kernel for particles of total volume, u, masses, q, coalescing with particles of total volumes, w, and masses, s. The delta functions are present to ensure conservation of volume and mass in a collision. In passing, we note that an opportunity exists here to allow for nonconservation of volume, due say to some packing effect. It would only be necessary to rewrite the delta function as $\delta(v_p - a(u_p + w_p))$ where a is somewhat greater than unity and must be derived experimentally. Of course the mass must always be conserved. The second term on the right hand side describes the rate at which particles are removed from (v,m) due to coagulation. Finally, the last term is an independent source.

It should be noted that many authors simplify Eqn. (3.69) by setting $R(v,m) = R(v)$, $I_p(v,m) = I_p(v)$ and $K(u,q|w,s) = K(u|w)$, i.e. dependence on total particle volume only. There are some physical cases where this is true, for example, Brownian coagulation and deposition, but in general such an assumption is unjustified. However, if the densities of the materials involved do not differ substantially, then the error in making the above assumption (using some average density) should not be too large and may well be overshadowed by the other assumptions made such as spherical particles and spatial homogeneity. For that special case, Eqn. (3.69) reduces to:

$$\frac{\partial}{\partial t} n(v,t) + R(v,t) n(v,t) + \sum_{p=1}^{N} \frac{\partial}{\partial v_p} \left[I_p(v,t) n(v,t) \right]$$

$$= \frac{1}{2} \int_0^\infty du \int_0^\infty dw\, n(u,t) n(w,t) K(u|w) \prod_{p=1}^{N} \delta(v_p - u_p - w_p)$$

$$- n(v,t) \int_0^\infty du\, K(u|v) n(u,t) + S(v,t) \tag{3.70}$$

This equation, in a slightly different form, was derived by Simons [1982].

If Eqn. (3.70) is multiplied by:

$$\delta\left(v - \sum_{p=1}^{N} v_p \right)$$

and integrated over v, it may be shown that it reduces to:

Aerosol Science: Theory and Practice

$$\frac{\partial}{\partial t}N(v,t) + R(v,t)\,N(v,t) + \frac{\partial}{\partial v}[I(v,t)\,N(v,t)]$$
$$= \tfrac{1}{2}\int_0^v du\, K(u|v-u)\,N(u,t)\,N(v-u,t) - N(v,t)\int_0^\infty du\, K(u|v)\,N(u,t) + S(v,t) \quad (3.71)$$

where:

$$I(v,t) = \sum_{p=1}^{N} I_p(v,t)$$

Eqn. (3.71) is the standard equation for the aerosol particle volume distribution regardless of species distribution.

If the distribution of mass and volume is desired, we must return to Eqn. (3.69) and define:

$$N(v,m,t) = \int dv \int dm\, n(v,m,t)\,\delta\!\left(v - \sum_{p=1}^{N} v_p\right)\delta\!\left(m - \sum_{p=1}^{N} m_p\right) \quad (3.72)$$

Then, after averaging over the above delta functions, we find:

$$\frac{\partial}{\partial t}N(v,m,t) + R(v,m,t)\,N(v,m,t) + \frac{\partial}{\partial v}[I(v,m,t)\,N(v,m,t)]$$
$$= \tfrac{1}{2}\int_0^v du \int_0^m ds\, N(u,s,t)\,N(v-u,m-s,t)\,K(u,s|v-u,m-s)$$
$$- N(v,m,t)\int_0^\infty du \int_0^\infty ds\, K(u,s|v,m)\,N(u,s,t) + S(v,m,t) \quad (3.73)$$

Details of this manipulation may be found in Simons [1982].

3.8 Two species equations

As an example of the general equation, let us assume that we have a two species aerosol, which can be described by the distribution function, $n(v_1,v_2,m_1,m_2,t)$. From Eqn. (3.69) we find:

$$\frac{\partial}{\partial t}n(v_1,v_2,m_1,m_2,t) + R(v_1,v_2,m_1,m_2,t)\,n(v_1,v_2,m_1,m_2,t)$$
$$+ \frac{\partial}{\partial v_1}[I_1(v_1,v_2)\,n(v_1,v_2,m_1,m_2,t)] + \frac{\partial}{\partial v_2}[I_2(v_1,v_2)\,n(v_1,v_2,m_1,m_2,t)]$$
$$= \tfrac{1}{2}\int_0^{v_1}du_1 \int_0^{v_2}du_2 \int_0^{m_1}dm_1' \int_0^{m_2}dm_2'\, n(u_1,u_2,m_1',m_2',t)\,n(v_1-u_1,v_2-u_2,m_1-m_1',m_2-m_2',t)$$
$$K(u_1+u_2,m_1'+m_2'|v_1+v_2-u_1-u_2,m_1+m_2-m_1'-m_2')$$
$$- n(v_1,v_2,m_1,m_2,t)\int_0^\infty du_1 \int_0^\infty du_2 \int_0^\infty dm_1' \int_0^\infty dm_2'$$
$$K(u_1+u_2,m_1'+m_2'|v_1+v_2,m_1+m_2)\,n(u_1,u_2,m_1',m_2',t) + S(v_1,v_2,m_1,m_2,t) \quad (3.74)$$

The Dynamic Equation for the Aerosol Distribution

As an example of where such a situation might be met, let us assume that we have gravitational settling and deposition. In this case:

$$R(v,m) = \frac{U_s}{H}$$

where the magnitude of the Stokes velocity, U_s, (see Section 2.9) is:

$$U_s = \frac{m g}{6\pi \mu a}$$

Assuming equivalent spherical particles, we can write this as:

$$U_s = \left(\frac{4\pi}{3}\right)^{1/3} \frac{m g}{6\pi \mu v^{1/3}} \tag{3.75}$$

where $m = m_1 + m_2$ and $v = v_1 + v_2$.

The corresponding coagulation kernel is (see Section 4.2):

$$K(u,m|v,m') = \left(\frac{3}{4\pi}\right)^{1/3} \frac{g}{6\mu} \left(u^{1/3} + v^{1/3}\right)^2 \left|\frac{m}{u^{1/3}} - \frac{m'}{v^{1/3}}\right| \tag{3.76}$$

The physical nature of the particles is dependent upon the source term. Thus, for example, if there is no continuous source but rather an initial burst, for a two component aerosol this would take the form:

$$n(v_1,v_2,m_1,m_2,0) = Q_1 \,\delta(v_1 - v_{10})\,\delta(v_2)\,\delta(m_1 - m_{10})\,\delta(m_2)$$
$$+ Q_2 \,\delta(v_2 - v_{20})\,\delta(v_1)\,\delta(m_2 - m_{20})\,\delta(m_1) \tag{3.77}$$

Then the total number of particles injected will be:

$$N_0 = \int dv_1 \int dv_2 \int dm_1 \int dm_2 \, n(v_1,v_2,m_1,m_2,0) = Q_1 + Q_2$$

The total volume of particulate is:

$$\phi_0 = \int dv_1 \int dv_2 \,(v_1 + v_2) \int dm_1 \int dm_2 \, n(v_1,v_2,m_1,m_2,0) = Q_1 v_{10} + Q_2 v_{20}$$

and the total mass of particulate is:

$$M_0 = \int dv_1 \int dv_2 \int dm_1 \int dm_2 \,(m_1 + m_2)\, n(v_1,v_2,m_1,m_2,0) = Q_1 m_{10} + Q_2 m_{20}$$

Therefore, the average volume of a particle is ϕ_0/N_0, the average mass is M_0/N_0, and the average density is M_0/ϕ_0.

There is some practical interest in knowing the volume of a species p in an aerosol particle of total volume, v, and total mass, m. For example, if one of the species is radioactive, then clearly the total activity will be related to the fraction of this species residing in an aerosol particle.

Let $M_p(v,m,t)$ be the volume of species p residing in an aerosol particle of total volume, v, and total mass, m. Then, by definition:

$$M_p(v,m,t) = \int dm\, \delta\left(m - \sum_{p=1}^{N} m_p\right) \int dv\, v_p\, \delta\left(v - \sum_{p=1}^{N} v_p\right) n(v,m,t) \qquad (3.78)$$

Operating on Eqn. (3.69) with the two delta functions and integrating leads, after following the procedure described by Simons [1982], to the following equation:

$$\frac{\partial}{\partial t}M_p(v,m,t) + R(v,m)\, M_p(v,m,t) + \frac{\partial}{\partial v}\left[I(v,m,t)\, M_p(v,m,t)\right]$$

$$= \int_0^m ds \int_0^v du\, K(u, v-u|s, m-s)\, M_p(u,s,t)\, N(v-u, m-s, t)$$

$$- M_p(v,m,t) \int_0^\infty ds \int_0^\infty du\, K(u, v|s, m)\, N(u,s,t) + I_p(v,m,t)\, N(v,m,t) + Q_p(v,m,t) \qquad (3.79)$$

where $N(v,m,t)$ is the solution of Eqn. (3.73).

This equation is interesting from several points of view but mostly because it is linear in the unknown variable, M_p. To be sure, it is a complicated integro-differential equation but its linearity opens up a range of powerful analytical and numerical techniques for its solution.

In the special case when K, R, and I are independent of m, or m can be related to v, we may integrate over m to obtain the reduced distribution:

$$\frac{\partial}{\partial t}M_p(v,t) + R(v)\, M_p(v,t) + \frac{\partial}{\partial v}\left[I(v,t)\, M_p(v,t)\right]$$

$$= \int_0^v du\, K(u, v-u)\, M_p(u,t)\, N(v-u, t) - M_p(v,t) \int_0^\infty du\, K(u,v)\, N(u,t)$$

$$+ I_p(v,t)\, N(v,t) + Q_p(v,t) \qquad (3.80)$$

Such an equation was derived by Simons [1982] who also suggested a practical way to modify it when the particles are composition dependent.

3.9 Radioactive aerosols

Eqn. (3.80) is particularly useful if one or more of the species is radioactive, for then the activity due to these rather special components can be isolated. Some very useful work in this area has been presented by Simons [1981, 1982] whose purpose was to estimate the error in the conventional method for calculating airborne radioactivity. Suppose, for example, that in an accident in a liquid metal cooled fast reactor there is an aerosol produced consisting of a mixture of sodium oxide particles (size ≈ 0.5 μm) and radioactive fuel particles (≈ 0.025 μm). The established procedure for dealing with such aerosols is to assume that the radioactive

The Dynamic Equation for the Aerosol Distribution

particles are rapidly captured by the larger sodium particles and that the subsequent history is for a 'single species' aerosol with the constituents given by the initial concentration. Thus, at any given time after the initial burst, the airborne activity is proportional to the mass of aerosol remaining. However, this neglects the fact that each species will coagulate and deposit differently. For example, the small fuel particles will undergo rapid Brownian motion and deposit on the surface of the sodium particles and hence the airborne radioactivity at least initially will depend on the surface area of particles rather than their volume. Eventually, as time proceeds, the formation of larger particles by the coalescence of these smaller surface activated components, leads to greater volume mixing. To put it otherwise, as the aerosol ages, the activity remaining in suspension changes from a surface area effect to a volume effect. Since the mass deposition mechanism is generally due to gravitational settling, which removes larger particles preferentially, the total airborne activity will be underestimated if it is assumed to depend on the total volume of particles suspended. In view of these complications, it is clearly necessary to deal properly with the two species described above by means of an extended distribution function, $n(v,s,t)$. Here, v is the total volume of the particle and s is the activity (in appropriate units) residing in that volume. Assuming that the radioactivity does not affect the coagulation or, more precisely, that the volume of the radioactive material is small compared with the nonradioactive such that $K(u,v)$ depends only on the total volume of the particle, the equation describing the situation is:

$$\frac{\partial}{\partial t} n(v,s,t) + R(v) n(v,s,t) = \tfrac{1}{2} \int_0^v du \int_0^s dr\, K(u,v-u)\, n(u,r,t)\, n(v-u,s-r,t)$$

$$- n(v,s,t) \int_0^\infty du \int_0^\infty dr\, K(u,v)\, n(u,r,t) + Q(v,s,t) \tag{3.81}$$

It should be noted that $n(v,s,t)$ is not precisely analogous to the $n(v_1,v_2,t)$ defined earlier. This is because in that definition the total volume, $v = v_1 + v_2$, whereas in the function, $n(v,s,t)$, v is the total volume and s is another property which does not affect v but which is still conserved in a collision, i.e. radioactivity. In this sense, then, Eqn. (3.81) is much simpler than Eqn. (3.73). We can define a kth moment of $n(v,s,t)$ as follows:

$$N_k(v,t) = \int_0^\infty ds\, s^k\, n(v,s,t) \tag{3.82}$$

Then $N_0(v,t)dv$ is the total number of particles present with volumes in the range, v to $v+dv$, and $N_1(v,t)dv$ is the total radioactivity of particles with volumes in the range, v to $v+dv$. If we set $N_0 = N$ and $N_1 = M$, we can readily find the equations that they obey. For $N(v,t)$ we find:

$$\frac{\partial}{\partial t} N(v,t) + R(v) N(v,t) = \tfrac{1}{2} \int_0^v du\, K(u,v-u)\, N(u,t)\, N(v-u,t)$$

$$-N(v,t) \int_0^\infty du\, K(u,v)\, N(u,t) + Q_0(v,t) \tag{3.83}$$

where Q_0 is the source of particles. For $M(v,t)$ we find:

$$\frac{\partial}{\partial t} M(v,t) + R(v) M(v,t) = \int_0^v du\, K(u,v-u)\, M(u,t)\, N(v-u,t)$$

$$-M(v,t) \int_0^\infty du\, K(u,v)\, N(u,t) + Q_1(v,t) \tag{3.84}$$

where Q_1 is the source of radioactivity. Higher moments than these may be generated if variances in the amount of radioactivity are required.

Eqns. (3.83) and (3.84) form a powerful tool for studying the effects of radioactive contamination and we shall discuss results and solutions in a later section. Other papers relevant to the problem of multispecies aerosols are Lushnikov [1976], Gelbard and Seinfeld [1978, 1980], and Friedlander [1970, 1971].

3.10 The effect of electrical charge on the dynamic equation

The importance of electrical charge on aerosol particles in determining their collision frequency was first recognized by meterologists. In such investigations, it was observed that atmospheric ions formed by cosmic rays or other means would diffuse onto cloud droplets and electrify them. An analogous situation would exist in the strong radiation field which would be present in the interior of the containment vessel of a nuclear reactor following an accidental release of radioactive material. The high ionization, due mainly to α and β particles, would lead to strong ionization of the background gas. There would exist, therefore, positive ions and electrons. However, the electrons because of their greater mobility will not only recombine with positive ions but also attach themselves to neutral atoms forming negative ions. The positive and negative ions and the electrons will diffuse onto the aerosol particles leading some to be positively and others negatively charged. In addition many aerosol particles will themselves be radioactive and it is necessary to calculate the residual charge on them by virtue of this activity.

The continuous creation of charged species by the background radiation and their destruction by recombination suggests that an equilibrium charge distribution will be established in the gas and hence on the aerosols themselves. Keefe et al. [1959] have applied Boltzmann's statistics to this problem and argue that the number of particles per unit volume having an energy E is:

$$N(E) = A \exp\left(-\frac{E}{kT}\right)$$

For a charged spherical particle of radius, a, and charge, $\pm qe$ (q is an integer), we can write:

$$E = E_0 + \frac{q^2 e^2}{2a}$$

where E_0 is the energy of the particle in the absence of charge and the second term is the electrostatic energy. Thus, if N_q is the number of particles having q unit charges, we may write:

$$N_q = N_0 \exp\left(-\frac{q^2 e^2}{2akT}\right)$$

Thus, the fraction having charge, q, is:

$$f(q) = \frac{\exp\left(-\dfrac{q^2 e^2}{2akT}\right)}{\displaystyle\sum_{q=-\infty}^{\infty} \exp\left(-\dfrac{q^2 e^2}{2akT}\right)}$$

The Dynamic Equation for the Aerosol Distribution

for which a reasonable approximation is:

$$f(q) \approx \left(\frac{e^2}{2\pi a k T}\right)^{1/2} \exp\left(-\frac{q^2 e^2}{2 a k T}\right) \tag{3.85}$$

This leads to an average number of unit charges per particle being:

$$\bar{q} = \int_{-\infty}^{\infty} |q| f(q) \, dq = \left(\frac{2 a k T}{\pi e^2}\right)^{1/2} \tag{3.86}$$

For example, the average number of unit charges on an aerosol particle of 0.25 μm radius, at T=293 K, is (1.67). In general, the number of charges increases as the square root of the radius.

Gunn [1955] has also made calculations of the charge distribution and finds that if more careful account is taken of the mobilities of the positive and negative ions the distribution is slightly skewed as follows:

$$f_G(q) = \left(\frac{e^2}{2\pi a k T}\right)^{1/2} \exp\left(-\frac{e^2}{2 a k T}\left[q - \frac{a k T}{e^2} \ln\left(\frac{\lambda_+}{\lambda_-}\right)\right]^2\right) \tag{3.87}$$

where λ_+ and λ_- are the electrical conductivities of the positive and negative ions, respectively. Experimental work carried out on cloud droplets confirms this asymmetry with an excess of 2.2 e being observed on droplets of 1.15 μm radius.

It should be stressed that the equilibrium charge distribution takes time to become established following the creation of the ions. This can be a fairly complex process and has been investigated by a number of authors (Flanagan and O'Connor [1961], Gunn [1954], and Cooper and Reist [1973]). The work of Flanagan and O'Connor leads to an expression for the time required for small aerosol particles (Aitken nuclei) to come into equilibrium with bipolar ions. According to Parker and Reist, however, Flanagan and O'Connor's procedure is not helpful for studying the approach to charge neutralization because the recombination terms are only appropriate for aerosols consisting of singly charged or uncharged particles. Thus, Cooper and Reist make use of Gunn's theory and some techniques of their own, both of which we will briefly describe.

Gunn's procedure assumes a stationary spherical particle (radius <5 μm) to which ions are diffusing in the presence of the electric field arising from the charges present on the particle. It is found that the free charge $Q(t)$ at time t is given in terms of the equilibrium charge Q_∞ as:

$$Q(t) = Q_\infty \left[1 - \exp\left[-4\pi e (N_+ B_+ + N_- B_-) t\right]\right] \tag{3.88}$$

where N_+ and N_- are the ion concentrations of positive and negative ions in the environment and B_+ and B_- are their corresponding mobilities. The equilibrium charge Q_∞ is seen from Eqn. (3.87) to be:

$$Q_\infty = \frac{a k T}{e^2} \ln\left(\frac{\lambda_+}{\lambda_-}\right) \tag{3.89}$$

where $\lambda_+ = N_+B_+$ and $\lambda_- = N_-B_-$. Q_∞ is to be distinguished from the average charge, q, defined in Eqn. (3.86). Q_∞ is interpreted as the most probable charge on a particle.

An important characteristic is the time constant:

$$\tau = \frac{1}{4\pi (N_+ B_+ + N_- B_-) e} \tag{3.90}$$

which, according to Cooper and Reist [1973], has a value $\tau = 0.28 \times 10^6 / N$ sec where N is the number of ions cm^{-3}. We would like τ to be short compared with any characteristic time constant of the overall aerosol behavior, e.g. settling and coagulation times, since otherwise it will be necessary to consider a time dependent coagulation process in which the charged state affects agglomeration. In order to avoid this, it is important to obtain a value for the average ion concentration. Cooper and Reist [1973] have made extensive studies of this problem. The balance equation used for the ion concentration is:

$$\frac{\partial N}{\partial t} + v \cdot \nabla N = 2Q - \frac{\alpha}{2} N^2 + D \nabla^2 N \tag{3.91}$$

Equal mobilities are assumed for the positive and negative ions although more detailed calculations could allow for the differences. In Eqn. (3.91) the second term on the left hand side is the drift term due to any mass motion arising from applied forces (e.g., electrical or gravitational). The first term on the right is the rate of production of ion pairs by the radioactive source, Q being the number of ions produced per unit volume per unit time. These ions are subject to recombination and the term $\alpha N^2/2$ accounts for this, with α being the recombination coefficient. Finally, the term $D\nabla^2 N$ allows for diffusion. The main thrust of the work of Cooper and Reist is to obtain approximate solutions of Eqn. (3.91). To do this they note that the critical parameter is the range, R, of the ionizing radiation. If the size of the system, L, is very much less than the range, R, then the spatial variation of N may be neglected and we can write:

$$\frac{dN}{dt} = 2Q - \frac{\alpha}{2} N^2 \tag{3.92}$$

This equation leads to an equilibrium ion density of:

$$N_\infty = 2 \left(\frac{Q}{\alpha} \right)^{1/2} \tag{3.93}$$

which is reached in a time $>> 1/(\alpha Q)^{1/2}$.

A second situation that can occur is when L>>R. Then, the space may be divided into two regions; r<R where the behavior is as described by Eqn. (3.92) and r>R where there is no production. In that case, it would be necessary to solve the equation:

$$\alpha N_s^2 = 2 D \nabla^2 N_s \tag{3.94}$$

where we have assumed a steady state and N_s refers to the region r>R. The boundary condition at r=R would be $N_s = N_\infty$. To obtain an estimate of how the ion density behaves in the region r>R, we assume a planar geometry with $N_s = N_\infty$ at x=0. Then, Eqn. (3.94) can be solved exactly to give:

The Dynamic Equation for the Aerosol Distribution

$$N_s(x) = N_\infty \left[1 + \left(\frac{\alpha N_\infty}{12 D}\right)^{1/2} x\right]^{-2} \tag{3.95}$$

Thus, at distances:

$$x \gg \left(\frac{12 D}{\alpha N_\infty}\right)^{1/2} = \left(\frac{6 D}{(\alpha Q)^{1/2}}\right)^{1/2} \tag{3.96}$$

the ion density will be negligible.

It remains now to calculate Q and R. Radioactive materials produce ionization through the emission of α and β particles and γ rays. The range of γ rays in air is so large (~100 m) that they produce very little ionization in containers. However, α and β particles have much smaller ranges and are major sources of ionization. Various range-energy relationships exist for charged particles. For example, Morgan and Turner [1967] give the range of α particles as:

$$\begin{aligned} R_\alpha \text{ (cm)} &= (0.56) \, E \, (\text{Mev}) & &; \quad E < 4 \text{ Mev} \\ &= (1.24) \, E - (2.62) & &; \quad 4 < E < 8 \text{ Mev} \end{aligned} \tag{3.97}$$

Since it takes 35.5 ev to produce an ion pair in air at STP, we may obtain the number of ion pairs per cm of range. For example, a 1 Mev α particle has a range of 0.56 cm and will produce 2.8×10^4 ion pairs or about 5×10^4 ion pairs per cm. The specific ionization is therefore $k_\alpha = (E/35.5)/R_\alpha$ ion pairs per cm per α.

For β-particles the process is more complicated since they do not have well defined ranges being emitted with a spectrum of energies. Instead, each emitter has a spectrum of ranges. However, the energy flux from a β source of strength 1 MeV can be represented fairly well by the relation:

$$I(x) = I_0 \exp(-\mu x)$$

The linear absorption coefficient, μ, is given by:

$$\mu \, (\text{cm}^{-1}) = 17 \, \rho \, E_{\max}^{-1.14}$$

where ρ is the air density and $E_{\max}$ is the maximum β energy. The energy dissipation per unit time per unit length is:

$$\frac{dI}{dx} = -\mu I = -\mu I_0 \exp(-\mu x)$$

Now, $I_0 = F E_{av}$, where F is the number of β particles emitted from the source per second per unit area, and E_{av} is the average energy. The β spectrum is such that $E_{av} = E_{\max}/3$, thus:

$$\frac{dI}{dx} = -\tfrac{17}{3} \rho E_{\max}^{-0.14} F \exp(-\mu x)$$

The specific ionization (ions produced per unit length per particle) is, therefore:

$$k_\beta(x) = \tfrac{17}{3} \rho \left[E_{max}^{-0.14} / (34 \text{ ev}) \right] \exp(-\mu x)$$
$$= 210 \, E_{max}^{-0.14} \exp(-\mu x) \left(\text{ion pairs cm}^{-1} \text{ in air} \right) \qquad (3.98)$$

where $r = 1.26 \times 10^{-3}$ gm cm^{-3} and 34 ev is expressed as 3.4×10^{-5} Mev. As an example, β particles from Kr-85, where $E_{max} = 0.67$ Mev, lead to $k_\beta = 222$ ion pairs cm^{-1} at $x = 0$.

In the case of α particles, which we take to have a well defined range, R_α, the total number of ions produced per unit volume per second is simply $Q = k_\alpha F$, where F is the number of α particles produced per unit area per unit time by the surface source. On the other hand, for β particles, there is an exponential decay of intensity and so the rate of ion creation also varies with distance from the source. Thus, the rate of creation of ions per unit volume per unit time at x is:

$$Q(x) = k_\beta(0) \exp(-\mu x) F \qquad (3.99)$$

For the situation prevailing in a reactor accident, we would also have volume sources arising from the particulate in suspension. If the α activity is uniformly distributed and is S_α disintegrations per unit volume per second, then the number of ion pairs produced per α is:

$$S_\alpha k_\alpha R_\alpha = S_\alpha E_\alpha / 35.5 \text{ (ev)}$$

Similarly, the number of ion pairs produced by β particles is $S_\beta E_{max} / 3 / 34$ (ev) per unit volume per second.

As far as the time constants are concerned, let us assume an activity of 1 Ci cm^{-3} of aerosol. For 1 Mev α particles this leads to:

$$Q_\alpha = 3.7 \times 10^{10} \times 10^6 / 35.5 = 10^{15} \text{ ion pairs cm}^{-3}\text{sec}^{-1}$$

Now the recombination coefficient, α, for ion pairs in air at STP is 3×10^{-6} cm^{3}sec^{-1}. Thus, from Eqn. (3.93), $N_\infty = 3.65 \times 10^{10}$ cm^{-3} and the time constant for ion equilibrium is $1/(\alpha Q)^{1/2} = 1.8 \times 10^{-5}$ sec. Using N_∞, we see from Eqn. (3.90) that $\tau = 7.7 \times 10^{-6}$ sec. Thus, the ions and aerosol charge rapidly come into equilibrium. Further work in this area may be found in Hoppel [1985] and Hoppel and Frick [1986] and also in the review by Whitby and Liu (Davies [1966]).

3.11 The effect of charge on coagulation

It is clear that charge will influence the way in which aerosol particles interact with each other. For example, unipolar charging (*i.e.* like charges) will surely lead to a reduced interaction and hence less coagulation. On the other hand, bipolar charging (unlike charges) will lead to enhanced coagulation. The problem remains of how to formulate the equation for such processes. We shall do this from an elementary point of view initially and then generalize the theory.

In order to calculate the degree of coagulation that takes place, we need to know the force between two charged spheres. This is not as simple as it seems at first. If the two spheres have the same sign of charge they will generally repel each other, but as they approach closer, the two particles, which are of finite size, induce a charge of opposite polarity so that a force of attraction is formed. When the particles are far apart this force is negligible but it may dominate

The Dynamic Equation for the Aerosol Distribution

the situation when they are very close together. It was shown by Russell [1922] that the force between two particles of equal radius, a, and interparticle distance, r, is given by:

$$F(r) = \left(1 + \frac{15\,a^6}{r^6} + \ldots\right)\frac{q_1 q_2}{r^2} - \left(\frac{2\,a^3}{r^3} + \frac{3\,a^5}{r^5} + \frac{4\,a^7}{r^7} + \ldots\right)\left(\frac{q_1^2 + q_2^2}{r^2}\right) \tag{3.100}$$

For r>>a, we regain the inverse square law but for sufficiently small separation the force becomes attractive. However, only by an exceptionally large difference in the magnitudes of the charges on each particle will the attractive force be significant, and this feature is generally neglected in the calculation of coagulation. Thus, we can assume that the particles obey the usual Coulomb's law and ignore the image force. However, for more details on image forces the reader should consult Keefe et al. [1959, 1961], Junge [1955], and Jacobi [1961].

In order to illustrate the effects of charge, we consider how Brownian motion is modified and recalculate the Brownian kernel. Consider two particles of radii, a_1 and a_2, which undergo Brownian diffusion but are subject to a force which acts between their centers. Then from Eqn. (4.273) we see that the rate of capture of particles 2 by test particle 1 is:

$$J_{12} = \frac{4\pi (D_1 + D_2) N_0}{\int_{a_1+a_2}^{\infty} \frac{dr}{r^2} \exp\left\{\frac{\Phi(r)}{kT}\right\}} \tag{3.101}$$

where N_0 is the unperturbed density of particles per unit volume. The coagulation kernel, $K(a_1,a_2)=J_{12}/N_0$, and can be written:

$$K(a_1, a_2) = 4\pi (D_1 + D_2)(a_1 + a_2)\,p \tag{3.102}$$

where:

$$\frac{1}{p} = \int_0^1 dx \exp\left\{\frac{1}{kT}\Phi\left(\frac{a_1 + a_2}{x}\right)\right\}$$

Thus, p includes the effect of the interparticle potential. When:

$$\Phi(r) = \frac{\nu\mu e^2}{r}$$

where ν and μ are integers representing the number of charges on each particle, we find:

$$p = \left(\frac{S_{\nu\mu}}{a_{12}}\right)\left(\exp\left(\frac{S_{\nu\mu}}{a_{12}}\right) - 1\right)^{-1} \tag{3.103}$$

where $S_{\nu\mu}=\nu\mu e^2/kT$ and $a_{12}=a_1+a_2$.

We are now in a position to formulate a balance equation for the charged aerosol particles. Let $n_+(t)$ be the total number of aerosol particles with the average positive charge and $n_-(t)$ the total number with the average negative charge, q. Then:

$$\frac{dn_+}{dt} = -\tfrac{1}{2} K_B \left\{n_+^2 p_s + n_- n_+ p_a\right\} + \left(\frac{\partial n_+}{\partial t}\right)_{\text{drift}} \tag{3.104}$$

and:

$$\frac{dn_-}{dt} = -\tfrac{1}{2} K_B \{n_-^2 p_s + n_- n_+ p_a\} + \left(\frac{\partial n_-}{\partial t}\right)_{drift} \qquad (3.105)$$

Here we have used the approximate Brownian kernel, $K_B = 16\pi Da$, where D and a are average diffusion coefficients and particle radii. Also, we have used an average value of p such that, for unlike charges:

$$p_a = \left(\frac{|S_{v\mu}|}{a_{12}}\right)\left(1 - \exp\left(-\frac{|S_{v\mu}|}{a_{12}}\right)\right)^{-1} = \frac{\lambda}{1 - \exp(-\lambda)} \qquad (3.106)$$

and for like charges:

$$p_s = \left(\frac{|S_{v\mu}|}{a_{12}}\right)\left(\exp\left(\frac{|S_{v\mu}|}{a_{12}}\right) - 1\right)^{-1} = \frac{\lambda}{\exp(\lambda) - 1} \qquad (3.107)$$

where $\lambda = \bar{q}^2 / 2akT$.

In view of the fact that the average charge on a particle is, from Eqn. (3.86), equal to $(2akT/\pi)^{1/2}$ we see that:

$$\lambda = \frac{2}{\pi} \frac{\sqrt{a_1 a_2}}{a_1 + a_2}$$

which has a maximum value of $1/\pi$ when $a_1 = a_2$. Thus, the extremum values of p_s and p_a are $p_{s,max} = (0.849)$ and $p_{a,max} = (1.168)$. The overall effect on coagulation is therefore relatively small.

The terms denoted by $(\partial n_\pm / \partial t)_{drift}$ refer to the movement of electrical charges due to electrostatic dispersion. Such dispersion occurs if the positive and negative charges in a unit volume do not balance. Then, because of the forces of repulsion between unipolar charged particles, they recede from each other with a velocity:

$$V = B \bar{E} \bar{q} \qquad (3.108)$$

where $\bar{E}$ is the field due to the charged particles and B is their mobility. It is recalled that $B = D/kT$. Thus, the flux of particles out of either the positive or negative population is:

$$\left(\frac{\partial n_+}{\partial t}\right)_{drift} = -\frac{D}{kT} \bar{q} n_+ \nabla \cdot E \qquad (3.109)$$

or:

$$\left(\frac{\partial n_-}{\partial t}\right)_{drift} = \frac{D}{kT} \bar{q} n_- \nabla \cdot E \qquad (3.110)$$

But, from Poisson's equation:

$$\nabla \cdot E = 4\pi Q$$

The Dynamic Equation for the Aerosol Distribution

where Q is the total charge per unit volume. Clearly:

$$Q = \bar{q}(n_+ - n_-) \tag{3.111}$$

and so:

$$\left(\frac{\partial n_+}{\partial t}\right)_{drift} = -4\pi \frac{D}{kT} \bar{q}^2 n_+ (n_+ - n_-) \tag{3.112}$$

and:

$$\left(\frac{\partial n_-}{\partial t}\right)_{drift} = 4\pi \frac{D}{kT} \bar{q}^2 n_- (n_+ - n_-) \tag{3.113}$$

The equation for the positive and negative particle populations are therefore:

$$\frac{\partial n_+}{\partial t} = -\tfrac{1}{2} K_B \{n_+^2 P_s + n_- n_+ P_a\} - 4\pi \frac{D}{kT} \bar{q}^2 n_+ (n_+ - n_-) \tag{3.114}$$

and:

$$\frac{\partial n_-}{\partial t} = -\tfrac{1}{2} K_B \{n_-^2 P_s + n_- n_+ P_a\} + 4\pi \frac{D}{kT} \bar{q}^2 n_- (n_+ - n_-) \tag{3.115}$$

In the case of unipolar charging, we can set $n_- = 0$ and $n_+ = n$, whence:

$$\frac{dn}{dt} = -\tfrac{1}{2} K_B n^2 P_s - 4\pi \frac{D}{kT} \bar{q}^2 n^2 = -8\pi D a n^2 \frac{\lambda e^\lambda}{e^\lambda - 1} \tag{3.116}$$

Since the factor involving λ is greater than unity, it is clear that unipolar charging leads to an increase in the removal rate of particles. This rather curious result stems from the effect of the electrostatic dispersion. In fact, if the charge correction term is separated into two parts:

$$\frac{\lambda}{e^\lambda - 1} + \lambda$$

then the first term is due to coagulation, which is less than unity, and the second term arises from electrostatic dispersion. These effects have been observed experimentally by Whytlaw-Gray and Patterson [1932].

Bipolar charging leads to a somewhat more complicated situation. To obtain an estimate of its effect, we add Eqns. (3.114) and (3.115) to get:

$$\frac{dn}{dt} = -8\pi D a n^2 \left\{ \frac{\lambda \exp(\lambda)}{\exp(\lambda) - 1} - \frac{2\lambda n_+ n_-}{n^2} \right\} \tag{3.117}$$

where $n = n_+ + n_-$. Now, if we set $n_+ = n_- = n/2$, then the enhancement factor in the curly brackets becomes a minimum and is equal to:

$$\frac{\lambda}{2} \frac{e^\lambda + 1}{e^\lambda - 1} \tag{3.118}$$

This factor is certainly greater than unity which indicates that symmetrical bipolar systems always increase the coagulation rate, although the effect is much less than that due to unipolar charging. An estimate of the enhancement factor can be obtained for a system which is in charge equilibrium, for then, we know from Eqn. (3.86) that $\lambda=1/\pi$. Hence, the enhancement factor is (1.007), which is a very small effect.

The major shortcoming of the above analysis is that it assumes nearly equal particles and does not allow for any statistical asymmetry of charge. Indeed, work carried out by Gillespie [1953] indicates that electric charge may have a significant effect on coagulation and surface loss. In his experiments, which employed porous silica powder and magnesium oxide, Gillespie was able to determine the distribution of charge which he found to be highly symmetrical (as expected from the equilibrium theory discussed above). Because the experiments were carried out in a finite container, the surface effects were also considered. Thus, the particle balance equation was written:

$$\frac{dn}{dt} = -k n^2 - \beta n$$

where k is the coagulation constant and β the surface loss constant. Some of the experimental results are shown in Figs. 3.5 and 3.6.

We see that the coagulation constant k is a function of time for three different initial conditions. These initial conditions correspond to different pressure air blasts used to charge the particles and therefore denote varying amounts of initial charge. It is clear that significant deviations from the uncharged state are present, but what is not clear is how much this difference depends on the various coagulation mechanisms present. For example, Gillespie includes in his analysis Brownian, gravitational, and shear coagulation and, in so far as his 'average-size' approximation will allow, writes:

$$k = \gamma k_0 \left(1 - \tfrac{1}{2} f_q^2\right) + A f_q^2 \bar{q}^2 \left(\frac{1}{a}\right)_{av}$$

where:

$$k_0 = \tfrac{4}{3} \frac{kT}{\mu} + 2\pi V_0 \bar{a}^2 + \tfrac{32}{3} \omega \bar{a}^3$$

In this formula, f_q denotes the fraction of particles which are charged and k_0 is the neutral coagulation kernel for Brownian, gravitational, and shear mechanisms, suitably averaged. V_0 is the sedimentation velocity differential and ω is the rate of shear. γ is a factor which describes variations in the rate of coagulation due to size distribution effects and possibly van der Waals forces.

A is a constant related to charge distribution:

$$A = \frac{\gamma}{3\mu} \left[1 + \frac{4}{\pi} + \left(1 - \exp\left\{-\frac{\bar{q}^2}{2kT}\left(\frac{1}{a}\right)_{av}\right\}\right)^{-1}\right]$$

Experiment shows that a plot of $k - k_0 \left(1 - \tfrac{1}{2} f_q^2\right)$ versus $\bar{q}^2 f_q^2 (1/a)_{av}$ leads to approximately a straight line. A is therefore a linear function of $\bar{q}^2$.

We have discussed this work for two reasons. First, it was a pioneering effort to examine charge effects and second, the theory is so rudimentary, compared with modern formulations, that one wonders whether the interpretation is correct. For this reason we will discuss below a consistent approach to the dynamic equation for charged aerosols.

The Dynamic Equation for the Aerosol Distribution

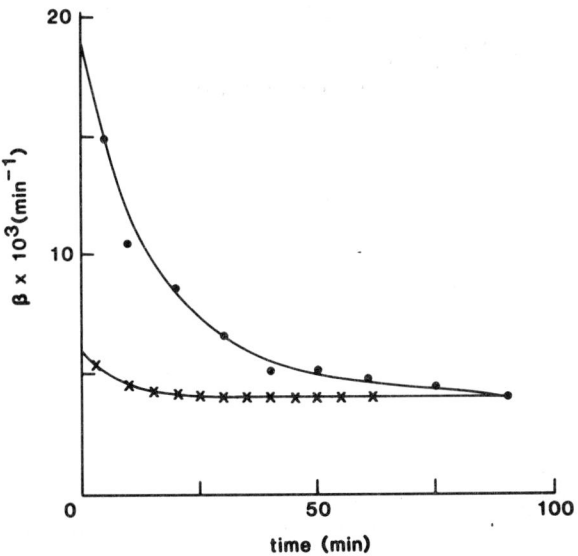

Figure 3.5: Variation of the surface loss constant with time. × and ● refer to Vycor aerosols produced by air blasts with pressures of 10 and 50 lb./in.², respectively. Adapted from Gillespie [1953] with permission of The Royal Society.

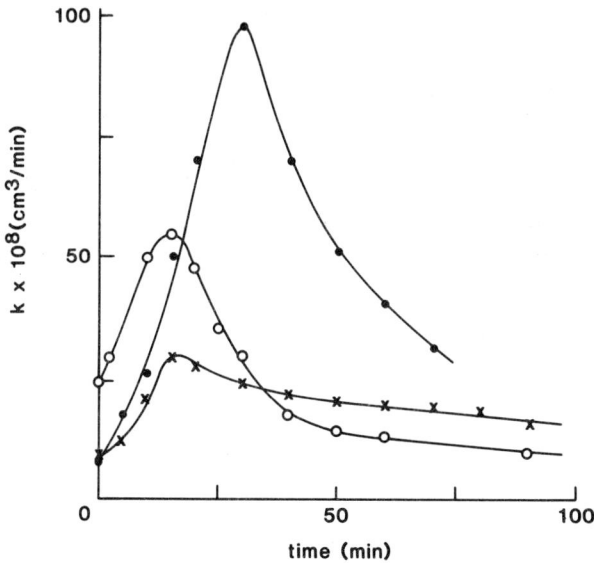

Figure 3.6: Variation of the coagulation constant with time. ×, O and ● refer to Vycor aerosols produced by air blasts with pressures of 10, 25 and 50 lb./in.², respectively. Redrawn from Gillespie [1953] with permission of The Royal Society.

In closing this section we mention some other experimental results that have been performed on the coagulation and scavenging of radioactive aerosols. That is the particles are not being charged simply from external radiation generating ions, but because they themselves are radioactive. This does, of course, lead to ionization of the surrounding air but there is the added complication of the intrinsic activity. This work was carried out by Rosinski et al. [1962] and used gold aerosols from exploding wires. Results were only measured over the time range for which Brownian motion was important. However, the deposition on the walls of the vessel was important. Rosinski et al. used the following equation to explain their results:

$$\frac{dn}{dt} = -kn^2 - \beta(t)n$$

where the wall surface loss time constant $\beta(t)$ was set equal to a−bt. This form was chosen to account for the change in particle size during coagulation and hence in the deposition rate, which is proportional to 1/(particle volume)$^{1/3}$. It was observed that $\beta(t)$ eventually became negligible. However, since a=6.86×10^{-5} sec^{-1} and b=9.53×10^{-9} sec^{-2}, the time scale has to be less than a/b=2 hr.

The conclusions of Rosinski et al. are interesting. For example, the coagulation constant of nonradioactive gold aerosols was found to be about five times larger than that of 'slightly' radioactive gold aerosols (50–900 mCi gm^{-1}). The difference was attributed to the presence of electrostatic charge on the nonradioactive particles. The average charge was calculated to be 0.94 e per particle and k=2.70×10^{-9} cm^{3}sec^{-1}. These authors also assume that there is no net electrostatic charge on radioactive aerosols because the residence time is very short compared to the time between disintegrations.

As the activity of the aerosols increased to around 3 Ci gm^{-1}, an unusual increase in the coagulation constant at early stages of coagulation was observed. This amounted to about twenty times the mean value determined for slightly radioactive systems. It seems, therefore, that the coagulation is enhanced by the presence of the highly ionized gas produced in the neighborhood of the particle. At later stages in the coagulation process, the coagulation constant decreases to the value for nonradioactive aerosols. These results were considered too complex to be explained by Rosinski et al. Further work in the paper on the scavenging of radioactive aerosols led to some interesting conclusions. One of these was that Brownian motion, in the presence of a water vapor concentration gradient around condensing droplets, is a most effective way to remove slightly radioactive aerosols.

Further relevant work in this area has been carried out by Kunkel [1950a,b]. He has studied the interactions of small particles suspended in air with the ions normally produced in air. Equations are established for the charge density as a function of time and, in particular, an exact solution for the equilibrium condition is derived which compares favorably with experiment. One conclusion is that multiply charged particles are present in coarse aerosols at all times, with charges up to ten electron units being not uncommon. The reader is also referred to the articles by Whitby and Liu [1966], and Bricard and Pradel [1966] (both in Davies [1966]).

3.12 The dynamic equation for a charged aerosol

In order to characterize an aerosol distribution with electric charge, we need to define an extended distribution function $n_v(v,t)$. Thus $n_v(v,t)dv$ is the number of aerosol particles with volumes in the range v to v+dv carrying v units of fundamental charge, e. We shall construct such an equation in stages following Zebel [1958b, 1966]. First, we neglect all processes

The Dynamic Equation for the Aerosol Distribution

except those of mutual electrostatic dispersion caused by a distributed array of charges in space. Thus, we can write:

$$\frac{\partial}{\partial t} n_v(v,t) = -\nabla \cdot [\mathbf{V} \, n_v(v,t)] \tag{3.119}$$

where **V** is the velocity with which unipolar electrically charged particles recede from one another. If **E** is the electric field due to the charges on the particles, then this velocity is:

$$\mathbf{V} = B(v) \, v e \mathbf{E} \tag{3.120}$$

where B(v) is the mobility of the particle of volume v, i.e. B=D/kT and ve is the charge on it. Thus, Eqn. (3.118) becomes:

$$\frac{\partial}{\partial t} n_v(v,t) = -v e B(v) \, n_v(v,t) \, \nabla \cdot \mathbf{E} \tag{3.121}$$

But, Poisson's equation is:

$$\nabla \cdot \mathbf{E} = 4\pi Q$$

where Q is the sum of all the particle charges in a unit volume:

$$Q = e \sum_{\mu=-\infty}^{\infty} \mu \int_0^\infty dv \, n_\mu(v,t) \tag{3.122}$$

Thus, Eqn. (3.121) reduces to:

$$\frac{\partial}{\partial t} n_v(v,t) = -4\pi B(v) \, e^2 \, v \, n_v(v,t) \sum_{\mu=-\infty}^{\infty} \mu \int_0^\infty dv \, n_\mu(v,t) \tag{3.123}$$

which is a nonlinear integro-differential equation. If all of the particles have the same size and charge, the equation simplifies to:

$$\frac{dN}{dt} = -4\pi \, B \, q^2 \, N^2 \tag{3.124}$$

The solution of which is:

$$N(t) = \frac{N_0}{1 + (t/\tau)} \tag{3.125}$$

where $\tau = 1/(4\pi B q^2 N_0)$. Thus, even with no coagulation, the electrostatic dispersion leads to a reduction in the number of particles in a unit volume. It should be noted, however, that these particles have to go somewhere and, in any practical situation, they will deposit on the walls.

Another situation that arises is when the particles have some positive and some negative charges. Then we have:

$$\frac{dN_+}{dt} = -4\pi\, B\, q_+\, N_+ \left(q_+ N_+ - q_- N_-\right) \tag{3.126}$$

and:

$$\frac{dN_-}{dt} = 4\pi\, B\, q_-\, N_- \left(q_+ N_+ - q_- N_-\right) \tag{3.127}$$

where N_+ is the number of aerosol particles carrying a charge q_+ and N_- the number carrying a charge q_-. These equations may be solved exactly to give:

$$\tau = \int_{N_+}^{N_{+0}} \frac{d\omega\, \omega^r}{\omega^{r+1}\left(\omega^r - r\,\kappa\right)} \tag{3.128}$$

$$N_- = \frac{\kappa}{N_+^r}$$

where $r = q_-/q_+$, $\kappa = N_{-0}(N_{+0})^r$ and $\tau = 4\pi B q_+^2 t$. If $q_+ = q_- = q$, this simplifies to:

$$N_+(t) = \kappa_0 \frac{N_{+0} + \kappa_0 + (N_{+0} - \kappa_0)\exp(-\tau_0)}{N_{+0} + \kappa_0 - (N_{+0} - \kappa_0)\exp(-\tau_0)}$$

where $\kappa_0^2 = N_{+0}\, N_{-0}$, $\tau_0 = 8\pi\, \kappa_0\, B q^2\, t$, and $N_-(t) = \kappa_0^2 / N_+(t)$. Thus:

$$N_+(\infty) = \left(N_{+0} N_{-0}\right)^{1/2} = N_-(\infty)$$

We now extend the analysis to include coagulation. Essentially, we employ the arguments used to deal with multispecies aerosols where charge can be regarded as a property to be conserved in addition to volume and mass. Thus, with the rate of change due to coagulation, we may write:

$$\frac{\partial}{\partial t} n_\nu(v,t) = \tfrac{1}{2} \sum_{\mu=-\infty}^{\infty} \int_0^v du\, K_{\mu,\nu-\mu}(u,v-u)\, n_\mu(u,t)\, n_{\nu-\mu}(v-u,t)$$

$$- n_\nu(v,t) \sum_{\mu=-\infty}^{\infty} \int_0^\infty du\, K_{\nu,\mu}(u,v)\, n_\mu(u,t) - 4\pi\, B(v)\, v e^2\, n_\nu(v,t) \sum_{\mu=-\infty}^{\infty} \mu \int_0^\infty du\, n_\mu(u,t)$$

$$- R_\nu(v)\, n_\nu(v,t) + S_\nu(v,t)$$

(3.128)

The first term on the right hand side of Eqn. (3.128) describes the formation of particles of volume, v, and charge, ν, from the coalescence of particles of volume, u and v−u, and charge, μ and ν−μ. The second term on the right is the number removed from (v,μ). The third term has already been discussed. The fourth term is due to losses from settling or diffusion or, indeed, from an applied field, although that would also enter the electrostatic term. Finally S_ν is a source term due to aerosol production. Eqn. (3.128) assumes that no ions are generated, i.e. once the initial condition or source have been specified, the total charge remains the same: there is no creation of charge, it is simply reapportioned between the particles or lost due to dispersion and deposition.

Zebel [1958b] has modified Eqn. (3.128) by introducing the simplifying assumption that the amount of charge on a particle is proportional to its volume. Thus, in the Brownian coagulation kernel described by Eqn. (3.102), the factor, $s_{v\mu}$, in Eqn. (3.103) is written:

The Dynamic Equation for the Aerosol Distribution

$$S_{\nu\mu} = \frac{\nu\mu e^2}{kT} = \frac{e^2 \sigma^2}{kT}\left(\frac{a_1}{a_0}\right)^3 \left(\frac{a_2}{a_0}\right)^3 = \frac{e^2 \sigma^2}{kT}\left(\frac{v}{v_0}\right)\left(\frac{u}{v_0}\right) \tag{3.129}$$

where σ is the average number of charges on the basic spherical unit of volume $v_0 = 4\pi a_0^3/3$. This seems a sensible assumption and does not conflict with Eqn. (3.86) because that result refers to the equilibrium value attained by aerosol particles residing in a 'bath' of ions. However, this matter does raise the question of how to model the situation when charge is being created by a radiation field. Such a situation would prevail in a nuclear reactor accident. Then, some mechanism must be introduced to allow the particles described by Eqn. (3.128) to increase their charge in ways other than by coagulation. It would seem that the process can be likened to condensation with the ions taking the place of vapor molecules. In that case, we require a term of the form:

$$\frac{\partial}{\partial \mu}\left[I_\mu(v,t)\, n_\mu(v,t)\right] \tag{3.130}$$

where I_μ is the charging rate, $d\mu/dt$. In writing this equation, we have assumed charge to be a continuous quantity. More precisely we should write:

$$(A_v + B_v)\, n_v(v,t) - B_{v-1}\, n_{v-1}(v,t) - A_{v+1}\, n_{v+1}(v,t) \tag{3.131}$$

where A_v is the number of singly charged positive ions gained per unit time by an aerosol particle with v units of charge and B_v the corresponding number of negative ions. A_v and B_v will be governed by their own time dependent equations which, in turn, will be functions of the radiation field. Terms of the type illustrated in Eqn. (3.131) have been discussed by Simons [1976a,b].

Returning to Eqn. (3.129), we see that by means of the above assumption, the dynamic equation for $n_v(v,t)$ now reduces to a simpler one for $n(v,t)$ as follows:

$$\frac{\partial}{\partial t} n(v,t) + R(v)\, n(v,t) = S(v,t) + \tfrac{1}{2}\int_0^v du\, K(u, v-u)\, n(u,t)\, n(v-u,t)$$

$$-n(v,t)\int_0^\infty du\, K(u,v)\, n(u,t) - 4\pi B(v)\, e^2\, \sigma^2\left(\frac{v}{v_0}\right)\int_0^\infty du\left(\frac{u}{v_0}\right) n(u,t) \tag{3.132}$$

Zebel [1958b] has solved this equation numerically for Brownian coagulation, neglecting the deposition term, and using an initial value $n(v,0)$. Then, the time history of the aerosol is studied for $\sigma=0, 1$, and 2. Zebel's results, displayed in terms of particle radius, are shown in Figs. 3.7-3.9. It is clear that the effect of increasing charge is to narrow the spread of sizes. That is, a unipolar charged aerosol inhibits the coagulation process and the electrostatic dispersion term dominates. This would imply greater wall deposition in a vessel of finite size. Note that Eqn. (3.119) can be more precisely written as:

$$\frac{\partial}{\partial t} n_v(v,t) = -ve\, B(v)\left[n_v(v,t)\, \nabla \cdot \mathbf{E} + \mathbf{E} \cdot \nabla n_v(v,t)\right] \tag{3.133}$$

If n_v is space dependent, there will exist a flow of particles in space with an effective diffusion coefficient, $veB(v)\mathbf{E}$.

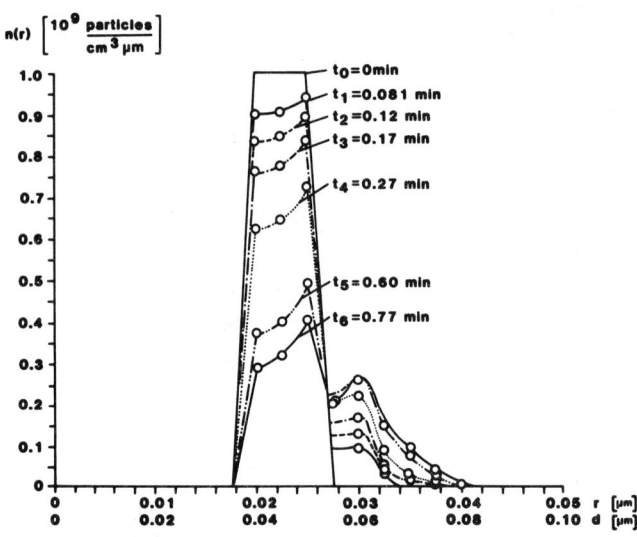

Figure 3.7: Aerosol evolution for weakly charged aerosol (σ=0). Adapted from Zebel [1958b] with permission of Steinkopff Verlag.

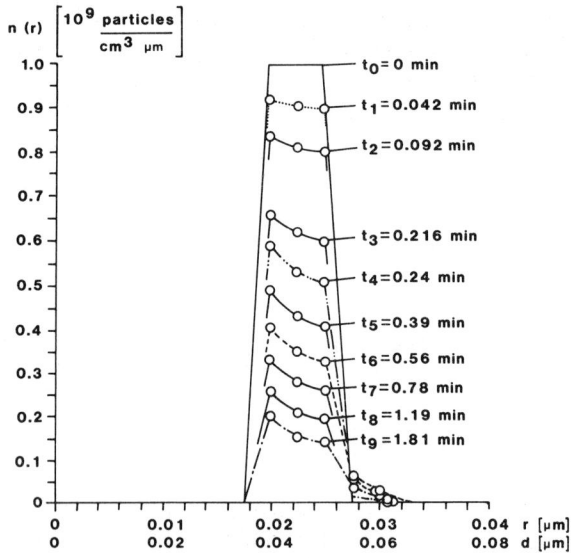

Figure 3.8: Aerosol evolution for moderately charged aerosol (σ=1). Adapted from Zebel [1958b] with permission of Steinkopff Verlag.

The Dynamic Equation for the Aerosol Distribution

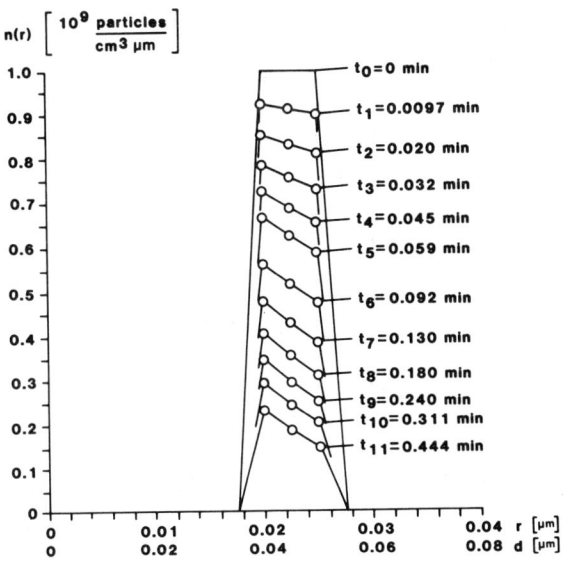

Figure 3.9: Aerosol evolution for strongly charged aerosol ($\sigma=2$). Adapted from Zebel [1958b] with permission of Steinkopff Verlag.

The major shortcoming of Eqn. (3.132) is that it applies only to unipolar aerosols. In practice, we have a range of positively and negatively charged particles that must be accounted for. We can develop the idea of Zebel by considering the aerosol population to be composed of $n^+(v,t)$ positively charged aerosols and $n^-(v,t)$ negatively charged ones. In that case, we can rewrite Eqn. (3.128) as two equations for the two groups n^+ and n^-. For n^+ we have:

$$\frac{\partial}{\partial t}n^+(v,t) + R^+(v)\,n^+(v,t)$$

$$= \tfrac{1}{2}\int_0^v du\, K_{++}(u, v-u)\, n^+(u,t)\, n^+(v-u,t) - n^+(v,t)\int_0^\infty du\, K_{++}(u,v)\, n^+(u,t)$$

$$- n^+(v,t)\int_0^\infty du\, K_{+-}(u,v)\, n^-(u,t) - \frac{4\pi D(v)}{kT}\, e^2\, \sigma^2\, g(v)\, n^+(v,t) \int_0^\infty du\, g(u)\,\{n^+(u,t) - n^-(u,t)\}$$

(3.134)

where $g(v) = (v/v_0)$ or any appropriate power of (v/v_0).

The physical meaning of the equation is the following. The left hand side denotes the rate of change of positively charged particles. The first and second terms on the right hand side are the formation and destruction terms for positively charged particles. The third term is a destruction term which occurs when a positive particle collides with a negative particle and has its mass changed from v. The fourth term is due to electrostatic dispersion and the fifth is due to deposition. The equation for n^- follows by the same arguments:

$$\frac{\partial}{\partial t}n^-(v,t)+R^-(v)\,n^-(v,t)$$

$$=\tfrac{1}{2}\int_0^v du\,K_{--}(u,v-u)\,n^-(u,t)\,n^-(v-u,t)-n^-(v,t)\int_0^\infty du\,K_{--}(u,v)\,n^-(u,t)$$

$$-n^-(v,t)\int_0^\infty du\,K_{+-}(u,v)\,n^+(u,t)$$

$$+\frac{4\pi D(v)}{kT}e^2\sigma^2\,g(v)\,n^-(v,t)\int_0^\infty du\,g(u)\{n^+(u,t)-n^-(u,t)\} \qquad (3.135)$$

The basic assumption behind these equations is that particles with the same volume have the same charge. The coagulation kernels K_{++}, K_{+-}, etc., refer to interactions between positive and positive particles, and positive and negative particles, etc., respectively. In the case of Brownian interaction we find:

$$K_{v\mu}(u,v)=\frac{2kT}{3\mu}\left[2+\left(\frac{u}{v}\right)^{1/3}+\left(\frac{v}{u}\right)^{1/3}\right]P_{v\mu}(u,v) \qquad (3.136)$$

where:

$$P_{v\mu}=\left(\frac{S_{v\mu}}{a_{12}}\right)\left(\exp\left(\frac{S_{v\mu}}{a_{12}}\right)-1\right)^{-1} \qquad (3.137)$$

and $a_{12}=a_1+a_2$ with $v=4\pi a_1^3/3$ and $u=4\pi a_2^3/3$. Then:

$$S_{--}=S_{++}=\frac{\sigma^2 e^2}{kT}g(v)\,g(u)=-S_{+-}=-S_{-+} \qquad (3.138)$$

In view of this we simplify our notation to $K_s=K_{++}=K_{--}$ and $K_a=K_{+-}=K_{-+}$.

There are two cases of interest: (1) where the total charge is conserved in a collision such that $g(v)=v/v_0$ and; (2) $g(v)=(v/v_0)^{1/6}$ which corresponds to the Boltzmann equilibrium situation in which the average charge is proportional to the square root of the radius (see Eqn. (3.86)). Case (1) is appropriate when no additional charges are being created. Case (2) applies when there exists a background equilibrium ion concentration arising, for example, from radiation.

For the case of $g(v)=v/v_0$, it is instructive to integrate Eqns. (3.134) and (3.135) over all volume, using Eqn. (3.12) to obtain:

$$\frac{dN^+}{dt}+\int_0^\infty dv\,R^+(v)\,n^+(v,t)$$

$$=-\tfrac{1}{2}\int_0^\infty du\int_0^\infty dv\,K_s(u,v)\,n^+(u,t)\,n^+(v,t)-\int_0^\infty du\int_0^\infty dv\,K_a(u,v)\,n^+(u,t)\,n^-(v,t)$$

$$-\frac{4\pi\,e^2\,\sigma^2}{kT\,v_0^2}\int_0^\infty dv\,D(v)\,v\,n^+(v,t)\int_0^\infty du\,u\{n^+(u,t)-n^-(u,t)\} \qquad (3.139)$$

and:

The Dynamic Equation for the Aerosol Distribution

$$\frac{dN^-}{dt} + \int_0^\infty dv\, R^-(v)\, n^-(v,t)$$

$$= -\tfrac{1}{2} \int_0^\infty du \int_0^\infty dv\, K_s(u,v)\, n^-(u,t)\, n^-(v,t) - \int_0^\infty du \int_0^\infty dv\, K_a(u,v)\, n^-(u,t)\, n^+(v,t)$$

$$+ \frac{4\pi\, e^2\, \sigma^2}{kT\, v_0^2} \int_0^\infty dv\, D(v)\, v\, n^-(v,t) \int_0^\infty du\, u\, \{n^+(u,t) - n^-(u,t)\} \tag{3.140}$$

where:

$$N^\pm(t) = \int_0^\infty dv\, n^\pm(v,t) \tag{3.141}$$

Similarly, if we define the total positive and negative charge:

$$\phi^\pm(t) = \pm\sigma e \int_0^\infty dv\, \frac{v}{v_0}\, n^\pm(v,t) \tag{3.142}$$

we find:

$$\frac{d\phi^+}{dt} = -\frac{\sigma e}{v_0} \int_0^\infty dv\, v \int_0^\infty du\, K_a(u,v)\, n^+(v,t)\, n^-(u,t) - \frac{\sigma e}{v_0} \int_0^\infty dv\, v\, R^+(v)\, n^+(v,t)$$

$$- \frac{4\pi\, e^2\, \sigma^2}{v_0^3\, kT} \int_0^\infty dv\, v\, D(v)\, v^2\, n^+(v,t) \int_0^\infty du\, u\, \{n^+(u,t) - n^-(u,t)\} \tag{3.143}$$

$$\frac{d\phi^-}{dt} = \frac{\sigma e}{v_0} \int_0^\infty dv\, v \int_0^\infty du\, K_a(u,v)\, n^-(v,t)\, n^+(u,t) + \frac{\sigma e}{v_0} \int_0^\infty dv\, v\, R^-(v)\, n^-(v,t)$$

$$+ \frac{4\pi\, e^2\, \sigma^2}{v_0^3\, kT} \int_0^\infty dv\, v\, D(v)\, v^2\, n^-(v,t) \int_0^\infty du\, u\, \{n^+(u,t) - n^-(u,t)\} \tag{3.144}$$

In the case of constant coagulation kernels and zero deposition these equations reduce to:

$$\frac{dN^+}{dt} = -\tfrac{1}{2} K_s \left(N^+\right)^2 - K_a\, N^+\, N^- - \frac{4\pi}{kT}\, \overline{D}\, \phi^+ \left(\phi^+ + \phi^-\right) \tag{3.145}$$

$$\frac{dN^-}{dt} = -\tfrac{1}{2} K_s \left(N^-\right)^2 - K_a\, N^+\, N^- - \frac{4\pi}{kT}\, \overline{D}\, \phi^- \left(\phi^+ + \phi^-\right) \tag{3.146}$$

$$\frac{d\phi^+}{dt} = -K_a\, N^-\, \phi^+ - \frac{4\pi}{kT}\, \overline{D}\, \overline{v}\, \phi^+ \left(\phi^+ + \phi^-\right) \tag{3.147}$$

and:

$$\frac{d\phi^-}{dt} = -K_a\, N^+\, \phi^- - \frac{4\pi}{kT}\, \overline{D}\, \overline{v}\, \phi^- \left(\phi^+ + \phi^-\right) \tag{3.148}$$

where $\overline{D}$ and $\overline{v}$ are given appropriate averages. Eqns. (3.145)-(3.148) will give some indication of the effect of charge on aerosol behavior.

3.13 A dynamic equation for nonspherical particles

We have seen that an equation for the population of spherical aerosol particles can be obtained without difficulty. We also know that most aerosols are not spherical. While some progress can be made in calculating the effective drag forces on aerosol particles, and to some extent the coagulation kernel, it requires far more effort to define a useful balance equation for such particles. In the case of a spherical particle, the only relevant parameter is the volume, which is conserved in a collision and leads to a new spherical particle. For nonspherical particles, additional parameters describing the shape are required. It would be most useful if these parameters, or some function of them, were to be conserved in a collision. Thus, let $s=(s_1,s_2,...)$ be set of 'descriptors' which define useful parameters characterizing the shape of the aerosol. In that case we may define (neglecting space dependence) $n(s,t)ds$ as the number of aerosol particles having descriptors in the range s to s+ds at time t. As examples of s we could set $s=(m,r_s)$ where m is the mass of a particle and r_s the Stokes radius. Another possibility is $s=(N,r_s)$ where N is the number of fundamental sub-units from which the aerosol particle is built. Other descriptors will suggest themselves according to the nature of the coagulation mechanisms and aerosol type. Thus we may define a generalized dynamic equation as follows:

$$\frac{\partial}{\partial t}n(s,t) + R(s)\,n(s,t)$$
$$= \tfrac{1}{2}\int ds' \int ds''\, K(s';s'')\,\delta(s-f(s',s''))\,n(s',t)\,n(s'',t)$$
$$- n(s,t)\int ds'\, K(s;s')\,n(s',t) + Q(s,t) \qquad (3.149)$$

Here, $R(s)$ is the removal rate per unit time for particles with descriptor s, $K(s,s')$ is the coagulation kernel, and Q is the source term. The delta function expresses some conservation relationships between s, s', and s''.

The simplest case of Eqn. (3.149) is when $s=(v)$ and the delta function expresses conservation of volume $v=v'+v''$. Then, we find the usual dynamic equation used in the earlier part of this chapter. As an example, which shows the usefulness of Eqn. (3.149), let us assume that $s=(N,r_s)$. Then, the distribution function is of the form $n(r_s,N,t)$ where $n(r_s,N,t)dr_s$ is the number of particles whose Stokes radius lies between r_s and r_s+dr_s and contains N primary sub-units. The collision terms in Eqn. (3.149) then become:

$$\tfrac{1}{2}\sum_{I,J}\int dr_s'\int dr_s''\, K(r_s',I;r_s'',J)\,n(r_s',I,t)\,n(r_s'',J,t)\,\delta_{N,I+J}\,\delta\!\left(r_e - \{r_e'^3 + r_e''^3\}^{1/3}\right)$$
$$- n(r_s,N,t)\sum_J \int dr_s'\, K(r_s,N;r_s',J)\,n(r_s',J,t) \qquad (3.150)$$

The conserved quantities are the number of primary sub-units and the equivalent volume defined by the radius r_e. Clearly, it is necessary to relate the Stokes radius to the equivalent volume radius. This is done experimentally through the dynamic shape factor, κ, defined by:

$$\kappa = r_e^2 / r_s^2$$

or:

$$r_e = \kappa^{1/2}\, r_s$$

Experiments are available relating $\kappa=\kappa(N)$ to the number of sub-units. For example, Kops [1976] has shown that:

The Dynamic Equation for the Aerosol Distribution

$$\kappa \propto N^{1/6} \quad ; \quad N < N^* \quad \text{and} \quad \kappa \propto N^{-1/6} \quad ; \quad N > N^*$$

where $N^* \approx 10^4$ for iron oxide.

In view of this result and after employing the properties of the delta function, we find that the dynamic equation can be written:

$$\frac{\partial}{\partial t} n(r_s, N, t) + R(r_s, N) \, n(r_s, N, t)$$

$$= \tfrac{1}{2} \sum_{I=1}^{N} \int_0^{L(\kappa)} dr_s' \, n(r_s', I, t) \, n\bigl(f(r_s, r_s'; N, I), N-I, t\bigr) K\bigl(f(r_s, r_s'; N, I), N-I; r_s', I\bigr) g(r_s, r_s'; N, I)$$

$$- n(r_s, N, t) \sum_{J=1}^{\infty} \int_0^{\infty} dr_s' \, K(r_s, N; r_s', I) \, n(r_s', I, t) + Q(r_s, N, t) \tag{3.151}$$

where:

$$L(\kappa) = r_s \left(\frac{\kappa(N)}{\kappa(I)} \right)^{1/2}$$

$$f = \left\{ r_s^3 \left[\frac{\kappa(N)}{\kappa(N-I)} \right]^{3/2} - r_s'^3 \left[\frac{\kappa(I)}{\kappa(N-I)} \right]^{3/2} \right\}^{1/3}$$

and:

$$g = \frac{r_s^2}{f^2} \left\{ \frac{\kappa(N)}{\kappa(N-I)} \right\}^{3/2}$$

This formulation assumes that the coagulation kernel can be adequately described by the Stokes radius. It may well be, however, that it contains other parameters that require different descriptors (*e.g.* mass, radius of gyration, *etc.*). A simplified form of Eqn. (3.151) has been employed by Okuyama *et al.* [1981]. These authors solved the equation numerically and then compared the results with those from a series of experiments involving carbon black, titanium oxide, and two kinds of iron oxide. The parameters used for comparison were the total number of particles and the geometric mean Stokes radius. Good agreement was found within the limitations of the experiment.

In Section 5.12 we elaborate further on the effect of shape on particle behavior.

References: Chapter 3

Bricard, J. and Pradel, J. [1966] "Electric Charge and Radioactivity of Naturally Occurring Aerosols" in Aerosol Science, Davies, C.N., editor (Academic, London, U.K.).

Cooper, D.W. and Reist, P.C. [1973] "Neutralizing charged aerosols with radioactive sources," *J. Colloid Interface Sci.* **45**, 17.

Davies, C.N., editor [1966] Aerosol Science (Academic, London, U.K.).

El-Hakeem, A.S. and Ryley, D.J. [1972] "The blow-down of a compressed air reservoir," *I.J.M.E.E.* **1**, 55.

Flanagan, V.P.V. and O'Connor, T.C. [1961] "Ionization equilibrium in aerosols," *Pure and Appl. Geophys.* **50**, 148.

Friedlander, S.K. [1970] "The characterization of aerosols distributed with respect to size and chemical composition," *J. Aerosol Sci.* **1**, 295.

Friedlander, S.K. [1971] "The characterization of aerosols distributed with respect to size and chemical composition - II. Classification and design of aerosol measurement devices," *J. Aerosol Sci.* **2**, 331.

Friedlander, S.K. [1977] Smoke, Dust and Haze (Wiley, New York, N.Y.).

Gelbard, F.M. and Seinfeld, J.H. [1978] "Coagulation and growth of a multicomponent aerosol," *J. Colloid Interface Sci.* **63**, 472.

Gelbard, F.M. and Seinfeld, J.H. [1980] "Simulation of multicomponent aerosol dynamics," *J. Colloid Interface Sci.* **78**, 485.

Gillespie, T. [1953] "The effect of the electric charge distribution on the ageing of aerosols," *Proc. Roy. Soc. (London)* **A216**, 569.

Goldman, D.E. [1950] Handbook of Aerosols (U.S. Government Printing Office).

Gunn, R. [1954] "Diffusion charging of atmospheric droplets by ions, and the resulting combination coefficients," *J. Meteor.* **11**, 339.

Gunn, R. [1955] "The statistical electrification of aerosols by ionic diffusion," *J. Colloid Sci.* **10**, 107.

Hoppel, W.A. [1985] "Ion-Aerosol attachment coefficients, ion depletion, and the charge distribution on aerosols," *J. Geophys. Res.* **90**, 5917.

Hoppel, W.A. and Frick, G.M. [1986] "Ion-aerosol attachment coefficients and the steady-state charge distribution on aerosols in a bipolar environment," *Aerosol Sci. and Tech.* **5**, 1.

Horlock, J.H. and Woods, W.A. [1966] "The thermodynamics of charging and discharging processes," *Proc. Inst. Mech. Engrs.* **180**, 16.

Jacobi, W. [1961] "Die anlagerung von naturlichen radionukliden an aerosolpartikel und niederschlagselemente in der atmosphare," *Pure and Appl. Geophys.* **50**, 260 (in German).

Jordan, H., Schumaker, P.M., Gieseke, J.A. and Lee, K.W. [1980] "Multiple Zone Aerosol Behaviour Model," Report # NUREG/CR-1294, BMI-2042/R7.

Junge, C. [1955] "The size distribution and aging of natural aerosols as determined from electrical and optical data on the atmosphere," *J. Meteor.* **12**, 13.

Keefe, D. and Nolan, P.J. [1961] "Influence of image forces on combination in aerosols," *Pure and Appl. Geophys.* **50**, 155.

Keefe, D., Nolan, P.J. and Rich, T.A. [1959] "Charge equilibrium in aerosols according to the Boltzmann law," *Proc. Roy. Irish Academy* **60A**, 27.

Kestin, J. and Glass, J.S. [1951] "The rapid discharge of gases from vessels," *Aircr. Eng.* **23**, 300.

Kops, J.A.M.M. [1976] "The Aerodynamic Diameter and Specific Surface Area of Branched Chain-like Aggregates" (Netherlands Energy Research Foundation) Report # ECN-5.

Kunkel, W.B. [1950a] "The static electrification of dust particles on dispersion into a cloud," *J. Appl. Phys.* **21**, 820.

Kunkel, W.B. [1950b] "Charge distribution in coarse aerosols as a function of time," *J. Appl. Phys.* **21**, 833.

Lushnikov, A.A. [1976] "Evolution of coagulating systems III. Coagulating mixtures," *J. Colloid Interface Sci.* **54**, 94.

Morgan, K.Z. and Turner, J.E. [1967] Principles of Radiation Protection (Wiley, New York, N.Y.).

Muller, H. [1928a] "Zur theorie der elektrischen ladung und der koagulation der kolloide," *Kolloidchem. Beih.* **26**, 257 (in German).

Muller, H. [1928b] "Zur allgemeinen theorie der raschen koagulation," *Kolloidchem. Beih.* **27**, 223 (in German).

Okuyama, K., Kousaka, Y. and Payakakes, A.C. [1981] "Evaluation of the effect of nonsphericity of fine aggregate particles in Brownian coagulation," *J. Colloid Interface Sci.* **81**, 21.

Rosinski, J., Werle, D. and Nagamoto, C.T. [1962] "Coagulation and scavenging of radioactive aerosols," *J. Colloid Sci.* **17**, 703.

Russell, A. [1922] "The problem of two electrified spheres," *Proc. Phys. Soc. (London)* **35**, 10.

Sakurai, A. and Takayama, F. [1983] "Spurt of gas from a hole into air," *J. Phys. Soc. Japan* **52**, 2963.

Schmidt, E. [1965] "Flow of gases from pressure vessels," *Chem.-Ing.-Tech.* **37**(11), 1091 (in German).

Schumann, T.E.W. [1940] "Theoretical aspects of the size distribution of fog particles," *Q. J. Roy. Meteor. Soc.* **66**, 195.

Shapiro, A.H. [1953] *The Dynamics and Thermodynamics of Compressible Fluid Flow*, Vol. I (Wiley, New York, N.Y.) Chapter 4.

Simons, S. [1976a] "The electric charge distribution on interstellar grains I. Calculation," *Astrophys. Space Sci.* **41**, 423.

Simons, S. [1976b] "The electric charge distribution on interstellar grains II. Some consequences," *Astrophys. Space Sci.* **41**, 435.

Simons, S. [1981] "The coagulation and deposition of radioactive aerosols," *Ann. Nucl. Energy* **8**, 287.

Simons, S. [1982] "The condensation, coagulation and deposition of a multicomponent radioactive aerosol," *Ann. Nucl. Energy* **9**, 473.

Simpson, D.R., Williams, M.M.R. and Simons, S. [1989] "Modelling of an aerosol in coupled chambers," *Nucl. Sci. and Eng.* **101**, 259.

Smoluchowski, M. von [1916] "Drei vortrage uber diffusion, Brownsche molekularbewegung und koagulation von kolloidteilchen." The title is translated as: "Three lectures on diffusion, Brownian motion and coagulation of colloid particles," *Phys. Z.* **17**, 557 (in German).

Smoluchowski, M. von [1917] "Versuch einer mathematischen theorie der koagulationskinetik kolloider losungen," *Z. f. Phys. Chemie* **92**, 129 (in German).

Twomey, S. [1977] *Atmospheric Aerosols* (Elsevier, New York, N.Y.).

van de Vate, J.F. [1980] "Investigations Into the Dynamics of Aerosols in Enclosures as Used for Air Pollution Studies" (Netherlands Energy Research Foundation) Report # ECN-86.

Whitby, K.T. and Liu, B.Y.H. [1966] "The Electrical Behaviour of Aerosols," in *Aerosol Science*, Davies, C.N., editor (Academic, London, U.K.).

Whytlaw-Gray, R. and Patterson, H.S. [1932] *Smoke: A Study of Aerial Disperse Systems* (Edward Arnold, London, U.K.).

Zebel, G. von [1966] "Coagulation of Aerosols," in *Aerosol Science*, Davies, C.N., editor (Academic, London, U.K.).

Zebel, G. von [1958a] Zur theorie des kogulation elektrisch ungeladener aerosole." The title is translated as: "On the theory of coagulation of electrically uncharged aerosols," *Kolloid Z.* **156**, 102 (in German).

Zebel, G. von [1958b] "Zur theorie des verhaltens elektrisch geladener aerosole." The title is translated as: "On the theory of electrically charged aerosols," *Kolloid Z.* **157**, 37 (in German).

CHAPTER 4

Coagulation Kernels

4.1 Introduction

The term 'coagulation' is used to describe the process of adhesion or fusion of two particles which takes place when they touch. Such a 'collision' takes place because of the relative velocity between the aerosol particles. This relative velocity can arise from a variety of physical causes. The most commonly known are:

a) Brownian motion
b) gravitational settling
c) turbulence

Another less common but feasible coagulation mechanism is due to acoustic forces arising from an applied sound wave. There also exists the possibility of second order effects due to temperature and density gradients. In general, any nonuniform disturbance in the fluid can lead to a velocity differential and therefore to coagulation.

It should be recognized that aerosol particles are not generally spherical. Indeed, they more often take on very complex shapes. For this reason it is difficult to quantify their interaction probability because, in order to do so, it is necessary to know the effective cross section presented by one particle to the other and their respective velocities. Both of these quantities depend critically on the shape. Unfortunately, there is no comprehensive theoretical treatment of particles of complex shape, although some work has been done on spheroids (Booth [1954]) which we discuss later. For this reason, our discussion of coagulation will assume that the particles entering into a coagulation are spherical and that the resulting compound particle is also spherical and equal to the sum of the two volumes. This would probably be true for small liquid droplets but almost certainly not for solids.

In order to quantify the coagulation process, let us assume that we have an aerosol cloud consisting of a single species which is described by the distribution function, $n(v)$, where $n(v)dv$ is the number of aerosol particles per unit volume of space whose individual volumes lie between, v and $v+dv$. Then the rate at which an aerosol particle of volume, u, coagulates with an aerosol particle of volume, v, *i.e.* the collision frequency, is:

$$K(u,v)\, n(u)\, n(v) \qquad (4.1)$$

$K(u,v)$, which has dimensions $[L^3 T^{-1}]$, is called the coagulation kernel or sometimes the agglomeration kernel. We shall use the former term.

In the rest of this chapter we will be concerned with developing methods to obtain analytical forms for $K(u,v)$ for a variety of physical processes. Before doing so, however, let us generalize Eqn. (4.1) to cover an aerosol containing N different species. In such a case, the distribution function itself must be generalized to:

$$n(v_1, v_2, \ldots, v_N)\, dv_1\, dv_2 \ldots dv_N \qquad (4.2)$$

which is the number of aerosol particles per unit volume of space which contain a volume, v_1 to v_1+dv_1, of species 1, v_2 to v_2+dv_2, of species 2, *etc.*

Coagulation Kernels

The total volume of the particle is:

$$v = \sum_{i=1}^{N} v_i \tag{4.3}$$

and the conventional aerosol particle distribution function for total volume is:

$$N(v) = \int n(v_1, v_2, ..., v_N) \, \delta\left(v - \sum_{i=1}^{N} v_i\right) dv_1 \, dv_2 \, ... \, dv_N \tag{4.4}$$

The rate at which aerosols containing u_1 to u_1+du_1 of species 1, u_2 to u_2+du_2 of species 2, *etc.*, coagulate with an aerosol containing v_1 to v_1+dv_1 of species 1, v_2 to v_2+dv_2 of species 2, *etc.*, is:

$$K(u_1, u_2, ..., u_N; v_1, v_2, ..., v_N) \, n(u_1, u_2, ..., u_N) \, n(v_1, v_2, ..., v_N) \tag{4.5}$$

$K(u_1, u_2, ..., u_N; v_1, v_2, ..., v_N)$ is then the generalized coagulation kernel. It is quite likely in practice that K(...) will not depend on the individual values of u_i and v_i, but on the sum of them and their average densities. We shall comment on this later.

A basic property of the coagulation kernel is its symmetry, $K(u,v)=K(v,u)$, and we shall use this property many times.

4.2 Brownian coagulation

4.2.1 Diffusion theory

Brownian motion is the name given to the irregular movement of small particles suspended in a fluid due to the random impact of molecules on their surfaces arising from thermal motion. The simultaneous random walk of such a large number of particles leads inevitably to collisions. Such a process is termed Brownian coagulation.

In order to calculate the trajectory of a Brownian particle, one follows Langevin in writing down the instantaneous equation of motion with an appropriate noise source term to account for molecular impacts:

$$m \frac{dv}{dt} = -f \, v + F(t) \tag{4.6}$$

Here, m is the particle mass, fv is the frictional retarding force or drag, and **F** is the random force term due to molecular bombardment. Although the frictional force is directly proportional to velocity, for the situations expected in aerosols, the calculation of the constant of proportionality, f, is, in general, very difficult. This is because the drag is a sensitive function of the ratio of the mean free path to the particle size, *i.e.* the Knudsen number. It means therefore that a detailed kinetic theory calculation would have to be performed. While such calculations have been carried out for single particles (Cercignani *et al.* [1968], Lea and Loyalka [1982], Law and Loyalka [1986]) they are not available for the associated Brownian problem.

In order to simplify the problem it is usual to consider limiting cases of particle size. If, for example, a particle is large compared with a mean free path of a gas atom in the surrounding fluid, then that fluid may be regarded as a continuum. Thus, we may employ hydrodynamic theory to calculate the drag on the sphere. As we have seen in Chapter 2, the drag force on a sphere moving with velocity, **v**, is:

$$F = 6\pi a \mu v \tag{4.7}$$

where a is the radius of the sphere and μ is the viscosity of the fluid. The friction coefficient, f, in Eqn. (4.6) is therefore:

$$f = 6\pi a \mu \tag{4.8}$$

According to Einstein [1956] and described by Chandrasekhar [1943], Brownian agitation leads to the diffusive motion of a large number of such particles. Such motion may be characterized by a diffusion equation of the form:

$$\frac{\partial}{\partial t} C(\mathbf{r},t) = D \nabla^2 C(\mathbf{r},t) \tag{4.9}$$

where D is the diffusion coefficient. The physical meaning of $C(\mathbf{r},t)$ can best be described if we assume that a burst of particles is released at the origin at time t=0. Then, $C(\mathbf{r},t)d\mathbf{r}$ is the number of particles to be found at **r** in d**r** at time, t, later.

The diffusion coefficient in Eqn. (4.9) also has a well defined physical meaning. Thus, if we write Eqn. (4.9) as:

$$\frac{\partial}{\partial t} C(x,y,z,t) = D \left[\frac{\partial^2}{\partial x^2} + \frac{\partial^2}{\partial y^2} + \frac{\partial^2}{\partial z^2} \right] C(x,y,z,t) \tag{4.10}$$

and integrate over x, y, and z $(-\infty,\infty)$ we find:

$$\int_{-\infty}^{\infty} dx \int_{-\infty}^{\infty} dy \int_{-\infty}^{\infty} dz\, C(x,y,z,t) = \text{constant} \tag{4.11}$$

Now, multiply Eqn. (4.10) by x^2, y^2, and z^2 in turn and integrate over all space to get:

$$\frac{d}{dt} \int_{-\infty}^{\infty} dx \int_{-\infty}^{\infty} dy \int_{-\infty}^{\infty} dz\, (x^2,y^2,z^2) C(x,y,z,t) = 2D \int_{-\infty}^{\infty} dx \int_{-\infty}^{\infty} dy \int_{-\infty}^{\infty} dz\, C(x,y,z,t) \tag{4.12}$$

Adding and using the result that:

$$\overline{x^2} + \overline{y^2} + \overline{z^2} = \overline{r^2} = \frac{\int dx \int dy \int dz (x^2 + y^2 + z^2) C(x,y,z,t)}{\int dx \int dy \int dz\, C(x,y,z,t)} \tag{4.13}$$

we obtain:

$$\overline{r^2} = 6Dt \tag{4.14}$$

Thus, D is related to the mean square distance of travel. Einstein, in his original analysis showed that, in the continuum regime and using Stokes' law for the resistance to motion:

Coagulation Kernels

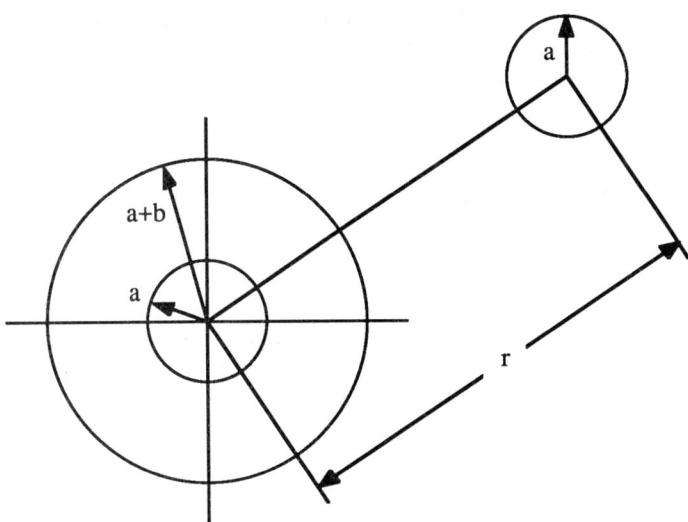

Figure 4.1: A polar coordinate system for two particle coagulation.

$$\overline{r^2} = \frac{kT}{\pi \mu a} t \qquad (4.15)$$

We see, therefore, by comparing Eqns. (4.14) and (4.15), that:

$$D = \frac{kT}{6\pi \mu a} \qquad (4.16)$$

More generally, Einstein showed that D=BkT where B is the mobility of the particle, *i.e.* the velocity per unit force. In order to apply this concept to coagulation, it is necessary to consider a test particle within the ensemble and to calculate the rate at which the other particles diffuse towards this test particle. If the test particle is of radius, b, and we wish to calculate the rate at which particles of radius, a, strike it, then we need to set up a polar coordinate system with the center of particle b at the origin, as shown in Fig. 4.1.

Since the diffusion process is isotropic in space, the diffusion equation reduces to:

$$\frac{\partial}{\partial t} C(r,t) = D_{ab} \left[\frac{\partial^2}{\partial r^2} C(r,t) + \frac{2}{r} \frac{\partial}{\partial r} C(r,t) \right] \qquad (4.17)$$

The boundary conditions associated with the problem are as follows:

(a) At r=a+b,   C(a+b,t)=0 $\qquad (4.18)$

follows from the fact that the spheres are impenetrable.

(b) At $r = \infty$, $C = C_\infty$ (a constant) $\qquad (4.19)$

implies that far away from the test particle we have an unperturbed uniform distribution.

The diffusion coefficient D_{ab} is the relative value for particles a and b since now both are moving. It is easy to see how D must be modified to take into account the relative motion. For then, the mean square distance of travel is the average of the distance between the two particles:

$$\overline{(r_a - r_b)^2} = \overline{r_a^2} + \overline{r_b^2} - 2\overline{r_a r_b} \tag{4.20}$$

Since there is no correlation between the positions of the two particles, $\overline{r_a r_b} = 0$ and we have:

$$D_{ab} = \frac{\overline{(r_a - r_b)^2}}{6t} = \frac{\overline{r_a^2} + \overline{r_b^2}}{6t} = \frac{kT}{6\pi\mu}\left(\frac{1}{a} + \frac{1}{b}\right) \tag{4.21}$$

Eqn. (4.17), subject to the boundary conditions of Eqns. (4.18) and (4.19), is readily solved and leads to:

$$C(r,t) = C_\infty \left[1 - \frac{a+b}{r} + \frac{a+b}{r} \mathrm{erf}\left(\frac{r-a-b}{\sqrt{4 D_{ab} t}}\right)\right] \tag{4.22}$$

Now, the rate at which test particle b captures particles a is given by the net current onto the test particle:

$$J = 4\pi (a+b)^2 D_{ab} \left.\frac{\partial C}{\partial r}\right|_{r=a+b} = 4\pi D_{ab} (a+b) C_\infty \left[1 + \frac{a+b}{\sqrt{\pi D_{ab} t}}\right] \tag{4.23}$$

The characteristic time:

$$\tau = \frac{(a+b)^2}{\pi D_{ab}} \tag{4.24}$$

is the time for diffusion over the distance of an aerosol diameter and is less than 10^{-5} sec for $a+b \leq 0.1$ μm and less than 10^{-2} sec for $a+b \leq 1.0$ μm. For times of interest, therefore, we can neglect the time transient in Eqn. (4.24) and obtain:

$$J = 4\pi D_{ab} (a+b) C_\infty \tag{4.25}$$

The collision frequency is the rate at which particles collide with the test particle per unit incident flux:

$$K(a,b) = \frac{J}{C_\infty} = 4\pi (D_a + D_b)(a+b) \tag{4.26}$$

Using the Stokes-Einstein value for the diffusion coefficient, we obtain:

$$K(a,b) = \frac{2kT}{3\mu}\left(\frac{1}{a} + \frac{1}{b}\right)(a+b) \tag{4.27}$$

Coagulation Kernels

which, in terms of the particle volumes, $u=4\pi a^3/3$ and $v=4\pi b^3/3$, leads to:

$$K(u,v) = \frac{2kT}{3\mu}\left(\frac{1}{u^{1/3}} + \frac{1}{v^{1/3}}\right)(v^{1/3} + u^{1/3}) = \frac{2kT}{3\mu}\left(2 + \left(\frac{u}{v}\right)^{1/3} + \left(\frac{v}{u}\right)^{1/3}\right) \tag{4.28}$$

4.2.2 Free molecular flow

Eqn. (4.28) is only valid when the particle is large compared with a mean free path. In the other limiting case, when the particle is small compared with a mean free path, a different approach is necessary. While the motion is still of a random nature, it is clear that, because of its small size, the aerosol particle can be considered like a large molecule and therefore travels in free unhindered flights between collisions. A Langevin equation of the type discussed above is therefore not appropriate. Instead, one calculates directly the number of collisions between the particle and the surrounding gas atoms by statistical mechanical averaging over the prevailing Maxwell-Boltzmann distribution. Thus, suppose the gas and aerosol are in thermodynamic equilibrium, then the aerosol velocity distribution function will be:

$$f_a(v) = n_a \left(\frac{m_a}{2\pi kT}\right)^{3/2} \exp\left\{-\frac{m_a v^2}{2kT}\right\} \tag{4.29}$$

where T is the temperature and m_a is the mass of an aerosol particle of radius, a. Consider now a mixture of two such aerosol species, one of mass, m_a, and the other of mass, m_b. The collision rate between these two aerosol 'clouds' will be (Hidy and Brock [1965], Williams [1971]):

$$R_{ab} = \int dv_a \int dv_b \, f_a(v_a) f_b(v_b) |v_a - v_b| \sigma(|v_a - v_b|) \tag{4.30}$$

where $\sigma(v)$ is the cross section for collision. Assuming hard-sphere collisions in which the cross section is simply the geometric cross section and independent of velocity:

$$\sigma(v) = \pi (a+b)^2 \tag{4.31}$$

we find:

$$R_{ab} = \pi (a+b)^2 \int dv_a \int dv_b \, f_a(v_a) f_b(v_b) |v_a - v_b| \tag{4.32}$$

Now, inserting Eqn. (4.29) into Eqn. (4.32) and using the following transformations:

$$v = v_a - v_b$$

and:

$$V = \frac{m_a v_a + m_b v_b}{m_a + m_b}$$

we obtain:

$$R_{ab} = (a+b)^2 n_a n_b \left[\frac{8\pi kT(m_a + m_b)}{m_a m_b}\right]^{1/2} \tag{4.33}$$

Thus, the collision frequency or coagulation kernel is:

$$K(a,b) = \frac{R_{ab}}{n_a n_b} = (a+b)^2 \left[\frac{8\pi kT(m_a+m_b)}{m_a m_b}\right]^{1/2} \quad (4.34)$$

If the particles are of the same material with the same density, then $m_a=4\pi a^3\rho/3$ and $m_b=4\pi b^3\rho/3$, and hence:

$$K(a,b) = \left(\frac{6kT}{\rho}\right)^{1/2}(a+b)^2\left(\frac{1}{a^3}+\frac{1}{b^3}\right)^{1/2} \quad (4.35)$$

In terms of particle volume, this can be written:

$$K(u,v) = \left(\frac{8\pi kT}{\rho}\right)^{1/2}\left(\frac{3}{4\pi}\right)^{2/3}\left(u^{1/3}+v^{1/3}\right)^2\left(\frac{1}{u}+\frac{1}{v}\right)^{1/2} \quad (4.36)$$

4.2.3 The slip flow regime

Eqns. (4.28) and (4.36) give the limiting values of the coagulation kernel for very large and very small particles, respectively (compared with a mean free path). In order to understand what this means in practice, let us calculate the mean free path, λ_g, in air at STP. Using a result given in Kennard [1938]:

$$\lambda_g = \frac{1}{\sqrt{2}\,\pi n\sigma^2} \quad (4.37)$$

where n is the number of air molecules per unit volume and σ^2 is the effective weighted cross section of air molecules. This result assumes hard-sphere interactions. The value of $\lambda_g=0.064$ µm for air. The typical size range of aerosol particles is from 50.0 µm down to 0.001 µm. Thus, it will be necessary to have a representation of K(a,b) over the whole range of Knudsen numbers.

Before considering this problem in more detail, let us note that a very useful semi-empirical method exists for extending the continuum results of Eqn. (4.28) to particle sizes which are comparable to a mean free path. The basis of this technique lies in the recognition that the diffusion coefficient is related to the friction coefficient through the Einstein formula D=BkT. Since:

$$V = BF \quad (4.38)$$

we see that for Stokes' law, $B=1/6\pi a\mu$. However, for a more general flow regime, Eqn. (4.38) is still valid and B itself will depend on the Knudsen number. To this end, Cunningham [1910] modified Stokes' law such that it could be written:

$$F = \frac{6\pi a\mu}{C}V \quad (4.39)$$

where C is a function of Knudsen number $Kn=\lambda_g/a$. In Cunningham's case the experimental results were fitted to:

$$C = 1 + A\,Kn \quad (4.40)$$

Coagulation Kernels

Later, Millikan [1923] repeated the experiment over a much wider range of Kn and found:

$$C = 1 + Kn\left[A + B\exp\left(-\frac{E}{Kn}\right)\right] \tag{4.41}$$

where A=(1.257), B=(0.400), and E=(1.10) (see also Allen and Raabe [1985]). In the limiting case of very small particles, Kn>>1, the modified Stokes' law becomes:

$$F = \frac{6\pi a^2 \mu}{\lambda_g (A+B)} V = (3.62)\pi a^2 \frac{\mu}{\lambda_g} V \tag{4.42}$$

This result agrees well with the exact theoretical result for perfectly accommodating spheres in which:

$$F = \tfrac{8}{3}\left(1 + \tfrac{1}{8}\pi\right)\frac{\pi a^2 \mu}{\lambda_g} V$$

then the coefficient (3.62) is replaced by (3.71).

Cercignani et al. [1968], Lea and Loyalka [1982], and Law and Loyalka [1986] have solved the BGK model of the Boltzmann equation over the complete range of Knudsen numbers and have verified their results against Millikan's data.

Phillips [1975] using the moments method has obtained a theoretical expression for F as:

$$F = -6\pi a \mu V Q$$

where:

$$Q = \frac{15 - 3c_1 Kn + c_2(8 + \pi\sigma)(c_1^2 + 2)Kn^2}{15 + 12 c_1 Kn + 9(c_1^2 + 1) + 18 c_2 (c_1^2 + 2) Kn^3}$$

This is exact in both limits, Kn→0 and Kn→∞. The result of these calculations leads therefore to a modified diffusion coefficient:

$$D_a = \frac{kT}{6\pi a \mu} C_a \tag{4.44}$$

From this, we may obtain a modified coagulation kernel:

$$K(a,b) = \frac{2 kT}{3\mu}(a+b)\left(\frac{C_a}{a} + \frac{C_b}{b}\right) \tag{4.45}$$

It should be noted that this formula has to be used with some care because, if we allow Kn→∞, then it reduces to:

$$K(a,b) = (A+B)\frac{2 kT \lambda_g}{\mu}(a+b)\left(\frac{1}{a^2} + \frac{1}{b^2}\right) \tag{4.46}$$

which does not agree with Eqn. (4.35). Thus, while the Cunningham correction accurately describes the change in drag with Knudsen number, it does not lead to the correct form of the coagulation kernel. Nevertheless, Eqn. (4.45) is used in practice for Brownian coagulation well into the noncontinuum region (Kn=0.25).

4.2.4 Fuchs' method

It is clearly of some importance to deal quantitatively with the transition regime between continuum and free molecular flow. Since it is very difficult to solve the Boltzmann equation in such a situation, various approximate techniques have been employed. One of the most interesting of these methods is due to Fuchs [1964] and involves the so-called Fuchs' jump distance. We shall describe Fuchs' technique below.

Fuchs assumes that the actual particle is enclosed by a spherical surface a distance, δ, from the real surface. The magnitude of δ depends on the radius of the particle and we shall say more about it below. If two particles are involved then each particle has its own value of δ and a combined δ, which we call, δ_{ab}, has to be considered.

We imagine, therefore, particles of radius, b, diffusing towards the test particle of radius, a. According to classical diffusion theory these particles would coalesce at r=a+b. Fuchs, however, stipulates that the classical diffusion theory solution is only valid for r>a+b+δ_{ab}. If the concentration at r=a+b+δ_{ab} is C', then the solution of the steady state diffusion equation may be written:

$$C(r) = C_\infty - (C_\infty - C')(a + b + \delta_{ab})/r \tag{4.47}$$

The current through the surface a+b+δ_{ab} is therefore:

$$J_{diff} = 4\pi (a + b + \delta_{ab})^2 (D_a + D_b) \frac{\partial C}{\partial r}\bigg|_{r=a+b+\delta_{ab}} \tag{4.48}$$

$$= 4\pi (a + b + \delta_{ab})(D_a + D_b)(C_\infty - C') \tag{4.49}$$

where D_a+D_b includes the Cunningham correction factors.

Fuchs now assumes that for r<a+b+δ_{ab}, the particles move according to the kinetic theory of gases as if in a vacuum. That is to say, the current of particles is given by $n\bar{v}/4$ where n is the concentration and $\bar{v}$ is the average thermal velocity (Kennard [1938]). In our notation, this means that the current of particles of type b striking the surface of particle a is:

$$J_{kin} = 4\pi (a + b)^2 \frac{\bar{v}_{ab}}{4} C' \tag{4.50}$$

where:

$$\bar{v}_{ab} = \left(\bar{v}_a^2 + \bar{v}_b^2\right)^{1/2} \tag{4.51}$$

with:

$$\bar{v}_a = \left(\frac{8kT}{\pi m_a}\right)^{1/2} \tag{4.52}$$

At steady state, the number of particles passing through the surface at a+b+δ_{ab} must equal the number incident on the surface, i.e.

Coagulation Kernels

$$J_{kin} = J_{diff} \tag{4.53}$$

This leads to a value for C' of:

$$C' = \frac{(a+b+\delta_{ab})(D_a+D_b)C_\infty}{(a+b+\delta_{ab})(D_a+D_b)+\frac{1}{4}(a+b)^2 \overline{v_{ab}}} \tag{4.54}$$

from which we obtain:

$$K(a,b) = \frac{J}{C_\infty} = \frac{4\pi(a+b)(D_a+D_b)C_\infty}{\dfrac{a+b}{a+b+\delta_{ab}} + \dfrac{4(D_a+D_b)}{(a+b)\overline{v_{ab}}}} \tag{4.55}$$

All of the parameters in this expression are defined except for δ_{ab}. Here we follow Fuchs who takes an average of the distance a particle will travel in the normal direction from the sphere surface after leaving in a random direction with a travel distance L. This travel distance is an apparent mean free path for the particle and can be explained in the following way. According to the theory of Brownian motion, a particle of mass, m, has a forward velocity with a mean value of $v=(8kT/\pi m)^{1/2}$. The surrounding fluid resists this motion and the instantaneous velocity is given by:

$$v = v_0 \exp(-t/\tau) \tag{4.56}$$

where τ can be considered as a mean free time between collisions. Thus, the apparent mean free path of a particle will be:

$$L = v\tau \tag{4.57}$$

Since $\tau = mB$, where B is the mobility, we see that:

$$L = vmB \tag{4.58}$$

But, $D = kTB$ and $v = (8kT/\pi m)^{1/2}$, thus:

$$L = \frac{8D}{\pi \overline{v}} \tag{4.59}$$

After averaging, Fuchs finds that:

$$\delta_a = \frac{1}{6aL_a}\left[(2a+L_a)^3 - (4a^2+L_a^2)^{3/2}\right] - 2a \tag{4.60}$$

and for the combined effect of two spheres of different radii, a and b, he assumes that:

$$\delta_{ab} = (\delta_a^2 + \delta_b^2)^{1/2} \tag{4.61}$$

on the basis that a thermal average is most appropriate.

Aerosol Science: Theory and Practice

The outcome of these calculations is that the coagulation kernel in Eqn. (4.55) spans the whole range of Knudsen numbers and therefore covers the transition regime where neither the hydrodynamic nor the free molecular results are valid. This result is self consistent and reasonable physically. Accurate results based on the model equations here have been reported by Loyalka [1976] and precise results based on the Boltzmann equation can be obtained by following the recent developments described in Chapter 6.

4.3 Gravitational coagulation

As particles settle under the action of gravity they acquire a constant velocity that is related to their mass and the drag of the surrounding fluid. Different size and mass particles will have varying settling velocities and so there will be a velocity differential. If two particles are within a certain distance of each other the faster one has the possibility of colliding with the slower one, thereby leading to coagulation. The calculation of the coagulation kernel for this problem is rather difficult but an elementary derivation is available which shows the general principles. We shall use the simple derivation and return to the more subtle issues later.

Consider a spherical test particle of radius, a, moving with a velocity, V_a. Then, consider a cloud of other particles of radius, b, velocity, V_b, and concentration, C_∞, moving in the same direction. The number of collisions per unit time will be equal to the product of the effective area, the relative velocity, and the concentration:

$$J = \pi (a+b)^2 |V_a - V_b| C_\infty \tag{4.62}$$

Thus, the coagulation kernel will be:

$$K(a,b) = \pi (a+b)^2 |V_a - V_b| \tag{4.63}$$

For small settling speeds, the velocity is available from:

$$m_a g = \frac{6\pi a \mu}{C_a} V_a \tag{4.64}$$

with a similar relation for V_b where we have included the Cunningham correction factor, C_a. Thus:

$$K(a,b) = \frac{g}{6\mu} (a+b)^2 \left| \frac{m_a C_a}{a} - \frac{m_b C_b}{b} \right| \tag{4.65}$$

If the particles are of the same material, then $m_a = 4\pi a^3 \rho/3$ and $m_b = 4\pi b^3 \rho/3$ and so:

$$K(a,b) = \frac{2 g \rho}{9 \mu} \pi (a+b)^2 |a^2 C_a - b^2 C_b| \tag{4.66}$$

In terms of particle volume, we may write this as:

$$K(u,v) = \frac{\rho g}{6\mu} \left(\frac{3}{4\pi} \right)^{1/3} \left(v^{2/3} + u^{2/3} \right) |v^{2/3} C_v - u^{2/3} C_u| \tag{4.67}$$

Coagulation Kernels

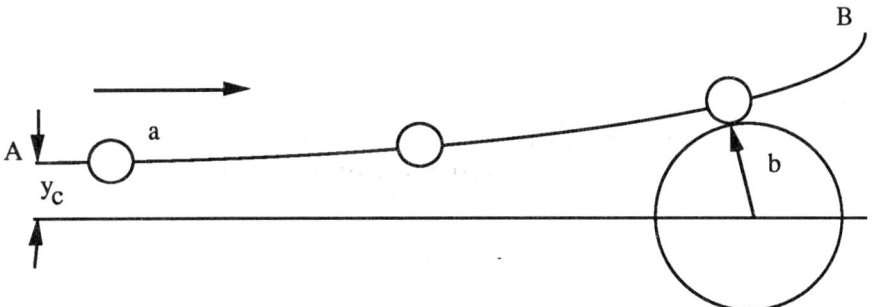

Figure 4.2: The geometry of a two particle collision.

where C_u and C_v are C_a and C_b written in terms of volume.

This formula suffers from a severe limitation. It ignores the fact that an interparticle force exists between the two particles which can lead to the particles not actually colliding, although they would seem able to do so on purely geometric grounds. Fig. 4.2 shows the situation that can occur.

On purely geometric grounds particle a would collide with particle b. However, due to the interparticle forces, particle a moves in a curved path, AB, which takes it past particle b. The critical trajectory in which particle a makes a grazing collision with particle b defines an impact parameter, y_c, and hence a collision efficiency:

$$\varepsilon = \left(\frac{y_c}{a+b}\right)^2 \tag{4.68}$$

All particles with $y<y_c$ strike b and all those with $y>y_c$ do not. Generally, $y_c<a+b$ and so the coagulation rate is less efficient than geometric considerations would imply.

In order to calculate y_c accurately, it is necessary to solve the associated equations of motion of the two particles with the fluid forces. This technique is described later. However, there are simpler if less accurate techniques available which enable estimates of ε to be obtained. A particularly clear demonstration of this has been given by Dunbar and Kirby [1983] and by Morlock [1986] which also highlights a discrepancy between two widely used forms of ε. In calculating ε, we make the following assumptions:

(I) The motion of the particles is dominated by the fluid flow, *i.e.* they are small enough for inertial effects to be neglected.

(II) The smaller particle is so small compared with the larger that it does not appreciably perturb the fluid about the larger particle.

(III) Stokes flow is valid.

Let us assume that the large particle b is at rest and that the fluid is moving over it with an upstream unperturbed velocity, U. The Stokes stream function then determines the flow field about the sphere and the small particle is assumed to be frozen into that flow field.

If the flow is directed in the positive z-direction, then the Stokes stream function takes the form:

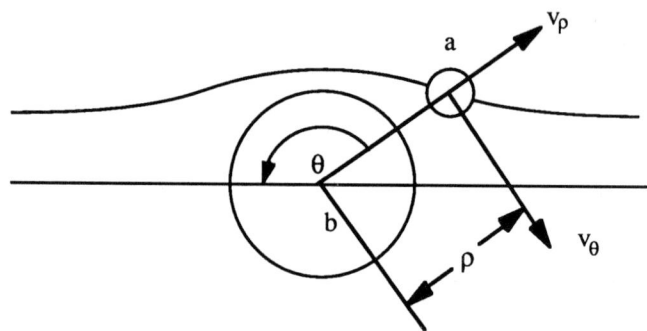

Figure 4.3: (ρ,θ) coordinates of two particle motion in a flow.

$$\psi(\rho,\theta) = -\tfrac{1}{2} U \rho^2 \sin^2(\theta)\left[1 - \frac{3b}{2\rho} + \frac{b^3}{2\rho^3}\right] \quad (4.69)$$

where $U=V_a-V_b$, the difference in the settling speeds and b is the radius of the sphere. The coordinates, (ρ,θ), are defined in Fig. 4.3.

The velocity components, $U_\rho(\rho,\theta)$ and $U_\theta(\rho,\theta)$, are obtained from the stream function as:

$$U_\rho(\rho,\theta) = U \cos(\theta)\left[1 - \frac{b}{2\rho} + \frac{b^3}{2\rho^3}\right] \quad (4.70)$$

and:

$$U_\theta(\rho,\theta) = -U \sin(\theta)\left[1 - \frac{3b}{4\rho} - \frac{b^3}{4\rho^3}\right] \quad (4.71)$$

The stream function is symmetric in θ about θ=π/2 and therefore its distance of closest approach to the sphere is at θ=π/2. We can define a critical stream line, y_c, as one which leads to a grazing collision between sphere a and sphere b in the equatorial plane, thus:

$$\psi_c = \psi(a+b, \pi/2) = -\tfrac{1}{2} U (a+b)^2 \left[1 - \frac{3}{2(1+\kappa)} + \tfrac{1}{2}\left(\frac{1}{1+\kappa}\right)^3\right] \quad (4.72)$$

where κ=a/b. For convenience we define:

$$f(\kappa) = 1 - \frac{3}{2(1+\kappa)} + \tfrac{1}{2}\left(\frac{1}{1+\kappa}\right)^3 \quad (4.73)$$

Thus:

$$\psi_c = -\tfrac{1}{2} U (a+b)^2 f(\kappa) \quad (4.74)$$

We wish to use this relation to obtain a critical impact parameter, y_c. The impact parameter is obtained as ρ→∞ and θ→0 such that ρ sin(θ)=y_c. Thus:

Coagulation Kernels

$$\psi = -\tfrac{1}{2} U y_c^2 \qquad (4.75)$$

Since the stream function is constant:

$$y_c = (a+b)^2 \, f(\kappa) \qquad (4.76)$$

Thus, from Eqn. (4.68) we find that:

$$\varepsilon = f(\kappa) \qquad (4.77)$$

Strictly speaking, this formula is only valid for $\kappa \ll 1$ because of assumption (II) above. Expanding $f(\kappa)$ for small κ leads to:

$$\varepsilon_F = \tfrac{3}{2}\left(\frac{\kappa}{1+\kappa}\right)^2 \qquad (4.78)$$

which agrees with a result obtained by Fuchs [1964].

A different approach to this problem has been followed by Pruppacher and Klett [1978]. These workers have noted that the Fuchs formula assumes that the particles settle with a relative speed, $V_a - V_b$, throughout the motion. This, however, is not the case because, as they pass near each other, the velocity is no longer in the downward direction but perturbed by the stream line motion. Hence, the velocity due to gravitational settling must be added vectorially to the fluid velocity.

In polar coordinates, the particle velocity has components:

$$v_\rho = -V_a \cos(\theta)$$

and:

$$v_\theta = V_a \sin(\theta)$$

which correspond to $v_x = 0$ and $v_z = -V_a$ in Cartesian coordinates.

The corresponding fluid velocity components are:

$$U_\rho = V_b \cos(\theta)\left[1 - \frac{b}{2\rho} + \frac{b^3}{2\rho^3}\right] \qquad (4.79)$$

$$U_\theta = -V_b \sin(\theta)\left[1 - \frac{3b}{4\rho} - \frac{b^3}{4\rho^3}\right] \qquad (4.80)$$

Adding the particle and fluid velocities leads to the correct relative velocity between the spheres, namely:

$$\omega_\rho = U_\rho + v_\rho = V_b \cos(\theta)\left[1 - \frac{V_a}{V_b} - \frac{3b}{2\rho} + \frac{b^3}{2\rho^3}\right] \qquad (4.81)$$

$$\omega_\theta = -U_\theta - v_\theta = V_b \sin(\theta)\left[1 - \frac{V_a}{V_b} - \frac{3b}{4\rho} - \frac{b^3}{4\rho^3}\right] \qquad (4.82)$$

In view of the relationship between velocity and stream function, it is clear that the paths of the smaller particles are described by the following modified stream function:

$$\psi_\infty = -\tfrac{1}{2} V_b \rho^2 \sin^2(\theta) \left[1 - \frac{V_a}{V_b} - \frac{3b}{2\rho} + \frac{b^3}{2\rho^3} \right] \tag{4.83}$$

If V_a and V_b are obtained from Stokes flow:

$$\frac{V_a}{V_b} = \frac{m_a C_a}{m_b C_b} \frac{b}{a} = \frac{a^2}{b^2} = \kappa^2 \quad \text{(if we set } C = 1\text{)} \tag{4.84}$$

Proceeding as before for the minimum impact parameter, we find:

$$\varepsilon_1 = f(\kappa) - \kappa^2 \tag{4.85}$$

which, for small κ, leads to $\varepsilon_1 \approx \kappa^2/2$ compared with the $3\kappa^2/2$ given by Fuchs.

A pictorial representation of the particle trajectory is shown in Fig. 4.4. The bold lines show the particle motion according to Eqn. (4.83) and the other lines show the stream lines. A value of $\kappa=0.5$ was used to exaggerate the effect.

It is not strictly correct to use the value of y_c obtained from the modified stream function directly in Eqn. (4.68). Rather, it is necessary to recalculate the coagulation kernel from first principles using the appropriate angle dependent relative velocities.

Let us recall that the coagulation kernel is obtained by calculating the current of particles of type a incident on the test particle, b. If we define a spherical surface S, of radius, a+b, around the larger particle, then we can see that a current of particles of type a is incident on only one hemisphere. Thus, the current of particles incident on the test particle will be:

$$J(\rho,\theta) = \omega(\rho,\theta) C_\infty \tag{4.86}$$

The total current incident on the hemisphere is then:

$$J_{tot} = -\int_S dS \cdot \omega(\rho,\theta) C_\infty \tag{4.87}$$

Since $K(a,b) = J_{tot}/C_\infty$, we obtain:

$$K(a,b) = -(a+b)^2 \int_{\pi/2}^{\pi} d\theta \sin(\theta) \int_0^{2\pi} d\phi \, \hat{\rho} \cdot \omega(a+b,\theta) \tag{4.88}$$

where $\hat{\rho}$ is the unit vector normal to the surface of the hemisphere. But:

$$\hat{\rho} \cdot \omega(a+b,\theta) = V_b \cos(\theta) \left[1 - \frac{V_a}{V_b} - \frac{3b}{2(a+b)} + \frac{b^3}{2(a+b)^3} \right] \tag{4.89}$$

$$= V_b \cos(\theta) \left[f(\kappa) - \kappa^2 \right] \tag{4.90}$$

Coagulation Kernels

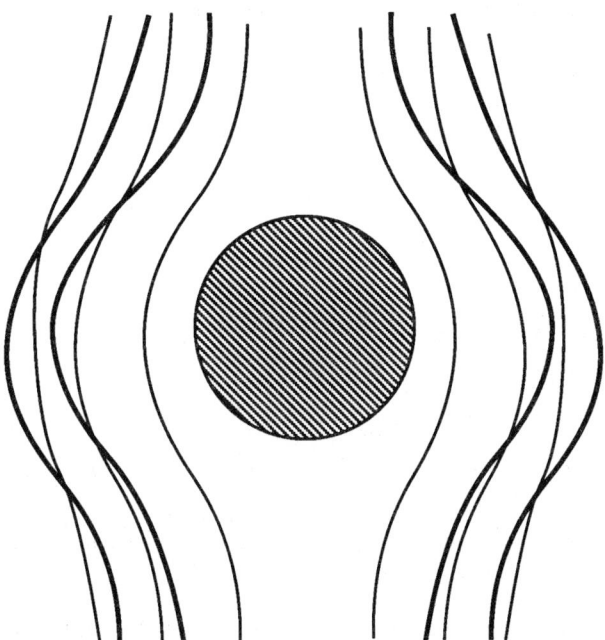

Figure 4.4: Stokes stream lines (normal) and particle trajectories (**bold**).

whence:
$$K(a,b) = -V_b (a+b)^2 \left[f(\kappa)-\kappa^2\right] \int_0^{2\pi} d\phi \int_{\pi/2}^{\pi} d\theta \sin(\theta) \cos(\theta)$$
$$= \pi (a+b)^2 V_a \left[f(\kappa)-\kappa^2\right]$$
$$= \pi (a+b)^2 (V_a - V_b) \frac{f(\kappa)-\kappa^2}{1-\kappa^2} \tag{4.91}$$

From this, we see that the collision efficiency can be defined as:
$$\varepsilon_2 = \frac{f(\kappa)-\kappa^2}{1-\kappa^2} \tag{4.92}$$

Expanding in κ^2 and keeping only the lowest order terms, we get:
$$\varepsilon_{PK} = \tfrac{1}{2}\left(\frac{\kappa}{1+\kappa}\right)^2 \tag{4.93}$$

which is the result obtained by Pruppacher and Klett and is to be preferred to that of Fuchs because it allows for a more accurate description of the particle trajectory.

It should be noted that $\varepsilon_2=0$ for $\kappa=(0.1915)$ and that no collisions are predicted for κ greater than this. Of course, this is absurd and can be explained by assumption (II) which demands that there is no perturbation of the fluid by the smaller particle. For this reason, all collision efficiencies obtained by these methods are restricted to small κ. Since in practice we wish to use ϵ for all κ between 0 and 1, it is necessary to carry out more detailed calculations which do not suffer from assumptions (I)–(III). Such matters are considered later in this chapter. It is worth noting here, however, that ϵ_{PK} is a reasonable approximation to the actual value of ϵ up to about $\kappa=0.5$. Thereafter, the collision efficiency tends to level off to a constant value about a factor of two less than ϵ_{PK}.

4.4 Laminar shear

It is necessary to know how the collision rate between aerosol particles is affected by the flow field. This is a complex matter, but one of the simplest flow fields arises when a constant shear force is applied and Couette flow occurs. The interaction of particles in such a flow field was first considered by Smoluchowski [1916]. The physical situation is shown in Fig. 4.5.

In order to calculate the collision rate, we consider the coordinate system shown in Fig. 4.6. The sphere is of radius, a+b.

Let the velocity gradient in the y-direction be a constant, G. Then, the relative velocity of the particles at y will be:

$$V = G y \, k \tag{4.94}$$

The current of type a particles, onto the test particle b, is:

$$J = -C_\infty V \tag{4.95}$$

and so the total current over the whole sphere is:

$$J_{tot} = -\int_S J \cdot dS \tag{4.96}$$

Now, setting:

$$y = (a+b)\sin(\theta)\sin(\phi)$$

and:

$$k \cdot dS = |\cos(\theta)| \, dS$$

$$dS = (a+b)^2 \sin(\theta) \, d\theta \, d\phi$$

we get:

$$J_{tot} = C_\infty G (a+b)^3 \int_0^\pi \sin(\phi) \, d\phi \int_0^\pi \sin^2(\theta)|\cos(\theta)| \, d\theta = \tfrac{4}{3} C_\infty G (a+b)^3 \tag{4.97}$$

The coagulation kernel is therefore:

$$K(a,b) = \tfrac{4}{3} G (a+b)^3 \tag{4.98}$$

or, in terms of particle volume:

Coagulation Kernels

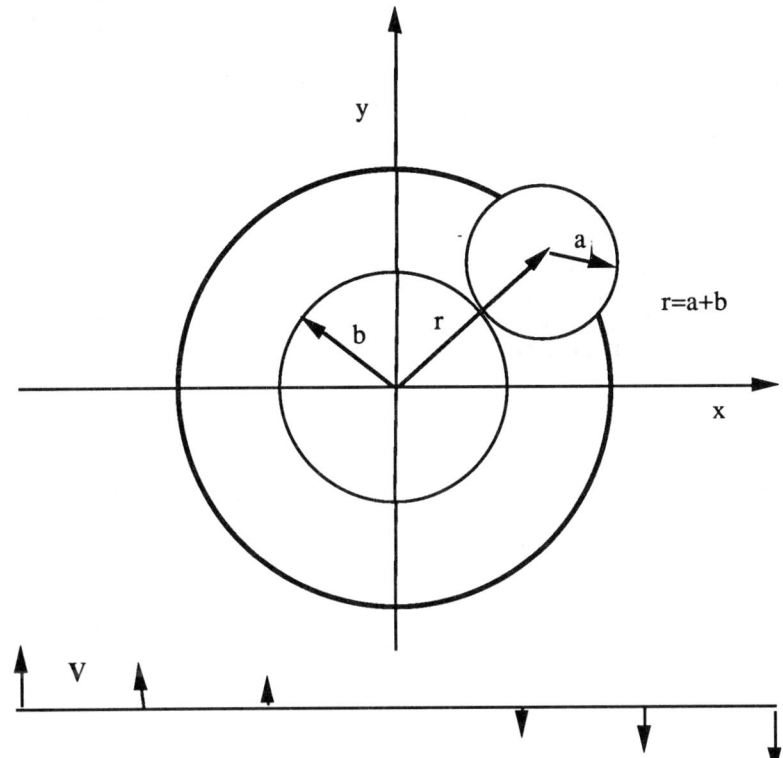

Figure 4.5: Two particle motion in laminar shear flow.

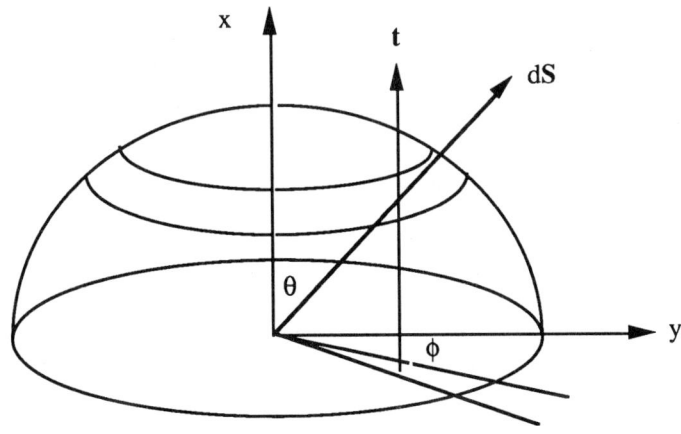

Figure 4.6: A coordinate system for laminar shear flow.

$$K(u,v) = \frac{G}{\pi}\left(u^{1/3} + v^{1/3}\right)^3 \tag{4.99}$$

This result assumes that the particles move with the unperturbed shear velocity and that any particle collides if it is within a cylinder of radius, a+b, with its axis parallel to the flow direction. In practice, as we have seen for gravitational coagulation, the cross sectional area will depend on the fluid forces between the particles and a collision efficiency must be introduced. Such calculations have been carried out by van de Ven and Mason [1977a,b], Arp and Mason [1976], Curtis and Hocking [1970], and Adler [1981] and will be discussed later.

4.5 Turbulent coagulation

In many practical situations, the flow field in a fluid is turbulent. That is, the fluid is in a state of random motion containing eddies of varying sizes and is moving in a more or less isotropic manner. The frequency of encounters between particles entrained in such flows will be substantially increased above that for a quiescent fluid. It will be shown that the increase in encounters in a turbulent fluid may be caused by two independent and essentially different mechanisms. One of these mechanisms, which is called the 'turbulent inertial effect,' is due to the large difference in the densities of the particle and the fluid; a situation that is particularly marked in a gas. Because of this density difference, the particles are not fully entrained by the turbulent eddies and an inertial effect forces particles out of one eddy into another. To put it another way, there are local accelerations arising in the gas which are transmitted to the particles. The velocities of colliding particles will therefore have little correlation and can lead to large relative velocities and collision rates. This mechanism was investigated by East and Marshall [1954] and we shall discuss it in Section 4.10.

There is another mechanism of coagulation which also depends on turbulence and it operates even for particles of equal size and mass. The term for this mechanism is 'turbulent diffusion.' The particles under consideration are usually smaller by at least an order of magnitude than the length scale of small eddies in the turbulence and thus, it will be the small scale motion that governs the relative motion of neighboring drops. Since the microscale of turbulence, even with very energetic mixing, is of order 100–500 μm and because many particles are usually much smaller than this, it is clear that these particles will become fully entrained in the eddies and the collisions that occur will arise from a form of diffusion. Thus, the particles are assumed to migrate through the fluid in a chaotic fashion rather like a 'macroscopic' Brownian motion. A physical explanation of the relative velocity that arises can be had from the fact that during the turbulent action there is a distortion of the fluid so that, for example, an initially spherical element of fluid becomes, after a period of time, ellipsoidal and hence the two particles will move closer together, eventually touching. Indeed, it is found that such motion may be described by a diffusion equation with a modified turbulent diffusion coefficient. In any practical situation, both turbulent mechanisms will coexist.

4.6 Turbulent diffusion

There are two seemingly unrelated methods for calculating the collision rate for turbulent diffusion. One depends on the use of a turbulent diffusion coefficient and the other on the use of statistically averaged fluid velocities. We shall describe both methods and attempt to make some connection between them.

The statistical averaging method is due to Saffman and Turner [1956]. Here, the particles are supposed to move with the fluid and so any statistical averaging of the fluid properties also applies to the particle properties. For two points close together in a turbulent fluid the relative

Coagulation Kernels

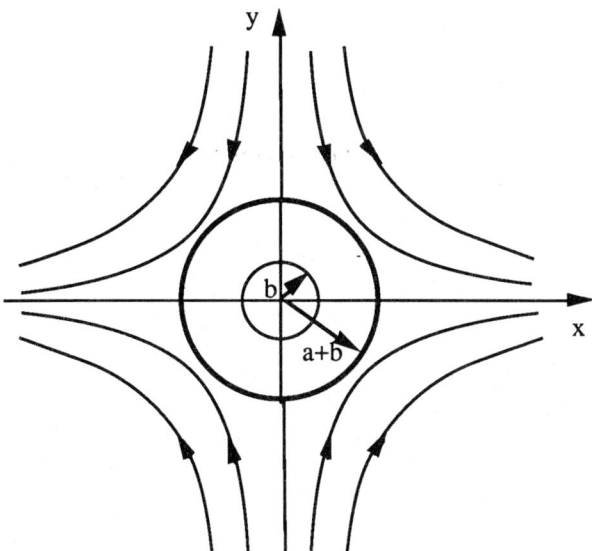

Figure 4.7: Stream lines of relative motion between two particles.

motion is that of uniform strain. Thus, if the center of one drop is taken as the origin, the stream lines of the relative motion are shown in Fig. 4.7.

The flux of type a particles onto the test particle, b, is obtained by noting that they move along the stream lines assuming, of course, no distortion of the stream lines (*i.e.*, a collision efficiency of unity).

If the particles are uniformly distributed, the collision rate of particles of radius, a, with a single particle of radius, b, is:

$$\left\langle -\int_{S(\omega_r<0)} \omega_r \, dS \right\rangle n_a \, n_b \tag{4.100}$$

where ω_r is the radial component of the relative velocity, S is the surface area of the sphere of radius, a+b, and the brackets, <...>, denote a statistical average over many realizations of the turbulence. n_a and n_b are the concentrations of particles a and b, respectively.

To evaluate this integral, we assume that the fluid is incompressible and that, as a result, the flux of fluid into the sphere will be equal to the flux out:

$$\int_{S(\omega_r<0)} \omega_r \, dS + \int_{S(\omega_r>0)} \omega_r \, dS = 0 \tag{4.101}$$

which also follows from the continuity equations. Thus:

$$-\int_{S(\omega_r<0)} \omega_r \, dS = \tfrac{1}{2} \int_{S(\text{all } \omega_r)} |\omega_r| \, dS \tag{4.102}$$

from which we may write:

$$\left\langle -\int_{S(\omega_r>0)} \omega_r \, dS \right\rangle = \tfrac{1}{2} \int_{S(\text{all }\omega_r)} \langle |\omega_r| \rangle \, dS \tag{4.103}$$

If the turbulence is isotropic compared with the length scale, a+b, then $\langle |\omega_r| \rangle$ is independent of position on the surface of the sphere. In that case, the integral over the sphere is given by:

$$\left\langle -\int_{S(\omega_r>0)} \omega_r \, dS \right\rangle = 2\pi (a+b)^2 \langle |\omega_r| \rangle \tag{4.104}$$

But, because of the isotropy, $\langle |\omega_r| \rangle = \langle |\omega_x| \rangle$ where ω_x is the component of velocity along the x-axis (which is arbitrary).

If a+b is small compared with the length scale of turbulence (*i.e.* small eddies) and if u_x is the component of the absolute velocity along the x-axis, then:

$$\omega_x = (a+b) \frac{\partial u_x}{\partial x} \tag{4.105}$$

Therefore:

$$\left\langle -\int_{S(\omega_r<0)} dS \right\rangle = 2\pi (a+b)^3 \left\langle \left|\frac{\partial u_x}{\partial x}\right| \right\rangle \tag{4.106}$$

Now, we would really like to relate the mean velocity gradient to a physical parameter of the fluid field, such as the turbulent energy dissipation rate per unit mass of fluid, ε_T.

Taylor [1935] has shown that, for isotropic eddies:

$$\left\langle \left(\frac{\partial u_x}{\partial x}\right)^2 \right\rangle = \frac{\varepsilon_T}{15 \nu} \tag{4.107}$$

where ν is the kinematic viscosity.

In order to relate the mean modulus to the mean square, it is necessary to postulate a distribution function for the gradient of the fluctuating velocity. Assuming this to be normally distributed, *e.g.*:

$$P(\chi) = \frac{1}{\sigma \sqrt{2\pi}} \exp\left(-\frac{\chi^2}{2\sigma^2}\right) \tag{4.108}$$

where $\chi = \partial u_x / \partial x$, we obtain:

$$\left\langle \left(\frac{\partial u_x}{\partial x}\right)^2 \right\rangle = \int_{-\infty}^{\infty} d\chi \, \chi^2 \, P(\chi) = \sigma^2 \tag{4.109}$$

Coagulation Kernels

and:
$$\left\langle \left|\frac{\partial u_x}{\partial x}\right|\right\rangle = \int_{-\infty}^{\infty} d\chi\, |\chi|\, P(\chi) = \sigma\sqrt{\frac{2}{\pi}} \tag{4.110}$$

Thus:
$$\left\langle \left(\frac{\partial u_x}{\partial x}\right)^2 \right\rangle = \sigma^2 = \frac{\pi}{2}\left\langle \left|\frac{\partial u_x}{\partial x}\right|\right\rangle^2 \tag{4.111}$$

and:
$$\left\langle \left|\frac{\partial u_x}{\partial x}\right|\right\rangle = \left(\frac{2\,\varepsilon_T}{15\pi\,\nu}\right)^{1/2} \tag{4.112}$$

It should be noted that there is a typographical error in the expression given by Saffman and Turner.

Using Eqn. (4.106) we find:

$$\left\langle \int_{S(\omega_r<0)} \omega_r\, dS \right\rangle = 2\pi\,(a+b)^3 \left(\frac{2\,\varepsilon_T}{15\pi\,\nu}\right)^{1/2} \tag{4.113}$$

and therefore the coagulation constant is:

$$K(a,b) = \left(\frac{8\pi\,\varepsilon_T}{15\,\nu}\right)^{1/2} (a+b)^3 = (1.29)\,(a+b)^3 \left(\frac{\varepsilon_T}{\nu}\right)^{1/2} \tag{4.114}$$

In terms of particle volume, this becomes:

$$K(u,v) = \frac{\tilde{G}}{\pi}\left(u^{1/3} + v^{1/3}\right)^3 \tag{4.115}$$

where:

$$\tilde{G} = \left(\frac{3\pi\,\varepsilon_T}{10\,\nu}\right)^{1/2} \tag{4.116}$$

We note that the turbulent diffusion coagulation kernel has the same functional dependence on u and v as does the laminar shear kernel with an effective velocity gradient of, $\tilde{G}$.

To put this mechanism into perspective, we need some estimate of the energy dissipation rate ε_T. An order of magnitude value can be obtained from the relation $\varepsilon_T \approx \bar{u}^3/L$, where $\bar{u}$ is the root mean square turbulent velocity and L is the length scale associated with the energy containing eddies. In clouds, where $L \approx 50$ m and $\bar{u} \approx 2$ m sec^{-1}, we find $\varepsilon_T \approx 0.16$ m^{2}sec^{-3}. On the other hand, in a closed container, it might be expected from convection alone to find $\bar{u} \approx 0.5$ m sec^{-1} and $L \approx 10$ m, from which, $\varepsilon_T \approx 0.0125$ m^{2}sec^{-3}. Some values of ε_T and the corresponding values of $\tilde{G}$ are given in Table 4.1 for air.

It should be stressed that an accurate determination of ε_T requires detailed knowledge of the geometry and local heat sources in a reactor containment. For example, following a breach in

Table 4.1: Typical energy dissipation rates, ε_T, and corresponding values of $\tilde{G}$, the effective velocity gradient.

ε_T	0.1	0.01	0.001	0.0001
$\tilde{G}$	77.0	24.5	7.7	2.45

the primary circuit, a jet of steam will be emitted into the vessel. The intensity of this jet is strongly space dependent and hence so also will be ε_T. Sometimes only a fraction of the containment volume is active in producing turbulence and, in such a case, the value of ε_T would have to be reduced accordingly and/or the volume of the containment divided into appropriate sub-volumes.

An alternative method of calculating the coagulation kernel for turbulent diffusion exists based upon classical diffusion concepts. This approach is discussed by Levich [1962]. It depends upon the size of the particle being very much smaller than the turbulent microscale. A measure of this microscale is $\lambda_0 \approx (\nu^3 / \varepsilon_T)^{1/4}$. Thus, with ν for air at STP and $\varepsilon_T = 0.0125$ m^{2}sec^{-3}, we find $\lambda_0 \sim 840$ μm. Since the particles under consideration in aerosol problems are generally less than 50 μm, we are clearly in a region where a<<λ_0. Therefore, we can assume that the particles are transported from place to place by isotropic turbulence in a motion analogous to that of Brownian motion. Such motion may be characterized by a turbulent diffusion coefficient D_T (Levich [1962]) and the aerosol cloud properties are then determined by a classical diffusion equation of the form:

$$\frac{1}{r^2}\frac{d}{dr}\left(r^2 D(r)\frac{d}{dr}C(r)\right) = 0 \quad (4.118)$$

In this case, the diffusion coefficient, $D(r)$, is written

$$D(r) = D_B + D_T(r) \quad (4.119)$$

where D_B is the Brownian diffusion coefficient ($D_a + D_b$).

The turbulent diffusion coefficient can be shown, by means of mixing length theory, to depend on the distance from a solid wall or, in our case, to depend on the distance between the particles centers (although this requires some justification). We find:

$$D_T(r) = \alpha_0\, \varepsilon_T^{1/3}\, r^{4/3} \quad ; \quad r > r_0 \quad (4.120)$$

$$= \beta_0 \left(\frac{\varepsilon_T}{\nu}\right)^{1/2} r^2 \quad ; \quad r < r_0 \quad (4.121)$$

where, according to Levich, $\beta_0 = (0.15)$ and r is the distance between particle centers. It is clear that, at very close separations, the net diffusion coefficient approaches $D_B + D_T(a+b)$. This is dominated by Brownian motion if:

Coagulation Kernels

$$\frac{D_T(a+b)}{D_B} = \beta_0 \left(\frac{\varepsilon_T}{\nu}\right)^{1/2} \frac{(a+b)^2}{D_B} \ll 1 \tag{4.122}$$

For $\varepsilon_T = (0.0125)$ m^{2}sec^{-3} and air at STP, this condition becomes:

$$a\,b\,(a+b) \ll 3 \tag{4.123}$$

where a and b are in microns. Thus, for particles of 1 μm or less, we expect Brownian diffusion to dominate turbulent diffusion.

An explicit expression for the coagulation kernel may be obtained by solving Eqn. (4.118). This is a two region problem, $r >$ or $< \lambda_0$. Let C_1 be the solution in $r < \lambda_0$ and C_2 be the solution in $r > \lambda_0$. Then:

$$r^2 D(r) \frac{d}{dr} C(r) = \text{constant} \tag{4.124}$$

For $r < \lambda_0$, $D(r) = D_1(r)$ and therefore, integrating Eqn. (4.124) and using $C_1(a+b) = 0$, we find:

$$C_1(r) = A_1 \int_{a+b}^{r} \frac{dw}{w^2 D_1(w)} \tag{4.125}$$

Similarly, in the region $r > \lambda_0$, we integrate to find:

$$C_2(r) = C_\infty - A_2 \int_{r}^{\infty} \frac{dw}{w^2 D_2(w)} \tag{4.126}$$

To find A_1 and A_2 we use the continuity condition:

$$C_1 = C_2 \quad ; \quad r = \lambda_0 \tag{4.127}$$

$$D_1 C_1' = D_2 C_2' \quad ; \quad r = \lambda_0 \tag{4.128}$$

from which we find:

$$\frac{1}{A_1} = \frac{1}{A_2} = \frac{1}{C_\infty} \left\{ \int_{a+b}^{\lambda_0} \frac{dw}{w^2 D_1(w)} + \int_{\lambda_0}^{\infty} \frac{dw}{w^2 D_2(w)} \right\} \tag{4.129}$$

The current onto the test sphere is given by:

$$J = 4\pi (a+b)^2 D_1(a+b) \left.\frac{dC_1}{dr}\right|_{r=a+b} \tag{4.130}$$

$$= 4\pi A_1$$

$$= 4\pi C_\infty \left\{ \int_{a+b}^{\lambda_0} \frac{dw}{w^2 D_1(w)} + \int_{\lambda_0}^{\infty} \frac{dw}{w^2 D_2(w)} \right\}^{-1} \tag{4.131}$$

and thus, the coagulation kernel is:

$$K(a,b) = 4\pi \left\{ \int_{a+b}^{\lambda_0} \frac{dw}{w^2 D_1(w)} + \int_{\lambda_0}^{\infty} \frac{dw}{w^2 D_2(w)} \right\}^{-1} \tag{4.132}$$

Since $(a+b)/\lambda_0 \ll 1$, it is readily verified that $K(a,b)$ can be written:

$$K(a,b) = 4\pi \left\{ \int_{a+b}^{\infty} \frac{dw}{w^2 D_1(w)} \right\}^{-1} \tag{4.133}$$

with an error of order $(a+b)^3/\lambda_0^3$.

Using the value of $D_1(w)$ in the above expression leads to:

$$K(a,b) = 4\pi D_B (a+b) g(\chi) \tag{4.134}$$

where:

$$\chi = (a+b) \left(\frac{k_e}{D_B} \right)^{1/2} \tag{4.135}$$

with:

$$k_e = \beta_0 \left(\frac{\varepsilon_T}{\nu} \right)^{1/2}$$

and:

$$g(\chi) = \left(1 - \frac{\pi \chi}{2} + \chi \tan^{-1}(\chi) \right)^{-1} \tag{4.136}$$

We observe that for small χ, when Brownian motion dominates turbulent motion, the correction factor, $g \to 1$, and we regain the classical Brownian coagulation kernel. On the other hand, for $\chi \to \infty$, $g(\chi) \to 3\chi^2$ and:

$$K(a,b) = (5.65)(a+b)^3 \left(\frac{\varepsilon_T}{\nu} \right)^{1/2} \tag{4.137}$$

This result can be compared with that of Saffman and Turner discussed above where the value of the constant is (1.29) rather than (5.65). Note that the structure of the result, i.e. its dependence on ε_T, ν, and particle sizes, is identical. The discrepancy, amounting to a factor of (4.38), can be explained by noting that the factor β_0 in the expression for D_T is only qualitative and, moreover, the statistical arguments used by Saffman and Turner are also approximate. Nevertheless, a deeper understanding of these two approaches is desirable.

The approach using diffusion theory does appear to have certain practical advantages. For example, we have been able to include in the coagulation kernel the simultaneous action of Brownian motion. While it may be possible to modify the Saffman and Turner approach to include such effects, the procedure is by no means as straightforward as in the method employing the diffusion equation. Our final result for the coagulation kernel allows the transition from Brownian to turbulent diffusion to be investigated in terms of the parameter, χ. The physical meaning of χ can be inferred from:

Table 4.2: The ratio, γ_{BT}, as a function of χ^2.

χ^2	γ_{BT}	χ^2	γ_{BT}	χ^2	γ_{BT}	χ^2	γ_{BT}
0.001	1.048	0.01	1.138	0.1	1.282	1.0	1.165
0.002	1.067	0.02	1.183	0.2	1.287	2.0	1.102
0.003	1.081	0.03	1.211	0.3	1.271	3.0	1.074
0.004	1.092	0.04	1.231	0.4	1.251	4.0	1.058
0.005	1.102	0.05	1.246	0.5	1.233	5.0	1.048
0.006	1.111	0.06	1.257	0.6	1.216	6.0	1.040
0.007	1.119	0.07	1.266	0.7	1.201	7.0	1.035
0.008	1.126	0.08	1.273	0.8	1.187	8.0	1.031
0.009	1.132	0.09	1.278	0.9	1.175	9.0	1.028

$$\chi^2 = \frac{k_e (a+b)^2}{D_B} = \frac{D_T^*}{D_B} \tag{4.138}$$

where D_T^* is the turbulent diffusion coefficient of touching spheres. Thus, χ^2 is the ratio of the local turbulent diffusion coefficient to the Brownian diffusion coefficient.

The conventional procedure adopted when Brownian and turbulent coagulation are present is to simply add the kernels, e.g.:

$$K_{BT}^*(a,b) = K_B(a,b) + K_T(a,b) \tag{4.139}$$

where K_T refers to Eqn. (4.137) and K_B to Eqn. (4.27). In view of Eqn. (4.134), we can assess the accuracy of the addition approximation. Thus, we compute the ratio:

$$\gamma_{BT} = \frac{K_{BT}(a,b)}{K_{BT}^*(a,b)} \tag{4.140}$$

where K_{BT} refers to Eqn. (4.134). In terms of χ, we can write:

$$\gamma_{BT} = \frac{g(\chi)}{1+3\chi^2} \tag{4.141}$$

Table 4.2 shows the results for a range of χ^2 values. Clearly, the addition approximation underestimates the combined effect of Brownian and turbulent coagulation with the maximum error of 29% occurring at $\chi^2=0.2$.

4.7 Generalized theory of aerosol coagulation

The methods described above for obtaining coagulation kernels use apparently unconnected procedures. It is possible, however, to develop a unified theory of coagulation that enables all of the various processes to be brought together in a consistent manner. Not only is this practically useful, but it also aids in understanding the ways in which various coagulation mechanisms interact. Essentially, we consider all mechanisms leading to coagulation as being describable by a diffusion process with a superimposed drift velocity. To see how this formalism arises, it is useful to return to Einstein's original discussion of Brownian motion and his derivation of the associated diffusion equation.

Einstein considered the situation where a hypothetical, steady, external force, derivable from a potential, $\psi(\mathbf{r})$, acts on a particle driving it towards a boundary. This force acts against the random thermal motion of the particle which tends to move it away from the boundary. In this state of thermodynamic equilibrium involving the particle and the fluid in which it is suspended at temperature, T, the probability density function, $P(\mathbf{r})$, for the position of the particle is given by the Boltzmann distribution:

$$P(\mathbf{r}) = P_0 \exp\left(-\frac{\psi}{kT}\right) \tag{4.142}$$

Alternatively, if there exists a noninteracting cloud of such particles, then the concentration $n(\mathbf{r})$ will be distributed as:

$$n(\mathbf{r}) = n_0 \exp\left(-\frac{\psi}{kT}\right) \tag{4.143}$$

To put the Einstein argument another way, in this equilibrium system, the mean particle flux due to the action of applied forces is just balanced by the Brownian diffusion down the concentration or probability gradient.

If the concentration of particles is very low and their motion is governed by slow viscous flow theory, the steady velocity, $\mathbf{U}$, of the particle derived from the force, $\mathbf{F}$ (where $\mathbf{F}=-\nabla\psi$), is:

$$\mathbf{U} = \mathbf{b} \cdot \mathbf{F} \tag{4.144}$$

where $\mathbf{b}$ is the particle mobility tensor and is, in general, a second rank tensor.

Now the particle flux is, by definition, $-\mathbf{D} \cdot \nabla n$ where $\mathbf{D}$ is the diffusion coefficient. This flux is just balanced by the flux due to the applied force, $\mathbf{U}n$, i.e.:

$$\mathbf{U}n - \mathbf{D} \cdot \nabla n = 0 \tag{4.145}$$

or, using Eqn. (4.144):

$$\mathbf{b} \cdot \nabla\psi\, n + \mathbf{D} \cdot \nabla n = 0 \tag{4.146}$$

From the Boltzmann relation, Eqn. (4.143), we get:

$$\nabla n = -\frac{\nabla\psi}{kT} n \tag{4.147}$$

from which:

$$\mathbf{D} = kT\,\mathbf{b} \tag{4.148}$$

Coagulation Kernels

This defines the diffusion coefficient. For rigid spheres of radius, a, the mobility tensor is isotropic and can be obtained from Stokes' law, leading to the well known result:

$$\mathbf{D} = \frac{kT}{6\pi a \mu} \mathbf{I} \tag{4.149}$$

I being the unit tensor.

This derivation by Einstein is very useful since it links the effective diffusive force to the concentration gradient. Thus, from Eqn. (4.143) we see that:

$$\psi = -kT \ln\left(\frac{n}{n_0}\right) \tag{4.150}$$

whence:

$$\mathbf{F} = -\nabla\psi = kT\,\nabla\ln(n) \tag{4.151}$$

or a force:

$$\mathbf{f} = -kT\,\nabla\ln(n) \tag{4.152}$$

acting on the particle. Of course, this is not a literal force but one which leads to the same effect as the concentration gradient. It is sometimes called the thermodynamic force.

These results hold if the particle under consideration is free from any interaction with other particles and if external forces are constant. When interactions exist, however, there will be forces acting on the particle that are functions of interparticle distance.

Thus, we might expect the statistical properties of the velocity to change. Since the stationary nature of the random velocity is an essential ingredient of classical Brownian theory, it is clearly important to investigate this point. Batchelor [1976] in a classic paper, has examined this and several other problems associated with a combination of Brownian diffusion and hydrodynamic interaction. He notes, for instance, that two particles which are very close have an inhibited relative Brownian motion and respond to thermal forces with the mass and hydrodynamic resistance of a rigidly joined pair. When they are far apart, however, each has a classical isolated Brownian motion. A complete picture of such an interaction would be very complex indeed. However, Batchelor as well as Deutch and Oppenheim [1971] and Murphy and Aguirre [1972] avoid much of the complexity by supposing that the change in particle configuration during the relaxation time, τ, is negligibly small. In these circumstances, the velocities of the interacting particles remain stationary, random functions of time over a sufficiently large time averaging period. To put it another way, the particle configuration is essentially constant during the time, τ, that characterizes the diffusion process.

Batchelor then proceeds to generalize Einstein's arguments to a group of particles in equilibrium in which a steady thermodynamic force acts with a potential, $\psi(\mathbf{r}_1,\mathbf{r}_2,...,\mathbf{r}_m)$, which depends only on the relative position vectors, $\mathbf{r}_2-\mathbf{r}_1,...,\mathbf{r}_m-\mathbf{r}_1$, and not on the location in space. However, there is also an interactive force between the particles arising from actual interparticle forces due, for example, to van der Waals or similar close range mechanisms. Thus, the probability density for particles can be written:

$$P(\mathbf{r}_1,\mathbf{r}_2,...,\mathbf{r}_m) = P_0 \exp\left(-\frac{\psi+\Phi}{kT}\right) \tag{4.153}$$

In the case of hard spheres, Φ is infinite if any two particles overlap and zero otherwise. We may also write this in terms of a concentration distribution as:

$$n(r_1, r_2, \ldots, r_m) = n_0 \exp\left(-\frac{\Psi + \Phi}{kT}\right) \tag{4.154}$$

Also, by analogy with the single particle case, we see that the force acting on particle k will be:

$$F_k = -kT \frac{\partial}{\partial r_k} \ln[n(r_1, r_2, \ldots, r_m)] \tag{4.155}$$

In order to obtain the diffusive flux, we consider two spheres only, assuming that the probability of three spheres being close together at any given moment is negligible. Thus, we need the diffusive flux of sphere 2 relative to sphere 1 (the test particle).

When two spheres are acted on by two external forces, F_1 and F_2, in a fluid which is at rest at infinity they acquire velocities, U_1 and U_2. Then, the diffusive flux will be:

$$(U_2 - U_1) n(r) \tag{4.156}$$

where $r = r_2 - r_1$. The thermodynamic forces are, respectively:

$$F_1 = kT \nabla \ln[n(r)] \tag{4.157}$$

and:

$$F_2 = -kT \nabla \ln[n(r)] \tag{4.158}$$

which are equal and opposite to maintain equilibrium.

In order to proceed, it is clear that we need a relationship between U_1 and U_2 and F_1 and F_2. Such relationships have been discussed in Chapter 2 in the context of slow viscous flow and lead to:

$$U_1 = b_{11} \cdot F_1 + b_{12} \cdot F_2 \tag{4.159}$$

and:

$$U_2 = b_{21} \cdot F_1 + b_{22} \cdot F_2 \tag{4.160}$$

where the mobility tensors b_{ij} are related to the viscosity of the fluid, the dimensions of the spheres, and their distance apart.

Solving for $U_2 - U_1$ leads to:

$$U_2 - U_1 = (b_{21} - b_{11}) \cdot F_1 + (b_{22} - b_{12}) \cdot F_2 \tag{4.161}$$

and then from Eqns. (4.157) and (4.158) we get:

$$U_2 - U_1 = -kT (b_{11} + b_{22} - b_{12} - b_{21}) \cdot \nabla \ln[n(r)]$$

or:

$$(U_2 - U_1) n(r) = -kT (b_{11} + b_{22} - b_{12} - b_{21}) \cdot \nabla n(r) \tag{4.162}$$

Coagulation Kernels

This leads to a definition of the relative diffusivity of the two spheres in the form:

$$D(r) = kT(b_{11} + b_{22} - b_{12} - b_{21}) \tag{4.163}$$

Now, b_{ij} can be written as follows:

$$b_{ij}(r) = \frac{1}{3\pi\mu(a_i + a_j)}\left\{A_{ij}(r)\frac{\mathbf{rr}}{r^2} + B_{ij}(r)\left(\mathbf{I} - \frac{\mathbf{rr}}{r^2}\right)\right\} \tag{4.164}$$

where the dimensionless quantities, A_{ij} and B_{ij}, depend on:

$$\rho = \frac{2r}{a_1 + a_2} \quad \text{and} \quad \lambda = \frac{a_2}{a_1}$$

The properties of A_{ij} and B_{ij} are discussed in Chapter 2. We see, therefore, that:

$$D(r) = D_0 \left\{G(r)\frac{\mathbf{rr}}{r^2} + H(r)\left(\mathbf{I} - \frac{\mathbf{rr}}{r^2}\right)\right\} \tag{4.165}$$

where:

$$D_0 = \frac{kT}{6\pi\mu}\left(\frac{1}{a_1} + \frac{1}{a_2}\right) \tag{4.166}$$

$$G(r) = G(\rho,\lambda) = \frac{\lambda A_{11} + A_{22}}{1+\lambda} - \frac{4\lambda A_{12}}{(1+\lambda)^2} \tag{4.167}$$

and:

$$H(r) = H(\rho,\lambda) = \frac{\lambda B_{11} + B_{22}}{1+\lambda} - \frac{4\lambda B_{12}}{(1+\lambda)^2} \tag{4.168}$$

It is readily seen that as $\rho \to \infty$, G and H $\to 1$, and therefore $D \to D_0\mathbf{I}$, the free particle value.

Eqn. (4.162) can be generalized to include the effect of a real (as opposed to the thermodynamic) force between the particles, e.g. due to London-van der Waals or electrostatic interactions. To do this, we note that, from Eqns. (4.154) and (4.155), we can write for a two body system:

$$F_1 = kT\,\nabla\ln(n) + \nabla\Phi \tag{4.169}$$

and:

$$F_2 = -kT\,\nabla\ln(n) + \nabla\Phi \tag{4.170}$$

Then:

$$U_2 - U_1 = -\mathbf{D}\cdot\left(\frac{\nabla n}{n} + \frac{\nabla\Phi}{kT}\right)$$

or:

$$(U_2 - U_1)n(r) = -\mathbf{D}\cdot\left(\nabla n + n\frac{\nabla\Phi}{kT}\right) \tag{4.171}$$

183

4.8 The balance equation

The balance equation for the concentration, n(r), is obtained from the continuity equation:

$$\frac{\partial}{\partial t}n(r,t) + \nabla \cdot \left[(U_2 - U_1)n(r,t)\right] = 0 \tag{4.172}$$

Using Eqn. (4.171) we find:

$$\frac{\partial n}{\partial t} = \nabla \cdot \left[D \cdot \nabla n + n D \cdot \frac{\nabla \Phi}{kT}\right] \tag{4.173}$$

which has to be solved with the boundary conditions:

$$n(a_1 + a_2, t) = 0 \tag{4.174}$$

and:

$$n(r,t) \to n_\infty \quad ; \quad |r| \to \infty \tag{4.175}$$

The collision rate is easily obtained from the expression for the net current onto the test particle:

$$J = -n_1 n_2 \int_{S(a_1+a_2)} n(U_2 - U_1) \cdot dS \tag{4.176}$$

$$= n_1 n_2 \int_S \left[D \cdot \left(\nabla n + n \frac{\nabla \Phi}{kT}\right)\right] \cdot dS \tag{4.177}$$

Returning to the result obtained in Eqn. (4.165), we can conclude that the classical calculation of the Brownian coagulation kernel is incorrect. In that calculation, it was assumed that the diffusion coefficient is independent of particle separation whereas in fact, because of hydrodynamic interaction, there is a strong dependence on position. Fig. 4.8 shows this dependence, for the case of spherically symmetric diffusion, for various values of λ and illustrates how the Brownian diffusion coefficient decreases rapidly as the spheres approach. In this figure, and also in Fig. 4.9, we note that $b/a = a_2/a_1$. Both figures are reproduced from Alam [1987].

As an illustration of this effect, we will recompute the coagulation kernel for Brownian motion with the more general diffusion coefficient and the interparticle potential, $\Phi(r)$.

Neglecting initial transients in the diffusion equation, we can reduce the problem to that of solving the steady state diffusion equation:

$$\nabla \cdot \left(D \cdot \nabla n + \frac{n}{kT} D \cdot \nabla \Phi\right) = 0 \tag{4.178}$$

Now we will choose a spherical polar coordinate system in which to work. This being so, we may write the tensor, $D(r)$, in terms of the following dyadic notation:

$$D = D_0 G(r) i_r i_r + D_0 H(r) i_\theta i_\theta + D_0 H(r) i_\varphi i_\varphi \tag{4.179}$$

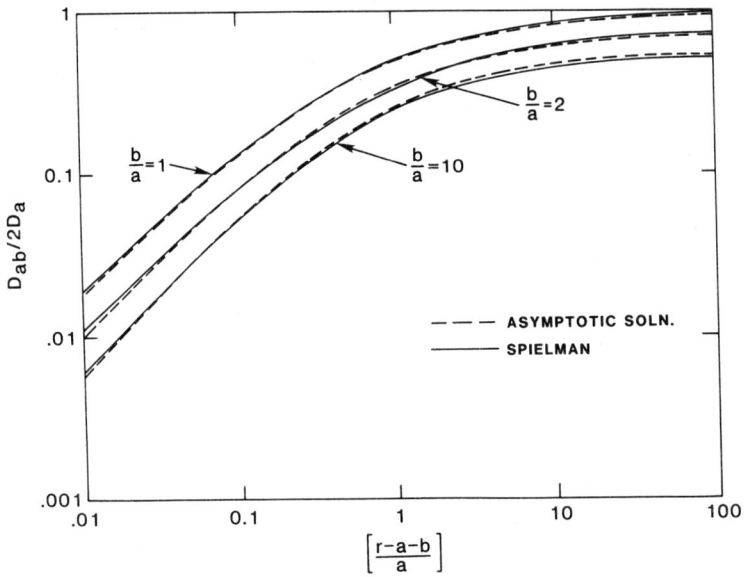

Figure 4.8: Dimensionless relative diffusion coefficient versus dimensionless particle separation for various radius ratios. The dashed curve corresponds to the asymptotic solution. Adapted from Alam [1987] with permission of the Elsevier Science Publishing Company.

An alternative representation is:

$$\mathbf{D} = D_0 \begin{bmatrix} G & 0 & 0 \\ 0 & H & 0 \\ 0 & 0 & H \end{bmatrix} \tag{4.180}$$

In terms of r, θ, and ϕ we obtain, with $n=n(r,\theta,\phi)$:

$$\nabla \cdot (\mathbf{D} \cdot \nabla n) = \frac{D_0}{r^2} \frac{\partial}{\partial r}\left[G(r) \, r^2 \frac{\partial n}{\partial r} \right]$$

$$+ \frac{D_0 \, H(r)}{r^2} \left[\frac{1}{\sin(\theta)} \frac{\partial}{\partial \theta}\left(\sin(\theta) \frac{\partial n}{\partial \theta} \right) + \frac{1}{\sin^2(\theta)} \frac{\partial}{\partial \phi}\left(\frac{\partial n}{\partial \phi} \right) \right] \tag{4.181}$$

and:

$$\nabla \cdot (n \, \mathbf{D} \cdot \nabla \Phi) = \frac{D_0}{r^2} \frac{\partial}{\partial r}\left[n \, G(r) \, r^2 \frac{\partial \Phi}{\partial r} \right]$$

$$+ \frac{D_0 \, H(r)}{r^2} \left[\frac{1}{\sin(\theta)} \frac{\partial}{\partial \theta}\left(n \sin(\theta) \frac{\partial \Phi}{\partial \theta} \right) + \frac{1}{\sin^2(\theta)} \frac{\partial}{\partial \phi}\left(n \frac{\partial \Phi}{\partial \phi} \right) \right] \tag{4.182}$$

Now, if the interparticle potential, $\Phi(\mathbf{r})$, is dependent only on the scalar, r, then we find that $n=n(r)$ and the diffusion equation becomes:

$$\frac{D_0}{r^2}\frac{d}{dr}\left[r^2 G(r)\left(\frac{dn}{dr}+\frac{n}{kT}\frac{d\Phi}{dr}\right)\right]=0 \tag{4.183}$$

This equation is readily integrated so that:

$$r^2 G(r)\left(\frac{dn}{dr}+\frac{n}{kT}\frac{d\Phi}{dr}\right)=C \tag{4.184}$$

Also, the coagulation rate is (with $n_1=n_\infty$ and $n_2=1$):

$$J=n_\infty(a_1+a_2)^2 D_0 \int_0^\pi d\theta \sin(\theta)\int_0^{2\pi}d\phi\left[G(a_1+a_2)\frac{\partial}{\partial r}n(r,\theta,\phi)+\frac{d\Phi}{dr}\frac{n(r,\theta,\phi)}{kT}\right]_{r=a_1+a_2} \tag{4.185}$$

although in our case $n(r,\theta,\phi)=n(r)$.

Solving Eqn. (4.184) subject to $n(\infty)=n_\infty$ and $n(a_1+a_2)=0$, we find:

$$n(r)=n_\infty \exp\left(-\frac{\Phi(r)}{kT}\right)-C\exp\left(-\frac{\Phi(r)}{kT}\right)\int_r^\infty \frac{dw}{w^2 G(w)}\exp\left(-\frac{\Phi(w)}{kT}\right) \tag{4.186}$$

where:

$$\frac{1}{C}=\frac{1}{n_\infty}\int_{a_1+a_2}^\infty \frac{dw}{w^2 G(w)}\exp\left(\frac{\Phi(w)}{kT}\right) \tag{4.187}$$

From this we may show that:

$$J=4\pi D_0 n_\infty C$$

or:

$$K(a,b)=\frac{J}{n_\infty}=4\pi D_0\left[\int_{a_1+a_2}^\infty \frac{dw}{w^2 G(w)}\exp\left(\frac{\Phi(w)}{kT}\right)\right]^{-1} \tag{4.188}$$

We note that some care should be exercised with Eqn. (4.185). This is because $G(a_1+a_2)=0$ and $n(a_1+a_2)=0$ and thus, on the face of it, J=0. However, $\partial n/\partial r$ is infinite at $r=a_1+a_2$ and together these two factors form a definite limit.

The simplest calculation would appear to be when:

$$\Phi=0 \quad ; \quad r>a_1+a_2$$
$$=-\infty \quad ; \quad r<a_1+a_2$$

i.e. hard-sphere interaction. In that case:

$$K(a,b)=4\pi D_0\left[\int_{a_1+a_2}^\infty \frac{dw}{w^2 G(w)}\right]^{-1} \tag{4.189}$$

Coagulation Kernels

Unfortunately, $G(w) \sim$ constant $(w-a_1-a_2)$ as $w \to a_1+a_2$ and this leads to a divergent integral and hence $K(a,b)=0$. Clearly, this is not acceptable physically and, in practice, what is occurring is that the resistance to motion becomes inversely proportional to the gap distance and thus the application of a finite force can never produce contact in a finite time. The anomalous behavior can be explained by arguing that, for all practical materials, there is always a short range attractive surface force that will overcome the hydrodynamic one. Alternatively, one can argue that when the gap separation becomes on the order of a mean free path, Stokes' theory fails and it is necessary to introduce some kinetic theory correction. Hocking [1973] has made the latter correction and developed a correction to the hydrodynamic force term. A full discussion of Hocking's work was given in Chapter 2. Here, we simply note the magnitude of the corrections that arise due to certain interparticle force laws.

As we have seen in Chapter 2, the factor $G(r)$ is rather complicated and some useful approximations to it have been proposed. One of these, for the case of equal size spheres, is given by Honig et al. [1971] in the form:

$$G(r) = \frac{6u^2 + 4u}{6u^2 + 13u + 2} \tag{4.190}$$

where $u=(r-2a)/a$.

For the more general case of unequal spheres, Alam [1987] gives:

$$G(r) = 1 + \frac{(2.6)ab}{(a+b)^2} \left\{ \frac{ab}{(a+b)(r-a-b)} \right\}^{1/2} + \frac{ab}{(a+b)(r-a-b)} \tag{4.191}$$

This expression gives numerical results that are within 1% of the exact values and, moreover, has the correct limiting behavior for small and large values of $r-a-b$. This behavior is illustrated in Fig. 4.8 and compared with the exact results of Spielman [1970].

As an example of the effect of viscous forces, we write the Brownian coagulation kernel in the form:

$$K(a,b) = 4\pi D_0 (a+b) W_c \tag{4.192}$$

where W_c, the enhancement factor, is:

$$\frac{1}{W_c} = (a+b) \int_{a+b}^{\infty} \frac{dw}{w^2 G(w)} \exp\left(-\frac{\Phi(w)}{kT}\right) \tag{4.193}$$

Φ is the van der Waals potential:

$$\Phi(r) = -\frac{A}{6} \left[\frac{2ab}{r^2-(a+b)^2} + \frac{2ab}{r^2-(a-b)^2} + \ln\left(\frac{r^2-(a+b)^2}{r^2-(a-b)^2}\right) \right] \tag{4.194}$$

and A is the Hamaker constant.

Fig. 4.9 shows the results with and without viscous forces for two size ratios. It is clear that significant differences arise in the value of W_c. We also note that Alam has introduced kinetic theory corrections in his work which enable the results to be used for very small particles.

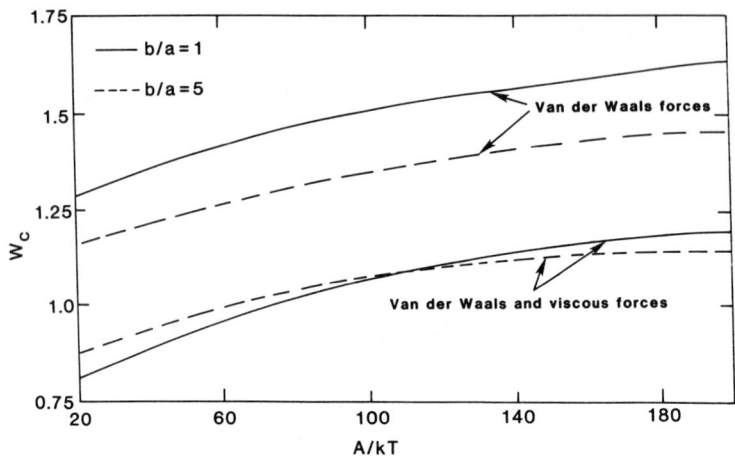

Figure 4.9: The enhancement factor for particles in the continuum regime (W) for the cases: 1) when only van der Waals forces operate and; 2) when both van der Waals forces and viscous forces operate. Adapted from Alam [1987] with permission of the Elsevier Science Publishing Company.

We shall return to Eqn. (4.193) for other potentials later. At the moment, we wish to generalize the procedure.

4.9 Diffusive and convective motion

The derivation of the diffusion equation given above for two particle interaction involved only diffusive motion. In many situations, there exist external forces acting on the particles which give rise to a net velocity. An obvious example is the force of gravity and yet another is nonuniform flow of the surrounding fluid due to turbulence or laminar shear.

In the case of gravitational settling, the force acting on a particle of mass m_i is $F_i = m_i g$. For an isolated particle moving in Stokes flow, the resistance will be $6\pi a_i \mu U_i^{(0)}$ and so, the asymptotic fall velocity is:

$$U_i^{(0)} = \frac{m_i g}{6\pi a_i \mu} = \frac{2 a_i^2 \rho_i}{9 \mu} g \tag{4.195}$$

where ρ_i is the density of the particle which we assume is much greater than that of the gas.

Suppose now we have two particles falling under gravity in a quiescent fluid. Their velocities are U_1 and U_2 and they are acted on by forces, F_1 and F_2, respectively. Then, as we have seen in Chapter 2, the velocities are

$$U_1 = b_{11} \cdot F_1 + b_{12} \cdot F_2 \tag{4.196}$$

and:

$$U_2 = b_{21} \cdot F_1 + b_{22} \cdot F_2 \tag{4.197}$$

Coagulation Kernels

where b_{ij} are the mobility tensors defined by Eqn. (4.164). The relative velocity is then

$$V_{12} = U_2 - U_1 = (b_{21} - b_{11}) \cdot F_1 + (b_{22} - b_{12}) \cdot F_2 \qquad (4.198)$$

where we have neglected the inertia of the particles. As the two spheres move apart, their relative velocity becomes:

$$V_{12}^{(0)} = U_2^{(0)} - U_1^{(0)} = (\lambda^2 \gamma - 1) U_1^{(0)} \qquad (4.199)$$

where $\gamma = \rho_2/\rho_1$. On substituting for the mobility tensors:

$$V_{12}(r) = V_{12}^{(0)} \left\{ \frac{rr}{r^2} L(r) + \left(I - \frac{rr}{r^2} \right) M(r) \right\} \qquad (4.200)$$

where:

$$L(r) = L(\rho, \lambda) = \frac{\lambda^2 \gamma A_{22} - A_{11}}{(\lambda^2 \gamma - 1)} + \frac{2(1 - \lambda^3 \gamma) A_{12}}{(1 + \lambda)(\lambda^2 \gamma - 1)} \qquad (4.201)$$

and:

$$M(r) = M(\rho, \lambda) = \frac{\lambda^2 \gamma B_{22} - B_{11}}{(\lambda^2 \gamma - 1)} + \frac{2(1 - \lambda^3 \gamma) B_{12}}{(1 + \lambda)(\lambda^2 \gamma - 1)} \qquad (4.202)$$

If we now consider the net relative velocity of the particles due to gravity, diffusion, and interparticle forces, we find:

$$V_{12}^*(r) = V_{12}(r) - D \cdot \left(\frac{\nabla n}{n} + \frac{1}{kT} \nabla \Phi \right) \qquad (4.203)$$

Thus, from the continuity equation, we find:

$$\frac{\partial}{\partial t} n(r,t) = -\nabla \cdot (V_{12}(r) n(r,t)) + \nabla \cdot \left[D \cdot \left(\nabla n(r,t) + \frac{1}{kT} \nabla \Phi \, n(r,t) \right) \right] \qquad (4.204)$$

Neglecting time transients, we find that the steady state flow of particles to a test particle is given by:

$$\nabla \cdot \left[D \cdot \left(\nabla n(r) + \frac{1}{kT} \nabla \Phi(r) n(r) \right) \right] - \nabla \cdot [V_{12}(r) n(r)] = 0 \qquad (4.205)$$

Subject to the boundary conditions:

$$n(r) = 0 \quad ; \quad r = a_1 + a_2 \qquad (4.206)$$

$$n(r) = n_\infty \quad ; \quad \rho \to \infty \qquad (4.207)$$

for all θ and ϕ.

The corresponding coagulation kernel is given by:

$$K(a_1,a_2) = \frac{1}{n_\infty}(a_1+a_2)^2 D_0 G(a_1+a_2) \int_0^{2\pi} d\phi \int_0^{\pi} d\theta \sin(\theta) \frac{\partial n}{\partial \theta}\bigg|_{r=a_1+a_2} \quad (4.208)$$

In terms of polar coordinates we may write:

$$\nabla \cdot (V_{12}\, n) = \frac{1}{r^2}\frac{\partial}{\partial r}(r^2 V_{12}^r\, n) + \frac{1}{r\sin(\theta)}\frac{\partial}{\partial \theta}(V_{12}^\theta \sin(\theta)\, n) + \frac{1}{r\sin(\theta)}\frac{\partial}{\partial \phi}(V_{12}^\phi\, n)$$

Eqns. (4.204) and (4.205) have been studied in some generality by Batchelor [1982]. In our work, however, we do not require such detail. Nevertheless, Eqn. (4.205) and its boundary conditions form the basis of any generalized approach to coagulation and we shall illustrate the power of the equation by means of several examples.

4.10 Brownian and gravitational coagulation

We have seen earlier in this chapter how coagulation kernels may be obtained for Brownian motion and for gravitational settling. In arriving at the expressions for K(a,b), it was assumed that these mechanisms acted alone. Although no indication was given as to what action to take when a system was simultaneously undergoing Brownian and gravitational coagulation, the currently accepted practice is to algebraically add the two kernels:

$$K_{BG}(a,b) = K_B(a,b) + K_G(a,b) \quad (4.209)$$

where subscripts B and G refer to Brownian and gravitational coagulation. In this case, therefore, the combined effect is given by:

$$K_{BG}(a,b) = \frac{2kT}{3\mu}\left(\frac{1}{a}+\frac{1}{b}\right)(a+b) + \frac{2g\rho\pi}{9\mu}(a+b)^2\,|a^2-b^2| \quad (4.210)$$

where we have neglected effects due to kinetic theory slip (*i.e.*, the Cunningham factor = 1).

The rationale behind the additive assumption is that when K_B is large, K_G is small and vice versa. This is certainly nearly true, for we know that Brownian motion dominates for small particles and gravitational settling for large ones. However, the precise nature of the approximation is unclear. In order to clarify this problem, we shall employ Eqn. (4.205) which incorporates the effects of both Brownian motion, via **D**, and gravitational settling, via V_{12}.

The example that we shall use will neglect a number of important aspects of the problem but will, nevertheless, lead to a quantitatively useful result. Thus, we assume that there are no interparticle forces and no hydrodynamic forces. In that case:

$$\mathbf{D} = D_0\, \mathbf{I}$$

and:

$$\mathbf{V}_{12} = \mathbf{V}_{12}^{(0)}$$

and the diffusion equation becomes:

Coagulation Kernels

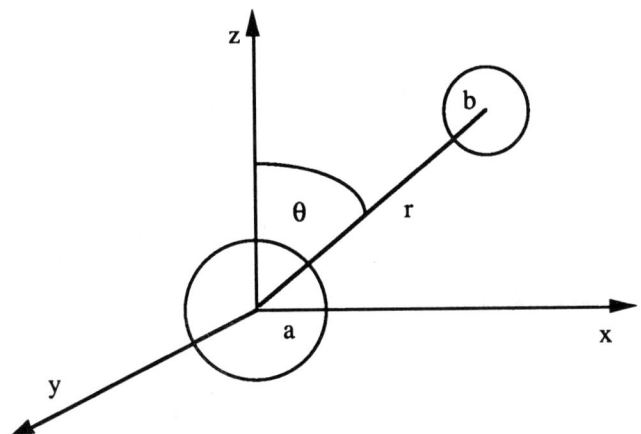

Figure 4.10: Spherical polar coordinates for the solution of Eqn. (4.211).

$$D_0 \nabla^2 n(r) - \mathbf{V}_{12}^{(0)} \cdot \nabla n(r) = 0 \tag{4.211}$$

Both D_0 and $\mathbf{V}_{12}^{(0)}$ are now independent of position because we no longer allow for the inhibiting effect of the fluid. In fact, this assumption was employed in arriving at Eqn. (4.210) and so we are consistent throughout. The use of Eqn. (4.211) enables us, therefore, to examine the accuracy of additivity.

In order to solve Eqn. (4.211) we cast it into spherical polar coordinates as shown in Fig. 4.10.

The sphere of radius, a, is the test particle and is located at the origin. The sphere of radius, b, has its center at the point (r,θ). We shall see that the problem is azimuthally symmetric and so the angle, ϕ, can be integrated out.

Now, from Eqns. (4.199) and (4.195) we see that

$$\mathbf{V}_{12}^{(0)} = \frac{2\rho g}{9\mu}(a^2 - b^2)\mathbf{k} \equiv -V\mathbf{k} \tag{4.212}$$

where $\mathbf{k}$ is the unit vector in the z-direction. Now:

$$\mathbf{k} \cdot \nabla n = \cos(\theta)\frac{\partial n}{\partial r} - \frac{\sin(\theta)}{r}\frac{\partial n}{\partial \theta} \tag{4.213}$$

and so the concentration of b type particles $n(r,\theta)$ is given by:

$$\frac{D_0}{r^2}\frac{\partial}{\partial r}\left(r^2 \frac{\partial n}{\partial r}\right) + \frac{D_0}{r^2 \sin(\theta)}\frac{\partial}{\partial \theta}\left(\sin(\theta)\frac{\partial n}{\partial \theta}\right) + V\cos(\theta)\frac{\partial n}{\partial r} - \frac{V\sin(\theta)}{r}\frac{\partial n}{\partial \theta} = 0 \tag{4.214}$$

In terms of $n(r,\theta)$, the coagulation kernel follows from Eqn. (4.185):

$$K_{BG}^{\bullet}(a,b) = 2\pi (a+b)^2 D_0 \int_0^\pi d\theta \sin(\theta) \frac{\partial n}{\partial r}\bigg|_{r=a+b} \quad (4.215)$$

The boundary conditions on Eqn. (4.214) are:

(1) At r=∞ ; n=n_∞ where n_∞ is the spatially constant number density of 'b' particles in the absence of coagulation.

(2) At r=a+b ; n=0. This is because no b particles can have centers inside the sphere. Thus, to prevent J from becoming infinite at r=a+b it is necessary that n=0.

(3) At θ=0, π ; ∂n/∂θ = 0. This is necessary to ensure that ∂n/∂θ is continuous everywhere.

Before solving Eqn. (4.214), we simplify the notation by setting x=r/(a+b), m(x,θ)=n(r,θ)/n_∞, and:

$$\beta = \frac{(a+b) V}{D_0} = \frac{4\pi \rho g a b}{3 kT} |a^2 - b^2| \quad (4.216)$$

Then, Eqn. (4.214) becomes:

$$\frac{1}{x^2}\frac{\partial}{\partial x}\left(x^2 \frac{\partial m}{\partial x}\right) + \frac{1}{x^2 \sin(\theta)}\frac{\partial}{\partial \theta}\left(\sin(\theta)\frac{\partial m}{\partial \theta}\right) + \beta \cos(\theta)\frac{\partial m}{\partial x} - \frac{\beta \sin(\theta)}{x}\frac{\partial m}{\partial \theta} = 0 \quad (4.217)$$

and:

$$K_{BG}^{\bullet}(a,b) = 2\pi (a+b) D_0 \int_0^\pi d\theta \sin(\theta) \frac{\partial m}{\partial x}\bigg|_{x=1} \quad (4.218)$$

As a measure of the accuracy of the sum kernel given in Eqn. (4.210), we calculate the ratio:

$$\gamma = \frac{K_{BG}^{\bullet}(a,b)}{K_{BG}(a,b)} = \frac{2}{4+\beta}\int_0^\pi \frac{\partial m}{\partial x}\bigg|_{x=1} \sin(\theta)\, d\theta \quad (4.219)$$

Eqn. (4.217) is readily solved by making the transformation, μ=cos(θ):

$$m(x,\mu) = g(x,\mu) \exp(-\tfrac{1}{2}\beta x \mu) \quad (4.220)$$

whereby the equation for g becomes:

$$\frac{1}{x^2}\frac{\partial}{\partial x}\left(x^2 \frac{\partial g}{\partial x}\right) + \frac{1}{x^2}\frac{\partial}{\partial \mu}(1-\mu^2)\frac{\partial g}{\partial \mu} - \tfrac{1}{4}\beta^2 g = 0 \quad (4.221)$$

subject to the boundary conditions:

$$g(1,\mu) = 0 \quad (4.222)$$

and:

$$\lim_{x\to\infty}[g(x,\mu)] = \exp(\tfrac{1}{2}\beta x \mu) \quad (4.223)$$

Coagulation Kernels

Using the separation of variables technique, the solution of Eqn. (4.221) can be written:

$$g(x,\mu) = \left(\frac{2\pi}{\beta x}\right)^{1/2} \sum_{n=0}^{\infty} \left[A_n K_{n+\frac{1}{2}}\left(\frac{\beta x}{2}\right) + B_n I_{n+\frac{1}{2}}\left(\frac{\beta x}{2}\right)\right] P_n(\mu) \qquad (4.224)$$

where I and K are the modified Bessel functions of half-integer order and $P_n(\mu)$ are Legendre polynomials.

Using $g(1,\mu)=0$ and the relationship:

$$\exp\left(\frac{\beta x \mu}{2}\right) = \left(\frac{\pi}{\beta x}\right)^{1/2} \sum_{n=0}^{\infty} (2n+1) I_{n+\frac{1}{2}}\left(\frac{\beta x}{2}\right) P_n(\mu) \qquad (4.225)$$

we see that the solution can be written:

$$m(x,\theta) = 1 - \exp\left(-\frac{\beta x \mu}{2}\right)\left(\frac{\pi}{\beta x}\right)^{1/2} \sum_{n=0}^{\infty} (2n+1) \frac{I_{n+\frac{1}{2}}(\beta/2)}{K_{n+\frac{1}{2}}(\beta/2)} K_{n+\frac{1}{2}}\left(\frac{\beta x}{2}\right) P_n(\mu) \qquad (4.226)$$

Noting that:

$$\left.\frac{\partial m}{\partial x}\right|_{x=1} = \left(\frac{\pi}{\beta}\right)^{1/2} \exp\left(-\frac{\beta \mu}{2}\right) \sum_{n=0}^{\infty} (2n+1) \frac{P_n(\mu)}{K_{n+\frac{1}{2}}(\beta/2)} \qquad (4.227)$$

and using the integral:

$$\int_{-1}^{1} d\mu\, P_n(\mu) \exp\left(-\frac{\beta \mu}{2}\right) = 2(-1)^n \left(\frac{\pi}{\beta}\right)^{1/2} I_{n+\frac{1}{2}}(\beta/2) \qquad (4.228)$$

we find:

$$\gamma = \frac{4\pi}{4+\beta} \frac{1}{\beta} \sum_{n=0}^{\infty} (-1)^n (2n+1) \frac{I_{n+\frac{1}{2}}(\beta/2)}{K_{n+\frac{1}{2}}(\beta/2)} \qquad (4.229)$$

This infinite sum is convergent and is shown in Table 4.3. It may be observed that $\gamma>1$ and hence that the sum kernel underestimates the true rate of coagulation by at most 30% although over most of the range of β, which is a measure of gravitational to diffusive forces, the deviation is much less.

We conclude, therefore, that the additivity assumption is a reasonable one within the limitations of the problem. Nevertheless, our exact kernel shows how the process gradually changes from Brownian to gravitational settling as the average particle size increases.

We can use these results to obtain a deeper understanding of the physical processes occurring during combined Brownian and gravitational coagulation. For example, when Brownian coagulation dominates the situation, an expansion in small values of β leads to:

$$K_{BG}(a,b) = K_B(a,b) + 2 K_G(a,b)$$

Table 4.3: Values of γ–1 as a function of β from Eqn. (4.229).

β	γ–1	β	γ–1	β	γ–1
0.0	0.0	1.0	0.15508	10.0	0.25913
0.1	0.02362	1.5	0.19326	15.0	0.23766
0.2	0.04473	2.0	0.21893	20.0	0.21822
0.3	0.06369	3.0	0.24871	25.0	0.20189
0.4	0.08079	4.0	0.26278	30.0	0.18822
0.5	0.09625	5.0	0.26874	35.0	0.17666
0.6	0.11103	6.0	0.27022	40.0	0.16676
0.7	0.12304	7.0	0.26914	45.0	0.15818
0.8	0.13469	8.0	0.26656	50.0	0.15067
0.9	0.14533	9.0	0.26310	60.0	0.13810

which implies that the effect of gravitational settling is relatively larger by a factor of 2 than the addition hypothesis predicts. Consider the deposition of small particles onto a large particle. In the conventional calculation of Brownian coagulation, the diffusion equation describes the depletion of the 'small' particle distribution in the neighborhood of the 'large' particle due to diffusive flow of small particles onto the large particle. If the large particle is falling through the small particles, then, apart from those with which it directly collides and coagulates, there will also be an additional flux of small particles into the region around the large particle which will partly offset the depletion in that region due to the diffusive flow mentioned above. This, in turn, will increase the diffusive flux of small particles onto the large one. The net effect is that the combined effect of Brownian and gravitational coagulation is likely to be greater than the sum of the separate effects. The result shown in Eqn. (4.229) confirms this. When the gravitational effect dominates, it is not obvious how to expand Eqn. (4.229) and therefore a detailed asymptotic analysis has been carried out for large β. The result is that:

$$\gamma - 1 \approx (4.5) \beta^{-2/3}$$

or:

$$K_{BG}(a,b) \approx K_B(a,b) + \left[1 + (4.5)\beta^{-2/3}\right] K_G(a,b) \approx K_G(a,b) + \left[1 + (1.125)\beta^{-2/3}\right] K_B(a,b)$$

Here, we can argue that the existence of a sedimentation flux increases the diffusional deposition by a proportional amount of $(1.125)\beta^{-2/3}$ (Simons et al. [1986]).

4.11 Generalized turbulent coagulation

We saw in Section 4.5 that motion due to turbulent diffusion could be adequately described by an effective diffusion coefficient $D_T(r)$. This is fully consistent with the formalism that will be developed in this section. It is not so clear, however, what approach to take with turbulent inertial motion in the context of the diffusion-convection equation. In this case it is V_{12} that contains the turbulent information. But we have arrived at Eqn. (4.205) on the basis that V_{12} is deterministic. A plausible modification to Eqn. (4.205) can be derived for turbulent motion if we write:

$$V_{12} = V_{12}^{(D)} + V_{12}^{(R)} \tag{4.230}$$

where $V_{12}^{(D)}$ is the deterministic component of the relative velocity and $V_{12}^{(R)}$ the random component. Then the diffusion-convection equation becomes:

$$\nabla \cdot D\left[\nabla n + \frac{\nabla \psi}{kT} n\right] - \nabla \cdot \left(V_{12}^{(R)} n\right) - \nabla \cdot \left(V_{12}^{(D)} n\right) = 0 \quad (4.231)$$

If this equation is statistically averaged over many turbulent realizations, we find:

$$\nabla \cdot D\left[\nabla \langle n \rangle + \frac{\nabla \psi}{kT} \langle n \rangle\right] - \nabla \cdot \left(V_{12}^{(D)} \langle n \rangle\right) - \left\langle \nabla \cdot \left(V_{12}^{(R)} n\right) \right\rangle = 0 \quad (4.232)$$

Now the average $\left\langle \nabla \cdot \left(V_{12}^{(R)} n\right) \right\rangle$ must be considered carefully. This average is to be taken over all orientations of the velocity with the concentration gradient and we can formally write:

$$\left\langle \nabla \cdot \left(V_{12}^{(R)} n\right) \right\rangle = \int dV_R \, P(V_R) \nabla \cdot (V_R n) \quad (4.233)$$

where $P(V_R)$ is the probability distribution of the velocity.

Let us assume that $\nabla \cdot V_R = 0$. Then, with $V_R = (V_R, \theta_R, \phi_R)$ and taking the arbitrary polar direction of the integrand as r, the integral can be written explicitly as:

$$2\pi \int_{\cos(\theta_R)<0} \sin(\theta_R) \, d\theta_R \int dV_R \, V_R^2 \, P(V_R, \theta_R) \cos(\theta_R) \frac{\partial}{\partial r}(V_R n)$$

where we have assumed azimuthal symmetry and hence no dependence on ϕ_R. θ_R is the angle between the radius vector and the turbulent velocity whose radial component is V_R. The integration is taken over the hemisphere corresponding to the flux of particles into the test sphere. We now write Eqn. (4.233) formally as:

$$\left\langle \nabla \cdot \left(V_{12}^{(R)} n\right) \right\rangle = -\frac{\partial}{\partial r}\langle V_R n \rangle \equiv -\frac{\partial J_R}{\partial r}$$

where J_R is a current. Now we can make use of the usual analogy between the kinetic theory of gases and mixing length theory. In mixing length theory, small pockets of fluid (or in this case groups of particles) are transported spatially in a random manner with a step length, L, analogous to a mean free path. In our case, however, L is determined by the scale of turbulence. Just as the current of molecules crossing a unit area per unit time is $n\bar{v}/4$, $\bar{v}$ being the root mean square velocity, in the case of turbulence we can write $J_R = \langle V_R \rangle n/4$, where:

$$\left\langle \nabla \cdot \left(V_{12}^{(R)} n\right) \right\rangle \equiv -\frac{\langle V_R \rangle}{4} \frac{\partial n}{\partial r} \quad (4.234)$$

This statistical approach is clearly open to criticism but, as we shall see, it leads to very useful results and is therefore worthy of further study.

The average turbulent velocity $<V_R>$ has yet to be defined and for this we appeal to the arguments put forward by Saffman and Turner [1956]. They set up a stochastic differential equation for the instantaneous velocity of a particle in a turbulent field in the following way. Let **c** be the instantaneous velocity of the particle and **u** the undisturbed velocity of the surrounding air. Then, the equation of motion of the drop is:

$$m_p \frac{d\mathbf{c}}{dt} = -6\pi a \mu (\mathbf{c}-\mathbf{u}) + m_a \frac{d\mathbf{u}}{dt} \qquad (4.235)$$

m_p is the mass of the particle and m_a is the mass of the same volume of air. The right hand side comprises the Stokes resistive force acting on the particle and the random force due to the acceleration of a pocket of air of mass, m_a. If ρ_p and ρ_a are particle and air densities, respectively, we find:

$$\frac{d\mathbf{c}}{dt} = -\frac{1}{\tau}(\mathbf{c}-\mathbf{u}) + \frac{\rho_a}{\rho_p}\frac{d\mathbf{u}}{dt} \qquad (4.236)$$

Setting **q**=**c**−**u**, this becomes:

$$\frac{d\mathbf{q}}{dt} + \frac{1}{\tau}\mathbf{q} = \left(\frac{\rho_a}{\rho_p}-1\right)\frac{d\mathbf{u}}{dt} \qquad (4.237)$$

where:

$$\tau = \frac{2 a^2 \rho_p}{9 \mu}$$

Now, if the relaxation time, τ, is small compared with the time scales of the smallest eddies, which is usually the case even for quite strong turbulence (Saffman and Turner [1956]), we can neglect the term d**q**/dt and obtain:

$$\mathbf{q} = -\tau \frac{d\mathbf{u}}{dt} \qquad (4.238)$$

where we have also noted that $\rho_a \ll \rho_p$.

If we consider the relative velocity of two such particles of radii, a and b, we find:

$$\left\langle (\mathbf{q}_a - \mathbf{q}_b)^2 \right\rangle = (\tau_a - \tau_b)^2 \left\langle \left(\frac{d\mathbf{u}}{dt}\right)^2 \right\rangle \qquad (4.239)$$

But Batchelor [1951] has shown, from arguments based on the theory of homogeneous turbulence, that:

$$\left\langle \left(\frac{d\mathbf{u}}{dt}\right)^2 \right\rangle = (3.9) \frac{\varepsilon_T^{3/2}}{\nu^{1/2}} \qquad (4.240)$$

Thus, the value of $\langle V_R \rangle$, which is the square root of $\langle (\mathbf{q}_a-\mathbf{q}_b)^2 \rangle$, is:

$$\langle V_R \rangle = (1.97)|\tau_a - \tau_b| \frac{\varepsilon_T^{3/4}}{\nu^{1/4}} \qquad (4.241)$$

Coagulation Kernels

To illustrate the methods described in this section, we consider the solution of Eqn. (4.232) for turbulent diffusion and turbulent inertial motion. Here, we must solve:

$$\frac{1}{r^2}\frac{d}{dr}\left(r^2 D_T(r)\frac{d}{dr}n(r)\right) + \frac{\langle V_R \rangle}{4}\frac{d}{dr}n(r) = 0 \tag{4.242}$$

subject to the boundary conditions $n(a+b)=0$ and $n(r\to\infty)=n_\infty$. The resulting coagulation kernel is found to be:

$$\frac{1}{K(a,b)} = \frac{1}{4\pi}\int_{a+b}^{\infty}\frac{dw}{w^2 D(w)}\exp\left\{-\frac{\langle V_R \rangle}{4}\int_{a+b}^{w}\frac{dw'}{D(w')}\right\} \tag{4.243}$$

To be more general, we include Brownian diffusion and write:

$$D(r) = D_0 + k_e r^2 \tag{4.244}$$

One of the integrals in Eqn. (4.243) may be carried out explicitly and leads to:

$$K(a,b) = \frac{\pi(a+b)^2 \langle V_R \rangle}{G(\chi,\beta)} \tag{4.245}$$

where:

$$\beta = (a+b)\langle V_R \rangle / D_0 \tag{4.246}$$

$$\chi = (a+b)(k_e / D_0)^{1/2} \tag{4.247}$$

and:

$$G(\chi,\beta) = 1 - 2\exp\left(\frac{\beta}{4\chi}\tan^{-1}(\chi)\right)\int_1^\infty \frac{dt}{t^3}\exp\left(-\frac{\beta}{4\chi}\tan^{-1}(t\chi)\right) \tag{4.248}$$

β is a measure of the ratio of inertial turbulent forces to Brownian forces and χ^2 is a measure of the ratio of turbulent diffusive forces to Brownian forces. Eqn. (4.245) incorporates into one formula the effects of Brownian diffusion, turbulent diffusion, and turbulent inertial motion. If we neglect the Brownian diffusion, i.e. allow $\beta\to\infty$, then:

$$K(a,b) = \frac{\pi\langle V_R \rangle^3}{32 k_e^2}\left(1 - \xi + \tfrac{1}{2}\xi^2 - \exp(-\xi)\right)^{-1} \tag{4.249}$$

where:

$$\xi = \frac{\beta}{4\chi^2} = \frac{\langle V_R \rangle}{4(a+b)k_e} \tag{4.250}$$

is a measure of turbulent inertial to diffusive forces.

It will be instructive to examine the limit of Eqn. (4.249) as $\xi\to 0$ and ∞. Thus, for $\xi\to\infty$, we find:

$$K(a,b) = 12\pi k_e (a+b)^3 = (5.65)(a+b)^3\left(\frac{\varepsilon_T}{\nu}\right)^{1/2} \tag{4.251}$$

which agrees, as we expect, with Eqn. (4.137).

If we allow $k_e \to \infty$, i.e. $\xi \to 0$, then:

$$K(a,b) = \pi (a+b)^2 \langle V_R \rangle \tag{4.252}$$

which, from Eqn. (4.241), leads to:

$$K(a,b) = (6.2)(a+b)^2 |\tau_a - \tau_b| \frac{\varepsilon_T^{3/4}}{\nu^{1/4}} \tag{4.253}$$

This result compares favorably with that of Saffman and Turner [1956] in which the factor is (5.65) rather than (6.2). We shall discuss the method of Saffman and Turner below. Before doing this, however, we use the relative velocity due to gravitational settling instead of $\langle V_R \rangle$:

$$V_G = \frac{2 \rho_p g}{9 \mu} |a^2 - b^2| \tag{4.254}$$

In that case, we see that the 'exact' result for gravitational coagulation is produced. This must be, to some extent, fortuitous but nevertheless it is very useful since it means that Eqn. (4.248) has the correct limiting behavior for all the major coagulation processes. All that is necessary is to replace $\langle V_R \rangle$ by $\langle V_R \rangle + V_G$. A further interesting result emerges if we allow $\chi \to 0$ in Eqn. (4.245) and set $\langle V_R \rangle = V_G$. This then corresponds to the combined effect of Brownian diffusion and gravitational settling. The result is (Williams [1988]):

$$K(a,b) = \frac{4\pi (a+b) D_0 \exp(-\beta/4)}{E_2(\beta/4)} \tag{4.255}$$

where $E_2(x)$ is an exponential integral.

But, we have an exact result for this problem obtained from Eqn. (4.229). These two results are compared in Table 4.4 where γ from Eqn. (4.229) is compared with γ^* obtained from Eqn. (4.255):

$$\gamma^* = \frac{4}{4+\beta} \frac{\exp(-\beta/4)}{E_2(\beta/4)} \tag{4.256}$$

We note that the error is small overall. For $\beta<2$, Eqn. (4.255) overestimates the correct value by about 5.3% and at worst, for $\beta>3$, the error changes sign increasing to a maximum of 7.7% at $\beta=30$, thereafter decreasing. The simple and accurate nature of Eqn. (4.255), compared with the slowly converging infinite sum which arises in the exact result, makes its use highly desirable in practical calculations. We must bear in mind, however, that these results do not include any interparticle forces or slip corrections, although the formalism is capable of including them.

4.12 Turbulent coagulation by the Saffman and Turner method

The use of a diffusion equation and a convective term for directed motion is a very powerful technique for computing coagulation rates. When the fluid is turbulent the diffusive effects

Table 4.4: The ratios, γ_{exact} and γ_{approx}, as functions of β.

β	γ_{exact}	γ_{approx}	$\Delta\%$	β	γ_{exact}	γ_{approx}	$\Delta\%$
0.1	1.0236	1.0609	−3.6	3.0	1.2487	1.2433	0.5
0.2	1.0447	1.0943	−4.7	4.0	1.2628	1.2387	1.9
0.3	1.0637	1.1190	−5.2	5.0	1.2687	1.2304	3.1
0.4	1.0808	1.1384	−5.3	6.0	1.2702	1.2209	3.9
0.5	1.0963	1.1543	−5.3	7.0	1.2691	1.2113	4.6
0.6	1.1110	1.1675	−5.1	8.0	1.2666	1.2019	5.1
0.7	1.1230	1.1787	−5.0	9.0	1.2631	1.1930	5.5
0.8	1.1347	1.1882	−4.7	10.0	1.2591	1.1846	5.9
0.9	1.1453	1.1964	−4.5	20.0	1.2182	1.1270	7.5
1.0	1.1551	1.2034	−4.2	50.0	1.1507	1.0649	7.5
1.5	1.1933	1.2267	−2.9	100.0	1.1072	1.0358	6.4
2.0	1.2189	1.2379	−1.6				

[$\Delta\% = (1 - \gamma_{approx}/\gamma_{exact}) \times 100$]

may be included by means of an effective diffusion coefficient and the inertial effects by a statistically averaged velocity. Nevertheless, as we saw in Section 4.4, Saffman and Turner obtained an expression for the turbulent diffusion coagulation rate from direct considerations based on turbulent fluctuations. A similar argument was used by these authors for the case of inertial turbulent coagulation. In such calculations it is assumed that the relative velocity of two particles, w, is distributed in some fashion, P(w), where P(w)dw is the probability that the relative velocity lies between w and w+dw. Then, the rate of collision between two particles of radii, a and b, with number densities, n_a and n_b, respectively, will be:

$$J = \pi (a+b)^2 n_a n_b \int w P(w) dw \qquad (4.257)$$

This type of averaging is used frequently in the kinetic theory of gases. In the case of turbulent motion it will be valid if the mean velocity of each drop is statistically independent of its relative velocity. Since the mean velocity is controlled by large, energy containing eddies and the relative velocity by small eddies, which are statistically independent for large turbulent Reynolds numbers, the condition is met.

The simplest function which corresponds to reality and also leads to a simple integration is the Gaussian form:

$$P(w) = \left(\frac{\beta}{\pi}\right)^{3/2} \exp(-\beta w^2) \qquad (4.258)$$

where β is chosen so that the variance corresponds to that of the particles.

We saw in Eqn. (4.239) that:

$$\text{var}(q_a - q_b) = (\tau_a - \tau_b)^2 \left\langle \left(\frac{du}{dt}\right)^2 \right\rangle \qquad (4.259)$$

But:
$$\text{var}(w) = \text{var}(c_a - c_b) = \text{var}(q_a - q_b) + \text{var}(u_a - u_b) \tag{4.260}$$

Also:
$$\text{var}(u_a - u_b) = (a+b)^2 \left\langle \left(\frac{du}{dx}\right)^2 \right\rangle$$

which, in view of the isotropy of the small eddies (Saffman and Turner [1956]) can be written:

$$\text{var}(u_a - u_b) = 5(a+b)^2 \left\langle \left(\frac{du}{dx}\right)^2 \right\rangle \tag{4.261}$$

Since (Taylor [1935]):

$$\left\langle \left(\frac{du}{dx}\right)^2 \right\rangle = \frac{\varepsilon_T}{15\,\nu} \tag{4.262}$$

we find that:

$$\text{var}(u_a - u_b) = \tfrac{1}{3}(a+b)^2 \frac{\varepsilon_T}{\nu} \tag{4.263}$$

In view of this:

$$\text{var}(w) = (3.9)(\tau_a - \tau_b)^2 \frac{\varepsilon_T^{3/2}}{\nu^{1/2}} + \tfrac{1}{3}(a+b)^2 \frac{\varepsilon_T}{\nu} \tag{4.264}$$

Now, from Eqn. (4.258), we see that $\text{var}(w) = 3/(2\beta)$, whence:

$$J = 2(a+b)^2 \, n_a \, n_b \left(\frac{\pi}{\beta}\right)^{1/2} \tag{4.265}$$

and:

$$K(a,b) = 2(a+b)^2 \left(\frac{\pi}{\beta}\right)^{1/2} \tag{4.266}$$

$$= 2(a+b)^2 \left(\frac{2\pi}{3} \text{var}(w)\right)^{1/2} \tag{4.267}$$

$$= 2\sqrt{2\pi}\,(a+b)^2 \left[(1.3)(\tau_a - \tau_b)^2 \frac{\varepsilon_T^{3/2}}{\nu^{1/2}} + \tfrac{1}{9}(a+b)^2 \frac{\varepsilon_T}{\nu}\right]^{1/2} \tag{4.268}$$

It is instructive to compare Eqn. (4.268) with the results obtained earlier.

First, allow the inertial effect to go to zero. Then we find:

$$K(a,b) = \tfrac{2}{3}\sqrt{2\pi}\left(\frac{\varepsilon_T}{\nu}\right)^{1/2}(a+b)^3 = (1.67)\left(\frac{\varepsilon_T}{\nu}\right)^{1/2}(a+b)^3 \tag{4.269}$$

Coagulation Kernels

This compares with the other result of Saffman and Turner in which the (1.67) is replaced by (1.29) (see Eqn. (4.114)). In the other limit, when the diffusive term goes to zero, we find:

$$K(a,b) = (5.71)(a+b)^2 |\tau_a - \tau_b| \frac{\varepsilon_T^{3/4}}{\nu^{1/4}} \qquad (4.270)$$

which compares favorably with the value we obtained from the diffusion-convection equation where the factor (5.71) is replaced by (6.2).

It is not obvious, *a priori*, whether the method of Saffman and Turner is better or worse than that of the diffusion equation. Both techniques have weaknesses. With Saffman and Turner it is the use of a Gaussian distribution and with the diffusion equation it is the adequacy of the statistical average of the convective term.

One further result follows from the work of Saffman and Turner. They assumed that gravitational forces may be included as an effective acceleration in the fluid. Thus, the stochastic forcing term in Eqn. (4.237) is augmented to:

$$\frac{du}{dt} + g$$

where **g** is the acceleration due to gravity. After statistical averaging and noting no correlation between **ù** and **g** we find that:

$$\left\langle \left(\frac{du}{dt}\right)^2 \right\rangle \Rightarrow \left\langle \left(\frac{du}{dt}\right)^2 \right\rangle + g^2$$

Thus, the end result for the coagulation kernel in Eqn. (4.268) is:

$$K(a,b) = 2\sqrt{2\pi}(a+b)^2 \left[(1.3)(\tau_a - \tau_b)^2 \frac{\varepsilon_T^{3/2}}{\nu^{1/2}} + \tfrac{1}{3}(\tau_a - \tau_b)^2 g^2 + (a+b)^2 \frac{\varepsilon_T}{9\nu} \right]^{1/2} \qquad (4.271)$$

If we neglect all turbulent motion, this result should lead to the gravitational coagulation kernel:

$$K(a,b) = (2.89)(a+b)^2 g |\tau_a - \tau_b| \qquad (4.272)$$

The true value has a factor, π, rather than (2.89) and so the Saffman and Turner method leads to an error of about 8%. This is small, but the result from diffusion theory is exact.

4.13 Electrostatic forces

Electrostatic and other forces acting between two bodies generally act along the line of centers (*i.e.*, they are central forces, in contrast to the hydrodynamic forces which are strongly noncentral). For central forces only, Eqn. (4.205) reduces to Eqn. (4.184) and thence to the coagulation kernel:

$$K(a,b) = 4\pi D_0 \left[\int_{a+b}^{\infty} \frac{dw}{w^2 G(w)} \exp\left(\frac{\Phi(w)}{kT}\right) \right]^{-1} \qquad (4.273)$$

Now suppose that the particles have electric charges, νe and μe, respectively. Then, from Coulomb's law [neglecting image forces]:

$$\Phi(r) = \frac{\nu \mu e^2}{r} \tag{4.274}$$

For convenience, we set $G=1$ and then:

$$K(a,b) = \frac{4\pi D_0 \nu \mu e^2}{kT} \left[\exp\left(\frac{\nu \mu e^2}{kT(a+b)}\right) - 1 \right]^{-1} \tag{4.275}$$

$$= 4\pi D_0 (a+b) Z(y) \tag{4.276}$$

where:

$$y = \frac{\nu \mu e^2}{kT(a+b)} \tag{4.277}$$

and:

$$Z(y) = \frac{y}{\exp(y) - 1} \tag{4.278}$$

Since ν and μ are integers that can vary between $-\infty$ and $+\infty$, we see that the correction factor, Z, due to charging varies from 0 to ∞. For example, if the particles both have like charges they will repel and inhibit coagulation. On the other hand, unlike charges attract and coagulation is enhanced. In Chapter 3 the distribution of charges is shown to depend on the time constants of the charging rate and the Boltzmann distribution law. It will, in fact, be necessary when employing the coagulation kernel for charged particles, to extend the definition of the distribution function to include not only the size but also the charge state.

We also note that, depending on the size and nature of the aerosol particles, the interaction may be governed by a modified Coulomb's law. Of the many approximate treatments used to represent electrostatic forces between two charged particles, Grover and Beard [1975] compared two methods which are based on opposite points of view. In the first method, the particles are represented by point charges fixed at their centers. In the second method, the particles are represented by two conducting spheres on whose surfaces the charges are completely free to redistribute themselves as the two particles approach. If the charges on the particles have opposite signs, the point charge approximation will underestimate the strength of the resulting attractive force, whereas the conducting spheres approximation will overestimate it. The reverse is true for charges of like sign. In practice, the actual electrostatic force will be intermediate between these two limits and so, therefore, will the corresponding coagulation kernel. The point charge approximation is very simple to use but the conducting spheres approximation is generally much more realistic. Grover and Beard concluded that the point charge approximation is acceptable for the majority of atmospheric aerosols provided their diameter is larger than about 200 μm. Since most nuclear aerosols are significantly smaller than this, the point charge approximation must be used with caution. Some further discussions of these matters may be found in Shahub and Williams [1988].

It is instructive to derive the modifications to the gravitational coagulation kernel that arise when the particles are charged. Then, Eqn. (4.205) leads to:

$$\nabla \cdot \left[D(r) \nabla n(r) + \frac{\nu \mu e^2 \hat{r}}{kT r^2} n(r) \right] - \nabla \cdot (V(r) n(r)) = 0$$

Coagulation Kernels

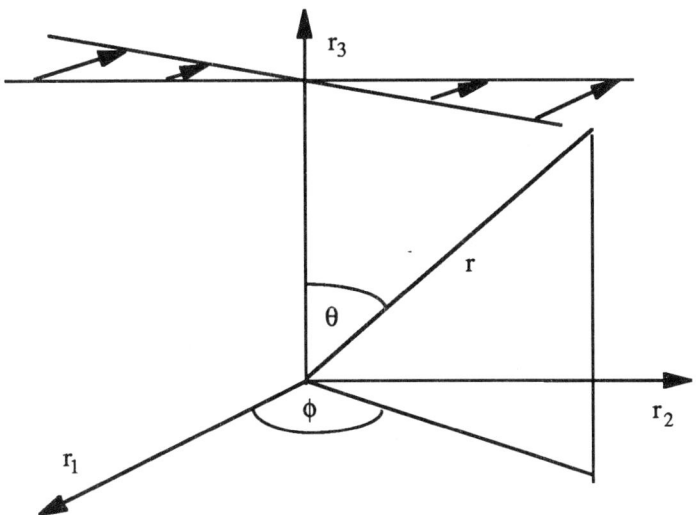

Figure 4.11: A coordinate system for Brownian and shear coagulation.

where $\hat{\mathbf{r}}$ is a unit vector in the r-direction and $\mathbf{V}(\mathbf{r})$ is the relative velocity between two particles falling under gravity and in their respective electric fields. There seems to be no easy solution to this equation so it will have to be evaluated numerically.

It is also important to account for deviations from hydrodynamic theory when the particles are smaller than, or comparable to, a mean free path. Such calculations have been carried out by Lassen [1961a,b] who used Fuchs' method.

4.14 Brownian and shear coagulation

We have seen above how laminar shear leads to coagulation and we have seen how Brownian motion leads to coagulation. We did not consider the simultaneous action of these two mechanisms at the time but now we do so. Shear and Brownian motion are extremely important in colloid science, particularly in the understanding of hydrosols which arise in various biological, technological, and environmental problems. Basic studies in this area have been carried out by Lin et al. [1970], Curtis and Hocking [1970], Zeichner and Schowalter [1977], van de Ven and Mason [1976, 1977a,b], and Batchelor and Green [1972].

Let us consider a fluid that is infinite in extent and whose velocity at infinity is a linear function of position. In this fluid are spheres which are undergoing Brownian motion. We wish to calculate the rate of collision of these spheres and to understand how this rate is enhanced by the shear effect and how the shear effect interacts with the Brownian motion.

We choose a coordinate system (see Fig. 4.11) in which one of the spheres is at the origin and at rest and the other spheres move up to it with a relative velocity, $\mathbf{V}$.

Let us write Eqn. (4.205) in this coordinate system:

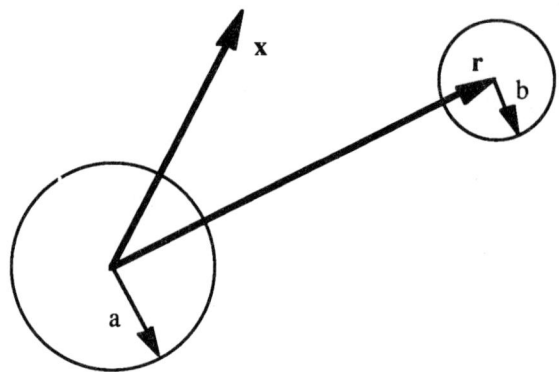

Figure 4.12: The relationship between the coordinates, **x** and **r**.

$$\frac{D_0}{r^2}\frac{\partial}{\partial r}\left(G(r)r^2\frac{\partial n}{\partial r}\right)+\frac{D_0 H(r)}{r^2}\left(\frac{\partial^2 n}{\partial \theta^2}+\frac{\cos(\theta)}{\sin(\theta)}\frac{\partial n}{\partial \theta}+\frac{1}{\sin^2(\theta)}\frac{\partial^2 n}{\partial \phi^2}\right)$$

$$+\frac{D_0}{r^2}\frac{\partial}{\partial \theta}\left(G(r)r^2 n\frac{\partial \Phi}{\partial r}\right)-\frac{1}{r^2}\frac{\partial}{\partial r}(r^2 V_r n)+\frac{1}{r\sin(\theta)}\frac{\partial}{\partial \theta}(V_\theta \sin(\theta) n)+\frac{1}{r\sin(\theta)}\frac{\partial}{\partial \phi}(V_\phi n)=0$$

(4.279)

where we have assumed that the interparticle potential is central.

We now consider the relative velocity, $\mathbf{V}=(V_r,V_\theta,V_\phi)$. For a linear flow field in which the asymptotic fluid velocity components far from the sphere are:

$$U_i(\mathbf{r})=E_{ij}\,r_j+\varepsilon_{ijk}\,\Omega_j\,r_k \tag{4.280}$$

Batchelor and Green [1972] show that the Cartesian components of the relative velocity are:

$$V_i(\mathbf{r})=U_i-\left\{A(r)\frac{r_i r_j}{r^2}+B(r)\left(\delta_{ij}-\frac{r_i r_j}{r^2}\right)\right\}r_k\,E_{jk} \tag{4.281}$$

In these equations, we use the summation convention and note that the rate of strain tensor elements are:

$$E_{ij}=\tfrac{1}{2}\left(\frac{\partial u_j}{\partial x_i}+\frac{\partial u_i}{\partial x_j}\right) \tag{4.282}$$

and the angular velocity is:

$$\Omega=\tfrac{1}{2}\nabla\times\mathbf{u} \tag{4.283}$$

A and B are functions of r and are discussed in Chapter 2. The coordinates, **x** and **r**, are related as shown in Fig. 4.12. Thus, **x** refers to an arbitrary position in the fluid and **r** refers to the interparticle orientation and separation. If the ambient flow is chosen to be:

$$U = (Kx_2, 0, 0) \tag{4.284}$$

we have:

$$E = \begin{bmatrix} 0 & K/2 & 0 \\ K/2 & 0 & 0 \\ 0 & 0 & 0 \end{bmatrix} \tag{4.285}$$

and:

$$\Omega = (0, 0, -K/2) \tag{4.286}$$

Now, in Cartesian coordinates:

$$V = i\,V_1 + j\,V_2 + k\,V_3 \tag{4.287}$$

but since:

$$\begin{aligned}i &= \sin(\theta)\cos(\phi)\,i_r + \cos(\theta)\cos(\phi)\,i_\theta - \sin(\phi)\,i_\phi \\ j &= \sin(\theta)\sin(\phi)\,i_r + \cos(\theta)\sin(\phi)\,i_\theta + \cos(\phi)\,i_\phi \\ k &= \cos(\theta)\,i_r - \sin(\theta)\,i_\theta \end{aligned} \tag{4.288}$$

we have:

$$\begin{aligned} V_r &= \sin(\theta)\cos(\phi)\,V_1 + \sin(\theta)\sin(\phi)\,V_2 + \cos(\theta)\,V_3 \\ V_\theta &= \cos(\theta)\cos(\phi)\,V_1 + \cos(\theta)\sin(\phi)\,V_2 - \sin(\theta)\,V_3 \\ V_\phi &= -\sin(\phi)\,V_1 + \cos(\phi)\,V_2 \end{aligned} \tag{4.289}$$

Since, from Eqn. (4.281), we can write:

$$V_1 = K\,r_2 \left\{ 1 - \frac{B}{2} - (A-B)\frac{r_1^2}{r^2} \right\} \tag{4.290}$$

$$V_2 = -K\,r_1 \left\{ \frac{B}{2} + (A-B)\frac{r_2^2}{r^2} \right\} \tag{4.291}$$

and:

$$V_3 = -K\,(A-B)\frac{r_1 r_2 r_3}{r^3} \tag{4.292}$$

where $r_1 = r\sin(\theta)\cos(\phi)$, $r_2 = r\sin(\theta)\sin(\phi)$, and $r_3 = r\cos(\theta)$. We have at once:

$$V_r = K\,r\,(1-A)\sin^2(\theta)\sin(\phi)\cos(\phi) \tag{4.293}$$

$$V_\theta = K\,r\,(1-B)\sin(\theta)\cos(\theta)\sin(\phi)\cos(\phi) \tag{4.294}$$

and:

$$V_\phi = -K\,r\sin(\theta)\left[\sin^2(\phi) + \tfrac{1}{2} B\left(\cos^2(\phi) - \sin^2(\phi)\right)\right] \tag{4.295}$$

Again, we note that these velocities neglect the inertial terms of the particles.

It is necessary to solve Eqn. (4.279) subject to the usual boundary conditions and then to compute the coagulation kernel from Eqn. (4.185). Such a calculation has been partially carried out by van de Ven and Mason [1977a,b] for spheres of equal radius, a. These authors employed perturbation theory together with matched inner and outer asymptotic expansions. The perturbation parameter was the Peclet number, Pe, defined as:

$$Pe = \frac{K a^2}{D_0} \tag{4.296}$$

The mathematical procedure is difficult and we simply quote the result for the coagulation kernel:

$$K(a,a) = 16\pi D_0 a \left[1 + (1.0272) Pe^{1/2}\right] + O(Pe) \tag{4.297}$$

where we have simplified the result by assuming no interparticle potential and no hydrodynamic forces. This is to be compared with the result obtained by adding the Brownian and laminar shear kernels of Eqns. (4.96) and (4.97), respectively, which leads to:

$$K(a,a) = 16\pi D_0 a \left[1 + \frac{2}{3\pi} Pe\right] \tag{4.298}$$

It is clear that there is no theoretical foundation for Eqn. (4.298) although it leads to the correct limits as Pe→0,∞.

It should be born in mind that these calculations neglect the inertia of the particle. This is a good approximation if the Reynolds number based on the Stokes velocity is much less than unity:

$$Re = \frac{a U_s}{\nu} \ll 1 \tag{4.299}$$

where:

$$U_s = \frac{2}{9}\left(\frac{\rho_p}{\rho_f} - 1\right)\frac{a^2}{\nu}$$

For colloids, where the fluid density, ρ_f, is close to the particle density, ρ_p, this is usually a very good approximation. However, for aerosols it is not so good as we may see from the case of water droplets in air. This leads to the condition:

$$Re = (6.7 \times 10^{-5}) a^3 \tag{4.300}$$

where a is in microns. Clearly, for a>20 μm, the approximation fails. We speak more about this problem in Chapter 2.

In spite of this restriction, the inertialess problem is very simple and instructive and merits some discussion since it leads to useful ideas about the trajectories of the particles as they pass one another. Batchelor and Green [1972] have examined the problem of two particles in shear flow when there is no Brownian motion and no interparticle forces. In such a case, the problem reduces to solving the differential equations:

$$\frac{\partial r}{\partial t} = V_r \tag{4.301}$$

$$r\frac{\partial \theta}{\partial t} = V_\theta \tag{4.302}$$

and:

$$r \sin(\theta) \frac{\partial \phi}{\partial t} = V_\phi \tag{4.303}$$

Dividing Eqn. (4.301) by Eqn. (4.302) leads to:

$$\frac{1}{r}\frac{\partial r}{\partial \theta} = \frac{1-A}{1-B} \tan(\theta) \tag{4.304}$$

and, from Eqns. (4.302) and (4.303):

$$\sin(\theta) \frac{\partial \phi}{\partial \theta} = -\frac{\sin^2(\phi) + \frac{1}{2} B \left(\cos^2(\phi) - \sin^2(\phi)\right)}{(1-B)\cos(\theta)\sin(\phi)\cos(\phi)} \tag{4.305}$$

Since $r_3 = r \cos(\theta)$, Eqn. (4.304) can be written:

$$\frac{1}{r_3}\frac{\partial r_3}{\partial r} = \frac{B-A}{1-A}\frac{1}{r} \tag{4.306}$$

and Eqn. (4.305) gives:

$$\frac{\partial}{\partial \theta}\left[\tan^2(\theta)\sin^2(\phi)\right] = -\frac{B}{1-B}\frac{\sin(\theta)}{\cos^3(\theta)} \tag{4.307}$$

which may also be written:

$$\frac{\partial}{\partial r}\left(\frac{r_2^2}{r_3^2}\right) = -\frac{B}{1-A}\frac{r}{r_3^2} \tag{4.308}$$

Eqns. (4.306) and (4.308) may be integrated and the constants of integration chosen to correspond to the values of r_2 and r_3 at infinity (say R_2 and R_3, respectively). We then find:

$$\frac{r_3}{R_3} = \exp\left\{\int_r^\infty \frac{A(r') - B(r')}{1 - A(r')} \frac{dr'}{r'}\right\} \tag{4.309}$$

and:

$$r_2^2 = \frac{r_3^2}{R_3^2}\left\{R_2^2 + \int_r^\infty \frac{B(r')}{1-A(r')} \frac{R_3^2}{r_3^2} r' \, dr'\right\} \tag{4.310}$$

These equations lead to a family of surfaces:

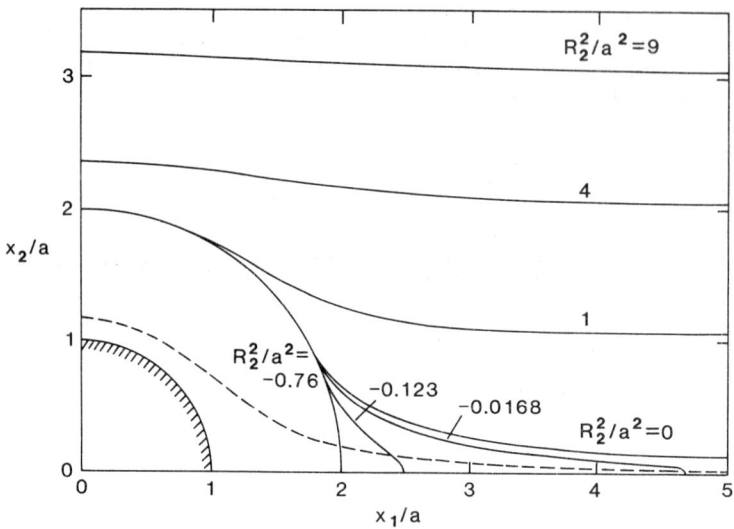

Figure 4.13: The trajectories, in the plane $x_3=0$, of the centre of one sphere relative to that of another of the same size in a steady simple shearing motion. The open trajectories ($R_2/a=1,2,3$) and the limiting trajectory ($R_2=0$) are taken from Lin et al. [1970]. Two closed trajectories ($R_2^2/a^2=-0.0168, -0.123$) have been calculated from Eqns. (4.309) and (4.310). The boundary of the region of closed streamlines in the steady flow around an isolated sphere is shown as a broken line. All these curves are symmetrical about both the x_1 and x_2-axes. Adapted from Batchelor and Green [1972] with permission of the Cambridge University Press.

$$r_2 = r\sin(\theta)\sin(\phi) = g(r) \quad \text{and} \quad r_3 = r\cos(\theta) = f(r)$$

which, when solved simultaneously, lead to a trajectory. Some typical trajectories are shown in Fig. 4.13 for equal spheres of radius, a, under various initial conditions.

It is found that certain conditions exist ($R_2^2<0$) where the particles capture each other and rotate together periodically as a permanent doublet. It is unlikely, however, that such a situation will be encountered in aerosol problems because of the inertia of the particles.

van den Ven and Mason [1977a,b] have solved Eqns. (4.301)-(4.303) numerically for equal size spheres and have obtained the collision efficiency. Further work along these lines has been carried out by Adler [1981] who also includes the influence of an electric field on particles which have a finite dielectric constant.

We note, in closing this topic, that Eqn. (4.279) becomes a standard second order partial differential equation analogous to, but somewhat more complicated than, Eqn. (4.204) when fluid forces are neglected ($G=H=1$) and V_r, V_θ, V_ϕ are independent of A and B. While we were able to find a complete solution to the latter equation, the one for shear motion does not seem readily amenable to analytical treatment.

Coagulation Kernels

4.15 Brownian and gravitational coagulation

We have seen in Section 4.7 that an analytical solution can be obtained for the coagulation kernel when Brownian diffusion and gravitational settling are present. This is only possible, however, when interparticle and fluid forces are neglected. The full equations, when these forces are present, can be written as:

$$\frac{D_0}{r^2}\frac{\partial}{\partial r}\left(G(r) r^2 \frac{\partial n}{\partial r}\right) + \frac{D_0 H(r)}{r^2}\left(\frac{\partial^2 n}{\partial \theta^2} + \cot(\theta)\frac{\partial n}{\partial \theta}\right)$$

$$\frac{D_0}{r^2}\frac{\partial}{\partial r}\left(G(r) r^2 \frac{1}{kT}\frac{\partial \Phi}{\partial r} n\right) - \frac{1}{r^2}\frac{\partial}{\partial r}[r^2 V_r n] - \frac{1}{r \sin(\theta)}\frac{\partial}{\partial \theta}[V_\theta \sin(\theta) n] = 0 \qquad (4.311)$$

with K(a,b) still given by Eqn. (4.208).

The velocity components, V_r and V_θ, can be obtained from Eqn. (4.200):

$$V_r = -V_\infty L(r) \cos(\theta) \qquad (4.312)$$

and:

$$V_\theta = V_\infty M(r) \sin(\theta) \qquad (4.313)$$

where $V_\infty = |\lambda^2 - 1| U_1^{(0)}$. Of course, for L=M=G=H=1, Eqn. (4.311) becomes Eqn. (4.214). We also note that, from symmetry, the azimuthal angle, ϕ, does not appear.

If we group together the third and fourth terms in Eqn. (4.311), then they become:

$$\frac{1}{r^2}\frac{\partial}{\partial r}\left[n r^2 \left(V_r - D_0 \frac{G(r)}{kT}\frac{\partial \Phi}{\partial r}\right)\right] \qquad (4.314)$$

Thus, the effective radial velocity is:

$$V_r^* = V_r - D_0 G(r)\frac{1}{kT}\frac{\partial \Phi}{\partial r} \qquad (4.315)$$

which is a result that we shall return to below.

Now, Melik and Fogler [1984a,b] have approximately solved Eqn. (4.311) by means of singular perturbation theory. They rewrite the equation in the form:

$$\frac{\partial}{\partial R}\left(R^2 G(R)\left\{\frac{\partial n}{\partial R} + \frac{1}{kT}\frac{\partial \Phi}{\partial R} n\right\}\right) + H(R)\left\{\frac{\partial^2 n}{\partial \theta^2} + \cot(\theta)\frac{\partial n}{\partial \theta}\right\}$$

$$= Gr\left\{\cos(\theta)\left[2R\, M(R)\, n - \frac{\partial}{\partial R}(R^2 L(R)\, n)\right] + R\, M(R)\sin(\theta)\frac{\partial n}{\partial \theta}\right\} \qquad (4.316)$$

where R=2r/(a+b) and the gravity number, Gr, is:

$$Gr = \frac{V_\infty (a+b)}{2 D_0} = \frac{\beta}{2} = \frac{2\pi\, g\rho\, a^4}{3 kT}\lambda(1-\lambda^2) \qquad (4.317)$$

Table 4.5: The gravity number, Gr, as a function of λ and a.

λ	a=0.5	a=0.7	a=1.0	λ	a=0.5	a=0.7	a=1.0
0.05	0.009	0.035	0.145	0.6	0.070	0.267	1.114
0.1	0.018	0.069	0.287	0.7	0.065	0.249	1.035
0.2	0.035	0.134	0.557	0.8	0.052	0.200	0.835
0.3	0.049	0.190	0.792	0.9	0.031	0.119	0.496
0.4	0.061	0.234	0.974	0.95	0.017	0.065	0.269
0.5	0.068	0.261	1.088				

with β being the same parameter used in Eqn. (4.217). The gravity number is shown in Table 4.5 for values of $\lambda = b/a$ from zero to unity, for $g=9.81$ m sec^{-2}, $T=298$ K, and $\rho=1000$ kg m^{-3}.

It is clear that if Gr is to be the perturbation parameter then $a < 0.5$ μm is required for reasonable convergence. The boundary conditions on Eqn. (4.316) are $n(2,\theta)=0$ and $n(\infty,\theta)=n_\infty$. Also, we require:

$$\frac{\partial n}{\partial \theta} = 0 \quad ; \quad \theta = 0, \pi$$

Before proceeding with a discussion of the method employed by Melik and Fogler, let us make a few observations on the existence of a steady state particle distribution which was implicitly assumed when we set $\partial n/\partial t=0$ in Eqn. (4.204).

In order for a steady state to exist, the net flux of particles crossing a surface S^* surrounding the test particle must be constant and independent of the choice of S^*. This follows from the relationship $(\nabla \cdot \mathbf{J})=0$ which implies that $\mathbf{J}$ is a constant over a closed surface surrounding the sink of particles. If this were not the case, there would be an accumulation or depletion of particles. This requirement demands that the two spheres touch in a finite time and this, in turn, depends on the nature of the interparticle potential, Φ, as $R \to 2$. For these very small interparticle distances, the dominant contribution to V_r^* will be:

$$V_r^* = \frac{dR}{dt} = -C_3 \frac{(R-2)}{kT} \frac{\partial \Phi}{\partial R} \tag{4.318}$$

In arriving at Eqn. (4.318), we have used the fact that as $R \to 2$:

$$L(R \to 2) \approx C_0 (R-2)$$

$$M(R \to 2) \approx C_1 + \frac{C_2}{\ln(R-2)}$$

and:

$$G(R \to 2) \approx C_3 (R-2)$$

Coagulation Kernels

where C_i are known constants.

If now we integrate Eqn. (4.318) from the surface out to a small interparticle distance, ω, then the time, Δt, to travel this distance is given by:

$$\Delta t = \frac{kT}{C_3} \int_2^{2+\omega} \frac{dR}{(R-2)\,\partial\Phi/\partial R} \tag{4.319}$$

Clearly, if the integral remains finite as $\omega \to 0$, then the two spheres will touch in a finite time. This will be true if the potential follows an inverse power law, e.g.:

$$\Phi(R) = \frac{A}{(R-2)^n} \tag{4.320}$$

The requirement corresponds, therefore, to the conditions:

$$\lim_{R \to 2}\left(\frac{1}{(R-2)\,\partial\Phi/\partial R}\right) = 0 \tag{4.321}$$

and:

$$\lim_{R \to 2}(\Phi) = 0 \tag{4.322}$$

These conditions are necessary ones for a steady state solution to exist. However, as Melik and Fogler point out, they are not sufficient. There do exist situations in which a large electrostatic barrier, coupled with a secondary minimum in the potential, can lead to instability. Such problems, however, are more important in colloids and generally do not enter into aerosol dynamics when the interparticle potentials are either electrostatic and/or London-van der Waals. As far as the present analysis is concerned, we will always assume that steady state solutions of Eqn. (4.316) exist.

The perturbation analysis of Melik and Fogler [1984a,b] employs asymptotic analysis and inner and outer expansions. Such details will not be discussed here, rather we will simply give the results obtained for the coagulation kernel:

$$K(a,b) = \frac{4\pi D_0 (a+b)}{W_{Br}}\left(1 + \frac{Gr}{W_{Br}}\right) + O(Gr^2 \ln(Gr)) \tag{4.323}$$

The presence of the logarithmic term is characteristic of singular perturbation theory.

The factor W_{Br} is given by:

$$W_{Br} = 2\int_2^\infty \frac{\exp(\Phi/kT)}{G(R)\,R^2}\,dR \tag{4.324}$$

If we assume the absence of interparticle and hydrodynamic forces, i.e. set $\Phi=0$ and $G=1$, we obtain $W_{Br}=1$ and hence:

$$K(a,b) = 4\pi D_0 (a+b)(1+\beta/2) + O(\beta^2 \ln(\beta)) \tag{4.325}$$

This result is consistent with the exact value for this simplified situation obtained in Section 4.7. We also note once again the inaccuracy inherent in the simple additive approximation with Brownian plus gravitational kernels.

4.16 Acoustic coagulation

Any mechanism that produces a relative velocity between aerosol particles has an associated coagulation rate and therefore leads to a coagulation kernel. The passage of sound waves in air leads to an oscillatory motion of the gas. Any entrained particles will also be subject to such a motion, although not necessarily in phase with it. A classical example of such motion can be observed in the Kundt's tube experiment in which standing sound waves in a closed pipe lead to the characteristic bunching of lycopodium particles along the pipe side.

The first serious experiments on acoustic coagulation were performed by Patterson and Cawood [1931, 1932] who made measurements on particles in systems very similar to Kundt's tube. These were followed by experiments in Germany (Brandt and Heidemann [1936]), Great Britain (Andrade [1936] and Parker [1936]), and the United States (St Clair [1938, 1949]). While efficiencies on the order of 50% were attainable in such situations as fog reduction and control of particulate emissions, the amount of energy required to drive the sonic sirens was such as to reduce commercial interest in the subject. Recent developments, however, have led to a resurgence of interest in this topic for pollution control and in certain types of scrubbers, filters, and electrostatic precipitators.

Aside from the commercial aspects of sonic coagulation, it is a very interesting and challenging physical problem and considerable work has been done in order to understand the mechanisms and forces that act between particles when they are situated in a sound field. We shall discuss these approaches and show how they can lead to the construction of a coagulation kernel.

As an introduction to the problem, let us ignore interparticle and fluid forces other than the Stokes drag which acts on any isolated sphere moving in a fluid. If a sound wave moves through a gas, then the local gas velocity, $U_g(t)$, varies sinusoidally with time:

$$U_g(t) = U_{g0} \sin(\omega t) \tag{4.325}$$

where ω is the frequency. According to acoustic theory, the maximum speed, U_{g0}, is given by:

$$U_{g0} = \left(\frac{2E}{\rho_g c_g} \right)^{1/2} \tag{4.326}$$

and the sound pressure by:

$$P_g = \rho_g c_g U_{g0}$$

where ρ_g and c_g are the gas density and speed of sound, respectively, and:

$$c_g = \left(\frac{P_s \gamma}{\rho_g} \right)^{1/2} \tag{4.327}$$

Also, γ is the ratio of the specific heats, P_s the static gas pressure, and E the sound intensity in W m^{-2}. As an example, we note that for E in the range 10^3–10^4 W m^{-2}, the maximum speed, U_{g0}, ranges from 2.2 to 7.0 m sec^{-1} in air.

If a small particle is now placed in the gas, it will undergo motion according to:

$$m_p \frac{du_p}{dt} = -6\pi a \mu (u_p - U_g) \tag{4.328}$$

where u_p, m_p, and a are the particle velocity, mass, and radius, respectively, and we assume that Stokes' law is valid. We could use the Cunningham correction for very small particles but will not complicate the present analysis with such refinements.

The solution of Eqn. (4.328) is:

$$u_p = \frac{U_{g0} \sin(\omega t - \phi)}{(1+\omega^2 \tau^2)^{1/2}} + \frac{\omega \tau U_{g0}}{(1+\omega^2 \tau^2)} \exp\left(-\frac{t}{\tau}\right) \tag{4.329}$$

where:

$$\tau = \frac{m_p}{6\pi a \mu} \tag{4.330}$$

and:

$$\phi = \tan^{-1}(\omega t) \tag{4.331}$$

The exponential term is an initial transient and rapidly disappears as the motion proceeds. We note, though, that there is a phase lag between the velocity of the particle and that of the gas. This is due to the particle's inertia and therefore its inability to respond instantly to changes in gas motion. In addition, the degree of entrainment of the particle in the vibration of the gas is:

$$\mu_p = (1+\omega^2 \tau^2)^{-1/2} \tag{4.332}$$

which physically denotes the ratio of the maximum amplitude of vibration of the particle motion to that of the gas.

Since there is a phase shift between the vibrations of the particle and those of the medium, the velocity, U_{gp}, at which the medium flows by the particle is:

$$U_{gp} = U_{g0} \sin(\omega t) - \mu_p U_{g0} \sin(\omega t - \phi) = \mu_g U_{g0} \cos(\omega t - \phi) \tag{4.333}$$

where:

$$\mu_g = \frac{\omega \tau}{(1+\omega^2 \tau^2)^{1/2}} \tag{4.334}$$

The maximum velocity of the gas flowing around the particle is therefore:

$$U_{gp} = \mu_g U_{g0} \tag{4.335}$$

where clearly $\mu_g^2 + \mu_p^2 = 1$.

It is instructive to calculate the amplitude of vibration of the particle, $A_p(t)$. This may be done by integrating Eqn. (4.329) (neglecting the exponential term), to get:

Aerosol Science: Theory and Practice

$$A_p(t) = \frac{U_{g0}\,\mu_p}{\omega}\left[\cos(\phi) - \cos(\omega t - \phi)\right] \tag{4.336}$$

Thus, the maximum amplitude is $U_{g0}\,\mu_p / \omega$. Taking the following values:

$$\begin{aligned}
E &= 1100 \text{ W m}^{-2}, & \mu &= 1.5 \times 10^{-5} \text{ kg m}^{-1}\text{sec}^{-1} \\
\rho_g &= 1 \text{ kg m}^{-3}, & \rho_p &= 1000 \text{ kg m}^{-3} \\
\gamma &= 5/3, & P_s &= 10^5 \text{ N m}^{-2}
\end{aligned}$$

we find c_g=408 m sec^{-1} and U_{g0}=2.3 m sec^{-1}. The maximum amplitude is then:

$$A_{p,max} = \frac{(3.7 \times 10^5)}{f}\left[1 + (8.6 \times 10^{-9})(fa)^2\right]^{1/2} \tag{4.337}$$

where $\omega = 2\pi f$ and a is in μm. If we choose a frequency of f=5 kHz, then for a=1, 10, and 20 μm we find that $A_{p,max}$=74, 16, and 8 μm, respectively. For an aerosol density of 10^{14} particles m^{-3}, we have an average interparticle distance of $(10^{-14})^{1/3}$=21 μm. It is clear that the rate of coagulation will be quite low for sufficiently high frequencies and radii, simply because the amplitude is substantially less than the interparticle distance, *i.e.* a particle will vibrate but there will be no partner within a distance $A_{p,max}$.

A further quantity of interest is the Reynolds number based on the maximum particle velocity. This will tell us in which flow regime the particle lies.

The Reynolds number is defined as:

$$\text{Re} = \frac{\mu_p\,U_{g0}\,a}{\nu} \tag{4.338}$$

which, for the data shown above, leads to:

$$\text{Re} = \frac{(0.15)\,a}{\left[1 + (8.6 \times 10^{-9})(fa)^2\right]^{1/2}} \tag{4.339}$$

where a in microns. For low frequencies such that $(8.6 \times 10^{-9})(fa)^2 \ll 1$, we see that particle sizes are confined to a<<1 μm. However, for higher frequencies the Reynolds number decreases and use of slow viscous flow theory becomes more accurate.

If we assume that two particles are present in the gas with velocities, U_{pa} and U_{pb}, respectively, then the instantaneous collision rate will be:

$$F = \pi(a+b)^2\,|U_{pa} - U_{pb}| \tag{4.340}$$

The total collision rate over half a cycle, π/ω, is:

$$\bar{F} = \pi(a+b)^2\,\frac{\omega}{\pi}\int_0^{\pi}dt\,|U_{pa} - U_{pb}| \tag{4.341}$$

Coagulation Kernels

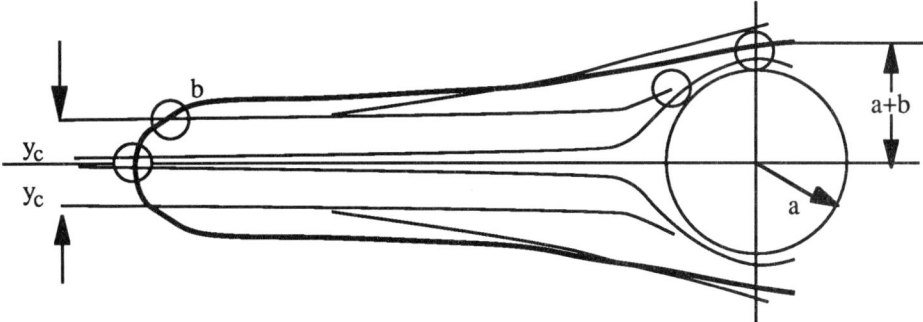

Figure 4.14: Trajectories of small particles as they oscillate about a large particle.

which is the coagulation kernel K(a,b). Now with:

$$\left|U_{pa} - U_{pb}\right| = U_{g0}\left\{\frac{\sin(\omega t - \phi_a)}{(1+\omega^2 \tau_a^2)^{1/2}} - \frac{\sin(\omega t - \phi_b)}{(1+\omega^2 \tau_b^2)^{1/2}}\right\} \quad (4.342)$$

we find:

$$K(a,b) = \frac{2 U_{g0} (a+b)^2 \omega}{(1+\omega^2 \tau_a^2)^{1/2} (1+\omega^2 \tau_b^2)^{1/2}} \left|\frac{1}{\tau_a} - \frac{1}{\tau_b}\right| \quad (4.343)$$

As far as we are aware, this result was first derived by Simons and Williams [1972]. It is interesting in that it has a maximum value when $\omega=1/(\tau_a\tau_b)^{1/2}$. Thus, for 1 μm particles colliding in air at STP we find $\omega_{max}=4000$ sec^{-1} or roughly 700 Hz. The physical explanation for the maximum is fairly straightforward. For small $\omega\tau$, the particles move virtually in phase with the gas and the average velocity differential between them is small. Thus the chance of collision is small. On the other hand, for large $\omega\tau$, particle inertia prevents them from responding to the motion of the gas and they remain more or less stationary. Again the chance of collision is small. At intermediate values of $\omega\tau$ there is an optimum phase and amplitude difference for a maximum collision rate. Of course, the best situation, as far as mass differential is concerned, is when one particle is large and the other small.

In the derivation of the coagulation kernel, the effect of fluid forces was neglected. Important details such as these have been thoroughly discussed by Mednikov [1965] in his classic treatise on acoustic coagulation. Mednikov has proposed that an important mechanism for particle coagulation is due to a so-called *orthokinetic* interaction. In this process, small particles are collected by larger ones because of the relative oscillating motion arising from the applied acoustic field. This is not dissimilar to the processes occurring during gravitational sedimentation. Thus, if we consider the trajectories of small particles as they oscillate about a larger particle, we have the situation depicted in Fig. 4.14.

The smaller particles will not move in straight lines because of the fluid forces and there is a collision efficiency, defined as:

$$\varepsilon = \left(\frac{y_c}{a+b}\right)^2 \quad (4.344)$$

215

There is also a critical trajectory, denoted by the bold line, which encloses a coagulation volume. The critical trajectory is one in which the smaller particle makes a grazing collision with the larger one. Any particle within this critical trajectory, *i.e.* inside the critical volume, will be captured. Once the critical volume is 'swept clean' it is refilled by parakinetic and attractive hydrodynamic interactions after every oscillation. Other mechanisms such as acoustic turbulence, also play a role in the filling mechanism.

Once the collision efficiency, ε, has been obtained, the coagulation kernel is rewritten as:

$$K_A^*(a,b) = \alpha \, \varepsilon \, K_A(a,b) \tag{4.345}$$

where α, called the refill factor, is due to the parakinetic and hydrodynamic mechanisms mentioned above. However, little is known about it and it is usually set equal to unity.

Some experimental work has been carried out by Brandt and Hiedemann [1936] and Neumann and Norton [1951] to verify the orthokinetic interaction. However, since the data are limited, no firm conclusions can be drawn although the results certainly point to qualitative agreement. In particular, it is found that K^* depends directly on U_{g0} and is therefore, by Eqn. (4.326), proportional to the square root of the sound intensity.

In addition to orthokinetic interactions, there are other mechanisms operating during acoustic coagulation. This must be so, otherwise the application of a sound field to a monodisperse aerosol would not lead to any significant coagulation, *i.e.* if $\tau_a = \tau_b$, then $K=0$. In practice this is not the case since experiments indicate that relatively monodisperse aerosols do, in fact, undergo substantial coagulation in both standing wave (Shirokova [1970]) and travelling wave (Shaw and Tu [1979]) fields. This is believed to be due to an enhancement of coagulation by diffusion.

An expression for the acoustic coagulation kernel has been obtained by Shaw and Rajendran [1979] by drawing an analogy with gravitational coagulation. The arguments are based on a dimensional analysis and the introduction of a hypothetical sphere, the radius of which gives the same gravitational settling velocity when falling in a gravitational field as does the actual particle of radius, a, when oscillating with average flow velocity, $\mu_p U_{g0}$. Analytical results are produced for the collision efficiency, but the formulae have not been tested experimentally and some of the arguments are open to criticism. Indeed, Mednikov [1965] advances strong arguments which suggest that the motion of a particle in a sound field is radically different from that in gravitational settling.

A consistent approach to sonic coagulation based upon Stokes flow is possible if we employ Eqns. (4.196) and (4.197). These equations relate the velocities of two interacting Stokesian particles subject to forces, $\mathbf{F}_1$ and $\mathbf{F}_2$. Bearing in mind the relative fluid motion of the sonic field, $\mathbf{U}_g(t)$, we see that the particle trajectories can be derived from:

$$\mathbf{U}_1 - \mathbf{U}_g = -m_1 \, \mathbf{b}_{11} \cdot \dot{\mathbf{U}}_1 - m_2 \, \mathbf{b}_{12} \cdot \dot{\mathbf{U}}_2 \tag{4.346}$$

and:

$$\mathbf{U}_2 - \mathbf{U}_g = -m_1 \, \mathbf{b}_{21} \cdot \dot{\mathbf{U}}_1 - m_2 \, \mathbf{b}_{22} \cdot \dot{\mathbf{U}}_2 \tag{4.347}$$

where the forces are simply the inertia terms and $\mathbf{U}_g(t) = k U_{g0} \sin(\omega t)$. For noninteracting particles, these equations reduce to Eqn. (4.328) but, for the general case, they require numerical solution. The coagulation kernel arises from an appropriate time average of the relative velocities.

Coagulation Kernels

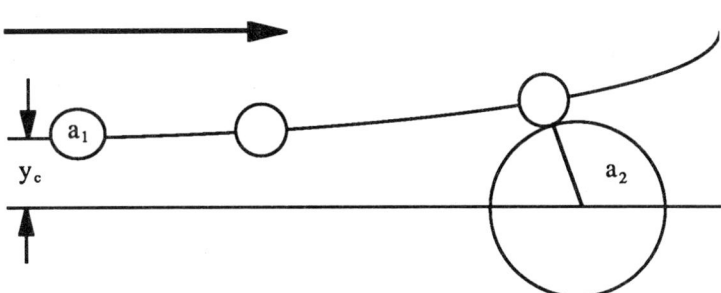

Figure 4.15: Trajectories of two spheres moving towards each other.

It is worth noting at this point that much of the early work on sonic coagulation, assumed that the particles interacted according to ideal fluid dynamics. Thus, using Bernoulli's law, it was possible to derive forces between the particles. This was done by Konig [1891a,b] using the theory of Bjerknes [1915] and the following results were obtained for the force components:

$$F_r = \frac{3\pi}{2\,r^4} \rho_g \, a_1^3 \, a_2^3 \, U_{gp}^2 \left[1 + 3\cos(2\,\theta)\right] \tag{4.348}$$

$$F_\theta = \frac{3\pi}{2\,r^4} \rho_g \, a_1^3 \, a_2^3 \, U_{gp}^2 \sin(2\,\theta) \tag{4.349}$$

where θ is the angle that the line of centers makes with the direction of the sound wave.

In fact, it is readily shown that the flow around particles in a sonic field is certainly not ideal and depends strongly on the viscous nature of the medium. It is evident, therefore, that Bernoulli forces are not appropriate. However, the criterion defined by Eqn. (4.338) regarding the Reynolds number should be born in mind since, if Re becomes too large, it may be necessary to modify the Stokes approximation.

4.17 Trajectory analysis

As we have noted above, due to the fluid forces generated by the motion of two particles as they pass by each other, there is a mutual deflection of their trajectories. This leads to a modification of the collision probability that would be expected on purely geometrical grounds. In order to incorporate this into the diffusion-advection equation or indeed to carry out any calculation of the collision dynamics, it is necessary to obtain the actual trajectories of the two particles. In this way the limiting condition of a grazing collision can be obtained. Fig. 4.15 shows the dynamical situation, with sphere 2 and sphere 1 moving towards each other. The impact parameter, y_c, defines the limiting situation and the collision efficiency, ε, is given by:

$$\varepsilon = \left(\frac{y_c}{a_1 + a_2}\right)^2 \tag{4.350}$$

Clearly, to obtain the grazing impact parameter, it is necessary to solve the equations of motion of the two particles.

This problem first assumed importance in the field of meteorology where it was necessary to calculate the rate of coalescence of raindrops. It is also important in filter design. Much of the pioneering work was carried out by meterologists and geophysicists. The early recognition and quantification of the collision efficiency was that of Langmuir [1948]. His result was crude and was soon superseded by improved methods (*e.g.*, Pearcey and Hill [1957], Shafrir and Neiburger [1963], Atkinson and Paluch [1968], Beard and Grover [1974], Pitter and Pruppacher [1974], Lin and Lee [1975], Schlamp et al. [1976], Wang et al. [1978], and Grover and Beard [1975]). These papers solved the equations of motion with various forms of interparticle force and sometimes with electrical forces present. Concurrent with these investigations, more detailed studies of the collision efficiency, employing improved interparticle fluid forces, were carried out by Hocking [1958, 1959], Davis and Sartor [1967], Hocking and Jonas [1970], Jonas [1972], Klett and Davis [1973], and Pertmer and Loyalka [1980].

Much of the early work in the calculation of trajectories, especially that in the meteorology area, made use of an approximation scheme as follows. It is assumed that the drag force on a sphere of radius, a, moving with a velocity, V, in a fluid is given by:

$$F_D = -6\pi a \mu (V - U) \qquad (4.351)$$

where U is the velocity that the fluid would have in the position of that sphere due to the other one if the first were not present. The fluid fields can be calculated to any desired degree of accuracy for isolated particles and can include effects arising from the nonlinear terms in the Navier-Stokes equations, *i.e.* deviations from Stokes flow. Nevertheless, the interaction effect is neglected.

Using this procedure, the equations of motion for the two spheres falling under gravity may be written:

$$m_1 \frac{dV_1}{dt} = m_1 g - 6\pi a_1 \mu (V_1 - U_2) \qquad (4.352)$$

and:

$$m_2 \frac{dV_2}{dt} = m_2 g - 6\pi a_2 \mu (V_2 - U_1) \qquad (4.353)$$

where g is the acceleration due to gravity.

Using a Cartesian coordinate system, these equations may be solved numerically with the additional equations:

$$V_1 = \frac{dr_1}{dt} \quad \text{and} \quad V_2 = \frac{dr_2}{dt} \qquad (4.354)$$

Trajectories are followed until grazing collisions take place. This determines y_c and hence the collision efficiency, $\varepsilon = (y_c/(a_1+a_2))^2$.

We shall not discuss the results of these early studies because most of them involve assumptions that can now be relaxed. In particular, as we have seen from the work in Chapter 2 on two sphere interactions, we now know the interparticle force law with some accuracy except perhaps for very small gaps and very small particles. Hocking and Jonas [1970] first used the complete set of interparticle forces derived above to calculate the trajectories of two spheres falling under gravity. To illustrate their procedure, we write the equations of motion for the two spheres of masses, m_1 and m_2, and radii, a_1 and a_2, as:

Coagulation Kernels

$$m_1 \dot{V}_1 = m_1 g - F_{D1} \tag{4.355}$$

and:

$$m_2 \dot{V}_2 = m_2 g - F_{D2} \tag{4.356}$$

where F_{Di} are the drag forces. Using Eqns. (2.358) and (2.359), the equations of motion become:

$$m_1 \dot{V}_1 = m_1 g - k_{11} \cdot V_1 - k_{12} \cdot V_2 \tag{4.357}$$

and:

$$m_2 \dot{V}_2 = m_2 g - k_{21} \cdot V_1 - k_{22} \cdot V_2 \tag{4.358}$$

We note in passing that, for particles sufficiently small, the inertia terms, $m_i \dot{V}_i$, may be neglected. In this case the equations simplify considerably and are the basis of calculations reported by Wacholder and Sather [1974] and Batchelor [1982]. In general, however, and especially for aerosols, this simplification is not valid. Indeed, we shall see below that even Stokes flow has its limitations for sufficiently large particles. For the moment, however, we consider Eqns. (4.357) and (4.358), where:

$$k_{ij} = 3\pi \mu (a_i + a_j) \left\{ S_{ij}(r) \frac{rr}{r^2} + T_{ij}(r) \left[I - \frac{rr}{r^2} \right] \right\} \tag{4.359}$$

Now, in order to solve Eqns. (4.357) and (4.358) numerically it is convenient to change to a Cartesian system of coordinates as shown in Fig. 4.16.

If we recall that r is the interparticle distance, then we may write:

$$r = i(x_1 - x_2) + j(y_1 - y_2)$$
$$\equiv i x + j y \tag{4.360}$$

Thus, $r^2 = x^2 + y^2$. We may also write the particle velocities as:

$$V_1 = i V_{1x} + j V_{1y}$$

and:

$$V_2 = i V_{2x} + j V_{2y}$$

Eqn. (4.357) then becomes:

$$m_1 \left(i \dot{V}_{1x} + j \dot{V}_{1y} \right) = m_1 g i - k_{11} \cdot \left(i V_{1x} + j V_{1y} \right) - k_{12} \cdot \left(i V_{2x} + j V_{2y} \right)$$

To obtain the equation for V_{1x}, we take the dot product with i to get:

$$m_1 \dot{V}_{1x} = m_1 g - i \cdot k_{11} \cdot i V_{1x} - i \cdot k_{11} \cdot j V_{1y} - i \cdot k_{12} \cdot i V_{2x} - i \cdot k_{12} \cdot j V_{2y}$$

But, from Eqns. (4.359) and (4.360), we find:

$$i \cdot k_{ij} \cdot i = 3\pi \mu (a_i + a_j) \left[S_{ij} \frac{x^2}{r^2} + T_{ij} \frac{y^2}{r^2} \right]$$

and:

$$i \cdot k_{ij} \cdot j = 3\pi \mu (a_i + a_j)[S_{ij} - T_{ij}] \frac{xy}{r^2}$$

This leads to:

$$m_1 \dot{V}_{1x} = m_1 g - 6\pi \mu a_1 \left(S_{11} \frac{x^2}{r^2} + T_{11} \frac{y^2}{r^2}\right) V_{1x} - 6\pi \mu a_1 (S_{11} - T_{11}) \frac{xy}{r^2} V_{1y}$$

$$-3\pi \mu (a_1 + a_2)\left(S_{12} \frac{x^2}{r^2} + T_{12} \frac{y^2}{r^2}\right) V_{2x} - 3\pi \mu (a_1 + a_2)(S_{11} - T_{11}) \frac{xy}{r^2} V_{2y}$$

(4.361)

Similar reasoning leads to:

$$m_1 \dot{V}_{1y} = -6\pi \mu a_1 (S_{11} - T_{11}) \frac{xy}{r^2} V_{1x} - 6\pi \mu a_1 \left(S_{11} \frac{y^2}{r^2} + T_{11} \frac{x^2}{r^2}\right) V_{1y}$$

$$-3\pi \mu (a_1 + a_2)(S_{12} - T_{12}) \frac{xy}{r^2} V_{2x} - 3\pi \mu (a_1 + a_2)\left(S_{12} \frac{y^2}{r^2} + T_{12} \frac{x^2}{r^2}\right) V_{2y}$$

(4.362)

$$m_2 \dot{V}_{2x} = m_2 g - 3\pi \mu (a_1 + a_2)\left(S_{21} \frac{x^2}{r^2} + T_{21} \frac{y^2}{r^2}\right) V_{1x} - 3\pi \mu (a_1 + a_2)(S_{21} - T_{21}) \frac{xy}{r^2} V_{1y}$$

$$-6\pi \mu a_2 \left(S_{22} \frac{x^2}{r^2} + T_{22} \frac{y^2}{r^2}\right) V_{2x} - 6\pi \mu a_2 (S_{22} - T_{22}) \frac{xy}{r^2} V_{2y}$$

(4.363)

and:

$$m_2 \dot{V}_{2y} = -3\pi \mu (a_1 + a_2)(S_{21} - T_{21}) \frac{xy}{r^2} V_{1x} - 3\pi \mu (a_1 + a_2)\left(S_{21} \frac{y^2}{r^2} + T_{21} \frac{x^2}{r^2}\right) V_{1y}$$

$$-6\pi \mu a_2 (S_{22} - T_{22}) \frac{xy}{r^2} V_{2x} - 6\pi \mu a_2 \left(S_{22} \frac{y^2}{r^2} + T_{22} \frac{x^2}{r^2}\right) V_{2y}$$

(4.364)

In addition we have:

$$\frac{dx}{dt} = V_{1x} - V_{2x} \quad \text{and} \quad \frac{dy}{dt} = V_{1y} - V_{2y}$$

(4.365)

These equations have been solved numerically by Hocking and Jonas [1970] using the following procedure. They start with a large value of the x-variable so that the particles are well separated and use successively smaller values of y. For each trajectory, the minimum distance of approach is found. A critical value, y_c, is found for which $r-a_1-a_2=\delta$, where δ is a small, but arbitrary, distance. It is necessary to use a finite value of δ because it is known that the theory for calculating the forces fails when the spheres approach very closely to one another. For the calculations reported by Hocking and Jonas, a value of $\delta=10^{-4} a_1$ was chosen, and Jonas showed that while the results are sensitive to the value of δ, its effect is largest for small collision efficiencies which are unlikely to lead to large numbers of collisions.

Coagulation Kernels

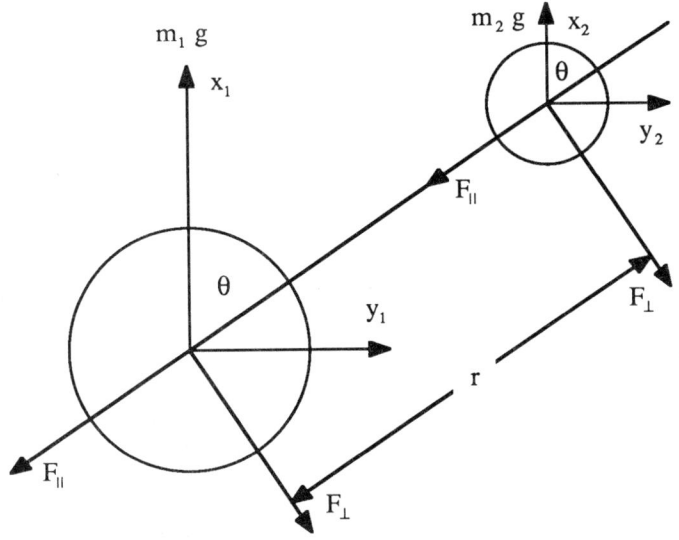

Figure 4.16: A Cartesian coordinate system for two particle motion.

Nevertheless, this is an unsatisfactory situation and in order to mitigate the error, Jonas [1972] has employed Hocking's [1973] slip corrected values for the forces as described in Chapter 2. This shows that for spheres approaching each other with a relative velocity, V_{12}, along their line of centers, the forces for very close separation are given by:

$$F_1 = 6\pi \mu \, a_1 \, V_{12} \, f_1$$

where:

$$f_1 = \frac{2 \, a_1 \, a_2^2}{(a_1 + a_2)^2 \, h} \frac{1}{\beta^2} \{(1+\beta)\ln(1+\beta) - \beta\} \tag{4.366}$$

$\beta = 6\lambda_g/h$, and h is the gap distance.

Now, for no slip, where $\beta \to 0$, $f = a_1 \, a_2^2 / \{(a_1 + a_2)^2 \, h\} = f_1^*$. Thus, for particle motion along the line of centers with no-slip, if the forces acting on the particles are:

$$F_1 = 6\pi \mu \, a_1 \, (k_1 \, U_1 + k_2 \, U_2) \tag{4.367}$$

and:

$$F_2 = 6\pi \mu \, a_2 \, (k_3 \, U_1 + k_4 \, U_2) \tag{4.368}$$

then we can modify these equations, for very close separation, to the form:

$$F_1 = 6\pi \mu \, a_1 \, \{U_1 \, (k_1 - f_1 + f_1^*) + U_2 \, (k_2 + f_1 - f_1^*)\} \tag{4.369}$$

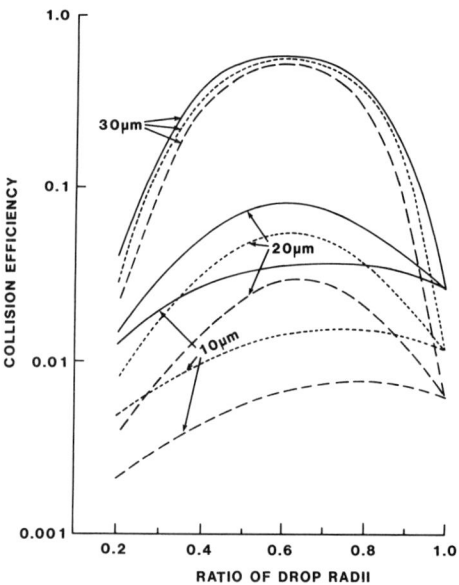

Figure 4.17: Calculated collision efficiencies for cloud droplets. The results are shown for several values of the larger droplet radius. The solid lines are those obtained by Jonas [1972] (with slip corrections) and the dashed and dotted lines are taken from the results of Hocking and Jonas [1970] for values of $\varepsilon=10^{-4}$ and 10^{-3} respectively ($\varepsilon=\delta$). Adapted from Jonas [1972] with permission of the Royal Meteorological Society.

and:
$$F_2 = 6\pi\mu a_2 \{U_1(k_3 + f_2 - f_2^*) + U_2(k_4 - f_2 + f_2^*)\} \tag{4.370}$$

In arriving at this result, we have noted that as $h \to 0$:

$$k_1 \to -f_1^* \quad , \quad k_2 \to f_1^* \quad , \quad k_3 \to f_2^* \quad \text{and} \quad k_4 \to -f_2^*$$

Thus, for large h, both f_i and f_i^* are negligible, but for small h, the singularity in k_i is cancelled by the appropriate value of f_i^* and the behavior near $h \to 0$ replaced by the more realistic behavior of f_i. This is not rigorous but is better than the pure hydrodynamic theory.

As far as the motion of spheres travelling perpendicular to their line of centers is concerned, the problem is less sensitive to the gap distance, but the calculation of the slip correction is more difficult. Hocking's results show that, up to a gap width of one mean free path, the slip and no-slip values of the force are very close. A linear interpolation was therefore used between a gap width of one and two mean free paths.

Using these modified expressions for the forces and a value of the mean free path of 0.1 μm, Jonas [1972] calculated the collision efficiencies of cloud droplets by a method similar to that of Hocking and Jonas [1970]. The results are shown in Fig. 4.17 for various size ratios.

Coagulation Kernels

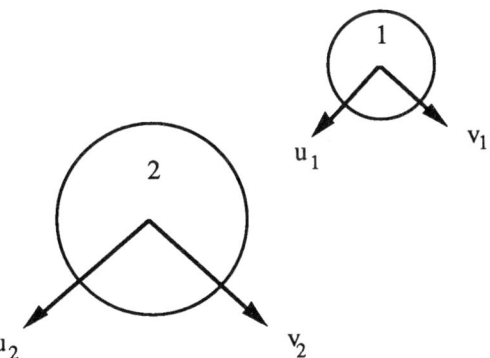

Figure 4.18: Two spheres moving with the velocities, u_i and v_i.

The sensitivity of the results to the value chosen for δ is clear. We also note that the collision efficiency becomes smaller for smaller droplets. It appears that inclusion of the slip effect leads to significantly higher values of collision efficiency. A semiempirical rule deduced by Jonas is that the no-slip equations can be employed provided that the minimum gap distance, $\delta=(1.3)\lambda_g$. This suggestion is in accordance with some ideas of Hocking [1973].

One of the limitations of the work described above is that it is restricted to very small Reynolds numbers, strictly speaking vanishingly small values, although in practice up to about Re=0.5 can be considered. Nevertheless, for large particles, there will be deviations from Stokes flow and it is therefore important to assess the errors involved.

In order to study this effect, Klett and Davis [1973] have made extensive use of the Oseen approximation in which the Navier-Stokes equations are replaced by:

$$\rho\, U_\infty \cdot (\nabla u) = -\nabla p + \mu\, \nabla^2 u \tag{4.371}$$

i.e., some aspects of fluid inertia are retained, with the fluid velocity at infinity playing an important role. In fact, Oseen's equation has been shown by Carrier [1953] to represent experiment better if U_∞ is replaced by cU_∞ where c is an empirical constant.

The argument for such a parameter is that since Stokes' theory neglects inertia altogether, while Oseen's theory overestimates it, at least in some regions of the flow, some compromise between the two limiting cases might be found, at least in an average sense, by introducing a semiempirical constant. It was found by Carrier that a value of c=(0.43) would lead to the vanishing of the integral of the difference between the exact and approximate forms of the convection terms over the whole field of flow. It was then found that the Oseen values of drag on spheres and cylinders could be used provided that the Reynolds number, Re, was replaced by cRe. The drag was then in good agreement with experiments in the range, 0.5<Re<20.0.

The drag forces for two spheres with the Carrier-modified Oseen approximation have been calculated by Klett and Davis [1973]. Fig. 4.18 shows two spheres moving with the velocities, u_i and v_i.

The components of the forces relative to the line of centers are found to be (after linearizing):

$$F_1^{\|} = 3\pi\, \mu\, a_1 \left[C_{11}(u_1+u_2) + C_{12}(u_1-u_2) + C_{13}(v_1+v_2) + C_{14}(v_1-v_2) \right] \tag{4.372}$$

Table 4.6: Comparison of force coefficients from Klett and Davis in the limit of zero Reynolds number with corresponding values (in parentheses) according to Stokesian hydrodynamics. Adapted from Klett and Davis [1973] with permission of the American Meteorological Society.

h/a	C_{11}	C_{12}	C_{23}	C_{24}
5.0	0.826 (0.826)	1.270 (1.270)	0.903 (0.902)	1.123 (1.123)
0.5	0.652 (0.653)	2.158 (2.772)	0.752 (0.754)	1.499 (1.512)
0.05	0.619 (0.648)	2.606 (2.711)	0.703 (0.714)	1.738 (1.886)

$$F_1^\perp = 3\pi \mu\, a_1 \left[C_{21}(u_1+u_2) + C_{22}(u_1-u_2) + C_{23}(v_1+v_2) + C_{24}(v_1-v_2) \right] \quad (4.373)$$

and:

$$F_2^\parallel = 3\pi \mu\, a_2 \left[C_{31}(u_1+u_2) + C_{32}(u_1-u_2) + C_{33}(v_1+v_2) + C_{34}(v_1-v_2) \right] \quad (4.374)$$

$$F_2^\perp = 3\pi \mu\, a_2 \left[C_{41}(u_1+u_2) + C_{42}(u_1-u_2) + C_{43}(v_1+v_2) + C_{44}(v_1-v_2) \right] \quad (4.375)$$

It is interesting to note that the forces along the line of centers depend on the velocities perpendicular to the line of centers as well as on those along the line of centers. Similarly, the forces perpendicular to the line of centers depend on the velocities along the line of centers as well as on those perpendicular to it. For pure Stokes flow, the cross terms C_{13}, C_{14}, C_{21}, C_{22} all vanish.

Table 4.6, which is a selection from Klett and Davis [1973] for $a_1=a_2$, shows the resistance terms C_{ij}, with the Stokes value in brackets. Of course, in general, the C_{ij} also depend on the Reynolds numbers for the two sphere problem. The values given here are in the limit of vanishingly small Re_i. we also note that for the case of equal spheres $C_{31}=C_{11}$, $C_{32}=-C_{12}$, $C_{43}=C_{23}$, and $C_{44}=-C_{24}$.

It is clear that very strong deviations arise at small separation especially for C_{12} and C_{32} which refer to equal anti-parallel motion along the line of centers.

Collision efficiencies based on these results for small Reynolds numbers were calculated and compared with the purely Stokesian results of Davis and Sartor [1967]. Fig. 4.19 shows the results and while the curves at $a_1=30$ µm are close, those for 10 µm and 20 µm are very different. Of course, these calculations do not include any slip effect and therefore must remain tentative.

The new forces derived from the Oseen calculation can also be employed in the small but nonzero range of Reynolds numbers. This also leads to a more realistic motion of two equal particles which according to Stokesian theory will remain in the same relative positions. However, Oseen's theory predicts a finite relative velocity, which is in accord with experiment (Jayaweera et al. [1964] and Hocking [1964]).

Using the nonzero Reynolds number drag coefficients, Klett and Davis computed collision efficiencies for particles in the size range $10<a_1<70$ µm (or equivalently $0.01<Re<2.0$)

corresponding to water drops in air at 900 mbar and 273 K. The results are shown in Fig. 4.20 together with the Davis and Sartor results which employed Stokesian drag. Although some of the results contain uncertain numerical errors (shown dotted), it is clear that there are significant differences. Also, none of the curves exhibit low mass ratio cutoffs, *i.e.* values of mass ratio below which the collision efficiency is zero. This spurious effect was predicted by Hocking's early analysis (Hocking [1958, 1959]) and was due to inadequate representation of the drag as a function of interparticle distance. The new collision efficiencies are larger than the Stokesian values for all droplet sizes, with a marked tendency for an increase (above unity) as the size ratio approaches unity. This is attributed to an inertial effect which leads to a closing velocity between nearly equal pairs of droplets.

A more recent and extensive series of calculations of collision efficiencies has been carried out by Pertmer and Loyalka [1980]. Their method is similar to that described above but employs different drag forces in different ranges of Reynolds number. This study, which is applicable to fast reactor safety, covers a mass range of 0.25 μm to 100 μm for sodium spheres. For Re<1 Stokes forces are used while for Re>1 Carrier-modified Oseen forces are employed. There are also corrections made for the particle density as the size increases to account for the lack of compactness of the aggregate.

The calculations solve the equations of motion, but since the actual values of the collision efficiency are to be used in an aerosol calculation, as described in Chapter 5, it would be inefficient to have to make such calculations for every a_1 and a_2. For this reason, the results are calculated at certain values and then a cubic spline fit is made to interpolate for intermediate values. Many numerical results are given by Pertmer and Loyalka but a useful summary can be obtained from Figs. 4.21 and 4.22 which are for Stokes forces and a range of particle radii from 0.5 μm to 60 μm. Fig. 4.23 shows the collision efficiency using the Oseen drag forces in one case, a_1=60 μm, for the Stokes drag. It is clear, as was also noted by Klett and Davis, that Stokesian dynamics leads to a significant underestimate of the collision efficiency and, moreover, does not show the marked increase near mass ratios of unity.

Pertmer and Loyalka also give some attention to the superposition method which is potentially valuable because of its simplicity. The fluid fields were obtained from the stream function of the Stokes approximation and, therefore, the radius of the larger particle was limited to less than 60 μm. Results are shown in Figs. 4.24 and 4.25 and, when compared with the 'exact' Stokesian results of Figs. 4.21 and 4.22, show a similar behavior. This gives some confidence in the superposition method although its extension to higher Reynolds numbers remains to be explored.

Finally, Pertmer and Loyalka compare their calculations with those of other workers, mainly from the atmospheric sciences. For example, Fig. 4.26 shows a comparison with the Stokesian regime of the work of Jonas [1972], and Hocking and Jonas [1970]. They note that good agreement is obtained for a_1>20 μm, but for smaller values noticeable differences occur. These differences are attributed to errors made by Hocking and Jonas in the numerical integration of the equations of motion. Gear's method, as used by Pertmer and Loyalka, is far superior to the techniques used by Hocking and Jonas due to the stiffness of the equations which becomes more pronounced for smaller radii.

Fig. 4.27 compares results using the Carrier-modified Oseen forces. Agreement is seen to be good and, in addition, Pruppacher and Klett [1978] have already shown that the theory is in good agreement with experimental data.

As a general conclusion, we note that the gravitational collision efficiency is very sensitive to the nature of the drag forces acting on the particles and it is important that these be known accurately. Clearly, the best drag force to use in the light of current knowledge is the Carrier-modified Oseen one.

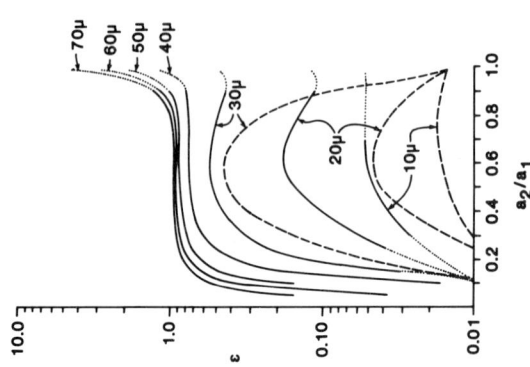

Figure 4.20: Collision efficiencies according to the method of Klett and Davis [1973] (solid and dotted lines) and the Stokesian treatment of Davis and Sartor [1967] (dashed lines). The dotted portions of the curves represent regions where the results are of doubtful accuracy, owing to the large number of integration steps required in the computations. Adapted from Klett and Davis [1973] with permission of the American Meteorological Society.

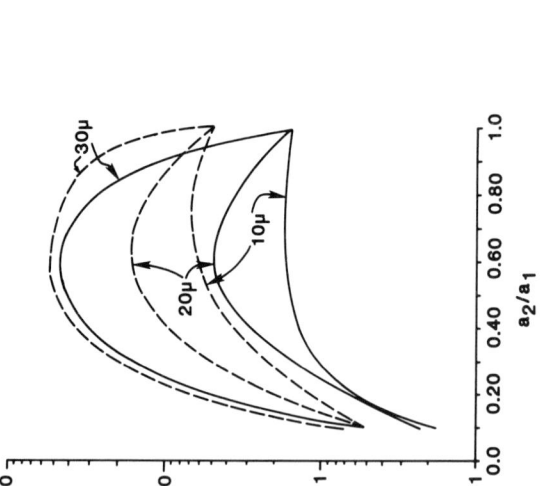

Figure 4.19: Comparison of collision efficiencies according to the formulation of Klett and Davis [1973] in the zero-Reynolds number limit with Davis and Sartor's [1967] Stokesian values. The full line denotes the Stokes flow and the dashed line the Re=0 limit of Oseen flow. Adapted from Klett and Davis [1973] with permission of the American Meteorological Society.

Coagulation Kernels

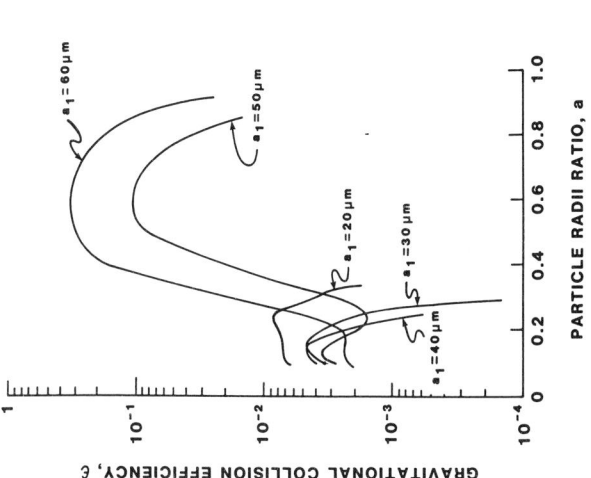

Figure 4.22: Gravitational collision efficiency, ε, versus a, Stokes drag forces $\rho_i = \rho(a_i)$. Adapted from Pertmer and Loyalka [1980] with permission of the American Nuclear Society.

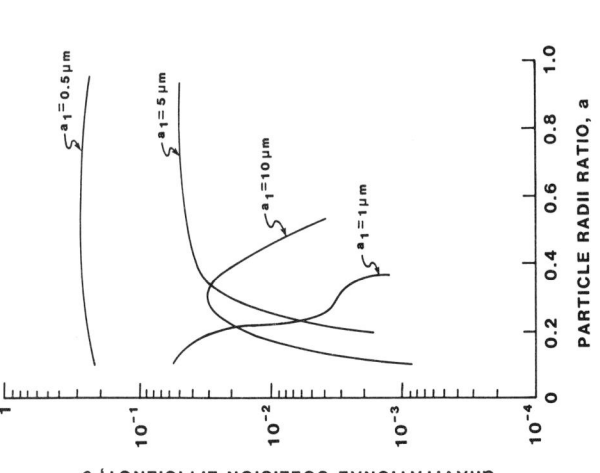

Figure 4.21: Gravitational collision efficiency, ε, versus a, Stokes drag forces $\rho_i = \rho(a_i)$. Adapted from Pertmer and Loyalka [1980] with permission of the American Nuclear Society.

Aerosol Science: Theory and Practice

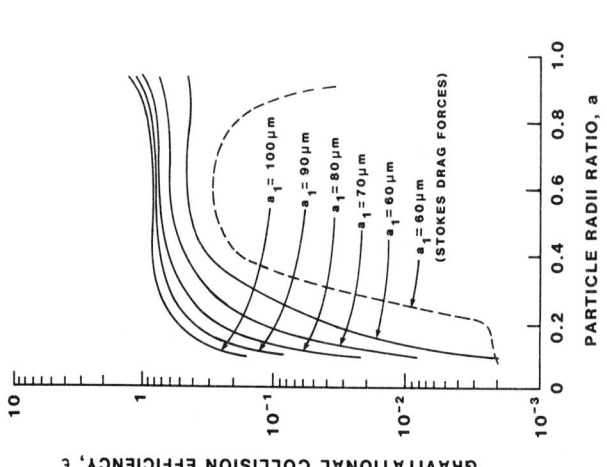

Figure 4.23: Gravitational collision efficiency, ε, versus a, Carrier-modified Oseen drag forces $\rho_i = \rho(a_i)$. Adapted from Pertmer and Loyalka [1980] with permission of the American Nuclear Society.

Coagulation Kernels

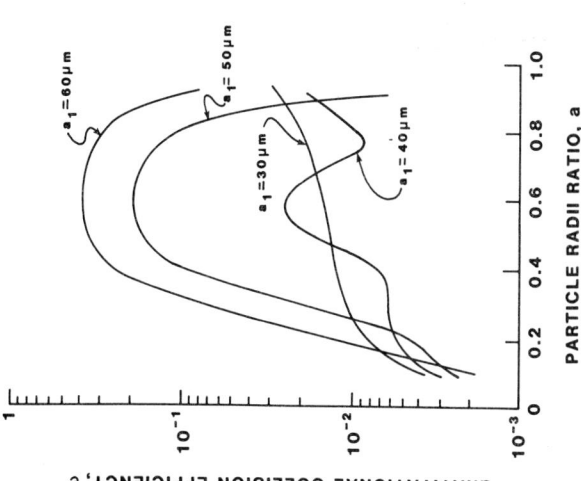

Figure 4.25: Gravitational collision efficiency, ε, versus a, superposition method $\rho_i = \rho(a_i)$. Adapted from Pertmer and Loyalka [1980] with permission of the American Nuclear Society.

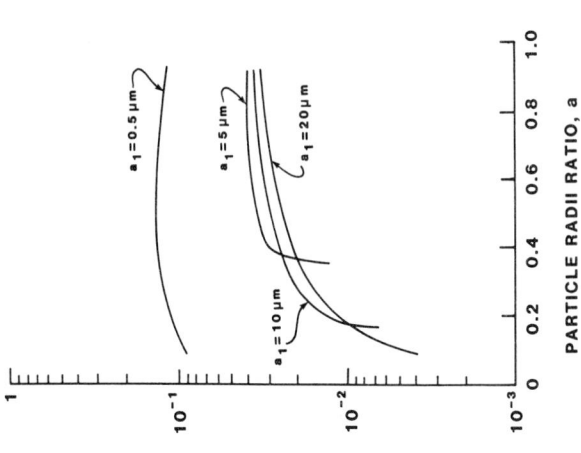

Figure 4.24: Gravitational collision efficiency, ε, versus a, superposition method $\rho_i = \rho(a_i)$. Adapted from Pertmer and Loyalka [1980] with permission of the American Nuclear Society.

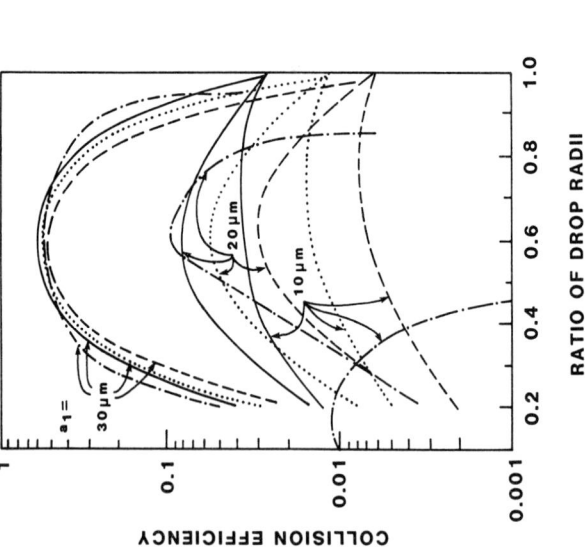

Figure 4.26: A comparison of the GCEFF results with the work of atmospheric sciences (Stokesian drag forces). Adapted from Pertmer and Loyalka [1980] with permission of the American Nuclear Society. Full line-Jonas [1972]; Dashed line-Hocking and Jonas [1970]; Dotted line-Hocking and Jonas [1970]; Dash dot line-Pertmer and Loyalka [1980].

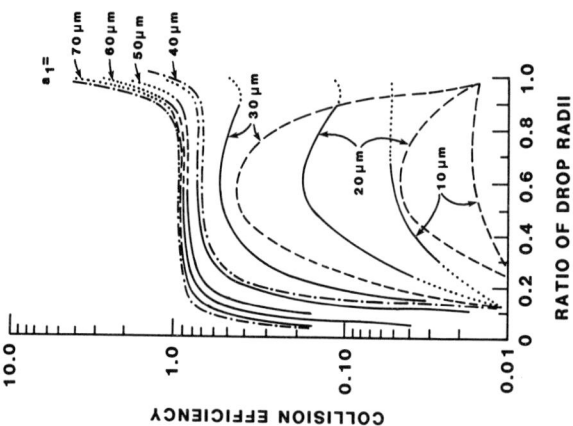

Figure 4.27: A comparison of the GCEFF results with the work of atmospheric sciences (Carrier-modified Oseen drag forces). Adapted from Pertmer and Loyalka [1980] with permission of the American Nuclear Society. Full line-Klett and Davis [1973]; Dashed line-Klett and Davis [1973]; Dotted line-Davis and Sartor [1967] (Stokesian); Dash dot line-Pertmer and Loyalka [1980].

Coagulation Kernels

4.18 The coagulation of nonspherical particles

We have seen how to calculate the coagulation kernels, K(a,b), for spherical particles of radii, a and b. In practice, however, aerosol particles are rarely spherical. It is therefore of some importance to investigate the influence of nonsphericity on the coagulation process. This is usually done by means of shape factors as discussed in Chapter 2. However, before commenting on this procedure, it is useful to outline an approach based on a more fundamental, although perhaps less practical, procedure.

Booth [1954] considered the problem of the Brownian coagulation of spheroidal particles. The procedure uses the diffusion equation as discussed in Section 4.5 but, because of the nonspherical symmetry, the boundary conditions and the diffusion equation become more difficult to solve. Moreover, the diffusion coefficient becomes a function of the orientation of the particles with respect to the direction of diffusion. Also, Brownian motion constantly changes orientation, an effect which occurs for spheres but can be ignored due to symmetry.

As a result of these complications, we do not expect a complete solution unless fairly severe approximations are made. Nevertheless, for aerosol units which differ considerably in size, quite reasonable results are obtainable. The main restriction is that the shapes considered fall into a category for which the diffusion equation is separable (the spheroid is one such case). Three cases have been studied by Booth; (i) coagulation of large spheroids with small spheres; (ii) coagulation of small spheroids with large spheres, and; (iii) coagulation of small and large spheroids. We note that Booth's calculations are for colloids, however the same principles can be used for aerosols.

4.18.1 Coagulation of large spheroids and small spheres

We consider prolate spheroids with long axis of length, A, and short axis of length, B. The spheres are of radius, S. It is assumed that A, B>>S and that the spheroid is stationary. The diffusion equation, $\nabla^2 C(r) = 0$, must be solved with the boundary conditions, $C = C_\infty$ at $r = \infty$, and with C=0 on a boundary which is the locus of the center of a sphere of radius, S, rolled over the surface of the spheroid. The analytical form of this surface is complicated, being determined by an equation of the fourth degree. In view of the size ratio of spheroid to sphere it is, however, close to a spheroid of surface S^* with axes A+S, B+S, and B+S (see Fig. 4.28).

In terms of the spheroidal coordinates, u and v, we may write (see Chapter 2 for definitions):

$$x = a\left(u^2 - 1\right)^{1/2} \left(1 - v^2\right)^{1/2} \quad \text{and} \quad z = a u v$$

The coordinates, u and v, are such that, $-1 < v < 1$ and $u_0 < u < \infty$. a and u_0 are given by:

$$a = \left[A^2 - B^2 + 2S(A - B)\right]^{1/2} \quad \text{and} \quad u_0 = \frac{A + S}{a}$$

u_0 defines the surface, S^*, of the extended spheroid and the diffusion equation takes the form:

$$D\left[\frac{\partial}{\partial u}\left(u^2 - 1\right)\frac{\partial C}{\partial u} + \frac{\partial}{\partial v}\left(1 - v^2\right)\frac{\partial C}{\partial v}\right] = 0 \tag{4.377}$$

The solution, subject to the boundary conditions, is:

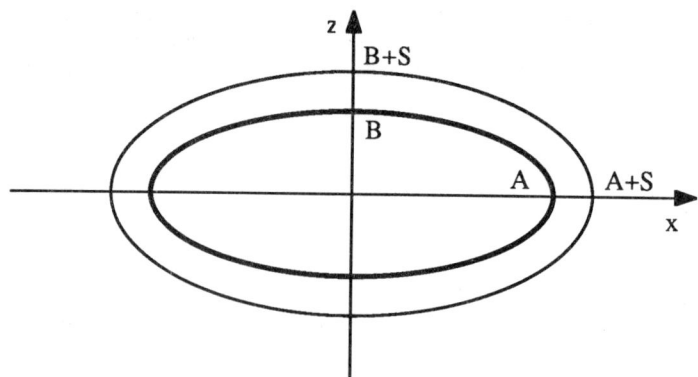

Figure 4.28: The boundary which is the locus of the center of a sphere of radius, S, rolled over the surface of a prolate spheroid with long axis of length, A, and short axes of length, B.

$$C = C_\infty \left[1 - \frac{Q_0(u)}{Q_0(u_0)}\right]$$

(4.378)

where:

$$Q_0(u) = \tfrac{1}{2} \ln\left(\frac{u+1}{u-1}\right)$$

(4.379)

The flux of spheres onto the spheroid is therefore:

$$J = -4\pi \int_{S^*} D \left.\frac{\partial C}{\partial n}\right|_{u_0} dS$$

(4.380)

which leads to:

$$J = \frac{4\pi a D C_\infty}{Q_0(u_0)}$$

(4.381)

The corresponding coagulation kernel is J/C_∞. The result is illuminating because it shows that the greater the deviation of spheroids, of fixed volume, from the spherical form, the more rapid is the coagulation. This may be demonstrated by calculating the ratio:

$$\frac{R_e}{R_0} = \frac{p^{2/3}}{\ln(2p)}$$

(4.382)

where p=A/B>>1. In this formula, R_e is the effective radius of the larger particle and is defined as the radius of a sphere which would give the same value of J as the spheroid. R_0 is the radius of a sphere with the same volume as the spheroid.

The result obtained in Eqn. (4.381) ignored the Brownian motion of the spheroid. The correct way to account for this would be to replace D in the diffusion equation by the sum of the coefficients for sphere and spheroid in the direction connecting (u,v) with the center of the

Coagulation Kernels

spheroid. This, however, makes the diffusion equation impossible to solve analytically in any useful manner, because D is then no longer a constant. The only practical solution is to add to D in the final result the value of the average diffusion coefficient for spheroids. That is (see Section 2.11):

$$D = \frac{kT}{6\pi\mu}\left[\frac{1}{S} + \frac{p\ln\left[p+(p^2-1)^{1/2}\right]}{A(p^2-1)^{1/2}}\right]$$
(4.383)

We have also neglected rotational Brownian motion. This has the effect of changing the orientation of the long axis of the spheroid as time proceeds. The condition that this may be neglected will be that the spheroid has rotated through a very small angle in a time during which $\partial C/\partial t$ is small. Quantitatively this means that:

$$\frac{S}{A} << \frac{a^2}{\pi D Q_0^2(u_0)}$$

which is generally true.

This concludes the discussion of case (i) except to note that the case of oblate spheroids can be dealt with in a similar fashion with the result:

$$J = \frac{4\pi C_\infty D a}{\tan^{-1}(u_0)}$$
(4.384)

where:

$$u_0 = (B+S)\left[A^2 - B^2 + 2S(A-B)\right]^{-1/2}$$

A is the length of the long axis, and B the length of the short one. If we wish to correct the diffusion coefficient we must write:

$$D = \frac{kT}{6\pi\mu}\left[\frac{1}{S} + \frac{p\tan^{-1}\left\{(1-p^2)^{1/2}/p\right\}}{A(1-p^2)^{1/2}}\right]$$
(4.385)

Booth [1954] shows how the above calculation may be extended to the case when an interparticle force is present.

4.18.2 Coagulation of large spheres and small spheroids

Consider now a large stationary sphere of radius, S, and spheroids diffusing towards it with two equal axes of length, b, and the other of length, a. The sizes are such that, S>>a, b. Since the coefficient of diffusion of the spheroids depends on their orientation with respect to the direction of diffusion, it is convenient to represent the concentration by C(r,θ). Here, r is the distance from the center of the sphere and θ is a well defined polar angle. In this case, the diffusion equation becomes:

$$\frac{D(\theta)}{r^2}\frac{\partial}{\partial r}\left[r^2\frac{\partial}{\partial r}C(r,\theta)\right] + \frac{kT}{C_b \sin(\theta)}\frac{\partial}{\partial \theta}\left[\sin(\theta)\frac{\partial}{\partial \theta}C(r,\theta)\right] = 0$$
(4.386)

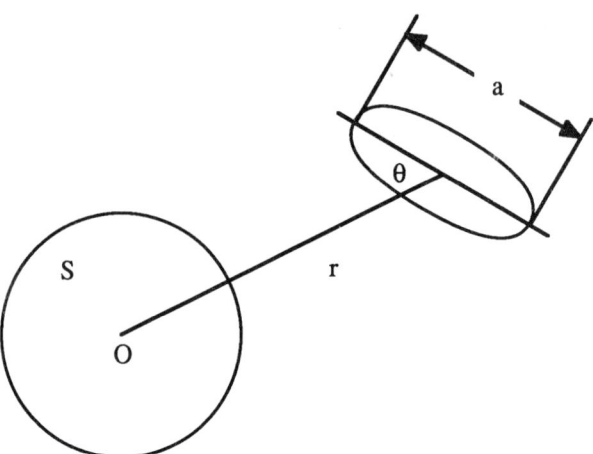

Figure 4.29: The angle, θ, relating the orientation of a spheroid to the direction of diffusion.

D(θ) is the coefficient of translational diffusion when the axis, a, of the spheroid makes an angle θ with the direction of diffusion as shown in Fig. 4.29.

D(θ) is given by:

$$\frac{1}{D(\theta)} = \frac{\cos^2(\theta)}{D_a} + \frac{\sin^2(\theta)}{D_b} \tag{4.387}$$

where D_a and D_b can be deduced from the results in Section 2.11. C_b is the rotational friction coefficient for rotation about the b-axis (Perrin [1934, 1936] and Sadron [1953]). The boundary conditions are:

$$C = \frac{C_\infty}{2\pi} \quad ; \quad \text{all } \theta, \quad r \to \infty$$

$$C = 0 \quad ; \quad \text{on} \quad r = S + \left[a^2 \cos^2(\theta) + b^2 \sin^2(\theta)\right]^{1/2} \tag{4.388}$$

A general solution to this problem is not available. A simple approximation, however, is to replace r in Eqn. (4.388) by its mean value, $\bar{r}$. C is now independent of θ and its value is:

$$C = C_\infty \left(1 - \frac{\bar{r}}{r}\right) \tag{4.389}$$

The coagulation rate is therefore:

$$J = 4\pi\, C_\infty\, \bar{r}\, \overline{D}(\theta) \tag{4.390}$$

where:

Coagulation Kernels

$$\overline{D}(\theta) = \frac{kT p \ln\left[p + (p^2 - 1)^{1/2}\right]}{6\pi \mu a (p^2 - 1)^{1/2}} \quad ; \quad p > 1 \tag{4.391}$$

$$= \frac{kT p \tan^{-1}\left[(1 - p^2)^{1/2} / p\right]}{6\pi \mu a (1 - p^2)^{1/2}} \quad ; \quad p < 1 \tag{4.392}$$

This solution should be approximately correct since, if we neglect the rotational Brownian term in Eqn. (4.386), the solution is:

$$C(r,\theta) = C_\infty \left[1 - \frac{1}{r}\left\{S + \left(a^2 \cos^2(\theta) + b^2 \sin^2(\theta)\right)^{1/2}\right\}\right] \tag{4.393}$$

which gives much the same value for J if S>>a, b. The effect of Brownian rotation will be, therefore, a smearing of the angular dependence which would lead C to lie somewhere between Eqns. (4.389) and (4.393). Booth shows how the above work may be extended to incorporate a radial potential.

4.18.3 Coagulation of large and small spheroids

This case may be developed by combining the methods of the other two cases. For example, (u,v) coordinates, fixed with respect to a large spheroid, are used to specify the centers of the smaller spheroids while their orientation is determined by the angle between the axes and the normal to the spheroidal surfaces, u=constant.

More recent contributions to coagulation between spheroids and spheres have been made by Tuttle and Loyalka [1985a,b,c]. In this work, the flow of fluid around an oblate spheroid is obtained numerically. This flow field is then used to calculate the collision efficiency between spheres and spheroids using the superposition method for the interparticle forces. More precisely, the forces on the sphere are obtained by assuming that its local velocity field is a superposition of that due to the sphere alone and that due to the spheroid alone. Similar calculations are made for the forces on the spheroid.

Results of these computations indicate that it is inappropriate to assume that agglomerates collide with the same efficiencies as spherical particles. A collision shape factor is introduced with which the spherical gravitational collision efficiencies must be modified. These factors are quite substantial in most cases, especially for highly distorted plate-like particles. The work of Tuttle and Loyalka follows closely that of Pitter and Pruppacher [1974] who examined the collision of supercooled water droplets with ice plates.

Recent results of Lee and Shaw [1984] show a strong shape effect for Brownian coagulation of fibrous-like particles. This work uses the concept of the effective collision diameter, d_{eff}. Then, assuming an equal orientation probability for the rod-like structures, a value of d_{eff} is obtained in terms of an average over $P(\theta,\phi)$, the orientation probability. The value of d_{eff} for the collision between two rods of lengths, l_1 and l_2, and diameters, d_1 and d_2, is:

$$d_{eff} = \frac{\left\{\pi L_2^2 \left[L_2 + 2 d_1 - d_1 (l_2 + d_2)/L_2\right] + \frac{3}{4} l_2^3 + \frac{1}{2} \pi^2 L_2^3 - 2\pi d_2 l_2 L_2\right\}}{\pi \left[4 L_2^2 + l_2^2 - (l_2 + d_2) L_2\right]} \tag{4.394}$$

where $L^2 = d^2 + l^2$. For a monodisperse aerosol where l>>d, we find $d_{eff} = (3/4)l$.

Lee and Shaw proceed by noting that according to Fuchs [1964], the calculation of the current of particles onto a body is analogous to the flow of electric charge (both of which are related to Laplace's equation). Thus, the coagulation kernel, K, is given by:

$$K = 2\pi (D_1 + D_2) C_E$$

where C_E is the electric capacitance of a conductor having the same shape as the absorbing body. The rods are regarded as prolate ellipsoids with a semi-major axis of length, $a = l_1/2 + d_{eff}$, and a semi-minor axis of length, $b = d_{eff}$. In this case:

$$C_E = \frac{\left(a^2 - b^2\right)^{1/2}}{\ln\left\{\left[a + \left(a^2 - b^2\right)^{1/2}\right]/b\right\}} \tag{4.395}$$

D_1 and D_2 correspond to the Brownian diffusion coefficients of randomly oriented rods. This result is clearly quite complex and would not fit conveniently into a formalism which requires spherical particles. Nevertheless, it is of some interest in dealing with coagulation between dendritic shapes.

References: Chapter 4

Adler, P.M. [1981] "Heterocoagulation in shear flow," *J. Colloid Interface Sci.* **83**, 106.

Alam, M.K. [1987] "The effect of van der Waals and viscous forces on aerosol coagulation," *Aerosol Sci. and Tech.* **6**, 41.

Allen, M.D. and Raabe, O.G. [1985] "Slip correction measurements of spherical solid aerosol particles in an improved Millikan apparatus," *Aerosol Sci. and Tech.* **4**, 269.

Andrade, E.N da C. [1936] "The coagulation of smoke by supersonic vibrations," *Trans. Faraday Soc.* **32**, 1111.

Arp, P.A. and Mason, S.G. [1976] "Orthokinetic collisions of hard spheres in simple shear flow," *Can. J. Chem.* **54**, 3769.

Atkinson, W.R. and Paluch, I. [1968] "Analytical approximation of numerically determined collision efficiencies of hydrometers," *J. Geophys. Res.* **73**, 2035.

Batchelor, G.K. [1951] "Pressure fluctuations in isotropic turbulence," *Proc. Camb. Philos. Soc.* **47**, 359.

Batchelor, G.K. [1976] "Brownian diffusion of particles with hydrodynamic interaction," *J. Fluid Mech.* **74**, 1.

Batchelor, G.K. [1982] "Sedimentation in a dilute polydisperse system of interacting spheres. Part 1. General theory," *J. Fluid Mech.* **119**, 379.

Batchelor, G.K. and Green, J.T. [1972] "The hydrodynamic interaction of two small freely-moving spheres in a linear flow field," *J. Fluid Mech.* **56**, 375.

Beard, K.V. and Grover, S.N. [1974] "Numerical collision efficiencies for small raindrops colliding with micron size particles," *J. Atmos. Sci.* **31**, 543.

Bjerknes, C.A. [1915] Hydrodynamisch Fernkrafte (Leipzig, Germany).

Booth, F. [1954] "The coagulation of non-spherical particles," *Faraday Society Discussions* **18**, 104.

Brandt, O. and Hiedemann, E. [1936] "The aggregation of suspended particles in gases by sonic and supersonic waves," *Trans. Faraday Soc.* **32**, 1101.

Carrier, G.F. [1953] "On slow viscous flow" (Brown University) Final report, Contract Nonr - 653 (00).

Cercignani, C. [1975] Theory and Application of the Boltzmann Equation (Scottish Academic Press, Edinburgh).

Cercignani, C., Pagani, C.D. and Bassanini, P. [1968] "Flow of a rarefied gas past an axisymmetric body II. Case of a sphere," *Phys. Fluids* **11**, 1399.

Chandrasekhar, S. [1943] "Stochastic problems in physics and astronomy," *Rev. Mod. Phys.* **15**, 1.

Cunningham, E. [1910] "On the velocity of steady fall of spherical particles through fluid medium," *Proc. Roy. Soc. (London)* **A83**, 357.

Curtis, A.S.G. and Hocking, L.M. [1970] "Collision efficiency of equal spherical particles in shear flow," *Trans. Faraday Soc.* **66**, 1381.

Davis, M.H. and Sartor, J.D. [1967] "Theoretical collision efficiencies for small cloud droplets in Stokes flow," *Nature Mag.* **215**, 1371.

Deutch, J.M. and Oppenheim, I. [1971] "Molecular theory of Brownian motion for several particles," *J. Chem. Phys.* **54**, 3547.

Dunbar, I.H. and Kirby, C.R. [1983] "Fast Reactor Aerosol Code Comparison Contract" (Safety and Reliability Directorate, UKAEA) Final report.

East, T.W.R. and Marshall, J.S. [1954] "Turbulence in clouds as a factor in precipitation," *Q. J. Roy. Meteor. Soc.* **80**, 26.

Einstein, A. [1956] Investigation on the Theory of the Brownian Movement (Dover, New York, N.Y.).

Fuchs, N.A. [1964] The Mechanics of Aerosols (Pergamon, New York, N.Y.).

Grover, S.N. and Beard, K.V. [1975] "A numerical determination of the efficiency with which electrically charged cloud drops and small raindrops collide with electrically charged spherical particles of various densities," *J. Atmos. Sci.* **32**, 2156.

Hidy, G.M. and Brock, J.R. [1965] "Some remarks about the coagulation of aerosol particles by Brownian motion," *J. Colloid Sci.* **20**, 477.

Hocking, L.M. [1958] Three-Dimensional Viscous Flow Problems Solved by the Stokes and Oseen Approximations (London University, London, U.K.) Ph.D. Thesis.

Hocking, L.M. [1959] "The collision efficiency of small drops," *Q. J. Roy. Meteor. Soc.* **85**, 44.

Hocking, L.M. [1964] "The behaviour of clusters of spheres falling in a viscous fluid, part 2. Slow motion theory," *J. Fluid Mech.* **20**, 129.

Hocking, L.M. [1973] "The effect of slip on the motion of a sphere close to a wall and of two adjacent spheres," *J. Eng. Math.* **7**, 207.

Hocking, L.M. and Jonas, P.R. [1970] "The collision efficiency of small drops," *Q. J. Roy. Meteor. Soc.* **96**, 722.

Honig, E.P., Roebersen, G.J. and Wiersema, P.H. [1971] "Effect of hydrodynamic interaction on the coagulation rate of hydrophobic colloids," *J. Colloid Interface Sci.* **36**, 97.

Jayaweera, K.O.L.F., Mason, B.J. and Slack, G.W. [1964] "The behaviour of clusters of spheres falling in a viscous fluid, 1. Experiment," *J. Fluid Mech.* **20**, 121.

Jonas, P.R. [1972] "The collision efficiency of small drops," *Q. J. Roy. Meteor. Soc.* **98**, 681.

Kennard, E.H. [1938] Kinetic Theory of Gases (McGraw Hill, New York, N.Y.).

Klett, J.D. and Davis, M.H. [1973] "Theoretical collision efficiencies of cloud droplets at small Reynolds numbers," *J. Atmos. Sci.* **30**, 107.

Konig, W. [1891a] "Hydrodynamisch - akustische untersuchungen," *Ann. Phys. Chem.* **42**, 353.

Konig, W. [1891b] "Hydrodynamisch - akustische untersuchungen," *Ann. Phys. Chem.* **42**, 549.

Langmuir, I. [1948] "The production of rain by a chain reaction in cumulus clouds at temperatures above freezing," *J. Meteor.* **5**, 175.

Lassen, L. [1961a] "Die anlagerung von zerfallsprodukten der naturlichen emanationen an elektrisch geladene aerosole (schwebstoffe)." The title is translated as: "The adsorption of decay products of natural emanations on electrically charged aerosols," *Z. Physik* **163**, 363 (in German).

Lassen, L. [1961b] "The attachment of natural radioactivity to aerosols," *Pure and Appl. Geophys.* **50**, 281.

Law, W.S. and Loyalka, S.K. [1986] "Motion of a sphere in a rarefied gas II: role of temperature variation in the Knudsen layer," *Phys. Fluids* **29**, 3886.

Lea, K.C. and Loyalka, S.K. [1982] "Motion of a sphere in a rarefied gas," *Phys. Fluids* **25**, 1550.

Lee, P.S. and Shaw, D.T. [1984] "Dynamics of fibrous-type particles: Brownian coagulation and the charge effect," *Aerosol Sci. and Tech.* **3**, 9.

Levich, V.G. [1962] Physicochemical Hydrodynamics (Prentice Hall, Englewood Cliffs, N.J.) (English translation).

Lin, C.J., Lee, K.J. and Sather, N.F. [1970] "Slow motion of two spheres in a shear field," *J. Fluid Mech.* **43**, 35.

Lin, C.L. and Lee, S.C. [1975] "Collision efficiency of water drops in the atmosphere," *J. Atmos. Sci.* **32**, 1412.

Loyalka, S.K. [1976] "Brownian coagulation of aerosol," *J. Colloid Interface Sci.* **57**, 578.

Mednikov, E.P. [1965] Acoustic Coagulation and Precipitation of Aerosols (Consultants Bureau, New York, N.Y.).

Melik, D.H. and Fogler, H.S. [1984a] "Gravity-induced flocculation," *J. Colloid Interface Sci.* **101**, 72.

Melik, D.H. and Fogler, H.S. [1984b] "Effect of gravity on Brownian flocculation," *J. Colloid Interface Sci.* **101**, 84.

Millikan, R.A. [1923] "The general law of fall of a small spherical body through a gas, and its bearing upon the nature of molecular reflection from surfaces," *Phys. Rev.* **22**, 1.

Morlock [1986] "The Collision Efficiency of Gravitational Coagulation of the Aerosol Particles in the Containment System of Fast Reactors" (Gessellschaft fur Reaktorsicherheit) Report # GRS-A-1246.

Murphy, T.J. and Aguirre, J.L. [1972] "Brownian motion of N interacting particles. I. Extension of the Einstein diffusion relation to the N-particle case," *J. Chem. Phys.* **57**, 2098.

Neumann, E.P. and Norton, J.L. [1951] "Application of sonic energy to commercial aerosol collection problems," *Chem. Eng. Prog. Symp. Ser. 1* **47**, 4.

Parker, R.C. [1936] "Experiments on coagulation by supersonic vibrations," *Trans. Faraday Soc.* **32**, 1115.

Patterson, H.S. and Cawood, W. [1931] "Phenomena in a sounding tube," *Nature Mag.* **127**, 667.

Patterson, H.S. and Cawood, W. [1932] "The reproducibility and rate of coagulation of stearic acid smokes," *Proc. Roy. Soc. (London)* **136A**, 538.

Patterson, H.S., Whytlaw-Gray, R. and Cawood, W. [1929] "The process of coagulation in smokes," *Proc. Roy. Soc. (London)* **124A**, 502.

Pearcey, T. and Hill, G.W. [1957] "A theoretical estimate of the collection efficiencies of small droplets," *Q. J. Roy. Meteor. Soc.* **83**, 77.

Perrin, F. [1934] "Mouvement Brownien d'un ellipsoide (I). Dispersion dielectrique pour des molecules ellipsoidales," *J. Phys. Radium* **5**, 497 (in French).

Perrin, F. [1936] "Mouvement Brownien d'un ellipsoide (II). Rotation libre et depolarisation des fluorescences. Translation et diffusion de molecules ellipsoidales," *J. Phys. Radium* **7**, 1 (in French).

Pertmer, G.A. and Loyalka, S.K. [1980] "Gravitational collision efficiency of post hypothetical core disruptive accident liquid metal fast breeder reactor aerosols: spherical particles," *Nucl. Tech.* **47**, 70.

Phillips, W.F. [1975] "Drag on a small sphere moving through a gas," *Phys. Fluids* **18**, 1089.

Pitter, R.L. and Pruppacher, H.R. [1974] "A numerical investigation of collision efficiencies of simple ice plates colliding with supercooled water drops," *J. Atmos. Sci.* **31**, 551.

Pruppacher, H.R. and Klett, J.D. [1978] Microphysics of Clouds and Precipitation (Reidel, New York, N.Y.).

Sadron, A. [1953] "Methods of determining the form and dimensions of particles in solution: a critical survey," *Prog. Biophys. and Biophys. Chem.* **3**, 237.

Saffman, P.G. and Turner, J.S. [1956] "On the collision of drops in turbulent clouds," *J. Fluid Mech.* **1**, 16.

Schlamp, R.J., Grover, S.N., Pruppacher, H.R. and Hamielec, A.E. [1976] "A numerical investigation of the effect of electric charges and vertical external electric fields on collision efficiency of cloud drops," *J. Atmos. Sci.* **33**, 1747.

Shafrir, U. and Neiburger, M. [1963] "Collision efficiencies of two spheres falling in a viscous medium," *J. Geophys. Res.* **68**, 4141.

Shahub, A.M. and Williams, M.M.R. [1988] "Brownian collision efficiency," *J. Phys. D Appl. Phys.* **21**, 231.

Shaw, D.T. and Rajendran, N. [1979] "Application of acoustic agglomerators for emergency use in liquid-metal fast breeder reactors," *Nucl. Sci. and Eng.* **70**, 127.

Shaw, D.T. and Tu, K.W. [1979] "Acoustic particle agglomeration due to hydrodynamic interaction between monodisperse aerosols," *J. Aerosol Sci.* **10**, 317.

Shirokova, N.L. [1970] "Aerosol Coagulation," in Physical Principles of Ultrasonic Technology, Vol. 2, Part X, Rozenberg, L.D., editor (Plenum Press, New York, N.Y.).

Simons, S. and Williams, M.M.R. [1971] unpublished but see Williams, M.M.R. [1976] "An introduction to the coagulation of aerosols," *J. Inst. Nuc. Engrs.* **17**, 83.

Simons, S., Williams, M.M.R. and Cassell, J.S. [1986] "A kernel for combined Brownian and gravitational coagulation," *J. Aerosol Sci.* **17**, 789.

Sitarski, M. and Seinfeld, J.H. [1977] "Brownian coagulation in the transition regime," *J. Colloid Interface Sci.* **61**, 261.

Smoluchowski, M. von [1916] "Drei vortrage uber diffusion, Brownsche molekularbewegung und koagulation von kolloidteilchen." The title is translated as: "Three lectures on diffusion, Brownian motion and coagulation of colloid particles," *Phys. Z.* **17**, 557 (in German).

Spielman, L.A. [1970] "Viscous interactions in Brownian coagulation, *J. Colloid Interface Sci.* **33**, 562.

St. Clair, H.W. [1938] "Sonic flocculation as a fume settler: theory and practice" (U.S. Bureau of Mines) Report # RI 3400, p. 51.

St. Clair, H.W. [1949] "Agglomeration of smoke, fog or dust particles by sonic waves," *Ind. Eng. Chem.* **41**, 2434.

Taylor, G.I. [1935] "Statistical theory of turbulence," *Proc. Roy. Soc. (London)* **A151**, 421.

Tuttle, R.F. and Loyalka, S.K. [1985a] "Gravitational collision efficiency of non-spherical aerosols I: definitions of shape factors," *Nucl. Tech.* **69**, 319.

Tuttle, R.F. and Loyalka, S.K. [1985b] "Gravitational collision efficiency of non-spherical aerosols II: motion of an oblate spheroid in a viscous fluid," *Nucl. Tech.* **69**, 327.

Tuttle, R.F. and Loyalka, S.K. [1985c] "Gravitational collision efficiency of non-spherical aerosols III: computer program NGCEFF and calculation of shape factors," *Nucl. Tech.* **69**, 337.

van de Ven, T.G.M. and Mason, S.G. [1976] "The microrheology of colloidal dispersions IV: pairs of interacting spheres in shear flow," *J. Colloid Interface Sci.* **57**, 505.

van de Ven, T.G.M. and Mason, S.G. [1977a] "The microrheology of colloidal dispersions VII," *Colloid and Polym. Sci.* **255**, 468.

van de Ven, T.G.M. and Mason, S.G. [1977b] "The microrheology of colloidal dispersions VIII," *Colloid and Polym. Sci.* **255**, 794.

van Kampen, N.G. [1981] <u>Stochastic Processes in Physics and Chemistry</u> (North Holland, Amsterdam, Netherlands).

Wacholder, E. and Sather, N.F. [1974] "The hydrodynamic interaction of two unequal spheres moving under gravity through a quiescent viscous fluid," *J. Fluid Mech.* **65**, 417.

Wang, P.K., Grover, S.N. and Pruppacher, H.R. [1978] "On the effect of electric charges on the scavenging of aerosol particles by clouds and small raindrops," *J. Atmos. Sci.* **35**, 1735.

Williams, M.M.R. [1971] <u>Mathematical Methods in Particle Transport Theory</u> (Butterworths, London, U.K.).

Williams, M.M.R. [1988] "A unified theory of aerosol coagulation," *J. Phys. D Appl. Phys.* **21**, 875.

Zeichner, G.R. and Schowalter, W.R. [1977] "Use of trajectory analysis to study stability of colloidal dispersions in flow fields," *A.I.Ch.E. J.* **23**, 243.

CHAPTER 5

Methods of Solving the Dynamic Equation

5.1 Introduction

The equation describing the behavior of aerosols is a nonlinear, integro-differential equation of considerable complexity. If all of the deposition, condensation, and coagulation mechanisms are included, it is clear that the only feasible method of solution is a numerical one. Indeed, in practical calculations this is the procedure and many computer codes exist which solve the dynamic equation using finite difference or finite element methods. We shall discuss the basis of such methods later in this chapter.

A major drawback of numerical methods is the vast amount of computer time necessary to obtain a useful survey of all the relevant parameters. This causes difficulty in gaining physical insight into the problem. For this reason a variety of approximate techniques have been developed which enable analytic solutions to be obtained. Such techniques use the complementary methods of kernel approximation and equation approximation or a combination of both. By kernel approximation we mean that the actual physical coagulation kernel, $K(u,v)$, is replaced by an approximate form which retains certain physical features of the 'exact' kernel but, at the same time, leads to a simple analytical solution of the dynamic equation. In the case of equation approximation, we retain the exact form of $K(u,v)$ but employ an approximate method to solve the resulting dynamic equation.

The purpose of this chapter is to discuss a number of exact solutions using simplified kernels, to study various approximate solution techniques, and finally, to elaborate on numerical procedures and how these are embodied in various computer codes.

5.2 Exact solutions

Exact solutions are useful since they provide benchmarks for verification of techniques that yield approximate or numerical solutions.

The earliest exact solution of the dynamic equation was given by Smoluchowski [1916] for the case when the particles are considered as discrete volumes. Then, the dynamic equation for pure coagulation becomes:

$$\frac{dn_i}{dt} = \tfrac{1}{2} \sum_{j=1}^{i-1} a_{j,i-j}\, n_j\, n_{i-j} - n_i \sum_{j=1}^{\infty} a_{i,j}\, n_j \tag{5.1}$$

where $n_i(0) = N\, \delta_{i,1}$.

Now, if the coagulation kernel $a_{i,j}$ is independent of particle size, we can write Eqn. (5.1) as:

$$\frac{dn_i}{d\tau} = \tfrac{1}{2} \sum_{j=1}^{i-1} n_j\, n_{i-j} - n_i \sum_{j=1}^{\infty} n_j \tag{5.2}$$

where $\tau = at$.

Eqn. (5.2) may be solved by first summing over all i to get:

$$N(t) = \sum_{i=1}^{\infty} n_i(t)$$

whence we find:

$$\frac{dN}{d\tau} = -\tfrac{1}{2} N^2$$

the solution of which is $N(\tau) = 2/(2+\tau)$. Now to find the n_i we solve Eqn. (5.2) recursively, e.g.:

$$\frac{dn_1}{d\tau} = -n_1 N$$

Thus:

$$n_1 = (1+\tau/2)^{-2}$$

$$n_2 = \frac{\tau}{2}(1+\tau/2)^{-3}$$

and:

$$n_i = \left(\frac{\tau}{2}\right)^{i-1}(1+\tau/2)^{-i-1}$$

Another case which may be solved exactly in the discrete sense is when $a_{i,j}=ij$. Then, Eqn. (5.1) becomes:

$$\frac{dn_i}{dt} = \tfrac{1}{2} i \sum_{j=1}^{i-1}(i-j) n_j n_{i-j} - i n_i \sum_{j=1}^{\infty} j n_j \qquad (5.3)$$

Multiplying by i and summing over all i leads to:

$$\sum_{i=1}^{\infty} i \frac{dn_i}{dt} = 0 \quad \text{or} \quad \sum_{i=1}^{\infty} i n_i = 1$$

Then, Eqn. (5.3) becomes:

$$\frac{dn_i}{dt} = \tfrac{1}{2} \sum_{j=1}^{i-1} j(i-j) n_j n_{i-j} - i n_i$$

This equation can be solved recursively to give:

$$n_1 = \exp(-\tau) \quad , \quad n_2 = \tfrac{1}{2}\tau \exp(-2\tau) \quad \text{and} \quad n_i = \frac{A_i}{i} \tau^{i-1} \exp(-i\tau)$$

where A_i are constants independent of τ.

However, because of the nature of this kernel, the second moment:

$$M_2 = \sum_{j=1}^{\infty} j^2 n_j$$

Methods of Solving the Dynamic Equation

is only convergent for $\tau < \tau_0$ where $\tau_0 < 1$ (McLeod [1962a,b]).

It is more useful to examine solutions of the dynamic equation in the continuous variable, v. In this respect, a solution of the equation:

$$\frac{\partial}{\partial t} n(v,t) = \tfrac{1}{2} \int_0^v du\, K(u, v-u)\, n(u,t)\, n(v-u,t) - n(v,t) \int_0^\infty du\, K(u,v)\, n(u,t) \qquad (5.4)$$

subject to $n(v,0) = n_0(v)$, was given by Schumann [1940] when K is a constant. Eqn. (5.4) becomes:

$$\frac{\partial}{\partial t} n(v,t) = -K\, n(v,t)\, N(t) + \tfrac{1}{2} K \int_0^\infty du\, n(u,t)\, n(v-u,t) \qquad (5.5)$$

Integrating this equation over all v leads to:

$$\frac{dN}{dt} = -\tfrac{1}{2} K\, N^2$$

whence:

$$N(t) = \frac{N_0}{1+pt}$$

where $p = K\, N_0 / 2$. Setting $\theta = 1/(1+pt)$ and:

$$\psi(v,\theta) = \frac{K}{2p}(1+pt)^2\, n(v,t)$$

Eqn. (5.5) reduces to:

$$-\frac{\partial}{\partial \theta}\psi(v,\theta) = \int_0^v du\, \psi(u,\theta)\, \psi(v-u,\theta)$$

Let us solve this equation by the Laplace transform technique. We define:

$$\overline{\psi}(s,\theta) = \int_0^\infty dv\, \exp(-s\, v)\, \psi(v,\theta)$$

which, by the convolution transform, leads to:

$$\psi(v,\theta) = \frac{1}{2\pi i} \int_L \frac{ds\, \exp(s\, v)}{\theta - 1 + [1/\overline{\psi}_0(s)]} \qquad (5.6)$$

Here, the initial condition is:

$$\overline{\psi}_0(s) = \int_0^\infty dv\, \psi(v,1)$$

243

Thus, the nature of the solution depends crucially on the initial conditions. This is in contrast to linear transport equations where the initial conditions generally only affect the overall amplitude of the solution.

As an initial condition, let us take the gamma distribution:

$$n(v,0) = \frac{N_0}{\hat{v}\,\Gamma(v+1)} \left(\frac{v}{\hat{v}}\right)^v \exp\left(-\frac{v}{\hat{v}}\right) \tag{5.7}$$

Inserting the Laplace transform of Eqn. (5.7) into Eqn. (5.6) we find:

$$\psi(v,\theta) = \frac{1}{\hat{v}} \exp\left(-\frac{v}{\hat{v}}\right) \frac{1}{2\pi i} \int_L \frac{dz\,\exp(vz/\hat{v})}{z^{v+1}-1+\theta} \tag{5.8}$$

If v is an integer, then the residue theorem leads to simple results (*e.g.* with v=0):

$$\psi(v,\theta) = \frac{1}{\hat{v}} \exp\left(-\frac{v\theta}{\hat{v}}\right)$$

or, from the definition of ψ:

$$n(v,t) = \frac{N_0}{\hat{v}(1+pt)^2} \exp\left\{-\frac{v}{\hat{v}(1+pt)}\right\} \tag{5.9}$$

In this case the distribution maintains its gamma-like form throughout, with the parameters, N_0 and $\hat{v}$, becoming time dependent, *i.e.* we can write Eqn. (5.9) as:

$$n(v,t) = \frac{N^2(t)}{\phi_0} \exp\left\{-\frac{v N(t)}{\phi_0}\right\} \tag{5.10}$$

where $\bar{v}(t) = v_0(1+pt)$ is the average particle volume at time, t. ϕ_0 is the volume of particulate per unit volume of space, *i.e.*:

$$\phi_0(t) = \int_0^\infty dv\, v\, n(v,t)$$

For v=1, the result becomes:

$$\psi(v,\theta) = \frac{1}{\hat{v}} \exp\left(-\frac{v}{\hat{v}}\right) \sinh\left\{\left(\frac{v}{\hat{v}}\right)(1-\theta)^{1/2}\right\}(1-\theta)^{-1/2}$$

which is not a gamma distribution. We discuss the general case of arbitrary v in Section 5.6, but results for some specific values of v may be found in Scott [1968] and Williams [1981]. A case which has some practical interest for naturally occurring atmospheric aerosols is one which obeys an initial size distribution of the form:

$$n(v,0) = \frac{a\, e^{-v/b}}{(v+c)^2}$$

Methods of Solving the Dynamic Equation

Using such a distribution, Mulholland and Baum [1980] have obtained the following result:

$$n(v,t) = \frac{a}{c^2}(1-\lambda)^2 \, \theta(\tilde{v},\lambda) \tag{5.11}$$

where $\tilde{v} = v/c$:

$$\lambda = \frac{K N_0 t/2}{(1+K N_0 t/2)\exp(c/b)\, E_2(c/b)}$$

and:

$$\theta(\tilde{v},\lambda) = \exp(-c\tilde{v}/b)\int_0^\infty \frac{\exp[-x(\tilde{v}+1)]\, x\, dx}{\{1-\lambda[1-x\exp(-x)E_i(x)]\}^2 + \{\pi\lambda x\exp(-x)\}^2} \quad ; \quad \lambda < 1$$

$$= 4a + \frac{\exp(-c\tilde{v}/b)\exp(x_0 \tilde{v})}{\lambda^2 \left[(1+x_0)\exp(x_0) E_1(x_0) - 1\right]} \quad ; \quad \lambda > 1$$

where x_0 is the root of the transcendental equation:

$$1 - \lambda\left[1 - x_0 \exp(x_0) E_1(x_0)\right] = 0$$

For sufficiently large λ, i.e. λ close to:

$$\lambda_\infty = \frac{\exp(-c/b)}{E_2(c/b)}$$

it may be shown that the solution reduces to:

$$n(v,t) = \frac{N^2(t)}{\phi_0}\exp\left\{-\frac{v N(t)}{\phi_0}\right\} \tag{5.12}$$

which is identical to the gamma distribution case for $v=0$ discussed earlier. Such asymptotic behavior will be useful in suggesting universal solutions at long times.

Let us now consider some exact solutions obtainable from other forms of the coagulation kernel. The first one we consider is, $K(u,v) = b(u+v)$. In this case, the dynamic equation becomes:

$$\frac{\partial}{\partial t} n(v,t) = \frac{bv}{2}\int_0^v du\, n(u,t)\, n(v-u,t) - b\, n(v,t)\int_0^\infty du\, (u+v)\, n(u,t) \tag{5.13}$$

Recalling the definitions of N and ϕ, Eqn. (5.13) can be written:

$$\frac{\partial}{\partial t} n(v,t) = \frac{bv}{2}\int_0^v du\, n(u,t)\, n(v-u,t) - b\, n(v,t)\left[\phi + vN\right] \tag{5.14}$$

We can obtain N(t) from the relation:

$$\frac{dN}{dt} = -\tfrac{1}{2}\int_0^\infty du \int_0^\infty dv\, K(u,v)\, n(u,t)\, n(v,t) \tag{5.15}$$

or:

$$\frac{dN}{dt} = -b\phi N$$

whence:

$$N(t) = N_0 \exp(-b\phi t)$$

It is more difficult to obtain the complete distribution. Thus, we define a function, $\psi(v,t)$, by:

$$n(v,t) = N_0\, \psi(v,t)\, f(t) \tag{5.16}$$

where $\psi(v,t)$ is determined by inserting Eqn. (5.16) into Eqn. (5.14):

$$f(t)\frac{\partial}{\partial t}\psi(v,t) + \psi(v,t)\frac{d}{dt}f(t)$$

$$= \frac{b}{2}v\, f^2(t)\, N_0 \int_0^v du\, \psi(u,t)\,\psi(v-u,t) - b\phi f(t)\,\psi(v,t) - b\, N(t)\, v\, f(t)\, \psi(v,t)$$

Now, define $f(t)$ by setting:

$$\frac{df}{dt} + b\phi f = 0$$

i.e. $f(t) = \exp(-b\phi t)$ and ψ obeys:

$$\frac{\partial}{\partial t}\psi(v,t) = \frac{b}{2} N_0\, v\, f(t) \int_0^v du\, \psi(u,t)\,\psi(v-u,t) - b\, N(t)\, v\, \psi(v,t)$$

Since $N(t) = N_0 f(t)$, we can define a new time variable:

$$\tau = b\phi \int_0^t dt'\, f(t') = 1 - \exp(-b\phi t)$$

whence:

$$\frac{\partial}{\partial \tau}\psi(v,\tau) = \frac{v}{v_0}\int_0^v du\, \psi(u,\tau)\,\psi(v-u,\tau) - \frac{v}{v_0}\psi(v,\tau)$$

where $v_0 = \phi/N_0$ is the average initial size of the particles. We now set:

$$\psi(v,\tau) = g(v,\tau)\exp(-v\tau/v_0)$$

leading to:

$$\frac{\partial}{\partial \tau}g(v,\tau) = \frac{v}{2v_0}\int_0^v du\, g(u,\tau)\, g(v-u,\tau) \tag{5.17}$$

Methods of Solving the Dynamic Equation

Defining the Laplace transform:

$$\bar{g}(p,\tau) = \int_0^\infty dv \, \exp(-pv) \, g(v,\tau)$$

and applying it to Eqn. (5.17), we find:

$$\frac{\partial}{\partial \tau} \bar{g}(p,\tau) = \frac{1}{v_0} \bar{g}(p,\tau) \frac{\partial}{\partial p} \bar{g}(p,\tau) \tag{5.18}$$

This is a partial differential and can be solved using the method of characteristics (Sneddon [1957]):

$$\bar{g}(p,\tau) = G\left[p - \frac{\tau}{v_0} \bar{g}(p,\tau)\right] \tag{5.19}$$

where:

$$G(p) = \bar{g}(p,0) \tag{5.20}$$

Eqn. (5.19) is a functional equation for $\bar{g}$ and may best be solved by using Lagrange's expansion (Abramowitz and Stegun [1973]). Then, we may write:

$$\bar{g}(p,\tau) = \sum_{k=0}^{\infty} \frac{(-1)^k}{(k+1)!} \left(\frac{\tau}{v_0}\right)^k \frac{d^k}{dp^k} G^{k+1}(p)$$

The Laplace inverse yields:

$$n(v,\tau) = N_0 (1-\tau) \exp(-v\tau/v_0) \sum_{k=0}^{\infty} \frac{1}{(k+1)!} \left(\frac{\tau v}{v_0}\right)^k L^{-1}\{G^{k+1}(p)\} \tag{5.21}$$

As an example, let us assume that the initial condition is:

$$n(v,0) = \frac{N_0}{v_0} \exp(-v/v_0)$$

from which $G(p) = N_0 / (1 + p v_0)$. It is then possible to sum the series in Eqn. (5.21) to get:

$$n(v,\tau) = N_0 (1-\tau) \exp(-(1+\tau) v/v_0) \frac{1}{v \tau^{1/2}} I_1\left(\frac{2 v \tau^{1/2}}{v_0}\right) \tag{5.22}$$

where $I_1(x)$ is a modified Bessel function.

It is interesting to note that, as t→∞, τ→1 and the volume spectrum assumes a time independent shape of the form:

$$n(v,\tau) = N_0 \exp(-b \phi t) \exp(-2 v/v_0) \frac{1}{v} I_1\left(\frac{2v}{v_0}\right) \tag{5.23}$$

We also note that the mean volume of a particle, $\bar{v}(t)$, obeys the relation, $\bar{v}(t) = v_0 \exp(b\phi t)$. Coagulation is therefore very rapid.

Another kernel of interest is:

$$K(u,v) = B u v \tag{5.24}$$

In this case, the dynamic equation can be written:

$$\frac{\partial}{\partial t} n(v,t) = \frac{B}{2} \int_0^v du \, u(v-u) \, n(u,t) \, n(v-u,t) - B v \phi \, n(v,t) \tag{5.25}$$

Setting:

$$n(v,t) = g(v,t) \exp(-B\phi v t)$$

Eqn. (5.25) becomes:

$$\frac{\partial}{\partial t} g(v,t) = \frac{B}{2} \int_0^v du \, u(v-u) \, g(u,t) \, g(v-u,t)$$

Taking the Laplace transform we find that:

$$\frac{\partial}{\partial t} \bar{g}(p,t) = \frac{B}{2} \bar{h}^2(p,t)$$

where:

$$\bar{g}(p,t) = \int_0^\infty dv \, \exp(-pv) \, g(v,t)$$

and:

$$\bar{h}(p,t) = \int_0^\infty dv \, v \exp(-pv) \, g(v,t) = -\frac{\partial}{\partial p} \bar{g}(p,t)$$

Therefore:

$$\frac{\partial}{\partial t} \bar{g}(p,t) = \frac{B}{2} \left(\frac{\partial}{\partial p} \bar{g}(p,t) \right)^2 \tag{5.26}$$

This equation may be solved by the method of Charpit (Sneddon [1957]) and leads to:

$$n(v,t) = N_0 \exp(-v\tau/v_0) \frac{1}{v} \sum_{k=0}^{\infty} \left(\frac{v\tau}{v_0} \right)^k \frac{1}{(1+k)!} L^{-1}\{-G'(p)\}^{k+1} \tag{5.27}$$

where $G(p) = \bar{g}(p,0)$.

For the special case of $n(v,0) = N_0 \exp(-v/v_0)/v_0$, as used above, we find:

$$n(v,t) = \frac{N_0}{v_0} \exp(-v(1+\tau)/v_0) \phi \sum_{k=0}^{\infty} \frac{y^k}{(k+1)!(2k+1)!} \tag{5.28}$$

where $y = \tau \phi (v/v_0)^3$ and $\tau = B f^2 t$.

The total number density is obtained easily from Eqns. (5.15) and (5.24) and leads to:

$$\frac{dN}{dt} = -\tfrac{1}{2} \phi^2 B$$

whence:

$$N(t) = N_0 \left(1 - \frac{\tau}{2}\right) \tag{5.29}$$

This result is interesting in that it predicts that the aerosol number will becomes zero at a finite time, $\tau=2$. This is clearly an unphysical result for an aerosol but does have some relevance to colloidal aggregation corresponding to the final stages of gelation (Ernst [1981]).

5.3 Growth and deposition

It is of some importance to study the effects of condensation and evaporation on an aerosol. In many practical instances vapors are present in containment and condense, thereby altering the size distributions of the aerosols. If such condensation phenomena are to be studied analytically, without too many simplifying assumptions, it is necessary to ignore coagulation. This is not an unphysical situation since, in many cases, the condensation can dominate the coagulation during certain time intervals.

Detailed calculations of the aerosol equation were made by Gelbard and Seinfeld [1979] who employed the radius distribution function, $n(r,t)$. They solved the dynamic equation in the form:

$$\frac{\partial}{\partial t} n(r,t) + \frac{\partial}{\partial r}[I(r,t) \, n(r,t)] + R(r,t) \, n(r,t) = S(r,t) \tag{5.30}$$

$I(r,t)$, which is actually dr/dt, assumes various functional forms determined by the environment and the materials involved. As an example, we note that for diffusion controlled growth, a law has been derived by Fuchs and Sutugin [1971] of the form:

$$I(r,t) = \frac{V_m \, D \, (r + \alpha \lambda)}{r^2 + b \lambda r + a \lambda^2} \left[N_\infty - N_+ \exp\left(\frac{2 \sigma V_m}{r k T}\right)\right] \tag{5.31}$$

where D is the diffusivity of the vapor molecules in the background gas. V_m is the molecular volume, σ is the surface tension, N_∞ is the gas phase concentration of condensing molecules far from the particle, and N_+ is the gas phase concentration of condensing molecules in equilibrium over a flat surface. The exponential factor is due to the Kelvin effect. The constants a, b, and α depend on the specific growth law and, as an example, Fuchs [1959] gives:

$$a = \frac{\alpha D}{\lambda} \left(\frac{2 \pi m}{k T}\right)^{1/2}$$

where λ is the mean free path of atoms in the background gas, $b = a/\alpha$, and $\alpha \approx 1$. The diffusion coefficient is given by the Knudsen value:

$$D = \frac{\lambda}{3}\left(\frac{8kT}{\pi m}\right)^{1/2}$$

Gelbard and Seinfeld [1979] give some exact solutions of Eqn. (5.30) for various growth laws with the deposition constant, R, independent of r. Also, they assume an initial value problem with no continuous source.

Another useful form for the condensation rate, given by Fuchs [1959] and modified by Pich et al. [1970], is $I(v,t) = B(S-1)v^{1/3}$ where the supersaturation, $S = p_v/p_s$, p_v being the actual pressure of the vapor and p_s the saturation vapor pressure. The constant, B, is given by:

$$B = 3^{1/3}(4\pi)^{2/3}\left[\frac{\rho L^2 M}{KRT^2} + \frac{\rho RT}{DMp_s}\right]^{-1}$$

with:

ρ = density of the liquid
L = latent heat
M = molecular weight of vapor molecule
K = thermal conductivity of the drop
R = gas constant
D = diffusion coefficient of the vapor.

We shall give a general derivation of the solution for arbitrary I, R, and S. For convenience, we employ the volume distribution function which obeys the equation:

$$\frac{\partial}{\partial t}n(v,t) + \frac{\partial}{\partial v}[I(v,t)n(v,t)] + R(v,t)n(v,t) = S(v,t) \tag{5.32}$$

This first order partial differential equation may be solved by the method of characteristics:

$$n(v,t) = \frac{1}{f(v)}f(F^{-1}[F(v)+G(0)-G(v)])n_0(F^{-1}[F(v)+G(0)-G(t)])$$

$$+ \int_0^t dt'\, S(v(t'),t')f(v(t'))\exp\left\{-\int_{t'}^t dt''\, R(v(t''),t'')\right\} \tag{5.33}$$

where:

$$F(v) = \int^v \frac{dv'}{f(v')}$$

$$G(t) = \int^t dt'\, H(t')$$

$$v(t') = F^{-1}[G(t')-G(t)+F(v)]$$

and $F^{-1}(x)$ is the solution of $x = F(v)$. Also, we have assumed that $I(v,t) = f(v)H(t)$ and the initial condition, $n(v,0) = n_0(v)$.

Methods of Solving the Dynamic Equation

If we are only concerned with an initial value problem, then S=0. On the other hand, for a source driven system, we may set $n_0(v)=0$. The steady state solution, when $S(v,t)=S(v)$, is readily obtained and leads to:

$$n(v) = \frac{1}{f(v)} \int_0^v dv' \, S(v') \exp\left\{-\int_{v'}^v \frac{R(v'')}{f(v'')} dv''\right\}$$
(5.34)

The complete solution for n(v,t) follows from Eqn. (5.33) by quadrature. However, we may consider some special cases.

First, we note that most deposition mechanisms can be written as:

$$R(v,t) = R_m(t) \, v^m$$

and we shall take R_m to be independent of time. It is also reasonable to take:

$$I(v,t) = F_\alpha(t) \, v^\alpha$$

where, again, we assume F_α is independent of time.

In the case of the initial value problem, and using these simple forms of R and I, we see that Eqn. (5.33) becomes:

$$n(v,t) = \left[1 - \frac{(1-\alpha) F_\alpha t}{v^{(1-\alpha)}}\right]^{\alpha/(1-\alpha)} n_0\left(\left[v^{(1-\alpha)} - (1-\alpha) F_\alpha t\right]^{\alpha/(1-\alpha)}\right)$$

$$\exp\left\{-\frac{R_m}{F_\alpha (m+1-\alpha)}\left[v^{m+1-\alpha} - \left(v^{(1-\alpha)} - (1-\alpha) F_\alpha t\right)^{[m/(1-\alpha)]+1}\right]\right\}$$
(5.35)

where $v^{(1-\alpha)} > (1-\alpha) F_\alpha t$. For $v^{(1-\alpha)} < (1-\alpha) F_\alpha t$, it should be noted that $n(v,t) = 0$. This discontinuity follows from the fact that, due to particle growth, the minimum particle size at time, t, is:

$$v_{min} = \left[(1-\alpha) F_\alpha t\right]^{1/(1-\alpha)}$$

If we choose a monodisperse initial condition, $n_0(v) = N_0 \, \delta(v - v_0)$, then the total aerosol density is:

$$N(t) = N_0 \exp\left\{-\frac{R_m}{F_\alpha (m+1-\alpha)}\left[\left\{v_0^{(1-\alpha)} + (1-\alpha) F_\alpha t\right\}^{[m/(1-\alpha)]+1} - v_0^{m+1-\alpha}\right]\right\}$$
(5.36)

and:

$$\phi(t) = \left[v_0^{(1-\alpha)} + (1-\alpha) F_\alpha t\right]^{1/(1-\alpha)} N(t)$$

Thus, the average volume of a particle increases with time according to:

$$\overline{v}(t) = \frac{\phi(t)}{N(t)} = \left[v_0^{(1-\alpha)} + (1-\alpha) F_\alpha t\right]^{1/(1-\alpha)}$$
(5.37)

If we have a steady source problem with the source written:

$$S(v,t) = S_0\, \delta(v-v_0)\, g(t)$$

Then, the solution becomes:

$$n(v,t) = \frac{S_0}{v^\alpha F_\alpha}\, g\!\left(t + \frac{1}{(1-\alpha)F_\alpha}\left\{v_0^{(1-\alpha)} - v^{(1-\alpha)}\right\}\right)\exp\!\left\{-\frac{R_m}{F_\alpha(m+1-\alpha)}\left(v^{m+1-\alpha} - v_0^{m+1-\alpha}\right)\right\}$$

(5.38)

where:

$$v_0 \le v \le \left[v_0^{1-\alpha} + (1-\alpha)F_\alpha t\right]^{1/(1-\alpha)}$$

and $n(v,t) = 0$ outside these limits.

This distribution differs from the initial value one in that the shape of the initial distribution is no longer preserved. The difference arises from the fact that once the aerosol has been released (in the initial value problem) it can only change by growth or by deposition. In the source problem, however, particles of initial volume, v_0, are continuously being produced leading to a distribution of volumes from v_0 up to the maximum value allowed by the growth law. Moreover, in the initial value problem, a monodisperse distribution will remain monodisperse with its volume equal to:

$$v = \left[(1-\alpha)F_\alpha t + v_0^{1-\alpha}\right]^{1/(1-\alpha)}$$

If the source is independent of time, with $g(t)=1$, the solution reverts to the steady state one:

$$n(v) = \frac{S_0}{v^\alpha F_\alpha}\exp\!\left\{-\frac{R_m}{F_\alpha(m+1-\alpha)}\left(v^{m+1-\alpha} - v_0^{m+1-\alpha}\right)\right\}$$

(5.39)

where $v \ge v_0$.

We observe that when deposition is negligible, $R_m = 0$, and:

$$n(v) = \frac{S_0}{v^\alpha F_\alpha}$$

which yields information about the growth law parameter, α. Numerical details may be found in Williams [1983], Brock [1983], Loyalka and Park [1988], and Gelbard and Seinfeld [1979].

These calculations have neglected coagulation and, as such, are limited in their application although they are important. It is not possible to perform analytical calculations with general forms of I and R if coagulation is included. However, there are some models of coagulation which do lead to tractable equations and, although they do not have any direct physical application, they show how the coagulation and condensation phenomena interact. They may also provide benchmark solutions against which approximate methods may be compared. A program of investigation along these lines has been pursued by Ramabhadran et al. [1976]. Before discussing this work, let us consider another problem raised by Ramabhadran et al. concerning the validity of describing condensation by a single derivative term:

$$\frac{\partial}{\partial v}[I(v)\, n(v)]$$

Methods of Solving the Dynamic Equation

In this connection, it is noted that for evaporating and condensing k-mers, the rate of accretion is given by:

$$-p_{k-1}\, n_{k-1} + p_k\, n_k - q_{k+1}\, n_{k+1} + q_k\, n_k$$

If this discrete representation is replaced by a continuum approximation, we have:

$$\frac{\partial}{\partial v}[\alpha_0(v)\, n(v)] - \frac{\partial^2}{\partial v^2}[\alpha_1(v)\, n(v)]$$

where:
$$\alpha_0(v) = (\Delta v)(p_k - q_k)$$

and:
$$\alpha_1(v) = \tfrac{1}{2}(\Delta v)^2 (p_k + q_k)$$

α_0 is the rate of change of the volume of a particle of size, $k\Delta v$. α_1 assumes the role of a diffusion coefficient in volume space. Both α_0 and α_1 may be computed from kinetic theory (Brock [1972]). Ramabhadran et al. [1976] show that, for most practical distributions, the diffusive term is negligible compared to the convective one and so justifies the conventional procedure.

But, to return to the problem of simultaneous coagulation and condensation, the equation to be solved is:

$$\frac{\partial}{\partial t} n(v,t) + \frac{\partial}{\partial v}[I(v)\, n(v,t)] = \left(\frac{\partial n}{\partial t}\right)_{coag} \tag{5.40}$$

subject to $n(v,0) = n_0(v)$.

The essential point is to choose convenient forms for I, K, and $n_0(v)$ such that Eqn. (5.40) can be solved analytically. It is found that the following combinations lead to tractable cases:

(1) $K = K_0 = $ constant , $I = \sigma_1 v$
(2) $K = K_1(u+v)$, $I = \sigma_1 v$
(3) $K = K_0$, $I = \sigma_0$

Two initial conditions are used:

$$n_0(v) = \frac{N_0}{v_0} \exp(-v/v_0) \tag{5.41}$$

and:

$$n_0(v) = \frac{N_0 v}{v_0^2} \exp(-v/v_0) \tag{5.42}$$

For each case, it is easy to obtain equations for the moments using Eqn. (5.15) and:

$$\frac{d\phi}{dt} = \int_0^\infty I(v)\, n(v,t)\, dv \tag{5.43}$$

These results assume that:

Aerosol Science: Theory and Practice

$$[I(v)\,n(v,t)]_0^\infty = 0 \quad \text{and} \quad [v\,I(v)\,n(v,t)]_0^\infty = 0$$

a matter that has been discussed by Simons [1983].

The following results are readily deduced:

(1) $\quad \dfrac{dN}{dt} = -\dfrac{K_0}{2} N^2 \quad ; \quad N = \dfrac{N_0}{1 + K_0\, N_0\, t/2}$

$\quad \dfrac{d\phi}{dt} = \sigma_1 \phi \quad ; \quad \phi = \phi_0 \exp(\sigma_1 t)$

(2) $\quad \dfrac{dN}{dt} = -K_1 N \phi \quad ; \quad N = N_0 \exp\left[-\dfrac{K_1 \phi_0}{\sigma_1}\left(\exp(\sigma_1 t) - 1\right)\right]$

$\quad \dfrac{d\phi}{dt} = \sigma_1 \phi \quad ; \quad \phi = \phi_0 \exp(\sigma_1 t)$

(3) $\quad \dfrac{dN}{dt} = -\dfrac{K_0}{2} N^2 \quad ; \quad N = \dfrac{N_0}{1 + K_0\, N_0\, t/2}$

$\quad \dfrac{d\phi}{dt} = \sigma_0 N \quad ; \quad \phi = \phi_0\left[1 + \dfrac{2\sigma_0}{K_0\, \phi_0}\ln\!\left(\dfrac{N_0}{N}\right)\right]$

Some instructive comments can be made on these results. In case (1), the total number of particles depends only on coagulation whereas the total volume of particulate depends only on the growth factor. In case (2), however, the total number of particles is influenced by both coagulation and growth because of the dependence of the coagulation process on volume. The total volume of particulate remains unaffected by coagulation. In case (3), the number of particles depends only on coagulation, but the total volume of particulate depends on both coagulation and growth. This is because the growth rate is independent of volume.

To obtain the volume distribution functions is more difficult and we only indicate the procedure for case (1). The equation to be solved is:

$$\dfrac{\partial}{\partial t} n(v,t) + \sigma_1 \dfrac{\partial}{\partial v}[v\,n(v,t)] = \tfrac{1}{2} K_0 \int_0^v du\, n(u,t)\, n(v-u,t) - K_0\, n(v,t)\, N(t) \tag{5.44}$$

Taking the Laplace transform:

$$\bar{n}(s,t) = \int_0^\infty dv\, \exp(-sv)\, n(v,t)$$

and defining:

$$\bar{g}(s,\tau) = \dfrac{2}{(1-\tau)^2}\, \bar{n}(s,t)$$

where $\tau = 1 - (N/N_0)$, we find:

Methods of Solving the Dynamic Equation

$$\frac{\partial \bar{g}}{\partial \tau} - \frac{2\Lambda s}{(1-\tau)^2} \frac{\partial \bar{g}}{\partial s} = \frac{\bar{g}^2}{2N_0} \tag{5.45}$$

where $\Lambda = \sigma_1 / (K_0 N_0)$.

Eqn. (5.45) may be solved by the method of characteristics and, for the two boundary conditions of Eqns. (5.41) and (5.42) respectively, leads to:

$$\bar{g}(s,\tau) = \frac{2 N_0 \exp\left(-\frac{2\Lambda\tau}{1-\tau}\right)}{v_0 + (1-\tau)\exp\left(-\frac{2\Lambda\tau}{1-\tau}\right)} \tag{5.46}$$

or:

$$n(v,t) = \frac{N_0}{v_0}(1-\tau)^2 \exp\left(-\frac{2\Lambda\tau}{1-\tau}\right) \exp\left[-\frac{v}{v_0}(1-\tau)\exp\left(-\frac{2\Lambda\tau}{1-\tau}\right)\right] \tag{5.47}$$

and:

$$\bar{g}(s,\tau) = \frac{2N_0}{\left[v_0 s \exp\left(\frac{2\Lambda\tau}{1-\tau}\right)+1\right]^2 - \tau} \tag{5.48}$$

or:

$$n(v,t) = \frac{N_0 (1-\tau)^2}{v\sqrt{\tau}} \exp\left(-\frac{2\Lambda\tau}{1-\tau}\right) \exp\left[-\frac{v}{v_0}\exp\left(-\frac{2\Lambda\tau}{1-\tau}\right)\right] \sinh\left[\frac{v}{v_0}\sqrt{\tau}\exp\left(-\frac{2\Lambda\tau}{1-\tau}\right)\right] \tag{5.49}$$

The other cases are dealt with by Ramabhadran et al. [1976].

It is interesting to compare Eqn. (5.47) with the solution for no growth, i.e. $\Lambda=0$, when:

$$n(v,t) = \frac{N_0}{v_0}(1-\tau)^2 \exp\left[-\frac{v}{v_0}(1-\tau)\right] \tag{5.50}$$

and also for the case of no coagulation, when:

$$n(v,t) = \frac{N_0}{v_0} \exp\left[-\left(\frac{v}{v_0}+1\right)\exp(-\sigma_1 t)\right] \tag{5.51}$$

The parameter, Λ, is a measure of the ratio of the condensation rate to the coagulation rate. A numerical study of Eqn. (5.47) shows that, for $\Lambda=0.1$, the coagulation dominates and the particles are spread widely over the range of volumes as time proceeds. On the other hand, for $\Lambda=10.0$, growth occurs mainly by condensation and the size distribution tends to narrow and does not exhibit the broadening effect of coagulation.

The numerical results presented by Ramabhadran et al. for various cases give an overall picture of the relative effects of condensation and coagulation. This picture is useful in the interpretation of situations involving more realistic coagulation processes and growth laws. Further exact solutions are given by Lushnikov [1973, 1974, 1976].

5.4 Space and time dependence

We have seen above that, in order to obtain analytical solutions and to some extent numerical solutions, it is necessary to employ the 'well-mixed' hypothesis. The result of this homogenization is an averaging out of spatial effects. For example, the losses due to gravitational settling reduce to a term of the form:

$$\frac{V_s(v)}{H} n(v,t)$$

where V_s is the Stokes velocity and H is the height of the vessel. H/V_s is the transit time of particles falling from the top of the vessel to the bottom. The accuracy of the well-mixed hypothesis is thus in question.

Some insight into the well-mixed hypothesis may be gained by the study of simple space dependent models. First, we consider the dynamic equation for no coagulation but allowing for space dependence. Then, in the one-dimensional case, we find (Williams [1984]):

$$\frac{\partial}{\partial t} n(v,x,t) + \frac{\partial}{\partial x}(U(v,x,t)\, n(v,x,t)) = 0 \tag{5.52}$$

subject to the initial condition $n(v,x,0) = n_0(v,x)$. Using the method of characteristics, the solution of Eqn. (5.52) is:

$$n(v,x,t) = n_0(v, x + U(v)\, t) \tag{5.53}$$

Let us compare this solution with the one from the well-mixed hypothesis where:

$$\frac{\partial}{\partial t} n(v,t) + \frac{U(v)}{H} n(v,t) = 0$$

is subject to $n(v,0) = n_0(v)$ and the solution is:

$$n(v,t) = n_0(v) \exp\left\{-\frac{U(v)\, t}{H}\right\} \tag{5.54}$$

We compare these two results by calculating the total number of particles in the container:

$$N_A(t) = \int_0^\infty dv\, n(v,t) \tag{5.55}$$

in the case of Eqn. (5.54), and:

$$N(t) = \int_0^\infty dv \int_0^H dx\, n(v,x,t) \tag{5.56}$$

in the case of Eqn. (5.53).

Let us take as the initial condition:

$$n_0(v,x) = N_0\, f(v)\, \delta(x - H) \tag{5.57}$$

Methods of Solving the Dynamic Equation

where we assume that the particles are located at x=H and fall onto a floor at x=0. Thus:

$$n(v,x,t) = N_0 \, f(v) \, \delta(x + U(v) \, t - H) \tag{5.58}$$

The spatial distribution of particles is then:

$$N_D(x,t) = N_0 \int_0^\infty dv \, f(v) \, \delta(x - H + Cv^{2/3} t) \tag{5.59}$$

where we have set $U(v) = C v^{2/3}$. After careful integration, we find:

$$N_D(x,t) = \tfrac{3}{2} \frac{N_0}{H \tau^3} \left(1 - \frac{x}{H}\right)^2 \exp\left[-\frac{1}{\tau^{3/2}}\left(1 - \frac{x}{H}\right)^{3/2}\right] \tag{5.60}$$

and:

$$N_D(t) = \tfrac{3}{2} \frac{N_0}{\tau^3} \int_0^1 dy \, (1-y)^2 \exp\left[-\left(\frac{1-y}{\tau}\right)^{3/2}\right] \tag{5.61}$$

where $\tau = Cv_0^{2/3} t / H$ is physically the ratio of the actual time to the time taken for particles of volume, v_0, to fall a distance, H.

Eqn. (5.61) can be evaluated analytically to give:

$$N_D(t) = N_0 \left[1 - \left(1 + \frac{1}{\tau^{3/2}}\right) \exp\left(-\frac{1}{\tau^{3/2}}\right)\right]$$

If we now use a constant source such that the particles are initially distributed uniformly between x=0 and x=H, then we find:

$$N_C(x,t) = \frac{N_0}{H} \left\{1 - \left[1 + \frac{1}{\tau^{3/2}}\left(1 - \frac{x}{H}\right)^{3/2}\right] \exp\left[-\frac{1}{\tau^{3/2}}\left(1 - \frac{x}{H}\right)^{3/2}\right]\right\} \tag{5.62}$$

and:

$$N_C(t) = N_0 \int_0^1 dy \left\{1 - \left[1 + \left(\frac{y}{\tau}\right)^{3/2}\right] \exp\left[-\left(\frac{y}{\tau}\right)^{3/2}\right]\right\} \tag{5.63}$$

The result from the well-mixed hypothesis is:

$$N_A(t) = \frac{3 N_0}{2 \tau^3} \int_0^\infty dy \, y^2 \exp(-y) \exp\left[-\left(\frac{y}{\tau}\right)^{3/2}\right] \tag{5.64}$$

Table 5.1 shows N_D, N_C, and N_A as functions of τ. It is clear that, apart from a minor deviation for the delta source at short times, the well-mixed hypothesis overestimates the rate of removal by sedimentation by a significant amount. In the case of the delta source located at the top of the vessel, the overestimate increases to a factor of six for long times and for a uniform source to a factor of 24.

The spatial distribution of the particles is interesting, especially for the delta source which exhibits a maximum in the density at the point:

Table 5.1: Particle densities, N_D, N_C, and N_A, as functions of dimensionless time, τ.

τ	N_C	N_D	N_A	N_A/N_C	N_A/N_D
0.01	0.985	1.0	0.985	1.0	0.99
0.02	0.970	1.0	0.971	1.0	0.97
0.05	0.925	1.0	0.928	1.0	0.93
0.1	0.850	1.0	0.863	1.02	0.86
0.2	0.699	1.0	0.748	1.07	0.75
0.5	0.308	0.774	0.500	1.62	0.65
1.0	7.89E–2	0.264	0.276	3.5	1.05
2.0	1.32E–2	4.96E–2	0.105	8.0	2.1
5.0	9.58E–4	3.77E–3	1.55E–2	16.0	4.1
10.0	1.23E–4	4.90E–4	2.53E–3	21.0	5.2
20.0	1.55E–5	6.20E–5	3.52E–4	23.0	5.7
50.0	9.99E–7	3.99E–6	2.36E–5	24.0	5.9
100.0	1.25E–7	5.00E–7	2.98E–6	24.0	6.0

(Note: We have corrected a few errors in Williams [1984]).

$$\left(\frac{x}{H}\right)_{max} = 1 - \left(\frac{4}{3}\right)^{2/3} \tau$$

Beyond $\tau = (3/4)^{2/3} = (0.825)$, the maximum disappears (or rather, remains at x=0).

5.5 Multicomponent aerosols

We have seen in Chapter 3 that when an aerosol is composed of a number of different species, it is necessary to consider an extended distribution function, $n(v_1,v_2,...,t)$, which describes the amount of each species contained within the total volume, v, of the particle. Solutions for such a case have been given by Gelbard and Seinfeld [1978a,b, 1979] under a variety of simplifying conditions. In particular, it is assumed that the coagulation kernel is independent of the particle volume and that the growth rate, $I_p(v_1,v_2,...)=\sigma_p v_p$. The first assumption approximates Brownian coagulation. The latter assumption indicates that the rate of growth of the component fraction of each particle depends on the current mass of that component in the particle. Strictly, this assumption has no physical foundation but it is indicative of what actually happens and has the advantage of leading to useful analytical solutions. The equation to be solved is:

$$\frac{\partial}{\partial t} n(v_1,v_2,...,v_N,t) + \sum_{p=1}^{N} \sigma_p \frac{\partial}{\partial v_p}\left[v_p n(v_1,v_2,...,v_N)\right]$$

$$= \tfrac{1}{2} K_0 \int_0^{v_1} du_1 \int_0^{v_2} du_2 ... \int_0^{v_N} du_N \, n(u_1,u_2,...,u_N,t) n(v_1-u_1, v_2-u_2,..., v_N-u_N, t)$$

$$- K_0 \, n(v_1,v_2,...,v_N,t) \int_0^{\infty} du_1 \int_0^{\infty} du_2 ... \int_0^{\infty} du_N \, n(u_1,u_2,...,u_N,t) \tag{5.65}$$

Methods of Solving the Dynamic Equation

subject to an initial condition:

$$n(v_1, v_2, \ldots v_N, 0) = n_0(v_1, v_2, \ldots v_N) \tag{5.66}$$

Eqn. (5.65) may be solved by defining an N-dimensional Laplace transform as follows:

$$\bar{n}(s_1, s_2, \ldots, s_N, t) = \int_0^\infty dv_1 \int_0^\infty dv_2 \cdots \int_0^\infty dv_N \exp\left\{-\sum_{p=1}^N s_p v_p\right\} n(v_1, v_2, \ldots, v_N, t) \tag{5.67}$$

To illustrate the method, we first set $I_p=0$ so that only coagulation is operating. Applying the generalized Laplace transform, we find:

$$\frac{d\bar{n}}{dt} = \tfrac{1}{2} K_0 \bar{n}^2 - K_0 \bar{n} N(t) \tag{5.68}$$

where the total particle density:

$$N(t) = \bar{n}(0, 0, \ldots 0, t) \tag{5.69}$$

To find $N(t)$, we set $\sigma_1 = \sigma_2 = \ldots = \sigma_N = 0$ in Eqn. (5.68) whence:

$$\frac{dN}{dt} = -\tfrac{1}{2} K_0 N^2$$

and:

$$N(t) = \frac{N_0}{1+\tau} \tag{5.70}$$

where $\tau = K_0 N_0 t / 2$. This result is the same as that for a single species aerosol and does not depend on the composition. Using Eqn. (5.70) in Eqn. (5.68), we readily find:

$$\bar{n} = \frac{\bar{n}_0}{(1+\tau)^2 - \tau(1+\tau)\bar{n}_0 / N_0} \tag{5.71}$$

To invert this Laplace transform it is necessary to specify the initial distribution. Gelbard and Seinfeld have examined various forms for n_0 but, for simplicity, we consider only two species and:

$$n_0(v_1, v_2) = \frac{N_0}{v_{10} v_{20}} \exp\left\{-\left(\frac{v_1}{v_{10}} + \frac{v_2}{v_{20}}\right)\right\} \tag{5.72}$$

where v_{10} and v_{20} are the average initial volumes of the particles. The Laplace transform of n_0 is:

$$\bar{n}_0 = \frac{N_0}{(1+s_1 v_{10})(1+s_2 v_{20})}$$

Inserting this into Eqn. (5.71) leads to a straightforward residue inversion and we find:

$$n(v_1, v_2, t) = \frac{N_0}{v_{10} v_{20} (1+\tau)^2} \exp\left\{-\left(\frac{v_1}{v_{10}} + \frac{v_2}{v_{20}}\right)\right\} I_0\left\{2\left[\frac{\tau v_1 v_2}{(1+\tau) v_{10} v_{20}}\right]^{1/2}\right\} \tag{5.73}$$

where $I_0(x)$ is a modified Bessel function.

When $I_p \neq 0$, we introduce a transformation such that:

$$T = \frac{\tau}{1+\tau} \quad \text{and} \quad \bar{p} = \frac{2\bar{n}}{(1-T)^2}$$

and set $\Lambda_p = \sigma_p / (K_0 N_0)$. Then, the Laplace transform gives:

$$\frac{\partial \bar{p}}{\partial \tau} - \sum_{p=1}^{N} \frac{2 \Lambda_p s_p}{(1-T)^2} \frac{\partial \bar{p}}{\partial s_p} = \frac{\bar{p}^2}{2 N_0}$$

This is a partial differential equation which can be solved by the method of characteristics. It is found that, for the same initial conditions as used above:

$$n(v_1, v_2, t) = \frac{N_0}{(1+\tau)^2} \frac{1}{v_{10} v_{20}} \exp\left\{-2(\Lambda_1 + \Lambda_2)\tau - \frac{v_1}{v_{10}} \exp(-2\Lambda_1 \tau) - \frac{v_2}{v_{20}} \exp(-2\Lambda_2 \tau)\right\}$$

$$I_0\left\{2\left[\frac{\tau v_1 v_2}{(1+\tau) v_{10} v_{20}} \exp[-2(\Lambda_1 + \Lambda_2)\tau]\right]^{1/2}\right\}$$

(5.74)

This result and the others given by Gelbard and Seinfeld [1978a,b] allow useful qualitative conclusions to be drawn about the relative influence of condensation and coagulation via the parameters, Λ_1 and Λ_2. Physically, Λ_i is the ratio of the characteristic time for growth of the ith particle, to the characteristic time for coagulation. Because the value of Λ_i depends on the chemical nature of the aerosol it is readily observed that this can have a marked influence on the evolution of the aerosol cloud. Another important use of these solutions is as a benchmark for assessing the accuracy of large aerosol computer codes.

5.6 Radioactive aerosols

Radioactivity is a property of an aerosol particle in much the same way as its volume, mass, or chemical composition. For this reason, we may describe the radioactive content of an aerosol by an extended distribution function. Suppose we have a multispecies aerosol whose components are radioactive. Since each component will have its own specific radioactive characteristics (e.g., α, β, γ decay), each component must have a radioactivity label. For example, we could describe a two component radioactive aerosol by the function:

$$n(v_1, v_2, s_1, s_2, t) \, dv_1 \, dv_2 \, ds_1 \, ds_2$$

where this is the number of aerosol particles with total volume, $v = v_1 + v_2$, and total radioactivity (in some convenient units), $s = s_1 + s_2$, with the volume of component 1 lying in the range, v_1 to $v_1 + dv_1$, and the radioactivity of component 1 lying in the range, s_1 to $s_1 + ds_1$. A similar description applies for component 2. The definition may be extended to any number of components.

Simons [1981] has used a simple example to show how radioactivity is shared among the aerosol particles as the cloud coagulates. To do this, it was assumed that the aerosol may be described by a distribution function, $n(v,s,t)$. As we have seen in Chapter 3, two useful moments of $n(v,s,t)$ are:

Methods of Solving the Dynamic Equation

$$N(v,t) = \int_0^\infty ds\, n(v,s,t)$$

which is the total number of particles of volume, v, and:

$$M(v,t) = \int_0^\infty ds\, s\, n(v,s,t)$$

which is the total radioactivity contained by particles of volume, v. Eqns. (3.83) and (3.84) describe N and M.

The important aspect of Eqn. (3.84) is that it is linear, although with a kernel that depends on $N(v,t)$. Moreover, $M(v,t)$ is physically a very important quantity since, from it, the total airborne radioactivity, J, may be found as a function of time:

$$J(t) = \int_0^\infty dv\, M(v,t) \tag{5.75}$$

In general, Eqns. (3.83) and (3.84) must be solved numerically but, under certain assumptions, an analytical solution may be obtained.

Let us take K and R to be independent of particle volume. Let us further take Laplace transforms of Eqns. (3.83) and (3.84) with respect to volume. Then, with an overbar denoting the Laplace transform, we find:

$$\frac{\partial S}{\partial t} = \tfrac{1}{2} K S^2 - R S \tag{5.76}$$

and:

$$\frac{\partial \overline{M}}{\partial t} = K S \overline{M} - R \overline{M} \tag{5.77}$$

where:

$$S(p,t) = \overline{N}(p,t) - \overline{N}(0,t) \tag{5.78}$$

Eqn. (5.76) is readily solved to give:

$$S(p,t) = \frac{\exp(-R t)\, S(p,0)}{1 - \dfrac{K}{2R}(1 - \exp(-R t))\, S(p,0)} \tag{5.79}$$

and then, Eqn. (5.77) leads to:

$$\overline{M}(p,t) = \frac{\overline{M}(p,0)}{\left[\exp\!\left(\dfrac{R t}{2}\right) - \dfrac{K}{R}\sinh\!\left(\dfrac{R t}{2}\right) S(p,0)\right]^2} \tag{5.80}$$

If:

$$f(v,t) = L^{-1}\left[\exp\!\left(\dfrac{R t}{2}\right) - \dfrac{K}{R}\sinh\!\left(\dfrac{R t}{2}\right) S(p,0)\right]^{-1}$$

then $M(v,t)$ may be expressed as:

$$M(v,t) = \int_0^v du\, f(v-u,t)\, M(u,0) \tag{5.81}$$

To illustrate the technique, we take:

$$N(v,0) = A \exp(-bv)$$

whence:

$$S(p,0) = -\frac{Ap}{b(p+b)}$$

and:

$$f(v,t) = \beta^2 \exp(-Rt)\left[\delta(v) + 2b(1-\beta)\exp(-b\beta v) + b^2(1-\beta)^2 v \exp(-b\beta v)\right]$$

where:

$$\beta = \left[1 + \frac{KA}{2Rb}(1-\exp(-Rt))\right]^{-1}$$

To obtain the amount of radioactivity, we need to specify the initial condition. Let us assume that, at t=0, the radioactivity is concentrated on monodisperse particles of volume, w. Then:

$$M(v,0) = B\,\delta(v-w)$$

Thus, from Eqn. (5.81):

$$M(v,t) = \begin{cases} 0 & , v < w \\ B\,f(v-w,t) & , v \geq w \end{cases}$$

For an alternative initial condition:

$$M(v,0) = B \exp(-cv)$$

it follows that:

$$M(v,t) = B\beta^2 \exp(-Rt)\left\{(1-\sigma)^2 \exp(-ct) + \sigma\left[2 - \sigma + b(1-\beta)v\right]\exp(-b\beta v)\right\}$$

where $\sigma = \beta(1-\beta)/(c - \beta\beta)$.

One of the uses to which the analytical solution developed above can be put, is to test the validity of the often used assumption that the radioactivity of a particle is directly proportional to its volume. Some insight into this matter may be gained if we write Eqn. (5.80), using Eqn. (5.79), as:

$$\overline{M}(p,t) = \exp(Rt)\left[\frac{S(p,t)}{S(p,0)}\right]^2 \overline{M}(p,0)$$

$$= \left[\Gamma(t)\,\overline{N}(p,t) + \Delta(t)\right]^2 \tag{5.82}$$

where:

Methods of Solving the Dynamic Equation

$$\Gamma(t) = \frac{K}{R} \sinh\left(\frac{Rt}{2}\right)$$

$$\Delta(t) = \exp\left(\frac{Rt}{2}\right) + n_0 \Gamma(t)$$

and n_0 is the total number of particles initially.

Performing a Laplace inversion on Eqn. (5.82) we obtain:

$$M(v,t) = \Delta^2(t) M(v,0) + 2 \Delta(t) \Gamma(t) \int_0^v du\, N(v-u,t) M(u,0)$$

$$+ \Gamma^2(t) \int_0^v du \int_0^u ds\, N(v-u,t) N(u-s,t) M(s,0) \qquad (5.83)$$

As t increases, the radioactivity is transferred from the initial distribution, $M(v,0)$, to the two distributions given by the second and third terms on the right hand side of Eqn. (5.83). For small values of t the transfer is mainly to the second term but, when $t \gg 1/n_0 K$, the transfer to the third term dominates.

Some physics may be introduced here if we consider that the aerosol being described consists of two species: sodium oxide, of average initial particle radius approximately 0.05 μm, mixed with a much smaller volume of radioactive fuel aerosol with particle radius around 0.025 μm.

With this situation in mind, let us take $M(v,0) = B\, \delta(v)$ which assumes that the initial radioactive particles are extremely small (actually zero). Then, from Eqn. (5.83), we obtain (Simons [1981]):

$$M(v,t) = B\left[\Delta^2(t)\, \delta(v) + 2 \Delta(t) \Gamma(t) N(v,t) + \Gamma^2(t) \int_0^v du\, N(u,t) N(v-u,t)\right] \qquad (5.84)$$

This expression has an interesting physical interpretation. The first term on the right hand side corresponds to the uncoagulated source particles being depleted by deposition and by coagulation. The second and third terms correspond to the radioactivity of those fuel particles which have coagulated with the sodium oxide. Initially the main contribution is due to the term proportional to N and means that the radioactivity of these composite particles is initially the same regardless of their size. As $t \to \infty$, the composite particle radioactivity is dominated by the last term in Eqn. (5.84). Here, we anticipate a result that will be proved later in this chapter; namely, that for Brownian coagulation (Wang and Friedlander [1967]):

$$N(v, t \to \infty) = A \exp(-bv)$$

Using this expression in Eqn. (5.84) we find:

$$M(v, t \to \infty) \propto v\, N(v, t \to \infty)$$

which means, physically, that for large times the radioactivity is proportional to the particle volume. To summarize, the initial radioactivity on a particle is size independent but as time proceeds it becomes more sensitive to particle size, eventually becoming proportional to volume. This latter behavior is not unexpected and, indeed, in most approximate calculations it is the basis of radioactivity calculations. However, we would expect some dependence on size even for small times; a dependence on particle surface area would not be unreasonable. Simons believes that this discrepancy is due to the assumption of a constant coagulation kernel. In any event, the procedure described above demonstrates clearly that deviations from the volume proportionality assumption do occur. We speak quantitatively about this effect in a later section.

Some further analytical examples are given by Simons [1981] and also an extension to condensation (Simons [1982]).

5.7 Self-preserving solutions

It is of considerable practical interest to know whether, after some time, the size distribution of aerosol particles tends to an asymptotic shape, or to put it another way, to what extent do the initial conditions affect the size distribution after a sufficiently long period of time?

Historically, we must credit Schumann [1940] for raising this question when, in the work cited above, he writes *"from the nature of the problem, one is inclined to surmise that there should be a tendency for the same distribution to be approached asymptotically after a sufficiently long time, no matter what the initial condition might be."* Schumann illustrated his hypothesis with a number of specific examples but was unable to furnish any general proof.

Further progress on this problem was made by Swift and Friedlander [1964] and we shall describe their work in some detail. We consider, initially, only the case of pure coagulation with no sources or sinks. We also recall the definitions of the total number of particles:

$$N(t) = \int_0^\infty n(v,t)\, dv \tag{5.86}$$

and the total volume fraction of particulate:

$$\phi(t) = \int_0^\infty v\, n(v,t)\, dv \tag{5.87}$$

which, for pure coagulation, is independent of time.

The basic equation for study is:

$$\frac{\partial}{\partial t} n(v,t) = \tfrac{1}{2} \int_0^v du\, K(u, v-u)\, n(u,t)\, n(v-u,t) - n(v,t) \int_0^\infty du\, K(u,v)\, n(u,t) \tag{5.88}$$

together with the auxiliary relation:

$$\frac{dN}{dt} = -\tfrac{1}{2} \int_0^\infty du \int_0^\infty dv\, K(u,v)\, n(u,t)\, n(v,t) \tag{5.89}$$

Subject to an initial condition, we can solve the above equation. However, we now pose the problem of whether a solution can be found in the form:

$$n(v,t) = g(t)\, \tilde{\psi}\!\left(\frac{v}{\tilde{v}(t)}\right) \tag{5.90}$$

Methods of Solving the Dynamic Equation

where g(t) and $\tilde{v}(t)$ are unknown functions of time, and $\tilde{\psi}(\eta)$ is independent of time, i.e., the coordinates are 'stretched' but the functional shape of $\tilde{\psi}$ remains constant as the aerosol ages. We refer to this as a self-preserving distribution.

Substituting Eqn. (5.90) into Eqns. (5.86) and (5.87) leads to:

$$g(t) = \frac{N}{\tilde{v}}\left[\int_0^\infty d\eta\, \tilde{\psi}(\eta)\right]^{-1} \tag{5.91}$$

and:

$$\tilde{v}(t) = \frac{\phi}{N}\frac{\int_0^\infty d\eta\, \tilde{\psi}(\eta)}{\int_0^\infty d\eta\, \eta\, \tilde{\psi}(\eta)} \tag{5.92}$$

If such a solution does exist, it will clearly be of considerable value in correlating experimental results and also for gaining physical insight into the problem.

We note that the integrals:

$$\tilde{\psi}_n = \int_0^\infty d\eta\, \eta^n\, \tilde{\psi}(\eta) \tag{5.93}$$

are dimensionless constants.

We can now write n(v,t) as:

$$n(v,t) = \frac{N^2}{\phi}\frac{\tilde{\psi}_1}{\tilde{\psi}_0^2}\tilde{\psi}\left(\frac{\tilde{\psi}_1}{\tilde{\psi}_0}\frac{Nv}{\phi}\right) \equiv \frac{N^2}{\phi}\psi\left(\frac{Nv}{\phi}\right) \tag{5.94}$$

which is a function of N and ϕ, and we note that ϕ/N is the average volume of a particle.

Substitution of Eqn. (5.94) into Eqns. (5.86) and (5.87) leads to ψ_0=1 and ψ_1=1 which are constraints on the form of $\psi(\eta)$.

The self-preserving distribution must also satisfy the integro-differential equation, Eqn. (5.88). Thus, we find:

$$\frac{1}{N^2}\frac{dN}{dt}\left[2\,\psi(\eta) + \eta\frac{\partial}{\partial\eta}\psi(\eta)\right]$$
$$= \tfrac{1}{2}\int_0^\eta d\eta'\, K\!\left(\frac{\phi\eta'}{N},\frac{\phi(\eta-\eta')}{N}\right)\psi(\eta')\,\psi(\eta-\eta') - \psi(\eta)\int_0^\infty d\eta'\, K\!\left(\frac{\phi\eta'}{N},\frac{\phi\eta}{N}\right)\psi(\eta') \tag{5.95}$$

and, from Eqn. (5.89):

$$\frac{1}{N^2}\frac{dN}{dt} = -\tfrac{1}{2}\int_0^\infty d\eta'\int_0^\infty d\eta\, K\!\left(\frac{\phi\eta}{N},\frac{\phi\eta'}{N}\right)\psi(\eta')\,\psi(\eta) \tag{5.96}$$

Now, what we wish to show is that a solution to Eqn. (5.95) exists which is independent of time. Clearly, because of the presence of the ratio, ϕ/N, in the argument of the coagulation kernel, this will not generally be true. However, for a class of homogeneous kernels such that:

$$K(au, av) = a^\alpha K(u, v) \tag{5.97}$$

the problem simplifies and we obtain:

$$\left(\frac{N}{\phi}\right)^\alpha \frac{1}{N^2} \frac{dN}{dt} \left[2\psi(\eta) + \eta \frac{\partial}{\partial \eta} \psi(\eta)\right]$$

$$= \tfrac{1}{2} \int_0^\eta d\eta'\, K(\eta', \eta-\eta')\, \psi(\eta')\, \psi(\eta-\eta') - \psi(\eta) \int_0^\infty d\eta'\, K(\eta, \eta')\, \psi(\eta') \tag{5.98}$$

and:

$$\left(\frac{N}{\phi}\right)^\alpha \frac{1}{N^2} \frac{dN}{dt} = -\tfrac{1}{2} \int_0^\infty d\eta' \int_0^\infty d\eta\, K(\eta, \eta')\, \psi(\eta')\, \psi(\eta) \tag{5.99}$$

Using Eqn. (5.99), the time dependent term in Eqn. (5.98) is seen to be a constant. Thus, ψ is indeed independent of time and it can be asserted that the self-preserving solution is a particular solution of Eqn. (5.88).

We now enquire about the usefulness of the homogeneous assumption in Eqn. (5.97). The kernels for the important coagulation mechanisms are, from Chapter 4:

Brownian (continuum):

$$K(u, v) = \frac{2kT}{3\mu} \left[2 + \left(\frac{u}{v}\right)^{1/3} + \left(\frac{v}{u}\right)^{1/3}\right] \tag{5.100}$$

whence $\alpha = 0$;

Brownian (Knudsen):

$$K(u, v) = \left(\frac{8\pi kT}{\rho}\right)^{1/2} \left(\frac{3}{4\pi}\right)^{2/3} \left(u^{1/3} + v^{1/3}\right)^2 \left(\frac{1}{u} + \frac{1}{v}\right)^{1/2} \tag{5.101}$$

whence $\alpha = 1/6$;

gravitational and turbulent inertial:

$$K(u, v) = K_s \left(v^{2/3} - u^{2/3}\right) \left|v^{2/3} - u^{2/3}\right| \tag{5.102}$$

where for gravitational coagulation:

$$K_s = \frac{\rho g}{6\mu} \left(\frac{3}{4\pi}\right)^{1/3}$$

Methods of Solving the Dynamic Equation

and for turbulent inertial:

$$K_s = (0.188) \frac{\varepsilon_T^{3/4} \rho_g^{1/4} \rho_p}{\mu^{5/4}}$$

whence $\alpha = 4/3$;

and laminar shear and turbulent diffusion:

$$K(u,v) = \frac{G}{\pi}\left(v^{1/3} + u^{1/3}\right)^3 \tag{5.103}$$

where in the case of laminar shear, G is the shear rate and in the case of turbulent diffusion:

$$G = \left(\frac{3\pi}{10} \frac{\varepsilon_T \rho_g}{\mu}\right)^{1/2}$$

whence $\alpha = 1$.

We observe that each of these kernels satisfies the homogeneity condition. We also observe that any linear combination of these kernels does not satisfy the homogeneity condition unless some of the parameters contained in the kernels (*e.g.*, the shear rate, G) are a special function of time. This may be seen easily if we consider Eqn. (5.96) for a combination of the Brownian kernel and the laminar shear kernel. Then we see that:

$$\frac{1}{N^2}\frac{dN}{dt} = -\frac{1}{2}\int_0^\infty d\eta' \int_0^\infty d\eta \left[K_B(\eta,\eta') + \frac{G\phi}{\pi N}\left(\eta^{1/3} + \eta'^{1/3}\right)^3\right]\psi(\eta)\,\psi(\eta') \tag{5.104}$$

There will be similar terms in Eqn. (5.95) and it is clear that for $\psi(\eta)$ to be independent of time we must have $G\phi/N$ equal to a constant or $G \propto N$. Similar artificial constraints arise for other combinations. These have been studied numerically by Wang and Friedlander [1967]. The method has been extended by Pich *et al.* [1970] to include condensation but, even in that case, similarity only exists if the supersaturation coefficient varies with time in a special way.

A further problem arises if there is a removal term, of the form $R(v)n(v,t)$, on the left hand side of Eqn. (5.88). Then, if:

$$R(bv) = b^\gamma R(v) \tag{5.105}$$

we find that the left hand side of Eqn. (5.95) becomes:

$$\frac{1}{N^2}\frac{dN}{dt}\left[2\psi(\eta) + \eta\frac{\partial}{\partial\eta}\psi(\eta)\right] + \frac{1}{N}\left(\frac{\phi}{N}\right)^\gamma R(\eta)\,\psi(\eta) - \frac{1}{N^2\phi}\frac{d\phi}{dt}\psi(\eta) \tag{5.106}$$

and the left hand side of Eqn. (5.96) becomes:

$$\frac{1}{N^2}\frac{dN}{dt} + \frac{1}{N}\left(\frac{\phi}{N}\right)^\gamma \int_0^\infty d\eta\, R(\eta)\,\psi(\eta) \tag{5.107}$$

We also have the additional equation:

$$\frac{d\phi}{dt} + N\left(\frac{\phi}{N}\right)^{\gamma+1} \int_0^\infty d\eta\, \eta\, \psi(\eta) = 0 \tag{5.108}$$

Even with homogeneity there is no way in which the equation for ψ can be made independent of time. It appears, therefore, that similarity is restricted to pure coagulation and homogeneous kernels. Further results for special cases are given by Drake [1972].

While we have shown that, for pure coagulation and homogeneous kernels, a self-preserving form develops, we have not examined how long it is necessary to wait before such a form is achieved. However, it is quite clear from the work in Section 5.1 on exact solutions that the waiting time depends sensitively on the initial conditions. In order to illustrate the manner in which the self-similar solution is approached let us consider in more detail the general solution given by Eqn. (5.8). This is the solution for a gamma distribution initial condition and leads to:

$$n(v,t) = \frac{N_0\, \theta^2}{\hat{v}} \exp\left(-\frac{v}{\hat{v}}\right) \frac{1}{2\pi i} \int_L \frac{dz\, \exp(vz/\hat{v})}{z^{\nu+1} - 1 + \theta} \tag{5.109}$$

To invert the Laplace transform, let us assume that ν is an integer. Then we find that there are $\nu+1$ poles corresponding to the roots of $z^{\nu+1} = 1 - \theta$. Defining:

$$\omega = \frac{z}{(1-\theta)^{1/(\nu+1)}} \tag{5.110}$$

we require the roots of $\omega^{\nu+1} = 1$ or:

$$\omega = \exp\left\{\frac{2k\pi i}{\nu+1}\right\}$$

where $k = 0, 1, 2, ..., \nu$.

Using the method of residues, we find:

$$n(v,t) = \frac{N_0\, \theta^2}{\hat{v}(\nu+1)} \frac{1}{(1-\theta)^{\nu/(\nu+1)}} \sum_{k=0}^{\nu} \exp\left\{\frac{2k\pi i}{\nu+1} + \Delta_k\, \eta\right\} \tag{5.111}$$

where:

$$\Delta_k = \frac{\nu+1}{\theta}\left[(1-\theta)^{1/(\nu+1)} \exp\left\{\frac{2k\pi i}{\nu+1}\right\} - 1\right]$$

Now, for small θ (i.e. $t \to \infty$) we can write this as:

$$n(v,t) \approx \frac{N_0\, \theta^2}{\hat{v}(\nu+1)}\left\{\exp(-\eta) + \sum_{k=1}^{\nu} \exp\left[-\frac{\nu+1}{\theta}\eta\left(1 - \exp\left(\frac{2k\pi i}{\nu+1}\right)\right)\right]\right\}\left\{1 + \frac{\nu\theta}{\nu+1} + ...\right\} \tag{5.112}$$

Thus, the transients in the summation die out very rapidly and the most slowly decaying term is $\nu\theta/(\nu+1)$ compared with unity. We may conclude then that:

Methods of Solving the Dynamic Equation

$$n(v,t) \to \frac{N^2(t)}{\phi} \exp(-\eta) \tag{5.113}$$

at a rate comparable to the decay of N(t) itself, irrespective of the value of v.

5.8 Brownian self-similarity

We illustrate the fact that $\psi(\eta)$ depends sensitively on the nature of the kernel by describing some work of Friedlander and Wang [1966] using the Brownian kernel of Eqn. (5.100) in Eqns. (5.98) and (5.99). Thus, we have:

$$\frac{1}{N^2}\frac{dN}{d\tau}\left[2\,\psi(\eta)+\eta\frac{\partial}{\partial\eta}\psi(\eta)\right]$$

$$=\int_0^\eta d\eta'\left[1+\left(\frac{\eta-\eta'}{\eta'}\right)^{1/3}\right]\psi(\eta')\,\psi(\eta-\eta')-\psi(\eta)\int_0^\infty d\eta'\left[2+\left(\frac{\eta}{\eta'}\right)^{1/3}+\left(\frac{\eta'}{\eta}\right)^{1/3}\right]\psi(\eta') \tag{5.114}$$

where we have used symmetry in the first term on the right hand side. Also, from Eqn. (5.99):

$$\frac{1}{N^2}\frac{dN}{d\tau}=-\tfrac{1}{2}\int_0^\infty d\eta\int_0^\infty d\eta'\left[2+\left(\frac{\eta}{\eta'}\right)^{1/3}+\left(\frac{\eta'}{\eta}\right)^{1/3}\right]\psi(\eta)\,\psi(\eta') \tag{5.115}$$

with $\tau = 2kTt/3\mu$. Now, defining:

$$a = \int_0^\infty d\eta\,\eta^{1/3}\,\psi(\eta)$$

and:

$$b = \int_0^\infty d\eta\,\eta^{-1/3}\,\psi(\eta)$$

we find:

$$\frac{1}{N^2}\frac{dN}{d\tau}=-(1+ab) \tag{5.116}$$

and hence that Eqn. (5.114) becomes:

$$(1+ab)\,\eta\frac{\partial}{\partial\eta}\psi(\eta)+\left(2ab-b\eta^{1/3}-a\eta^{-1/3}\right)\psi(\eta)$$

$$+\int_0^\eta \psi(\eta')\psi(\eta-\eta')\left[1+\left(\frac{\eta-\eta'}{\eta'}\right)^{1/3}\right]d\eta' = 0$$

The further transformations:

$$\zeta = (1+ab)^3\,\eta/a^3 \quad \text{and} \quad \chi(\zeta) = a^3\,\psi(\eta)/(1+ab)^4$$

Aerosol Science: Theory and Practice

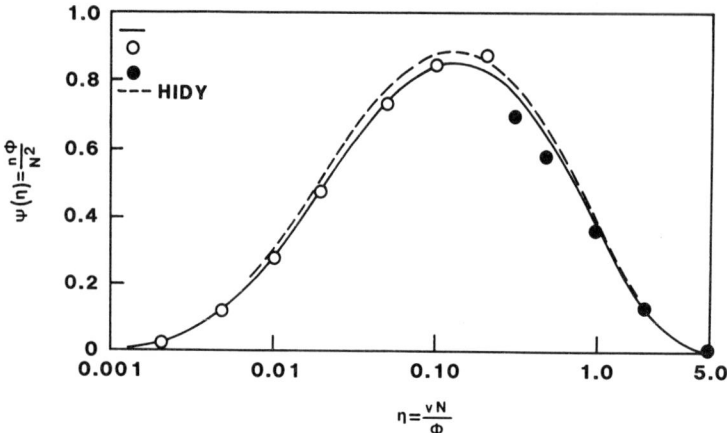

Figure 5.1: Self-preserving particle size distribution for Brownian coagulation. Adapted from Friedlander and Wang [1966] with permission of the Academic Press. Solid line-Friedlander and Wang [1966] numerical solution to Eqn. (5.117); O - Eqn. (5.117b); ● - Eqn. (5.117c); Dashed line-Hidy [1965] numerical simulations.

yield:

$$\zeta \frac{d}{d\zeta}\chi(\zeta) + \left[2\alpha - \alpha(1-\alpha)\zeta^{1/3} - \zeta^{-1/3}\right]\chi(\zeta) + \int_0^\zeta \chi(\zeta')\chi(\zeta-\zeta')\left[1+\left(\frac{\zeta-\zeta'}{\zeta'}\right)^{1/3}\right]d\zeta' = 0$$

where $\alpha = ab/(1+ab)$. (5.117)

Eqn. (5.117) has been solved numerically by Friedlander and Wang [1966] and the result is shown in Fig. 5.1. Also shown are some numerical simulations of Hidy [1965] who solved the discrete equations of aerosol coagulation (Eqn. (3.9)) for up to 600 particles. Friedlander and Wang also derived some analytical results for small and large values of η in the form:

$$\psi(\eta) = \frac{(0.5086)}{\eta^{(1.06)}} \exp\left[(1.758)\eta^{1/3} - (1.275)\eta^{-1/3}\right]$$

(5.117b)

for small η and:

$$\psi(\eta) = (0.915)\exp[-(0.95)\eta]$$

(5.117c)

for large η. These results are also shown Fig. 5.1. The value of ab≈(1.129).

We conclude that the ψ function for this case is very different from that found using the constant kernel. In particular, $\psi(\eta) \to 0$ as $\eta \to 0$, in contrast to the case of $\exp(-\eta)$ which arises for the constant kernel.

Another form of $\psi(\eta)$ has been derived by Lai *et al.* [1972] for the Knudsen form of the Brownian kernel, and will be discussed in Section 5.10.

Methods of Solving the Dynamic Equation

5.9 The moments method

We have seen that, under suitably restrictive conditions, a self-preserving form describes the aerosol distribution function. Thus, for example, in the case described by Eqn. (5.99) we can obtain the behavior of N(t) without solving the complete equation. In that case, we have:

$$\frac{dN}{dt} = -\tfrac{1}{2} \overline{K} \phi^\alpha N^{2-\alpha} \tag{5.118}$$

whence:

$$N(t) = N_0 \left\{ 1 + \tfrac{1}{2}(1-\alpha) \overline{K} N_0^{1-\alpha} \phi^\alpha t \right\}^{1/(\alpha-1)} \tag{5.119}$$

where $\overline{K}$ is the double integral on the right hand side of Eqn. (5.99). This simple technique is very efficient and avoids considerable numerical difficulty and expense, but it does appear to be restricted. Nevertheless, it is worthwhile considering the possibility of extending the method beyond its strict region of applicability with the addition of more parameters. The key to the problem is to choose a physically realistic and mathematically tractable function. Two such functions have been examined in this respect: (1) the log-normal distribution and; (2) the gamma distribution. These functions can be written as:

log-normal:

$$n(v) = \frac{N}{\sqrt{2\pi}\sigma} \frac{1}{v} \exp\left[-\frac{1}{2\sigma} \ln^2\left(\frac{v}{\hat{v}}\right) \right] \tag{5.120}$$

where N, $\hat{v}$, and σ are adjustable, and:

gamma:

$$n(v) = \frac{N}{\Gamma(v+1)} \frac{1}{\overline{v}} \left(\frac{v}{\overline{v}}\right)^v \exp\left(-\frac{v}{\overline{v}}\right) \tag{5.121}$$

where N, $\overline{v}$, and v are adjustable. In fact, the gamma distribution may be generalized to the form:

$$n(v) = \frac{N\beta}{\Gamma((v+1)/\beta)} \frac{1}{\overline{v}} \left(\frac{v}{\overline{v}}\right)^v \exp\left[-\left(\frac{v}{\overline{v}}\right)^\beta\right] \tag{5.122}$$

where N, v, β, and $\overline{v}$ are adjustable. In this form, the gamma distribution is called the Wiebull or modified gamma distribution.

There is some physical evidence to suggest that these functional forms will be appropriate. For example, it is observed, and can be predicted theoretically, that during the process of grinding the particle volume distribution function takes on a log-normal shape (Aitchison and Brown [1957]). In addition, Deirmendjian [1969] has shown in his work on electromagnetic scattering in polydispersions that the distribution of particle sizes in haze, rain, hail, and various cloud types can be closely described by a modified gamma distribution.

The rationale behind the moments method of solution of the dynamic equation is that the functional forms of log-normal or modified gamma remain valid throughout the coagulation and deposition history of the aerosol. The change in distribution is characterized by the time dependence of the adjustable parameters. Equations for these parameters are obtained by inserting them into the dynamic equation and taking an appropriate number of volume moments. This leads to a set of first order, nonlinear differential equations which can sometimes be solved analytically but which in any case are very simple to solve numerically. We will now illustrate the technique by some examples, but first let us note that both log-normal and gamma distributions can be put into a pseudo-self-similar form:

log-normal:

$$\psi(\eta) = \frac{1}{\sqrt{2\pi}\sigma}\frac{1}{\eta}\exp\left[-\frac{1}{2\sigma}\left(\ln\eta + \frac{\sigma}{2}\right)^2\right]$$
(5.123)

modified gamma:

$$\psi(\eta) = \frac{\beta\,\gamma(2)}{\gamma(1)^2}\left[\frac{\eta\,\gamma(2)}{\gamma(1)}\right]^{\nu}\exp\left[-\left(\frac{\eta\,\gamma(2)}{\gamma(1)}\right)^{\beta}\right]$$
(5.124)

where:

$$\gamma(n) \equiv \Gamma\left(\frac{\nu+1}{\beta}\right)$$
(5.125)

$\Gamma(x)$ being the gamma function.

In Eqns. (5.123) and (5.124), $\eta = v\,M_0 / M_1$, where the moments, M_n, are defined by:

$$M_n = \int_0^\infty dv\, v^n\, n(v)$$

Thus, for the log-normal:

$$M_n = N\,\hat{v}^n\,\exp\left(\frac{n^2\sigma}{2}\right)$$
(5.126)

In terms of these parameters, we then find:

$$N = M_0$$

$$\phi = M_1 = N\,\hat{v}\,\exp\left(\frac{\sigma}{2}\right)$$

and:

$$\mathrm{var}(n) = M_2 - \frac{M_1^2}{M_0} = \frac{\phi^2}{N}(\exp(\sigma) - 1)$$

Thus, the three moments, M_0, M_1, and M_2, serve to provide N, ϕ, and σ.

For the modified gamma distribution:

$$M_n = N\,\hat{v}^n\,\frac{\gamma(n+1)}{\gamma(1)}$$
(5.127)

where:

Methods of Solving the Dynamic Equation

$N = M_0$

$\phi = M_1 = N \bar{v} \gamma(2)/\gamma(1)$

and we need two additional moments, M_2 and M_3, to obtain the remaining parameters, v and β.

5.10 Coagulation with deposition, condensation, and a source

If the method of moments is to be useful, then it must be applicable to the actual physical situation including deposition, condensation, and an imposed source term. Thus, we shall apply it directly to Eqn. (3.31).

Multiplying Eqn. (3.31) by a function, $\xi(v)$, and integrating over $v(0,\infty)$, leads to:

$$\frac{d}{dt}\int_0^\infty dv\, \xi(v)\, n(v,t) + \int_0^\infty dv\, \xi(v)\, R(v)\, n(v,t) - \int_0^\infty dv\, I(v)\, n(v,t)\frac{\partial}{\partial v}\xi(v)$$

$$= \tfrac{1}{2}\int_0^\infty du \int_0^\infty dv\, [\xi(v+u) - \xi(v) - \xi(u)]\, K(u,v)\, n(u,t)\, n(v,t) + \int_0^\infty dv\, \xi(v)\, S(v,t) \tag{5.128}$$

where we have used Eqn. (3.11) and assumed that:

$$[I(v)\, n(v,t)\, \xi(v)]_0^\infty = 0$$

The first integral on the right hand side of Eqn. (5.128) can also be written in another way which is sometimes more convenient. Thus, if we split up the inner integral over v into two parts, namely $(0,v)$ and (v,∞), then we find, after using the symmetry of K, that:

$$\tfrac{1}{2}\int_0^\infty du \int_0^\infty dv\, [\xi(u+v) - \xi(u) - \xi(v)]\, K(u,v)\, n(u)\, n(v)$$

$$= \int_0^\infty du \int_0^u dv\, [\xi(u+v) - \xi(u) - \xi(v)]\, K(u,v)\, n(u)\, n(v)$$

If we now choose $\xi(v) = v^n$, $n = 0,1,2,...,N$, set $R(v) = R_m\, v^m$ and $I(v) = I_\alpha\, v^\alpha$, we find, using the log-normal distribution for $n(v,t)$, the following equations:

$$\frac{dN}{dt} = -R_m\, \phi^m\, N^{1-m}\, \exp\left[\frac{m(m-1)\sigma}{2}\right] - \tfrac{1}{2} J_0(\hat{v},\sigma)\, N^2 \tag{5.129}$$

$$\frac{d\phi}{dt} = -R_m\, \frac{\phi^{m+1}}{N^m}\, \exp\left[\frac{m(m+1)\sigma}{2}\right] + I_\alpha\, \phi^\alpha\, N^{1-\alpha}\, \exp\left[\frac{\alpha(\alpha-1)\sigma}{2}\right] \tag{5.130}$$

and:

$$\frac{d\sigma}{dt} = R_m\, \frac{G\phi^m}{N^m}\left\{2\exp\left[\frac{m(m+1)\sigma}{2}\right] - \exp\left[\frac{m(m-1)\sigma}{2}\right] - \exp\left[\frac{m(m+3)\sigma}{2}\right]\right\}$$

$$+ 2 I_\alpha\, \frac{\phi^{\alpha-1}}{N^{\alpha-1}}\left\{\exp\left[\frac{(\alpha-1)(\alpha+2)\sigma}{2}\right] - \exp\left[\frac{\alpha(\alpha-1)\sigma}{2}\right]\right\} + J_3(\hat{v},\sigma,N) \tag{5.131}$$

where:
$$J_3 = -\tfrac{1}{2} J_0 N + N J_2 \exp(-2\sigma) \tag{5.132}$$

$$\phi = N \hat{v} \exp(\sigma/2) \tag{5.133}$$

$$J_0 = \frac{1}{2\pi\sigma} \int_{-\infty}^{\infty} dp \int_{-\infty}^{\infty} dq\, K(\hat{v}\, e^p, \hat{v}\, e^q) \exp\left(-\frac{p^2+q^2}{2\sigma}\right) \tag{5.134}$$

and:
$$J_2 = \frac{1}{2\pi\sigma} \int_{-\infty}^{\infty} dp \int_{-\infty}^{\infty} dq\, K(\hat{v}\, e^p, \hat{v}\, e^q) \exp\left(-\frac{p^2+q^2}{2\sigma} + p + q\right) \tag{5.135}$$

Eqns. (5.129)-(5.131) constitute a closed set of first order nonlinear equations for N, ϕ, and σ. It is an easy matter to solve these numerically. As an example, let us assume that we have pure Brownian coagulation. In that case, only Eqns. (5.129) and (5.131) are relevant and, with:
$$J_0 = 2 K_B (1 + \exp(\sigma/9)) \tag{5.136}$$
and:
$$J_3 = K_B N (1 + \exp(\sigma/9))(2 \exp(-\sigma) - 1) \tag{5.137}$$

we find:
$$\frac{dN}{dt} = -N^2 (1 + \exp(\sigma/9)) \tag{5.138}$$
and:
$$\frac{d\sigma}{dt} = -N(1 + \exp(\sigma/9))(1 - 2\exp(-\sigma)) \tag{5.139}$$

where $\tau = N_0 K_B t$ and N is normalized to N_0.

An interesting physical observation can be made about Eqn. (5.138), namely, that now the effective coagulation rate depends on the dispersivity, σ, of the distribution. Thus, the more polydisperse the aerosol, the greater the reduction in particle number by coagulation.

Eqns. (5.138) and (5.139) are solved analytically to give:
$$N = \frac{\exp(\sigma) - 2}{\exp(\sigma_0) - 2} \tag{5.140}$$
and:
$$\tau = (\exp(\sigma_0) - 2) \int_\sigma^{\sigma_0} \frac{ds\, \exp(-s)}{(1 - 2 \exp(-s))^2 (1 + \exp(\sigma/9))} \tag{5.141}$$

In the case of a constant coagulation kernel, with $K(u,v) = 4 K_B$, we find:
$$N = \frac{1}{1 + 2\tau} \tag{5.142}$$

Methods of Solving the Dynamic Equation

and:

$$\sigma = \ln\left[2 + \frac{\exp(\sigma_0) - 2}{(1 + 2\tau)^{1/4}}\right] \tag{5.143}$$

It is interesting to see that in both cases, the exact kernel and the constant one, the value of $\sigma \to \ln(2)$ as $\tau \to \infty$. Thus, the pseudo-self-similar form in Eqn. (5.123) tends to a true self-similar form as τ increases.

In another example, we can use the shear kernel of Eqn. (5.103), with $\tau = Gt/\pi$, we find:

$$\frac{dN}{d\tau} = -N[1 + 3\exp(-2\sigma/9)]$$

and:

$$\frac{d\sigma}{d\tau} = 1 + 3\exp(-2\sigma/9)$$

These equations are readily solved to give:

$$N(t) = \left\{[1 + 3\exp(-2\sigma_0/9)]\exp(2\tau/9) - 3\exp(-2\sigma_0/9)\right\}^{-9/2} \tag{5.144}$$

and:

$$\sigma(t) = \tfrac{9}{2}\ln\left[(3 + \exp(2\sigma_0/9))\exp(2\tau/9) - 3\right] \tag{5.145}$$

In this case, $\sigma(t) \to t$ as $t \to \infty$, and self-similarity in the normal sense does not exist.

Detailed expressions are given in Williams [1986] for J_0 and J_2 for various coagulation mechanisms.

The only case which is not amenable to treatment, because the integrals, J_n, cannot be carried out explicitly, is that of free molecular Brownian coagulation. In this case, Lee et al. [1984] have developed an accurate approximation.

5.11 Modified gamma distribution

In the introduction, we noted another 'ansatz' for the volume distribution function, namely the modified gamma distribution. Following the same procedure as in the previous section, only using Eqn. (5.122), we find:

$$\frac{dN}{dt} + R_m N \bar{v}^m \frac{\gamma(m+1)}{\gamma(1)} = F_0(N, v, \beta, \bar{v}) \tag{5.146}$$

$$\frac{d\phi}{dt} + R_m N \bar{v}^{m+1} \frac{\gamma(m+2)}{\gamma(1)} = I_\alpha N \bar{v}^\alpha \frac{\gamma(\alpha+1)}{\gamma(1)} \tag{5.147}$$

$$\frac{dM_2}{dt} + R_m N \bar{v}^{m+2} \frac{\gamma(m+3)}{\gamma(1)} - 2 I_\alpha N \bar{v}^{\alpha+1} \frac{\gamma(\alpha+1)}{\gamma(1)} = F_2(N, v, \beta, \bar{v}) \tag{5.148}$$

and:
$$\frac{dM_3}{dt} + R_m\, N\, \bar{v}^{m+3}\, \frac{\gamma(m+4)}{\gamma(1)} - 3 I_\alpha\, N\, \bar{v}^{\alpha+2}\, \frac{\gamma(\alpha+3)}{\gamma(1)} = F_3(N,v,\beta,\bar{v}) \tag{5.149}$$

where:
$$\bar{v} = \frac{\phi}{N}\,\frac{\gamma(1)}{\gamma(2)} \tag{5.150}$$

The terms, F_n, are given in terms of the coagulation kernel as:

$$F_n = -\frac{\varepsilon_n N^2 \bar{v}^n}{\gamma(1)^2} \int_0^\infty dx\, x^{[(v+n)/\beta]-1}\, \exp(-x) \int_0^\infty dy\, y^{[(v+n)/\beta]-1}\, \exp(-y)\, K\!\left(\bar{v}\, x^{1/\beta},\bar{v}\, y^{1/\beta}\right) \tag{5.151}$$

where:
$$\varepsilon_0 = 1/2\,,\quad \varepsilon_2 = 1\quad \text{and}\quad \varepsilon_3 = 3 \tag{5.152}$$

Expressions for F_n may be found in Williams [1986].

Now it is possible to write:

$$\dot{M}_2 = A_1\, \dot{v} - A_2\, \dot{\beta} - A_3\, \dot{N} + A_4\, \dot{\phi} \tag{5.153}$$

and:
$$\dot{M}_3 = B_1\, \dot{v} - B_2\, \dot{\beta} - B_3\, \dot{N} + B_4\, \dot{\phi} \tag{5.154}$$

where the A_i and B_i are given below:

$$A_1 = \frac{\phi^2}{N\beta}\left[\tilde{\psi}(3) + \tilde{\psi}(1) - 2\tilde{\psi}(2)\right]\frac{\gamma(1)\,\gamma(3)}{\gamma(2)^2}$$

$$A_2 = \frac{\phi^2}{N\beta^2}\,\frac{\gamma(1)\,\gamma(3)}{\gamma(2)^2}\left[\tilde{\psi}(3)(v+3) + \tilde{\psi}(1)(v+1) - 2\tilde{\psi}(2)(v+2)\right]$$

$$A_3 = \frac{\phi^2}{N^2}\,\frac{\gamma(1)\,\gamma(3)}{\gamma(2)^2}$$

$$A_4 = \frac{2\phi}{N}\,\frac{\gamma(1)\,\gamma(3)}{\gamma(2)^2}$$

$$B_1 = \frac{\phi^3}{N^2\beta}\left[2\tilde{\psi}(1) + \tilde{\psi}(4) - 3\tilde{\psi}(2)\right]\frac{\gamma(1)^2\,\gamma(4)}{\gamma(2)^3}$$

$$B_2 = \frac{\phi^3}{N^2\beta^2}\,\frac{\gamma(1)^2\,\gamma(4)}{\gamma(2)^3}\left[2\tilde{\psi}(1)(v+1) + \tilde{\psi}(4)(v+4) - 3\tilde{\psi}(2)(v+2)\right]$$

$$B_3 = \frac{2\phi^3}{N^3} \frac{\gamma(4)\,\gamma(1)^2}{\gamma(2)^3}$$

and:

$$B_4 = \frac{3\phi^2}{N^2} \frac{\gamma(4)\,\gamma(1)^2}{\gamma(2)^3}$$

with:

$$\tilde{\psi}(n) \equiv \psi\left(\frac{v+n}{\beta}\right)$$

where $\psi(z)$ is the digamma function.

Using Eqns. (5.146)-(5.150) in Eqns. (5.153) and (5.154), it is clear that we can obtain equations in the form:

$$\dot{N} = F_0'(N, v, \beta, \phi) \tag{5.155}$$

$$\dot{\phi} = F_1'(N, v, \beta, \phi) \tag{5.156}$$

$$\dot{v} = \frac{B_2\left(F_2' - A_4\,F_1' + A_3\,F_0'\right) - A_2\left(F_3' - B_4\,F_1' + B_3\,F_0'\right)}{B_2\,A_1 - A_2\,B_1} \tag{5.157}$$

and:

$$\dot{\beta} = \frac{B_1\left(F_2' - A_4\,F_1' + A_3\,F_0'\right) - A_1\left(F_3' - B_4\,F_1' + B_3\,F_0'\right)}{B_2\,A_1 - A_2\,B_1} \tag{5.158}$$

where:

$$F_0' = F_0 - R_m\,N\,\overline{v}^m\,\frac{\gamma(m+1)}{\gamma(1)}$$

$$F_1' = I_\alpha\,N\,\overline{v}^\alpha\,\frac{\gamma(\alpha+1)}{\gamma(1)} - R_m\,N\,\overline{v}^{m+1}\,\frac{\gamma(m+2)}{\gamma(1)}$$

$$F_2' = F_2 + 2 I_\alpha\,\overline{v}^{\alpha+1}\,\frac{\gamma(\alpha+2)}{\gamma(1)} - R_m\,N\,\overline{v}^{m+2}\,\frac{\gamma(m+3)}{\gamma(1)}$$

and:

$$F_3' = F_3 + 3 I_\alpha\,N\,\overline{v}^{\alpha+2}\,\frac{\gamma(\alpha+3)}{\gamma(1)} - R_m\,N\,\overline{v}^{m+3}\,\frac{\gamma(m+4)}{\gamma(1)}$$

Thus, we have a set of four coupled first order, nonlinear differential equations for N, ϕ, v, and β. In some cases it may be worthwhile setting $\beta=1$ or some other constant value. Then the equation for β would be absent and the equation for v becomes:

$$\dot{v} = \left(F_2' + A_3\,F_0' - A_4\,F_1'\right)/A_1 \tag{5.159}$$

Let us apply this technique to the case of Brownian coagulation with no deposition as we did in the last section. The equations for the modified gamma distribution are algebraically tedious to write down, but we show the equations for the reduced distribution with $\beta=1$:

Table 5.2: Particle density and variance as a function of time.

τ	gamma N	gamma Var	log-normal N	log-normal Var	mod-gamma N	mod-gamma Var	AEROSIM N	AEROSIM Var
0.0	1.0	0.333	1.0	0.333	1.0	0.333	1.0	0.333
0.5	0.490	1.34	0.494	1.36	0.490	1.47	0.486	1.56
1.0	0.322	2.34	0.327	2.39	0.323	2.48	0.321	2.78
1.5	0.240	3.33	0.245	3.42	0.240	3.48	0.239	4.02
2.0	0.191	4.32	0.195	4.46	0.191	4.48	0.190	5.25
2.5	0.158	5.31	0.162	5.49	0.159	5.48	0.158	6.49
3.0	0.135	6.30	0.139	6.53	0.136	6.47	0.135	7.72
3.5	0.118	7.28	0.122	7.56	0.118	7.47	0.118	8.95
4.0	0.105	8.27	0.108	8.59	0.105	8.46	0.105	10.2
4.5	0.094	9.24	0.097	9.64	0.0943	9.46	0.0940	11.4
10.0	0.0443	20.06	0.0460	21.07	0.0445	20.37	0.0442	21.9

$$\frac{dN}{dt} = -N^2 \left\{ 1 + \frac{\Gamma(v+\frac{4}{3})\Gamma(v+\frac{2}{3})}{\Gamma(v+1)^2} \right\} \tag{5.160}$$

$$\frac{dv}{dt} = -N \left\{ v(v+1) + (v^2+v-\tfrac{2}{9})\frac{\Gamma(v+\frac{4}{3})\Gamma(v+\frac{2}{3})}{\Gamma(v+1)^2} \right\} \tag{5.161}$$

where $\Gamma(x)$ is the gamma function. While it is possible to obtain analytical solutions of these equations, in practice it is more efficient to evaluate them numerically by a Runge-Kutta method.

5.12 A comparative study of the moments method

In order to illustrate the above technique, we shall carry out some numerical calculations and discuss the results. First consider Eqns. (5.138) and (5.139) and Eqns. (5.160) and (5.161). For initial conditions, we set $\sigma_0 = \ln(4/3)$ and $v_0 = 2$ which ensures that the initial variances of the log-normal and gamma distributions are equal. Results are shown in Table 5.2 for the particle density and the variance as a function of time, τ. The results in this column headed 'AEROSIM' are from a computer code developed by the United Kingdom Atomic Energy Authority. We regard these results as the benchmark.

We hasten to add that we do not claim AEROSIM to be 'exact', simply that it produces values which should be reasonably accurate. However, as we shall see later, there is some question regarding its convergence for certain coagulation processes. Comparisons of log-normal, gamma and modified gamma distributions with AEROSIM show excellent agreement and none can be said to be superior. However, the inclusion of a removal term in the equation significantly alters these conclusions and we shall discuss the deviations below. First, it is

Methods of Solving the Dynamic Equation

useful to consider the consistency of the gamma distribution with self-similarity. We recall that for $\beta=1$, $\psi(\eta)$ is given by:

$$\psi(\eta) = \frac{\nu+1}{\Gamma(\nu+1)} \left[(\nu+1)\eta\right]^{\nu} \exp\left[-(\nu+1)\eta\right] \qquad (5.162)$$

and thus, only if $\nu(t)$ is independent of time will similarity be obeyed. We illustrate these matters in several ways. For example, Friedlander and Wang [1966] have carried out numerical calculations for $\psi(\eta)$ as we explained in Section 5.7. They note that for Brownian coagulation:

$$\frac{dN}{d\tau} = -(1+ab)N^2$$

where $ab=(1.129)$. Our calculations using the gamma distribution show an interesting behavior for $\nu(t)$. Fig. 5.2 illustrates the point with ν given for various values of the initial distribution parameter, ν_0. We observe that regardless of the value of ν_0, $\nu(\infty)$ is tending to a constant value such that $\nu(\infty) \sim (0.113)$. The value of ab is shown in Fig. 5.3 and there it is seen to tend to a constant value of (1.175) which is close to the value (1.129) calculated by Friedlander and Wang [1966]. This gives strong support to a self-similarity developing as time proceeds.

The actual form of $\psi(\eta)$ is also of interest and in Fig. 5.4 we show this for one initial condition at various times. Superimposed on the curves is the value of $\psi(\eta)$ calculated by Friedlander and Wang and also the value of $\psi(\eta)$ at infinite time as calculated from the modified gamma distribution, *i.e.* with variable $\beta(t)$. The results are striking. The relaxation of the gamma distribution to an asymptotic slope is evident and all final forms of ψ appear to be close together. This is especially true of Friedlander and Wang and the modified gamma distribution.

As a further test of similarity we have calculated $\psi(\eta)$ from Eqn. (5.162) using the free molecular form of the Brownian kernel. The results are shown in Figs. 5.5 and 5.6 for two initial conditions at various times, and we again note the tendency of the results to the exact value of ψ calculated by Lai et al. [1972], which is shown superimposed. We note that for free molecular coagulation, it may be shown that:

$$\frac{dN}{d\tau} = -\frac{\alpha}{2} N^{11/6}$$

where $\tau = K_f N_0 \bar{v}_0^{1/6} t$ and:

$$K_f = \left(\frac{8\pi kT}{\rho_p}\right)^{1/2} \left(\frac{3}{4\pi}\right)^{2/3}$$

Now, Lai et al. [1972] obtain $\alpha=(6.67)$. Fig. 5.7 shows $\alpha(t)$ deduced from the gamma distribution for various initial conditions and it is clearly tending to values which, while not identical, seem to cluster about $\alpha=(6.67)$. This is confirmation of the acceptable nature of the gamma distribution for representing size spectra. Further examples of the moments method using other coagulation mechanisms may be found in Williams [1986].

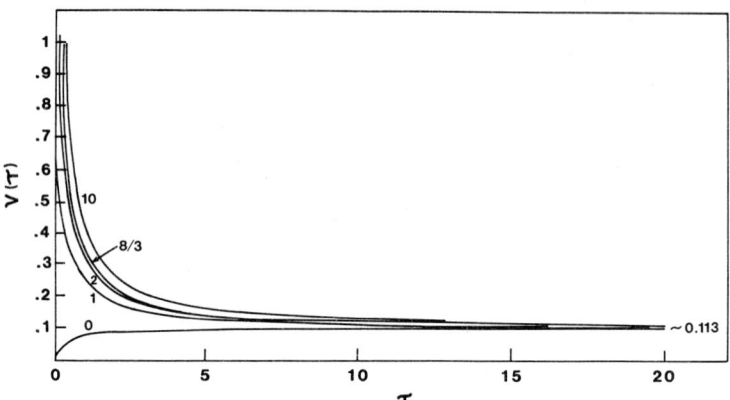

Figure 5.2: v for various initial values for the hydrodynamic coagulation kernel. Adapted from Williams [1986].

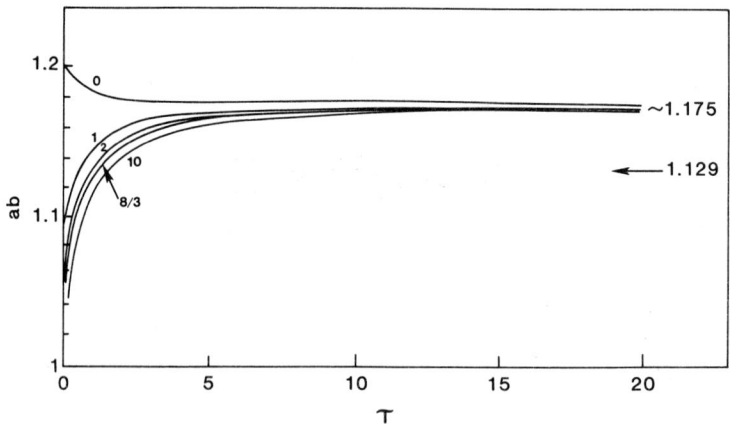

Figure 5.3: The coefficient 'ab' for the hydrodynamic coagulation kernel. Adapted from Williams [1986].

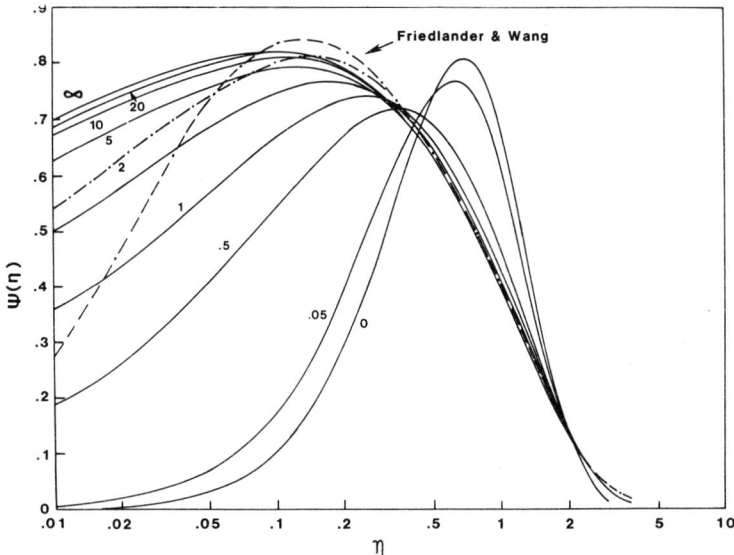

Figure 5.4: The similarity function at different times for the hydrodynamic coagulation kernel for constant $\beta=1$. The dashed-crossed line denotes the results of Friedlander and Wang [1966] and the dashed-dotted line denotes those from the modified gamma distribution at infinite time ($v_0=2, \beta_0=1$). Adapted from Williams [1986].

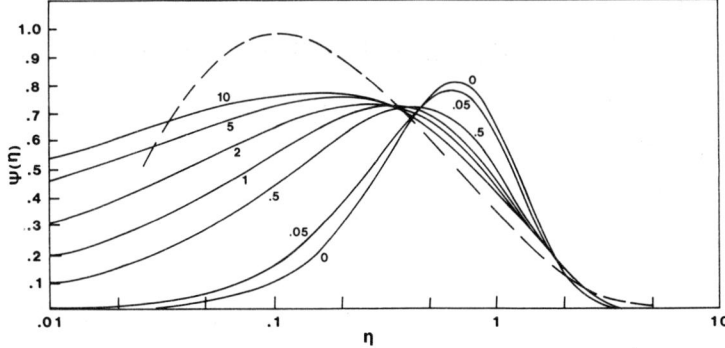

Figure 5.5: The similarity function at different times for the free molecule coagulation kernel. The normal Gamma distribution is used with $v_0=2$. The dashed line is from the paper of Lai et al. [1972]. Adapted from Williams [1986].

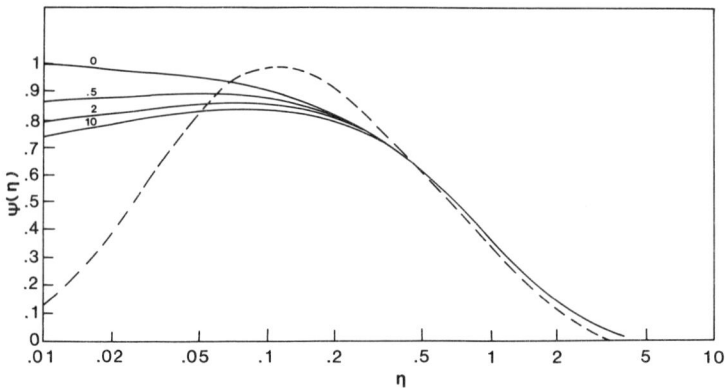

Figure 5.6: The similarity function at different times for the free molecule coagulation kernel. The normal Gamma distribution is used with $v_0=1$. The dashed line is from the paper of Lai et al. [1972]. Adapted from Williams [1986].

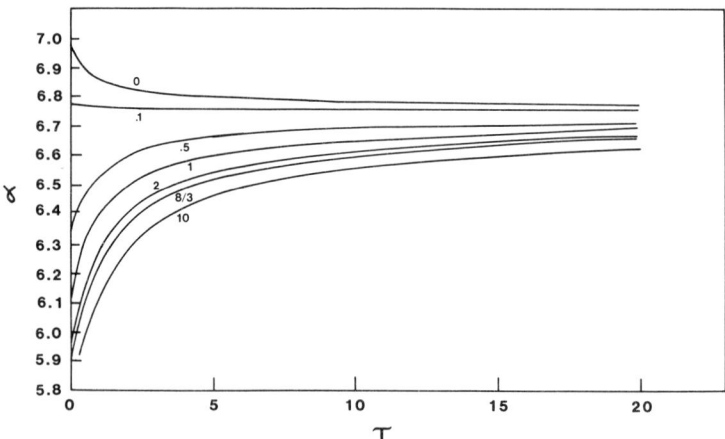

Figure 5.7: The value of α for the free molecule coagulation kernel for various initial conditions. Adapted from Williams [1986].

Methods of Solving the Dynamic Equation

5.13 Shape dependence

As we have seen in Section 3.13, a formalism for including the shape dependence of particles may be introduced by means of descriptors. In this section we wish to illustrate that technique and at the same time show how useful the moments method can be to obtain simple first order results for rather complex problems.

In order to describe shape dependence, we note that in the classical formulation of the coagulation terms in the dynamic equation it is assumed that the two particles of volumes, v' and v", coalesce to form a new particle of volume, v, where v=v'+v". Let us assume, however, that due to incomplete packing, the new composite volume is larger than the sum of the components. We may describe this situation mathematically by writing $v_s = v'_s + v''_s$ where the subscript, s, denotes the volume of the solid in the particle. According to Mountain and Mulholland [1984] (see Eqn. (2.225)) the effective radius, R, of a particle is related to v_s by:

$$v_s = v_0 \left(\frac{R}{R_0}\right)^{1/\alpha}$$

The effective volume, v, is then given by $v = 4\pi R^3 / 3$, whence:

$$v_s = \left(\tfrac{4}{3}\pi R_0^3\right)^{1-(1/3\alpha)} v^{(1/3\alpha)}$$

Thus, the conservation condition in terms of effective volume is:

$$v^{(1/3\alpha)} = v'^{(1/3\alpha)} + v''^{(1/3\alpha)} \tag{5.163}$$

where we call α the 'fractal index' which has been discussed in Section 2.18. Clearly, for $\alpha > 1/3$, the packing is incomplete. Indeed, Monte Carlo calculations indicate that $\alpha \approx 1/(1.78) = (0.56)$. It may be useful therefore, to study the behavior of the aerosol distribution for α in the range, $1/3 \leq \alpha \leq 1$. Physically, it seems that $\alpha = 2/3$ corresponds to addition of surface areas of the two components and $\alpha = 1$ to addition of linear dimensions. Thus, a larger α implies a more porous aggregate. In practice, one might expect α to be slightly dependent on aerosol age, but we shall ignore that matter here and simply examine how the volume of particulate and the spectral shape depend on α.

In order to carry out this task, we note that the aerosol balance equation for pure coagulation may be written from Eqn. (3.149) (using Eqn. (5.163)) as:

$$\frac{\partial}{\partial t} n(v,t) = \tfrac{1}{2} \int dv' \int dv'' \, K(v',v'') \, \delta\!\left[v - \left(v'^{(1/3\alpha)} + v''^{(1/3\alpha)}\right)^{3\alpha}\right] n(v',t)\, n(v'',t)$$

$$- n(v,t) \int_0^\infty dv' \, K(v',v)\, n(v',t) \tag{5.164}$$

Using the properties of the delta function, we can write the first integral in Eqn. (5.164) as:

$$\tfrac{1}{2} \int_0^v dv' \left[1 - \left(\frac{v'}{v}\right)^{(1/3\alpha)}\right]^{3\alpha - 1} K\!\left(v', \left[v^{(1/3\alpha)} - v'^{(1/3\alpha)}\right]^{3\alpha}\right) n(v',t)\, n\!\left(\left[v^{(1/3\alpha)} - v'^{(1/3\alpha)}\right]^{3\alpha}, t\right)$$

However, in order to apply the moments method this reduction is unnecessary and all that need be done is to multiply Eqn. (5.164) by v^n and integrate over all $v(0,\infty)$. The result is:

$$\frac{d}{dt}M_n(t) = \tfrac{1}{2}\int_0^\infty dv' \int_0^\infty dv'' \, K(v',v'') \left\{ \left[v'^{(1/3\alpha)} + v''^{(1/3\alpha)} \right]^{3\alpha n} - v'^n - v''^n \right\} n(v',t) \, n(v'',t)$$

(5.165)

M_n being the nth moment of $n(v,t)$.

Now, if we assume that the distribution follows the log-normal form:

$$n(v,t) = \frac{N(t)}{\sqrt{2\pi\sigma(t)}} \frac{1}{v} \exp\left[-\frac{1}{2\sigma(t)} \ln^2\left(\frac{v}{\hat{v}(t)}\right) \right]$$

then after some algebra we obtain:

$$\frac{d}{dt}\left[N \hat{v}^n \exp\left(\frac{n^2 \sigma}{2}\right) \right] = \frac{N^2}{4\pi\sigma} \hat{v}^n \, F_n(\hat{v},\sigma)$$

where:

$$F_n = \int_{-\infty}^{\infty} dx \int_{-\infty}^{\infty} dy \, K(\hat{v}\,e^x, \hat{v}\,e^y) \exp\left(-\frac{x^2+y^2}{2\sigma}\right) \left\{ \left[\exp\left(\frac{x}{3\alpha}\right) + \exp\left(\frac{y}{3\alpha}\right) \right]^{3\alpha n} - e^{nx} - e^{ny} \right\}$$

For illustrative purposes, we use the Brownian kernel:

$$K(u,v) = K_B \left[2 + \left(\frac{u}{v}\right)^{1/3} + \left(\frac{v}{u}\right)^{1/3} \right]$$

where $K_B = 2kT/3\mu$.

Then, setting n=0, 1/(3α), and 2/(3α) leads to:

$$\frac{dN}{dt} = -N^2 \left[1 + \exp(\sigma/9) \right]$$

(5.166)

$$\frac{d}{dt}\left[N \hat{v}^{(1/3\alpha)} \exp\left(\frac{\sigma}{18\alpha^2} \right) \right] = 0$$

(5.167)

and:

$$\frac{d}{dt}\left[N \hat{v}^{(2/3\alpha)} \exp\left(\tfrac{2}{9}\frac{\sigma}{\alpha^2} \right) \right] = 2 N^2 \hat{v}^{(2/3\alpha)} \exp\left(\tfrac{1}{9}\frac{\sigma}{\alpha^2} \right) \left[1 + \exp\left(\frac{\sigma}{9}\right) \right]$$

(5.168)

It should be noted that Eqn. (5.167) corresponds to conservation of solid volume as we expect. However, we do not have conservation of effective volume as we shall show below.

Using Eqns. (5.166) and (5.167) in Eqn. (5.168) we obtain:

$$\frac{d\sigma}{dt} = -9\alpha^2 \, N \left[1 + \exp\left(\frac{\sigma}{9}\right) \right] \left[1 - 2\exp\left(-\frac{\sigma}{9\alpha^2} \right) \right]$$

(5.169)

Methods of Solving the Dynamic Equation

where $\tau = N_0 K_B t$ and we have normalized N to N_0, the initial value.
Eqns. (5.166) and (5.169) are readily solved analytically and lead to

$$N = \frac{\exp\left(\frac{\sigma}{9\alpha^2}\right) - 2}{\exp\left(\frac{\sigma_0}{9\alpha^2}\right) - 2} \tag{5.170}$$

and:

$$\tau = \frac{1}{9\alpha^2}\left[\exp\left(\frac{\sigma_0}{9\alpha^2}\right) - 2\right] \int_\sigma^{\sigma_0} \frac{ds \exp\left(-\frac{s}{9\alpha^2}\right)}{\left[1 + \exp\left(\frac{s}{9}\right)\right]\left[1 - 2\exp\left(-\frac{s}{9\alpha^2}\right)\right]^2} \tag{5.171}$$

The volume of particulate, ϕ, is given by:

$$\phi = N^{1-3\alpha} \exp\left\{\frac{\sigma_0 - \sigma}{2}\left(\frac{1}{3\alpha} - 1\right)\right\} \tag{5.172}$$

and the effective particle density (relative to the initial value) is, $\rho = 1/\phi$. The volume of solid particulate, ϕ_s, is given by:

$$\phi_s = \int_0^\infty dv_s \, v_s \, n(v_s)$$

where:

$$n(v_s) = \int dv \, \delta\left(v - \left(\frac{4\pi R_0^3}{3}\right)^{1-(1/3\alpha)} v^{(1/3\alpha)}\right) n(v)$$

whence:

$$\phi_s = \left(\frac{4\pi R_0^3}{3}\right)^{1-(1/3\alpha)} M_{(1/3\alpha)}$$

which, from Eqn. (5.165), is indeed constant in time.

We have evaluated these quantities numerically for $\alpha = 1/3$, (0.56), 2/3, and 1. Fig. 5.8 shows the variance, σ, of the distribution. It is clear that this increases with time but also increases as α decreases. Indeed, σ is very sensitive to α. We also note that $\sigma(\infty)$ tends to a constant value as $\tau \to \infty$ and that this can be obtained from Eqn. (5.170) by setting N=0, whence:

$$\sigma(\infty) = \ln(2) = (0.693) \quad : \quad \alpha = \tfrac{1}{3}$$
$$= \frac{9}{(1.78)^2} \ln(2) = (1.97) \quad : \quad \alpha = (0.56)$$
$$= 4\ln(2) = (2.77) \quad : \quad \alpha = \tfrac{2}{3}$$
$$= 9\ln(2) = (6.24) \quad : \quad \alpha = 1$$

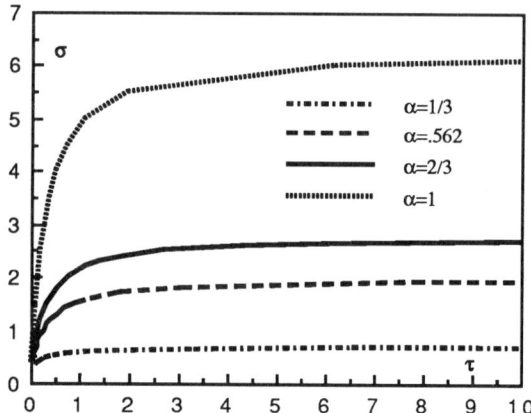

Figure 5.8: The variance, σ, as a function of the fractal index, α.

As σ increases with time, examination of Eqn. (5.166) shows that the effective coagulation constant $(1+\exp(\sigma/9))$ also increases with time and thus, for increasing α, the density reduces more rapidly. This behavior is illustrated in Fig. 5.9 where we plot $1/N$ versus τ (the results for $\alpha=2/3$ and (0.56) cannot be distinguished visually). Thus, a fractal structure leads to a more rapid decrease in particle density.

Finally, in Fig. 5.10, we examine the effective particle density, $1/\phi$, as a function of τ and α. For $\alpha=1/3$ there is no variation with time since in this case volume is conserved. On the other hand, as α increases, the density variation becomes more marked, demonstrating incomplete packing. This is an oversimplified picture but does illustrate the effect of shape on particle dynamics.

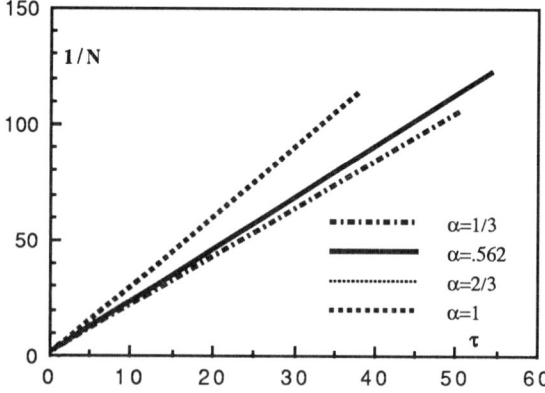

Figure 5.9: The inverse particle density, $1/N$, as a function of the fractal index, α.

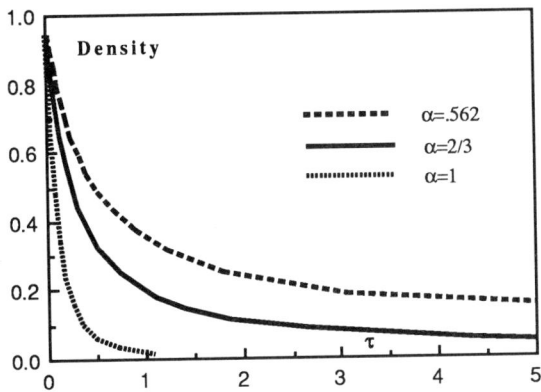

Figure 5.10: The effective particle density, ϕ, as function of the fractal index, α.

A modification of the above method to account for variation in packing structure as the aerosol ages would be to assign different values of α to each coagulation mechanism. For example, it is known that Brownian coagulation dominates for small particles and gravitational for large ones. Thus, if the α-values are different the packing will change as the particles grow. Experimental evidence may suggest what values of α to use. Quantitatively, however, one would write for the net coagulation kernel:

$$K(u,v) = \sum_i K_i(u,v;\alpha_i) \tag{5.173}$$

5.14 Numerical methods

The methods discussed so far in this chapter are very helpful in providing exact or approximate solutions for various representations of the aerosol rate processes. The exact solutions serve as benchmarks, and together with the approximate solutions, are useful in parametric studies as well as for providing insight into the nature of aerosol evolution. Their utility for practical use is limited for several reasons.

a) Exact solutions are feasible only for special, and not necessarily realistic, representations of the rate processes and for certain initial and boundary conditions.

b) Aerosol evolution, especially if there are abrupt changes such as source reinforcement, may not follow a prescribed form of distribution. Often bimodal, trimodal, or more complicated distributions may evolve, and these cannot be dealt with in the context of log-normal, gamma, or other forms.

c) In many applications, those in nuclear industry in particular, particle evolution is strongly coupled to the environment. Thus, the dynamical equation must be solved simultaneously with equations that describe the thermal hydraulic and chemical conditions of the gaseous/vaporous environment of the particles.

Aerosol Science: Theory and Practice

For practical applications, one must resort to numerical methods. From a numerical viewpoint, at least in principle, the dynamical equation is not too difficult to solve. Since methods for solving partial differential equations have been studied extensively and many algorithms and computer programs are available, the preferred approach here has been to reduce the dynamical equation first to a system of partial (ordinary) differential equations by employing some approximation to the integral of the coagulation terms.

The integral can be evaluated (approximated) in various ways. The approximations are likely to be most successful if they preserve the basic properties of coagulation. One such property is the conservation of mass or volume in coagulation. The various quadratures must either be accurate enough to preserve this property or one must preserve it forcibly by evaluating the integrals in a special manner or by applying corrective steps to the calculations.

Within the past years, several promising methods have emerged. For clarity, we shall first discuss the methods used with single species and spatial homogeneity. Then, we shall discuss the generalizations of these approaches to multispecies aerosols and spatially inhomogeneous situations.

5.15 The J-space transform and cubic spline method

This method has been used by several authors (Berry [1967], Suck and Brock [1979], Emami and Loyalka [1981]). In particular, following the development of Middleton and Brock [1976], to describe particle size variations over several decades in size, one introduces the transformations:

$$v(J) = v_0 \, a^{b(J-1)} \tag{5.174}$$

and:

$$Y(J,t) = n(v,t) \, v(J) \, b \ln(a) \tag{5.175}$$

where v_0 is the smallest aerosol size corresponding to $J=1$, and a and b are some arbitrary constants. J is an integer and varies from 1 to a fairly large number that would allow consideration of all sizes of interest. With this transformation we have:

$$\frac{\partial}{\partial t} Y(J,t) = \int_1^{J_d} dJ' \, \frac{v(J)}{v(J_c)} Y(J_c,t) \beta(J_c,J') Y(J',t) - \int_1^\infty dJ' \, Y(J,t) \beta(J,J') Y(J',t)$$

$$- \frac{\partial}{\partial J} \left[\frac{I(v(J))}{v(J) \, b \ln(a)} Y(J,t) \right] + \tilde{S}(v(J),t) \tag{5.176}$$

where:

$$v(J_c) = v(J) - v(J')$$

$$J_c = J + \frac{1}{b \ln(a)} \ln\left(1 - a^{(J'-J)b}\right)$$

and:

$$J_d = J - \frac{\ln(2)}{b \ln(a)}$$

Eqn. (5.176) is converted to a system of ordinary differential equations by use of quadrature formulas for the integrals and cubic splines for the derivative on J and one writes:

Methods of Solving the Dynamic Equation

$$\frac{d}{dt}Y(J,t) = f_J\left(Y(1,t), Y(2,t), Y(3,t), ..., Y(m,t), \tilde{S}(J,t)\right) \quad (5.177)$$

together with the associated initial conditions. Here, $J = 1,2,3,...,m$, where m is chosen sufficiently large so that all particles of interest are covered. The method can yield fairly accurate results, but the computational costs can be quite high because of the repeated cubic spline fits. Note that one usually takes a=2 and b=1/2

5.16 Finite element method

We write the GDE in the form:

$$\frac{d}{dt}n(v_i,t) = \int_0^\infty \int_0^\infty du\, dw\, \overline{K}(v_i;u,w)\, n(u,t)\, n(w,t) - \frac{\partial}{\partial v}(I(v)\, n(v,t))\bigg|_{v=v_i} + S(v_i,t) \quad (5.178)$$

where:

$$\overline{K}(v;u,w) = \tfrac{1}{2}\left[\delta(v-(u+w)) - \delta(v-u) - \delta(v-w)\right] K(u,w) \quad (5.179)$$

$n(v,t)$ can be expressed by an interpolation formula in terms of $n(v_i,t)$:

$$n(v,t) = \sum_{i=1}^{I} \phi_i(v)\, n(v_i,t) \quad (5.180)$$

where $\phi_i(v)$ are some known basis functions. Eqn. (5.180) can be substituted into Eqn. (5.178) and the integral and the derivative on the right hand side can be expressed in terms of the basis functions. This process leads to a system of equations of the form:

$$\frac{d}{dt}n(v_i,t) = \sum_j \sum_k \beta_{i;j,k}\, n(v_j,t)\, n(v_k,t) - \sum_j \gamma_{ij}\, n(v_j,t) + S(v_i,t) \quad (5.181)$$

where:

$$\beta_{i;j,k} = \int_0^\infty \int_0^\infty du\, dw\, \overline{K}(v_i,u,w)\, \phi_j(u)\, \phi_k(w) \quad (5.182)$$

and:

$$\gamma_{ij} = \frac{\partial}{\partial v}\left[I(v)\, \phi_j(v)\right]\bigg|_{v=v_i} \quad (5.183)$$

With $n(v_i,0)$ known, Eqn. (5.181) can be solved with available algorithms or standard computer programs (for example, Gear's Method, the Runge-Kutta method, or the program LSODE).

The accuracy of this approach depends on the choice of $\phi_i(v)$ and Eqn. (5.180). Since aerosol sizes can span several decades, the v_i are chosen on some geometric scale.

It has been verified that with proper choice of Eqn. (5.180), numerical quadrature and differentiation, the approach yields results that agree well with several of the exact solutions described in the Sections (5.1)-(5.11).

There are, however, several areas where the above approach requires improvement. First, if the aerosol distribution shifts to larger (because of coagulation/condensation) or smaller (because of deposition/evaporation) sizes, the grid must be shifted (moved). Second, the calculation of $\beta_{i;j,k}$ consumes considerable computational time. Strictly speaking, these would be time dependent but retention of time dependence, within a given time interval, makes the calculations quite expensive. Thus, the preferred approach is to calculate the coefficients once at the start of the calculations, and later as needed. Accurate evaluation is necessary to ensure mass conservation in the case of coagulation alone.

It is important to note that Eqn. (5.178) is hyperbolic and, in the absence of coagulation, its numerical solution poses the difficulties of dispersion and diffusion. We discuss these aspects in some detail in Section 5.18.

5.17 The group (sectional) method

In nuclear reactor physics, one of the more successful approaches for solving the transport (diffusion) equations consists of resorting to the multiple energy group equations. Here one writes conservation equations for neutron density in each group using 'transfer' cross sections. Similar ideas can be used to treat the integral in the dynamical equation. Next, we describe this approach as developed by Gelbard and Seinfeld [1980].

The basic ideas behind the approach are easily understood. Suppose we divide the size spectrum into m groups, $v_{\ell-1} \leq v < v_\ell$. The total volume of particles in group ℓ is then given by:

$$Q_\ell(t) = \int_{v_{\ell-1}}^{v_\ell} dv\, v\, n(v,t)$$

(5.184)

Coagulation creates particles whose volumes will lie in this group and also removes particles from this group. Thus, we could reduce the equation (in the absence of condensation/evaporation) to the form:

$$\frac{d}{dt} Q_\ell(t) = \tfrac{1}{2} \sum_{i=1}^{\ell-1} \sum_{j=1}^{\ell-1} \int_{v_{i-1}}^{v_i} dv \int_{v_{j-1}}^{v_j} dv'\, \eta(v_{\ell-1} < v+v' < v_\ell)(v+v') C$$

$$+ \sum_{i=1}^{\ell-1} \int_{v_{i-1}}^{v_i} dv \int_{v_{j-1}}^{v_j} dv'\, \eta(v+v' < v_\ell)\, v\, C$$

$$- \sum_{i=1}^{\ell-1} \int_{v_{i-1}}^{v_i} dv \int_{v_{\ell-1}}^{v_\ell} dv'\, \eta(v+v' > v_\ell)\, v'\, C$$

$$- \tfrac{1}{2} \int_{v_{\ell-1}}^{v_\ell} dv \int_{v_{\ell-1}}^{v_\ell} dv'\, \eta(v+v' > v_\ell)(v+v') C$$

$$- \sum_{i=\ell+1}^{m} \int_{v_{\ell-1}}^{v_\ell} dv \int_{v_{\ell-1}}^{v_\ell} dv'\, v'\, C + \int_{v_{\ell-1}}^{v_\ell} dv\, v\, S(v,t)$$

(5.185)

where:

$$C = K(v,v')\, n(v,t)\, n(v',t)$$

(5.186)

Also, η is the function:

$\eta(x) = 1$; if x is true
$ = 0$; otherwise

The equation is self explanatory. The first two terms on the right hand side are creation terms and the next three terms are removal terms.

Methods of Solving the Dynamic Equation

To close this set of equations in terms of $Q_\ell(t)$, it is necessary to relate $n(v,t)$ to this quantity. This can be accomplished by the use of Eqn. (5.184) and any appropriate approximation for n that simplifies the evaluation of the integrals in Eqn. (5.185). In general, one could write:

$$n(v,t) = \sum_{\ell=1}^{m} \eta(v_{\ell-1} < v < v_\ell) n_\ell(t) f_\ell(v) \tag{5.187}$$

where $n_\ell(t)$ are unknown and $f_\ell(v)$ are some prescribed functions (constants, 1/v, piecewise linear, *etc.*). If one takes $f_\ell(v)=1/v$, then, using Eqn. (5.187) in Eqn. (5.184), we find:

$$n_\ell(t) = \frac{Q_\ell(t)}{v_\ell - v_{\ell-1}} \tag{5.188}$$

which represents the total number of particles (of the same average volume) in the ℓth section at time, t.

Substituting Eqns. (5.187) and (5.188) into Eqn. (5.186), one obtains a set of differential equations:

$$\frac{d}{dt} Q_\ell(t) = \tfrac{1}{2} \sum_{i=1}^{\ell-1} \sum_{j=1}^{\ell-1} \beta^{(1)}_{ij\ell} Q_i(t) Q_j(t) + Q_\ell(t) \sum_{i=1}^{\ell-1} \beta^{(2)}_{i\ell} Q_i(t) - Q_\ell(t) \sum_{i=1}^{\ell-1} \beta^{(3)}_{i\ell} Q_i(t)$$

$$- \tfrac{1}{2} \beta^{(4)}_{\ell\ell} Q^2_\ell(t) - Q_\ell(t) \sum_{i=l+1}^{m} \beta^{(5)}_{i\ell} Q_i(t) + S_\ell(t) \tag{5.189}$$

where:

$$\beta^{(1)}_{ij\ell} = \int_{v_{i-1}}^{v_i} dv \int_{v_{j-1}}^{v_j} dv' \, \eta(v_{\ell-1} < v+v' < v_\ell)(v+v') g_{ij}(v,v')$$

$$\beta^{(2)}_{i\ell} = \int_{v_{i-1}}^{v_i} dv \int_{v_{\ell-1}}^{v_\ell} dv' \, \eta(v+v' < v_\ell) v \, g_{i\ell}(v,v')$$

$$\beta^{(3)}_{i\ell} = \int_{v_{i-1}}^{v_i} dv \int_{v_{\ell-1}}^{v_\ell} dv' \, \eta(v+v' > v_\ell) v' \, g_{i\ell}(v,v')$$

$$\beta^{(4)}_{\ell\ell} = \int_{v_{\ell-1}}^{v_\ell} dv \int_{v_{\ell-1}}^{v_\ell} dv' \, \eta(v+v' > v_\ell)(v+v') g_{\ell\ell}(v,v')$$

$$\beta^{(5)}_{i\ell} = \int_{v_{i-1}}^{v_i} dv \int_{v_{\ell-1}}^{v_\ell} dv' \, v' \, g_{i\ell}(v,v') \tag{5.190}$$

and:

$$g_{ij}(v,v') = \frac{K(v,v')}{(v_i - v_{i-1})(v_j - v_{j-1}) v v'} \tag{5.191}$$

A computer program can now be set up for solving the above equations in a fairly general format.

We have considered above the conservation of volume in each section but one could develop similar equations for any quantity of importance (*e.g.*, mass). Equations thus obtained preserve the conservation of important quantities in the description of aerosol dynamics, and thus lead to physically meaningful descriptions even with low order (group) approximations.

The method can be used for condensation/evaporation also, but then it has serious limitations (Park and Loyalka [1989], Tsang and Rao [1988]) because of numerical diffusion. These

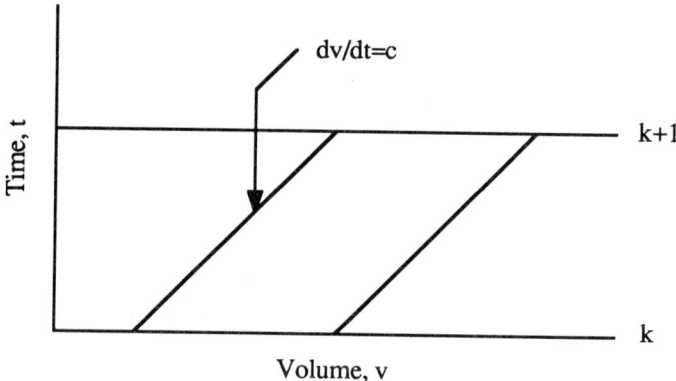

Figure 5.11: Integration along a constant characteristic.

problems can be partially solved by using 'moving' sections, but then, in the presence of coagulation, it would become necessary to evaluate β at each computational step. Thus, here the method does not appear to lead to any advantages over the methods of the previous section.

Since condensation/evaporation can be a very important process in aerosol evolution, we discuss the numerical aspects in the next several sections.

5.18 Numerical solutions for condensation/evaporation without coagulation

To elucidate the difficulties associated with numerical solutions of the GDE in the presence of condensation/evaporation, we examine the equation in the absence of coagulation. Thus, we consider the equation:

$$\frac{\partial}{\partial t} n(v,t) = -\frac{\partial}{\partial v}[I(v) n(v,t)] \qquad (5.192)$$

In general, I is a function of both the particle size and time, but for the moment we will assume it to be a constant, c. Then, Eqn. (5.192) can be written as:

$$\frac{\partial}{\partial t} n(v,t) = -c \frac{\partial}{\partial v} n(v,t) \qquad (5.193)$$

which has the analytical solution:

$$n(v,t) = n_0(v - ct) \quad ; \quad t > v/c$$
$$= 0 \quad ; \quad \text{otherwise} \qquad (5.194)$$

where $n_0(v)$ is the initial distribution. We recognize that Eqn. (5.193) has the characteristic:

$$v - ct = \text{constant} \qquad (5.195)$$

along which particles 'move' with a constant rate, c. In other words, along the characteristic (see Fig. 5.11):

Methods of Solving the Dynamic Equation

$$\frac{dn}{dt} = 0 \tag{5.196}$$

Use of Euler's forward difference method with respect to time and a second order central difference on size in Eqn. (5.193) gives:

$$\frac{n_j^{k+1} - n_j^k}{\Delta t} = \frac{c}{2\,\Delta v}\left(n_{j+1}^k - n_{j-1}^k\right) \tag{5.197}$$

where the subscript, j, indicates current size and the superscript, k, indicates current time. To examine the stability of the method, we use von Neumann stability analysis, and note that Eqn. (5.193) has the solution:

$$n(v,t) = \exp(i\ell v)\exp(-i\ell c t) \tag{5.198}$$

while the difference equation has the solution:

$$n_j^k = \exp(i\,\ell\,j\,\Delta v)\left[1 - \frac{i\,c\,\Delta t}{\Delta v}\sin(\ell\,\Delta v)\right]^k \tag{5.199}$$

That is:

$$n_j^k - n(v,t)\Big|_{v=v_j,\,t=t_k} = \exp(i\,\ell\,v)\left\{\left[1 - \frac{i\,c\,\Delta t}{\Delta v}\sin(\ell\,\Delta v)\right]^k - \left[\exp(-i\,\ell\,c\,\Delta t)\right]^k\right\} \tag{5.200}$$

We note that the numerical solution differs substantially from the exact solution in that:

i) The exact solution is propagated without a change in amplitude, by the original Eqn. (5.193).

ii) The finite difference Eqn. (5.197) yields a solution, Eqn. (5.199), whose amplitude increases in time (as k increases) for any values of the parameters, and the method is unconditionally unstable!

To understand the second point we note that the substitution:

$$n(v,t) = \exp(i\ell v)\,\varphi(t) \tag{5.201}$$

in Eqn. (5.193) leads to:

$$\varphi'(t) = \alpha\,\varphi(t) \tag{5.202}$$

where:

$$\alpha = i\ell c \tag{5.203}$$

is a purely imaginary number. It is known that with α imaginary, Euler's method is unstable for Eqn. (5.202), and so the result is not surprising. We note that Eqn. (5.193) imposes some very stringent requirements on the positive nature of the solution, n(v,t), and 'forbidden signals.' By the latter, we imply that at point (v,t) the domain of dependence is well defined, and any numerical method that distorts this significantly will likely lead to serious problems. The central differencing above distorts the domain of dependence (by extending it to the right of the point, (j,k+1)) and causes difficulty.

A better approach would be to backward difference in both size and time. The simplest approach of this type leads to:

$$\frac{n_j^{k+1} - n_j^k}{\Delta t} = -\frac{c}{\Delta v}\left(n_j^k - n_{j-1}^k\right) \tag{5.204}$$

which is known as the upwind scheme. A von Neumann analysis of this method leads to:

$$n_j^{k+1} = \left[1 - \frac{c\,\Delta t}{\Delta v}\left(1 - \exp(-i\ell\Delta v)\right)\right]n_j^k \tag{5.205}$$

and this method is stable. That is, the method does not increase the magnitude of the solution with increasing time if:

$$C_0 = \frac{c\,\Delta t}{\Delta v} \leq 1 \tag{5.206}$$

where C_0 is the Courant number.

Although in the above example $C_0=1$ leads to the exact solution (at the stability limit), this is a result of special circumstances. As such, Eqn. (5.204) is only first order accurate in space and time. But, we note that (as with the central difference approach) use of higher order formulas here has the full potential of introducing spurious solutions that can lead to both dispersion (solution oscillating to negative values) and diffusion (distortion in the shape, or undesired spread of the peak). These arise from a large number of Fourier components in the solution of the difference equations. Thus, it becomes imperative that the difference equations preserve the basic properties of the original equation (solution) as closely as possible. The desirable requirement is to integrate along the characteristics but this is not always easy as the function, I, changes continually.

In actual aerosol computations, the upwind method requires very small time steps as both the aerosol distribution, and the growth rate, I, change by orders of magnitude over the range of v. Since Δv also must be kept small (to avoid excessive truncation errors), modifications of the upwind schemes are sought. We discuss in the next section one particularly useful modification of the method.

5.19 The Smolarkiewicz method

Smolarkiewicz [1983] studied the problem of advection in fluid flows but his method applies directly to the problem of aerosol growth. We note that the upwind difference approximation to Eqn. (5.192), in the conservative form, gives:

$$n_j^{k+1} = n_j^k - \frac{\Delta t}{\Delta v}\left(F_{j+1/2}^k - F_{j-1/2}^k\right) \tag{5.207}$$

where:

$$F_{j+1/2}^k = \tfrac{1}{2}\left\{\left(I_{j+1/2}^k + |I_{j+1/2}^k|\right)n_j^k + \left(I_{j+1/2}^k - |I_{j+1/2}^k|\right)n_{j+1}^k\right\} \tag{5.208}$$

with:

$$I_{j+1/2}^k = \tfrac{1}{2}\left(I_j^k + I_{j+1}^k\right) \tag{5.209}$$

Methods of Solving the Dynamic Equation

This formula is easily implemented on a computer. But, despite the fact that it meets the conservation requirement, because of its first order accuracy it leads to the problem of numerical 'diffusion.' To examine this, Smolarkiewicz, using Taylor series expansions in Eqn. (5.193), noted that the formula actually is a close approximation to the equation:

$$\frac{dn}{dt} = -\frac{\partial}{\partial v}(In) + \frac{\partial}{\partial v}\left(K\frac{\partial n}{\partial v}\right) \tag{5.210}$$

where K, the 'diffusivity,' is expressed as:

$$K = \tfrac{1}{2}\left(|I|\,\Delta v - I^2\,\Delta t\right) \tag{5.211}$$

and is thus responsible for diffusion in the solution. Since one cannot choose K to be zero, the diffusion is inherent to the method. Also, this term is important to the stability of the method, and cannot be simply subtracted. The argument given by Smolarkiewicz was that Eqn. (5.207) be used as a predictive step only, and that a corrective step then be applied to reverse the effects of diffusion to the extent possible by using the equation:

$$\frac{\partial n}{\partial t} = \frac{\partial}{\partial v}\left(K\frac{\partial n}{\partial v}\right) \tag{5.212}$$

in a reverse manner. Thus, we write Eqn. (5.192) in the form:

$$\frac{\partial n}{\partial t} = -\frac{\partial}{\partial v}(I_d\, n) \tag{5.213}$$

where:

$$I_d = \begin{cases} -\dfrac{K}{n}\dfrac{\partial n}{\partial v} & \text{if} \quad n > 0 \\ 0 & \text{if} \quad n = 0 \end{cases} \tag{5.214}$$

and I_d is referred to as the 'diffusion velocity.' An 'antidiffusion velocity,' $\tilde{I}$, is then defined as:

$$\tilde{I} = \begin{cases} -I_d & \text{if} \quad n > 0 \\ 0 & \text{if} \quad n = 0 \end{cases} \tag{5.215}$$

The Smolarkiewicz method then leads to the scheme:

i) Predictive Step:

$$n_j^* = n_j^k - \frac{\Delta t}{\Delta v}\left(F_{j+1/2}^k - F_{j-1/2}^k\right) \tag{5.216}$$

ii) Corrective Step:

$$n_j^{k+1} = n_j^* - \frac{\Delta t}{\Delta v}\left(F_{j+1/2}^* - F_{j-1/2}^*\right) \tag{5.217}$$

where, in F^*, $\tilde{I}$ is calculated as:

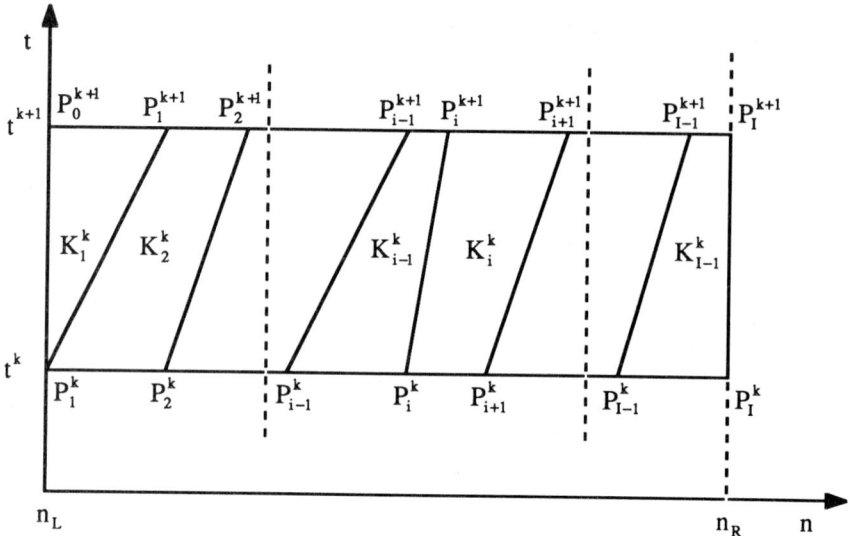

Figure 5.12: Finite elements in space and time at a typical time step. Adapted from Varoḡlu and Finn [1980] with permission of the Academic Press.

$$\tilde{I}_{j+1/2} = K_{j+1/2} \frac{1}{\frac{1}{2}\left(n_j^* + n_{j+1}^* + \varepsilon\right)} \frac{n_{j+1}^* - n_j^*}{\Delta x} = \frac{\left(\left|I_{j+1/2}\right|\Delta v - I_{j+1/2}^2 \Delta t\right)}{\frac{1}{2}\left(n_j^* + n_{j+1}^* + \varepsilon\right)} \frac{n_{j+1}^* - n_j^*}{\Delta x} \qquad (5.218)$$

and ε is a small quantity (10^{-15}). Note that the corrective step can be used repeatedly. For stability here, one requires:

$$\max_j \left(\frac{\left|\tilde{I}_{j+1/2}\right|\Delta t}{\Delta v}\right) \leq 1 \qquad (5.219)$$

which, with the upwind scheme criterion, is easily satisfied. Thus, the method is stable and consistent.

5.20 A moving grid characteristic based finite element method

Neither the upwind nor Smolarkiewicz methods overcome the problem of diffusion to complete satisfaction and additional methods are needed. Tsang and Brock [1983], and later Tsang and Rao [1988, 1990] and Tsang and Huang [1990], have explored one promising method that essentially derives from the work of Varoḡlu and Finn [1980] in fluid dynamics. Although the details of this method are tedious, the basic ideas are easily followed.

Suppose $\hat{n}(v,t)$ is an approximation to the actual distribution $n(v,t)$. Then, we solve the problem of Eqn. (5.192) in a weak sense by requiring:

Methods of Solving the Dynamic Equation

$$\int_{t_1}^{t_2} dt \int_{v_0}^{v_{max}} dv\ \varphi(v,t) \left(\frac{\partial \hat{n}}{\partial t} + \frac{\partial}{\partial v}(I\hat{n}) \right) = 0 \tag{5.220}$$

where $\varphi(v,t)$ is a continuous function defined in the interval of interest. We consider a computational grid in the size-time domain, with $t_1 = t^k$ and $t_2 = t^{k+1}$, where the superscript, k, indicates the kth step (and known level) and k+1 indicates the next level. The domain of integration is discretized by the spatial-temporal elements, $K_0^n, K_1^n, ..., K_{I-1}^n$, as illustrated in Fig. 5.12.

Note that while the first element is triangular, all others are trapezoidal. The lines, $P_i^k\ P_i^{k+1}$, are chosen to be first order approximations to the characteristics of the dynamical equation (e.g., for I=constant, these will be exactly the straight lines). Further, the trapezoidal elements are transformed from the v–t global coordinate system to an η–ξ local coordinate system such that the transformed element (in the η–ξ local system) is a unit square. The function, $\hat{n}$, is defined as a piecewise polynomial based on each element, K_i^n, in the form:

$$\hat{n}(\eta,\xi) = (1-\eta)(1-\xi)\, n_i^k + (1-\eta)\,\xi\, n_i^{k+1} + \eta(1-\xi)\, n_{i+1}^k + \eta\,\xi\, n_{i+1}^{k+1} \tag{5.221}$$

The functions, $\varphi(v,t)$, are chosen to be the basis functions for the space, V, of all continuous functions defined on the finite elements, K_i^n, and are simply the hat functions.

The integral in Eqn. (5.220) is now integrated by parts with substitution of Eqn. (5.221), use of the Jacobian of transformation, and low order quadratures (for example, second order Newton-Cotes). This process gives a banded matrix for the evaluation of the unknown values, n_i^{k+1}, which can be found quite efficiently.

5.21 Numerical treatment of multicomponent aerosols

The methods discussed above should all, in principle, apply to multicomponent aerosols with similar limitations and effectiveness. Thus, if we consider the equation:

$$\frac{\partial}{\partial t} n(v_1, v_2, ..., v_m, t)$$

$$= \int_0^\infty du_1\, du_2 \ldots du_m \int_0^\infty dw_1\, dw_2 \ldots dw_m\ \overline{K}(v_1, v_2, ..., v_m; u_1, u_2, ..., u_m, w_1, w_2, ..., w_m)$$

$$n(u_1, u_2, ..., u_m, t)\, n(w_1, w_2, ..., w_m, t)$$

$$+ \sum_i \frac{\partial}{\partial v_i} \left(I_i(v_1, v_2, ..., v_m)\, n(v_1, v_2, ..., v_m, t) \right) + S(v_1, v_2, ..., v_m, t) \tag{5.222}$$

then corresponding to the J-space transform method, we will consider a $J_1, J_2, ..., J_m$ transform, and the number of coupled ordinary differential equations that we would need to solve would increase m fold from those for single species aerosols. Quite clearly, the evaluation of the integrals and the time derivative would become much more expensive in computational costs. Similarly, for the application of the finite element method, one would generalize Eqn. (5.180) to a basis on m dimensions.

Extensive numerical efforts to solve Eqn. (5.222) are currently in progress. For the problem of condensation/evaporation alone, Tsang and Huang [1990] have successfully analyzed a binary aerosol using the method of Section 5.20. The details of their development, like for most finite element methods of this level, are quite tedious, and we will not discuss them here.

Quite a different approach to the problem has been taken in the group (sectional) method. Gelbard and Seinfeld [1978a,b, 1979, 1980], assume that:

$$\overline{K}(v_1, v_2, ..., v_m; u_1, u_2, ..., u_m, w_1, w_2, ..., w_m) \Rightarrow \hat{\overline{K}}(v; u, w) \tag{5.223}$$

where:
$$v = \sum_i v_i, \text{ etc.} \tag{5.224}$$

and $\hat{\overline{K}}$ is some coagulation kernel that is independent of particle composition. Next, the distribution function, $n(v_1, v_2, ..., v_m, t)$, is related to a distribution, $\tilde{n}(v, t)$, of the total mass via the assumption that in each section, ℓ, the particle composition is independent of particle size. There are also some other auxiliary assumptions. These assumptions allow Eqn. (5.222), with coagulation alone, to be written in the form:

$$\frac{d}{dt} Q_{\ell k}(t) = \frac{1}{2} \sum_{i=1}^{\ell-1} \sum_{j=1}^{\ell-1} \left(\overline{\beta}_{ij\ell}^{(1a)} Q_{jk}(t) Q_i(t) + \overline{\beta}_{ij\ell}^{(1b)} Q_{ik}(t) Q_j(t) \right) + \sum_{i=1}^{\ell-1} \overline{\beta}_{i\ell}^{(2)} Q_\ell(t) Q_{ik}(t)$$

$$- \sum_{i=1}^{\ell-1} \overline{\beta}_{i\ell}^{(3)} Q_i(t) Q_{\ell k}(t) - \frac{1}{2} \overline{\beta}_{\ell\ell}^{(4)} Q_\ell^2(t) - Q_\ell(t) \sum_{i=\ell+1}^{m} \overline{\beta}_{i\ell}^{(5)} Q_i(t) + S_{\ell k}(t) \tag{5.225}$$

where $Q_{\ell k}$ is the mass concentration of species k in the ℓth section and:

$$Q_\ell = \sum_k Q_{\ell k} \tag{5.226}$$

The coefficients, $\overline{\beta}$, are similar to those for the single species aerosols, and are given by:

$$\overline{\beta}_{ij\ell}^{(1a)} = \int_{v_{i-1}}^{v_i} dv \int_{v_{j-1}}^{v_j} dv' \, \eta(v_{\ell-1} < v + v' < v_\ell) \, v \, g_{ij}(v, v')$$

$$\overline{\beta}_{ij\ell}^{(1b)} = \int_{v_{i-1}}^{v_i} dv \int_{v_{j-1}}^{v_j} dv' \, \eta(v_{\ell-1} < v + v' < v_\ell) \, v' \, g_{ij}(v, v')$$

$$\overline{\beta}_{i\ell}^{(2)} = \int_{v_{i-1}}^{v_i} dv \int_{v_{\ell-1}}^{v_\ell} dv' \, \eta(v + v' < v_\ell) \, v \, g_{i\ell}(v, v')$$

$$\overline{\beta}_{i\ell}^{(3)} = \int_{v_{i-1}}^{v_i} dv \int_{v_{\ell-1}}^{v_\ell} dv' \, \eta(v + v' > v_\ell) \, v' \, g_{i\ell}(v, v')$$

$$\overline{\beta}_{\ell\ell}^{(4)} = \int_{v_{\ell-1}}^{v_\ell} dv \int_{v_{\ell-1}}^{v_\ell} dv' \, \eta(v + v' > v_\ell) (v + v') \, g_{\ell\ell}(v, v')$$

$$\overline{\beta}_{i\ell}^{(5)} = \int_{v_{i-1}}^{v_i} dv \int_{v_{\ell-1}}^{v_\ell} dv' \, v' \, g_{i\ell}(v, v') \tag{5.227}$$

The functions, g, are same as those defined earlier, and there is little difference from the coefficients defined for a single component aerosol. This circumstance is a direct consequence of a rather severe approximation to the physics of the problem.

Methods of Solving the Dynamic Equation

We should also note here that Kim and Seinfeld [1990] have extended the moving sectional method to mixed aerosol condensation/evaporation problem. They have presented interesting results retaining several simplifying assumptions.

5.22 Spatial inhomogeneities

In the absence of strong turbulence, the assumption of spatial homogeneity is not justified. This is especially true for time scales of localized aerosol sources in large chambers. Thus, it is important to consider the spatial terms in the GDE. The task can be carried out in several different ways:

> i) Zonal Methods: Here the spatial domain is divided into fairly large subvolumes (say 4 to 5 'zones') and the aerosol currents among the various zones are allowed to interflow. This approach has been used by Jordan *et al.* [1980] and Simpson *et al.* [1989]. Several useful results have been obtained.

> ii) Finite Differencing or Finite Elements on Space: Here, using any of the methods of the previous sections, the GDE is reduced to a set of partial differential equations (instead of ODEs). These are then solved by standard finite difference or finite element methods (*e.g.*, Park and Loyalka [1989]).

In principle, these extensions are quite straightforward. The calculations, however, are both tedious and computationally quite expensive. But, for many applications such computations will be quite necessary.

5.23 Computer programs and benchmarking

A number of computer programs have been written that attempt to solve various versions of the dynamical equations by using some of the methods described above. Many of these programs have undergone considerable evolution over the past several years. In general, for prescribed forms of the rate processes, and for single component aerosols, some of these programs do provide results with good numerical accuracy. We review briefly the benchmark results first, and then we comment on the areas where considerable caution must be exercised in the use of these programs.

The existing programs can provide useful results when one species is dominant, the particles are nearly spherical, the aerosol is homogeneous, temporal variations are not large, and the coupling with the surroundings is weak. Early assessment of the J-space transform technique has been given by Middleton and Brock [1976] who showed that for weak condensation and coagulation this method worked quite well. Similar conclusions are evident from the work of Gelbard and Seinfeld [1978a,b, 1979, 1980] regarding the sectional method. Emami and Loyalka [1980] applied the J-space method to a wider range of benchmark problems and again found that for weak processes this method worked quite well. Park and Loyalka [1988a] developed a computer program (AEROMECH) that provides for use of both the J-space and sectional methods and allows different models for rate processes. They found that the J-space transform method worked better for condensation dominated problems while the sectional method had advantages in coagulation dominated problems. Thus, it was suggested that a hybrid of the two might be advantageous in some circumstances. Seigneur *et al.* [1986] compared results based on the J-space method (COAGUL and CONFEMM programs), the sectional method (MAEROS and ESMAP programs), and the moments method (AGRO program), but did not verify these against known analytical results.

Tsang and Rao [1988] considered the condensation problem with:

Table 5.3: Comparison of computing time, number and mass concentration between different numerical schemes. Adapted from Tsang and Rao [1988] with permission of the Elsevier Science Publishing Company.

Numerical scheme	Case 1 $a=10^{-10}$, $M_1=5\times10^{-9}$ gm/cm³ $\sigma_g=1.15$, $d_g=10^{-5}$ cm			Case 2 $a=10^{-13}$, $M_1=10^{-9}$ gm/cm³ $\sigma_g=2.24$, $d_g=6\times10^{-5}$ cm		
	$M_0\times10^{-6}$	$M_1\times10^9$	C.P.U.	$M_0\times10^{-6}$	$M_1\times10^9$	C.P.U
Initial	8.7957	5.0		6.3873	1.0	
Analytical	8.7956	70.609		6.3873	1.342	
Section. (40)	8.8178	69.659	5.8	6.4033	1.3411	2.4
Section. (20)	8.8840	68.912	2.9	6.4513	1.3413	1.9
Smolar. (40)	8.7957	67.186	82.6	6.3873	1.3429	8
Smolar. (20)	8.7957	66.154	33.3	6.3873	1.3463	4.4
Upwind (40)	8.7957	72.346	7.2	6.3873	1.3526	3
Upwind (20)	8.7957	74.624	4.2	6.3873	1.3654	2
CONFEMM (40)	8.7847	70.522	5	6.3888	1.3421	6.1
CONFEMM (20)	8.7517	70.259	2.8	6.3933	1.3431	3.2

Numbers in parentheses refer to number of grid intervals per decade of particle diameter. The units of M_0 and M_1 are #/cm³ and g/cm³, respectively. C.P.U. time is in seconds on an IBM 3081.

and:
$$I(v,t) = a\, v^{1/3} \tag{5.228}$$

$$n(d_p,0) = \frac{N}{\sqrt{2\pi}\,\ln(\sigma_g)} \exp\left[-\frac{\left(\ln(d_p)-\ln(d_g)\right)^2}{2\left(\ln(\sigma_g)\right)^2}\right]\frac{1}{d_p} \tag{5.229}$$

They studied both the accuracy and computational time requirements of four different methods for both strong and weak condensation. The problem parameters, the results for total number and mass concentrations, and the cpu times are shown in Table 5.3. The computed distributions are shown in Figs. 5.13 and 5.14. It is evident that, for strong condensation, only a moving grid method that is based on the use of characteristics provides acceptable results for the distribution. Accuracy in the computation of the moments does not assure accuracy in the computation of the distribution itself.

As we noted earlier, the sectional method can be modified so that the sections are moved along the characteristics. This seems to be an idea inherent also in the work of Neiburger and Chien [1960]. The modified method (Gelbard [1990]) is then quite suitable for problems of condensation alone.

Use of the available programs for analysis or design of experiments or natural/industrial aerosol processes must be undertaken with considerable care. First, it is important to select a method and rate process models appropriate to the problem as the computed results are dependent on these (Loyalka [1983] and Buckley and Loyalka [1990]). Next, little trust can be placed in the available multicomponent aerosol codes because of the rather drastic approximations that have been employed, both with respect to the physical processes and the numerics. No doubt, the situation is being rectified and one can be optimistic about new understandings as well as emergence of a new generation of software, especially with the recent advances in hardware, theoretical and experimental methods, and computational fluid mechanics.

Methods of Solving the Dynamic Equation

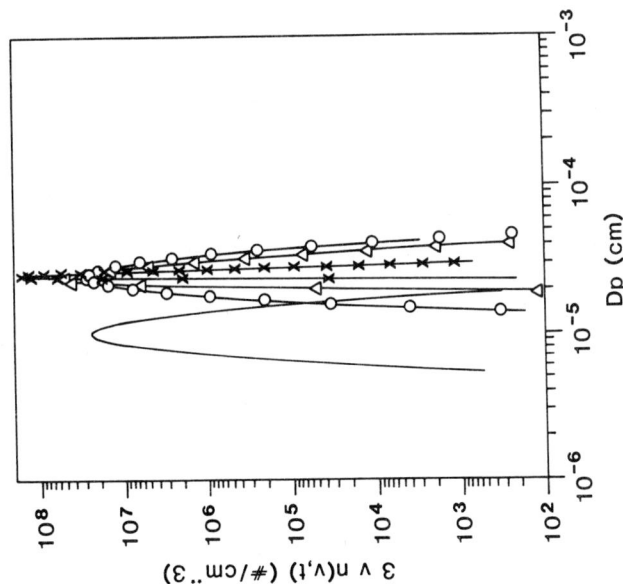

Figure 5.14: The number distribution function, $3vn(v)$ for different numerical schemes after 10,000 seconds for Case 2. The solid curve on the left is the initial size distribution. –x–x– CONFEMM and analytical solution (x x : analytical solution, line: CONFEMM); –△–△– Smolarkiewicz solution; –O–O– upwind differencing and sectional method (O O : upwind differencing, line: sectional method). Adapted from Tsang and Rao [1988] with permission of the Elsevier Science Publishing Company.

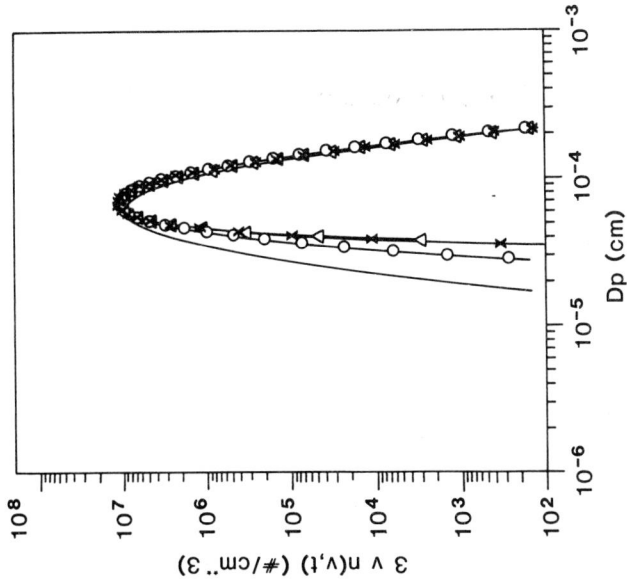

Figure 5.13: The number distribution function, $3vn(v)$ for different numerical schemes after 5 seconds for Case 1. The solid curve on the left is the initial size distribution. –x–x– CONFEMM and analytical solution (x x : analytical solution, line: CONFEMM); –△–△– Smolarkiewicz solution; –O–O– upwind differencing and sectional method (O O : upwind differencing, line: sectional method). Adapted from Tsang and Rao [1988] with permission of the Elsevier Science Publishing Company.

References: Chapter 5

Abramowitz, M. and Stegun, I.A., editors [1973] Handbook of Mathematical Functions (Dover, New York, N.Y.).

Aitchison, J. and Brown, J.A.C. [1957] The Log-Normal Distribution (Cambridge University Press, Cambridge, U.K.).

Berry, E.X. [1967] "Cloud droplet growth by condensation," *J. Atmos. Sci.* **24**, 688.

Brock, J.R. [1972] "Condensational growth of atmospheric aerosols," *J. Colloid Interface Sci.* **39**, 32.

Brock, J.R. [1983] "New aspects of aerosol growth processes," *Aerosol Sci. and Tech.* **2**, 109.

Buckley, R.L. and Loyalka, S.K. [1990] "Aerosol evolution in reactor containments: role of gravitational collision efficiency," *Trans. Am. Nucl. Soc.* **61**, 254.

Deirmendjian, D. [1969] Electromagnetic Scattering on Spherical Polydispersions (Elsevier, New York, N.Y.).

Drake, R.L. [1972] "A General Mathematical Survey of the Coagulation Equation," in Topics in Current Aerosol Research, Vol. 3, part 2, Hidy, G.M. and Brock, J.R., editors (Pergamon, Oxford, U.K.).

Emami, A. [1980] "Effect of Condensation on Aerosol Behavior in Post-hcda LMFBR Containment" (University of Missouri-Columbia) Ph.D. Thesis.

Emami, A. and Loyalka, S.K. [1981] "Role of condensation in aerosol source term for liquid-metal fast breeder reactor containment," *Nucl. Tech.* **52**, 162.

Ernst, M.H. [1981] "Nonlinear model-Boltzmann equations and exact solutions," *Phys. Rep.* **78**, 1.

Ferziger, J.H. [1981] Numerical Methods for Engineering Applications (Wiley, New York, N.Y.).

Friedlander, S.K. and Wang, C.S. [1966] "The self-preserving particle size distribution for coagulation by Brownian motion," *J. Colloid Interface Sci.* **22**, 126.

Fuchs, N.A. [1959] Evaporation and Droplet Growth in Gaseous Media (Pergamon, New York, N.Y.).

Fuchs, N.A. and Sutugin, A.G. [1971] "High-Dispersed Aerosols," in Topics in Current Aerosol Research, Vol. 2, Hidy, G.M. and Brock, J.R., editors (Pergamon, Oxford, U.K.).

Gelbard, F.M. [1990] "Modeling multicomponent aerosol particle growth by vapor condensation," *Aerosol Sci. and Tech.* **12**, 399.

Gelbard, F.M. and Seinfeld, J.H. [1978a] "Numerical solution of the dynamic equation for particulate systems," *J. Comput. Phys.* **28**, 357.

Gelbard, F.M. and Seinfeld, J.H. [1978b] "Coagulation and growth of a multicomponent aerosol," *J. Colloid Interface Sci.* **63**, 472.

Gelbard, F.M. and Seinfeld, J.H. [1979] "Exact solutions of the general dynamic equation for aerosol growth by condensation," *J. Colloid Interface Sci.* **68**, 173.

Gelbard, F.M. and Seinfeld, J.H. [1980] "Simulation of multicomponent aerosol dynamics," *J. Colloid Interface Sci.* **78**, 485.

Gieseke, J.A., Lee, K.W. and Reed, L.D. [1978] HAARM-3 User's Manual (Battelle Memorial Institute) Report # BMI-NUREG-1991.

Hidy, G.M. [1965] "On the theory of the coagulation of noninteracting particles in Brownian motion," *J. Colloid Sci.* **20**, 123.

Hubner, R.S., Vaughn, E.V. and Banume, L. [1973] HAA-3 User's Report (Atomics International) Report # AI-AEC-13038.

Jordan, H., Schumaker, P.M., Gieseke, J.A. and Lee, K.W. [1980] "Multiple Zone Aerosol Behaviour Model," Report # NUREG/CR-1294, BMI-2042/R7.

Kim, Y.P. and Seinfeld, J.H. [1990] "Simulation of multicomponent aerosol condensation by moving sectional method," *J. Colloid Interface Sci.* **135**, 185.

Lai, F.S., Friedlander, S.K., Pich, J. and Hidy, G.M. [1972] "The self-preserving particle size distribution for Brownian coagulation in the free-molecule regime," *J. Colloid Interface Sci.* **39**, 395.

Lee, K.W., Chen, H. and Gieseke, J.A. [1984] "Log-normally preserving size distribution for Brownian coagulation in the free-molecule regime," *Aerosol Sci. and Tech.* **3**, 53.

Lee, K.W., Gieseke, J.A. and Reed, L.D. [1979] "Sensitivity Analysis of the HAARM-3 Code" (Battelle Memorial Institute) Report # BMI-2008, NUREG/CR-0527.

Loyalka, S.K. [1983] "Mechanics of aerosols in nuclear reactor safety: a review," *Prog. Nuc. Energy* **12**, 1.

Loyalka, S.K. and Park, J.W. [1988] "Aerosol growth by condensation: a generalization of Mason's formula," *J. Colloid Interface Sci.* **125**, 712.

Lushnikov, A.A. [1973] "Evolution of coagulating systems," *J. Colloid Interface Sci.* **45**, 549.

Lushnikov, A.A. [1974] "Evolution of coagulating systems II, asymptotic size distributions and analytical properties of generating functions," *J. Colloid Interface Sci.* **48**, 400.

Lushnikov, A.A. [1976] "Evolution of coagulating systems III, coagulating mixtures," *J. Colloid Interface Sci.* **54**, 94.

McLeod, J.B. [1962a] "On an infinite set of nonlinear differential equations," *Q. J. Math.* **13**, 119.

McLeod, J.B. [1962b] "On an infinite set of nonlinear differential equations (II)," *Q. J. Math.* **13**, 193.

Middleton, P. and Brock, J.R. [1976] "Simulation of aerosol kinetics," *J. Colloid Interface Sci.* **54**, 249.

Mountain, R.D. and Mulholland, G.W. [1984] "Stochastic Dynamics Simulation of Particle Aggregation," in <u>Kinetics of Aggregation and Gelation</u> (Elsevier, New York, N.Y.) pp. 83-86.

Mulholland, G.W. and Baum, H.R. [1980] "Effect of initial size distribution on aerosol coagulation," *Phys. Rev. Lett.* **45**, 761.

Neiburger, M. and Chien, C.W. [1960] "Computations of the Growth of Cloud Drops by Condensation Using an Electronic Digital Computer," in <u>Physics of Precipitation</u>, Weickemann, H. and Smith, W.E., editors (Geographical Monograph Series) Vol. 5, p. 191.

Park, J.W. and Loyalka, S.K. [1988a] "Vapor condensation on aerosols: numerical simulation," *Trans. Am. Nucl. Soc.* **56**, 287.

Park, J.W. and Loyalka, S.K. [1988b] "Kinetic theory of gelation: numerical simulation and comparison with analytical results," *J. Colloid Interface Sci.* **125**, 615.

Park, J.W. and Loyalka, S.K. [1989] "Role of spatial inhomogeneities in source term aerosol dynamics," *Nucl. Sci. and Eng.* **101**, 269.

Pich, J., Friedlander, S.K. and Lai, F.S. [1970] "The self-preserving particle size distribution for coagulation by Brownian motion - III. Smoluchowski coagulation and simultaneous Maxwellian condensation," *J. Aerosol Sci.* **1**, 115.

Ramabhadran, T.E., Peterson, T.W. and Seinfeld, J.H. [1976] "Dynamics of aerosol coagulation and condensation," *A.I.Ch.E. J.* **22**, 840.

Ramsdale, S.A. [1986] <u>AEROSIM-S User Manual</u> (Safety and Reliability Directorate, UKAEA) Report # SRDR-832.

Schumann, T.E.W. [1940] "Theoretical aspects of the size distribution of fog particles," *Q. J. Roy. Meteor. Soc.* **66**, 195.

Scott, W.T. [1968] "Analytic studies of cloud droplet coalescence I," *J. Atmos. Sci.* **25**, 54.

Seigneur, C., Hudishweskyj, A.B., Seinfeld, J.H., Whitby, K.T., Whitby, E.R., Brock, J.R. and Barnes, H.M. [1986] "Simulation of aerosol dynamics: a comparative review of mathematical models," *Aerosol. Sci. and Tech.* **5**, 205.

Simons, S. [1981] "The coagulation and deposition of radioactive aerosols," *Ann. Nucl. Energy* **8**, 287.

Simons, S. [1982] "The condensation, coagulation and deposition of a multicomponent radioactive aerosol," *Ann. Nucl. Energy* **9**, 473.

Simons, S. [1983] "On the conservation of volume during particle coagulation," *J. Phys. A Math. Nucl. Gen.* **16**, L81.

Simpson, D.R., Williams, M.M.R. and Simons, S. [1989] "Modeling of an aerosol in coupled chambers," *Nucl. Sci. and Eng.* **101**, 259.

Smolarkiewicz, P.K. [1983] "A simple positive definite advection scheme with small implicit diffusion," *Monthly Weather Rev.* **111**, 479.

Smoluchowski, M. von [1916] "Drei vortrage uber diffusion, Brownsche molekularbewegung und koagulation von kolloidteilchen." The title is translated as: "Three lectures on diffusion, Brownian motion and coagulation of colloid particles," *Phys. Z.* **17**, 557 (in German).

Sneddon, I.N. [1957] <u>Elements of Partial Differential Equations</u> (McGraw Hill, New York, N.Y.).

Suck, S.H. and Brock, J.R. [1979] "Evolution of atmospheric aerosol particle size distributions via Brownian coagulation: numerical simulation," *J. Aerosol Sci.* **10**, 581.

Swift, D.L. and Friedlander, S.K. [1964] "The coagulation of hydrosols by Brownian motion and laminar shear flow," *J. Colloid Sci.* **19**, 621.

Tsang, T.H. and Brock, J.R. [1983] "Simulation of condensation aerosol growth by condensation and evaporation," *Aerosol Sci. and Tech.* **2**, 311.

Tsang, T.H. and Huang, L.K. [1990] "On a Petrov-Galerkin finite element method for evaporation of polydisperse aerosols," *Aerosol Sci. and Tech.* **12**, 578.

Tsang, T.H. and Rao, A. [1988] "Comparison of different numerical schemes for condensational growth of aerosols," *Aerosol Sci. and Tech.* **9**, 271.

Tsang, T.H. and Rao, A. [1990] "A moving finite element method for the population balance equation," *Int. J. Numer. Methods Fluids* **10**, 001.

Varoḡlu, E. and Finn, W.D.L. [1980] "Finite elements incorporating characteristics for one-dimensional diffusion-convection equation," *J. Comput. Phys.* **34**, 371.

Wang, C.S. and Friedlander, S.K. [1967] "The self-preserving particle size distribution for coagulation by Brownian motion, II," *J. Colloid Interface Sci.* **24**, 170.

Williams, M.M.R. [1981] "Some exact and approximate solutions of the nonlinear Boltzmann equation with applications to aerosol coagulation," *J. Phys. A Math. Nucl. Gen.* **14**, 2073.

Williams, M.M.R. [1983] "The time-dependent behaviour of aerosols with growth and deposition," *J. Colloid Interface Sci.* **93**, 252.

Williams, M.M.R. [1984] "On some exact solutions of the space- and time-dependent coagulation equation for aerosols," *J. Colloid Interface Sci.* **101**, 19.

Williams, M.M.R. [1986] "Some topics in nuclear aerosol dynamics," *Prog. Nuc. Energy* **17**, 1.

CHAPTER 6

Condensation and Evaporation

6.1 Introduction

The most important aspect of aerosol particles is, perhaps, their interaction with gas molecules. Under standard conditions, a current of 10^{23} gas molecules/cm^{2}sec impinges on a particle and, in equilibrium, a current of the same magnitude is returned back from the particle's surface to the surrounding gas. Generally, the molecules incident on a particle's surface can react with the particle constituents, they can be absorbed or adsorbed, or they can be scattered back into the gas. The problems of acid rain (in which sulphur and nitrogen oxides react with aqueous particles or other fine and coarse particle constituents), growth of rain drops from early cluster formation to later larger sizes, and the absorption of I and Cs vapors by particles during a nuclear reactor accident are all examples of particle interactions with gas constituents.

The magnitudes of the rate processes in an aerosol will be governed by the partial pressures of the gas constituents, their molecular properties, the particle constituents, the particle sizes, and the ambient temperature. If the partial pressure of a vapor is several times its saturation value then this highly nonequilibrium situation will lead to the formation of clusters of vapor molecules (in the absence of other particles on which the vapor could condense). The growth of these clusters through vapor condensation will progress to equilibrium in which the incoming and outgoing molecular currents at the particle (cluster) surfaces are the same (assuming that the vapor supply is limited). Conversely, rapid expansion of an aerosol, or heating of the particles (say by a laser beam) can lead to rapid vaporization of particles.

In general, treatment of particle growth/evaporation/reactions requires consideration of heat and mass transfer between a particle and the ambient gas mixture. For most problems of interest in aerosol mechanics, the vapor or reactant species concentration is quite small compared to that of the host gas and the mass transfer part can be treated by regarding the gas as dilute (vapor/reactant molecules transport through their collisions with the host molecules and aerosol particles only with vapor-vapor interactions ignored). The heat transfer, which affects the mass transfer through its impact on temperature, is then largely governed by the collisions of the host gas molecules amongst themselves. Convective heat and mass transfer, or radiative heat transfer will also play significant roles under some conditions.

In this chapter, we focus upon rate expressions for the condensation and evaporation of a particle in a dilute mixture where mass and heat transport are governed by diffusion and conduction. We will first consider spherical particles. Starting with some elementary considerations, we will eventually describe recent progress in solving the Boltzmann equation. Finally, we will consider the case of nonspherical particles.

Before proceeding with the above plan, we note that under most conditions of interest (see Seinfeld [1986], Hidy and Brock [1971], or Twomey [1977]), particle growth kinetics are such that rate expressions can be determined by solving the steady state equations of molecular transport in the gas. Also, the equilibrium (apparent saturation) density of the vapor over the particle (droplet) surface is a function of the particle size (the Kelvin effect), the particle constituents, and the gas-surface interaction properties. We will not discuss these aspects and only assume that the "equilibrium" density, n_s, can be expressed as:

$$n_s = n_s(T, a, c_i, ...) \qquad (6.1)$$

where T is the surface temperature, a is the particle radius, c_i is the concentration of species i, *etc.* In much of what follows we will assume that the functional form of n_s is known and can be determined through the expressions available in the literature (*e.g.*, Twomey [1977], Mason [1962], Davies [1966], Pruppacher and Klett [1978]) or through experiments. We note that for condensation of water vapor, which is an important factor in light water reactor accidents, the hygroscopic nature of the aerosols would reduce the value of n_s to levels such that aerosol growth would occur even in subsaturated environments. Thus, it is imperative that n_s be known, both experimentally and theoretically, to a fairly high degree of accuracy.

6.2 Basic equations

We consider a spherical particle situated in a dilute gas mixture in which a vapor or a gas (component 1) is diffusing in a background gas (component 2). The number density, n_1, of the vapor is much less than that of the background gas, n_2. Far from the sphere, the vapor is at a density, $n_\infty > n_s$. The vapor will condense on the particle and the temperature of the particle will be higher than that of the gas to enable transfer of heat. One might also encounter a situation where a particle is hot due to internal heat generation (*e.g.*, radioactive particles) or radiative absorption. Most often, one is concerned about conductive (molecular) heat transfer, although radiative and convective heat transport might also be of interest.

If the concentration and thermal gradients are small, then the basic equations of mass and heat transfer can be considered separately instead of as coupled calculations which we postpone to a later stage (temperature affects saturation density, n_s, through the Clausius-Clapeyron equation). Thus, let us first consider the isothermal mass transport of the vapor. The mass condensation rate, $\tilde{u}$ (mass/time), of vapor condensing on a sphere, can be expressed as:

$$\tilde{u} = 4\pi \tilde{a}^2 \, m_v \, (J_- - J_+) \qquad (6.2)$$

where J_- is the current (#/cm²sec) of the vapor molecules directed towards the particle and J_+ is the outward current of the vapor molecules emitted from the particle. These radial currents are defined as:

$$J_- = \int_{(\tilde{c}\cdot n_r)<0} d\tilde{c} \, |\tilde{c} \cdot n_r| \, f_v^-(\tilde{a},\tilde{c}) \qquad (6.3)$$

and:

$$J_+ = \int_{(\tilde{c}\cdot n_r)>0} d\tilde{c} \, |\tilde{c} \cdot n_r| \, f_v^+(\tilde{a},\tilde{c}) \qquad (6.4)$$

where **r** is the position vector, **c** is the molecular velocity, and f_v is the molecular distribution of the vapor. The superscripts + and − indicate the distributions for outgoing and incoming molecules, respectively. Note that n_r is a unit normal at **r**, directed from the sphere into the vapor-gas mixture.

For diffuse reflection, f_v^+ is simply a Maxwellian corresponding to n_s and the ambient temperature. f_v^-, however, must generally be determined by solving the linear Boltzmann equation. It is convenient to introduce here the Knudsen number, Kn, defined as:

$$Kn = \ell_D / \tilde{a} \qquad (6.5)$$

in which ℓ_D is the vapor molecular mean free path:

Condensation and Evaporation

$$\ell_D = 2D\left(\frac{m_v}{2kT}\right)^{1/2} \tag{6.6}$$

and D is the diffusion coefficient of the vapor in the gas. For Kn>>1, the incident distribution, f_v^-, seen by the particle is essentially a Maxwellian and we have:

$$J_- = \frac{n_\infty \bar{v}}{4} \tag{6.7}$$

where:

$$\bar{v} = \left(\frac{8kT}{\pi m_v}\right)^{1/2} \tag{6.8}$$

is the mean speed of the vapor molecules. Since:

$$J_+ = \frac{n_s \bar{v}}{4} \tag{6.9}$$

we have:

$$\lim_{Kn \gg 1}(\tilde{u}) = 4\pi \tilde{a}^2 \, m_v \frac{\bar{v}}{4}(n_\infty - n_s) \tag{6.10}$$

Denoting this limiting value by, $\tilde{u}_{fm}$, the free molecular condensation rate, we have:

$$\tilde{u}_{fm} = \pi \tilde{a}^2 \, (\rho_\infty - \rho_s)\left(\frac{8kT}{\pi m_v}\right)^{1/2} \tag{6.11}$$

in which $\rho = m_v n$ is the vapor density.

In the continuum limit, Kn<<1, one writes:

$$\tilde{u}_{cont} = -4\pi \tilde{a}^2 \, m_v \, J \tag{6.12}$$

where J is the net radial current:

$$J = J_+ - J_- \tag{6.13}$$

and is given by Fick's law as:

$$J = -D\frac{dn}{dr} \tag{6.14}$$

n(r) in Eqn. (6.14) is determined from the equation:

$$\frac{1}{r^2}\frac{d}{dr}\left(r^2 \frac{d}{dr}\right)n(r) = 0 \tag{6.15}$$

with the boundary conditions:

and:
$$n(r) = n_s \; ; \; r = \tilde{a}$$
$$\lim_{r \to \infty}[n(r)] = n_\infty$$

One obtains:

$$\tilde{u}_{cont} = 4\pi D \tilde{a} (\rho_\infty - \rho_s) \qquad (6.16)$$

For all intermediate Kn, however, $\tilde{u}$ must be obtained by solving the Boltzmann equation (Note that actually D is also obtained from the Chapman-Enskog solution to this equation). In the near continuum limit (Kn~0.1) however, fairly accurate results are obtained by considering the equations:

$$\frac{1}{r^2}\frac{d}{dr}\left(r^2\frac{d}{dr}\right)n(r) = 0$$

$$n(r) - n_s = \tilde{\zeta}\frac{dn}{dr} \; ; \; r = \tilde{a}$$

and:
$$\lim_{r \to \infty}[n(r)] = n_\infty \qquad (6.17)$$

where $\tilde{\zeta}$ is known as the jump coefficient (it is analogous to the velocity slip of Chapter 2). This coefficient is obtained by solving the Boltzmann equation and its evaluation has been discussed by Loyalka [1982b] and Loyalka, Hamoodi, and Tompson [1989a].

Solving Eqn. (6.17) for n(r), and using Eqns. (6.12) and (6.14), we find that:

$$\tilde{u}_{jump} = 4\pi D \tilde{a} (\rho_\infty - \rho_s)\left(1 - \frac{\tilde{\zeta}}{\tilde{a}}\right) \qquad (6.18)$$

That is, the condensation rate does not follow the prescription of Eqn. (6.16) but is reduced from the continuum value by a factor that accounts for the discontinuity of the molecular velocity distribution. We can write:

$$\tilde{u} = \tilde{u}_{cont}(1 - Kn\,\zeta) \qquad (6.19)$$

where $\zeta = \tilde{\zeta}/\ell_D$ is the dimensionless jump coefficient which varies from about (0.94) to (1.01) as the mass ratio of the background to vapor gas molecules:

$$\beta^2 = m_b/m_v \qquad (6.20)$$

varies from $\beta^2 \gg 1.0$ to $\beta^2 \ll 1.0$.

Very similar results are obtained for the conductive heat transfer in the background gas. The rate of heat transfer, $\tilde{q}$ (energy/time) is expressed as:

Condensation and Evaporation

$$\tilde{q} = 4\pi \tilde{a}^2 (Q_+ - Q_-) \tag{6.21}$$

where Q_+ is the conductive heat current (energy/area time) directed away from the particle and Q_- is the inward heat current. These are expressed as:

$$Q_- = \int_{(\tilde{c} \cdot n_r)<0} d\tilde{c} \, |\tilde{c} \cdot n_r| \tfrac{1}{2} m \tilde{c}^2 \, f_b^- (\tilde{a}, \tilde{c}) \tag{6.22}$$

and:

$$Q_+ = \int_{(\tilde{c} \cdot n_r)>0} d\tilde{c} \, |\tilde{c} \cdot n_r| \tfrac{1}{2} m \tilde{c}^2 \, f_b^+ (\tilde{a}, \tilde{c}) \tag{6.23}$$

subject to the condition that the background gas has no net flow, specifically:

$$\int d\tilde{c} \, (\tilde{c} \cdot n_r) \, f_b(r, c) = 0 \tag{6.24}$$

We consider the case of diffuse reflection at the particle-gas interface. In the free molecular limit:

$$f_b^- = n_{b\infty} \left(\frac{m_b}{2\pi k T_\infty} \right)^{3/2} \exp\left(-\frac{m_b \tilde{c}^2}{2 k T_\infty} \right)$$

and:

$$f_b^+ = n_{bw} \left(\frac{m_b}{2\pi k T_w} \right)^{3/2} \exp\left(-\frac{m_b \tilde{c}^2}{2 k T_w} \right) \tag{6.25}$$

where n_{bw} is determined from Eqn. (6.24). One finds that:

$$n_{bw} \approx n_{b\infty} \left(1 - \tfrac{1}{2} \frac{T_w - T_\infty}{T_\infty} \right) \tag{6.26}$$

and we can write:

$$f_b^+ = f_b^- \left(1 - \frac{T_w - T_\infty}{T_\infty} \right) \tag{6.27}$$

Thus, in the free molecular limit, the heat transfer rate is given by:

$$\tilde{q}_{fm} = 4\pi \tilde{a}^2 \tfrac{1}{2} p_\infty \left(\frac{8 k T_\infty}{\pi m_b} \right)^{1/2} \frac{T_w - T_\infty}{T_\infty} \tag{6.28}$$

Again, using continuum theory, we find that in this limit:

$$\tilde{q}_{cont} = 4\pi \tilde{a} \lambda (T_w - T_\infty) \tag{6.29}$$

where λ is the thermal conductivity of the gas. Note that in the jump limit:

$$T_w - \tilde{T}_{asy}(0) = \tilde{\zeta}_{TJ} \frac{d}{d\tilde{r}} \tilde{T}_{asy}(\tilde{r}) \tag{6.30}$$

where $\tilde{\zeta}_{TJ}$ is the temperature jump distance (Loyalka [1989]).

As before, one finds:

$$\tilde{q}_{jump} = \tilde{q}_{cont}\left(1 - \frac{\tilde{\zeta}_{TJ}}{\tilde{a}}\right) = \tilde{q}_{cont}\left(1 - Kn\,\zeta_{TJ}\right) \tag{6.31}$$

where now, the Knudsen number is defined as:

$$Kn = \ell_t / \tilde{a} \tag{6.32}$$

and ℓ_t is a mean free path:

$$\ell_t = \frac{4}{5} \frac{\lambda T_\infty}{p_\infty}\left(\frac{m_b}{2kT_\infty}\right)^{1/2} \tag{6.33}$$

Several theoretical results have been reported for both $\tilde{u}$ and $\tilde{q}$ in the transition regime. These include those of Loyalka [1973, 1982b], Williams [1975], Cercignani and Pagani [1967], Monchick and Blakemore [1988], Tompson and Loyalka [1988], and Pazooki and Loyalka [1988].

Recently, significant progress has been made in the direct numerical solution of the Boltzmann equation. We will discuss this work in the next section. Before doing so, we note that for arbitrary Kn it is possible to develop some simple interpolation formulas (Loyalka [1983]):

$$\frac{\tilde{u}}{\tilde{u}_{cont}} = f_c(Kn_c) \quad \text{and} \quad \frac{\tilde{q}}{\tilde{q}_{cont}} = f_h(Kn_h) \tag{6.34}$$

where:

$$f_c(Kn_c) = \left[1 + Kn_c \frac{\frac{\tilde{u}_{cont}}{\tilde{u}_{fm}}\xi_c + \zeta_c}{Kn_c\,\xi_c + 1}\right]^{-1} \tag{6.35}$$

and:

$$f_h(Kn_h) = \left[1 + Kn_h \frac{\frac{\tilde{q}_{cont}}{\tilde{q}_{fm}}\xi_h + \zeta_h}{Kn_h\,\xi_h + 1}\right]^{-1} \tag{6.36}$$

with:

$$Kn_c = \frac{\ell_D}{\tilde{a}} \quad \text{and} \quad Kn_h = \frac{\ell_t}{\tilde{a}} \tag{6.37}$$

The other quantities in Eqns. (6.35) and (6.36) depend on the intermolecular and/or gas-surface interactions. For diffuse reflection, we have:

$$\frac{\tilde{u}_{cont}}{\tilde{u}_{fm}} = \sqrt{\pi} \, Kn_c \quad \text{and} \quad \frac{\tilde{q}_{cont}}{\tilde{q}_{fm}} = \frac{5\sqrt{\pi}}{4} Kn_h \tag{6.38}$$

while the other quantities are of order 1. For the BGK model, these quantities, the dimensionless jump distances, ζ, and the auxiliary factors, ξ, are given as:

$$\zeta_c = (1.0161) \quad ; \quad \zeta_h = (1.1759) \tfrac{5}{8} \pi^{1/2} \quad ; \quad \xi_c = (1.3333) \quad ; \quad \xi_h = (1.9234) \tag{6.39}$$

An expression for the mass transfer jump coefficient, which is dependent on the mass ratio, has been given by Loyalka, Hamoodi and Tompson [1989a]:

$$\zeta_c = (0.9769) - (0.0518)\, z + (0.0018)\, z^2 + (0.0196)\, z^3 \quad ; \quad z = \log_{10}(\beta^2) \quad ; \quad 0.1 \le \beta^2 \le 10.0 \tag{6.40}$$

and may be used in preference to the value (1.0161) given above. Note that the black sphere value:

$$\zeta_c = \left(\frac{3\sqrt{\pi}}{4}\right)(0.7104) \approx (0.94)$$

of the often used Fuchs-Sutugin formula is meaningful only for the diffusion of a very light species in a heavy background gas and, as such, is not appropriate for use in most aerosol applications. Also, as discussed by Tompson and Loyalka [1986], the formulas above do not have correct near free molecular limits, and their derivatives can be substantially in error. Thus, they should be used with caution.

The heat transfer expression given above is most useful only if the vapor occurs in trace amounts and/or if it is a light species. It is known that heavy contaminants lower the heat transfer in a gas substantially. In such cases the heat transfer expression above should be modified, and a result appropriate for gas mixtures should be used. Such an expression has been given by Loyalka [1982b] for plane parallel plates, and can be simply adapted to the case of a spherical droplet by noting that:

$$\tilde{q}_{fm} = 4\pi \tilde{a}^2 \, \tfrac{1}{2} \, \frac{T_w - T_\infty}{T_\infty} \sum_i p_{i\infty} \left(\frac{8 k T}{\pi m_{ib}}\right)^{1/2} \tag{6.41}$$

where i indicates the ith species.

6.3 Transition regime heat and mass transfer

For arbitrary Knudsen number, we consider here the linearized Boltzmann equation for the vapor molecular distribution, f_v:

$$\tilde{c} \cdot \frac{\partial}{\partial \tilde{r}} f_v(\tilde{r},\tilde{c}) = \int \{f_v(\tilde{r},\tilde{c}')\, f_b(\tilde{r},\tilde{c}_1') - f_v(\tilde{r},\tilde{c})\, f_b(\tilde{r},\tilde{c}_1)\} |\tilde{c} - \tilde{c}_1| \, \tilde{\sigma}(|\tilde{c} - \tilde{c}_1|, \theta) \sin(\theta) \, d\theta \, d\varepsilon \, d\tilde{c}_1 \tag{6.42}$$

where f_b is a Maxwellian, $\tilde{\sigma}$ is the collision cross-section, θ is the scattering angle, ε is the angle between the planes $(\tilde{c},\tilde{c}')$ and $(\tilde{c}-\tilde{c}',\tilde{c}-\tilde{c}_1)$, and $\tilde{r}$ and $\tilde{c}$ are the dimensional molecular position and velocity vectors, respectively. We write:

$$f_b(\tilde{r},\tilde{c}) = n_b \left(\frac{m_b}{2\pi kT}\right)^{3/2} \exp\left(-\frac{m_b \tilde{c}^2}{2kT}\right)$$

and:

$$f_v(\tilde{r},\tilde{c}) = n_v \left(\frac{m_v}{2\pi kT}\right)^{3/2} \exp\left(-\frac{m_v \tilde{c}^2}{2kT}\right)(1+h(\tilde{r},\tilde{c}))$$

(6.43)

where $n_v = n_s$ is the saturation density (or the density of the molecular distribution evaporating from the particle surface). Eqn. (6.42) can be written as:

$$c \cdot \frac{\partial}{\partial r} h(r,c) = \varepsilon \, Lh(r,c)$$

(6.44)

where:

$$c = \left(\frac{m_v}{2kT}\right)^{1/2} \tilde{c} \quad \text{and} \quad r = \frac{\tilde{r}}{\ell}$$

(6.45)

and:

$$\varepsilon = \frac{n_b \, \tilde{\sigma} \, \ell}{\sqrt{\pi} \, \beta}$$

(6.46)

$\ell = \ell_D$ is a mean free path and $\tilde{\sigma}$ has dimensions of a collision cross-section.

If we also assume that the molecules interact as rigid spheres, then:

$$\tilde{\sigma} = \frac{\pi}{4}\left(\tilde{d}_b + \tilde{d}_v\right)^2$$

(6.47)

where $\tilde{d}_b$ and $\tilde{d}_v$ are the diameters of the background gas and the vapor molecules, respectively. Now, the dimensionless, linearized, Boltzmann collision operator can be expressed as:

$$Lh(r,c) = -v(c)\, h(r,c) + \int dc_1 \, \exp(-c_1^2) K(c,c_1) h(r,c_1)$$

(6.48)

where:

$$v(c) = \left(2\beta c + \frac{1}{\beta c}\right)\frac{\sqrt{\pi}}{2} \, \text{erf}(\beta c) + \exp(-\beta^2 c^2)$$

(6.49)

and:

$$K(c,c') = \frac{1}{4\pi}\left(\beta + \frac{1}{\beta}\right)^2 \frac{1}{|c-c'|} \exp\left(c'^2 - \frac{\beta^2}{4|c-c'|^2}\left[(c^2-c'^2) - \frac{|c-c'|^2}{\beta^2}\right]^2\right)$$

(6.50)

The error function is defined as:

$$\text{erf}(x) = \frac{2}{\sqrt{\pi}} \int_0^x dt \, \exp(-t^2)$$

(6.51)

The diffusion coefficient, D, is obtained in terms of the Chapman-Enskog function, $\hat{\phi}_d(c)$, which is the solution of the equation:

Condensation and Evaporation

$$L(c\, \hat{\phi}_d(c)) = c \tag{6.52}$$

specifically:

$$D = -\left(\frac{4\beta}{3\, n_b\, \tilde{\sigma}^2}\right)\left(\frac{2\,kT}{m_v}\right)^{1/2} \int_0^\infty dc\, c^4 \exp(-c^2)\, \hat{\phi}_d(c) \tag{6.53}$$

Also, ε can be expressed as:

$$\varepsilon = -\frac{8}{3\sqrt{\pi}} \int_0^\infty dc\, c^4 \exp(-c^2)\, \hat{\phi}_d(c) \tag{6.54}$$

For the future, it is convenient to define a function:

$$\phi_d(c) = \frac{1}{\varepsilon} \hat{\phi}_d(c) \tag{6.55}$$

Solutions of Eqn. (6.52) by Sonine polynomial expansion have been obtained (see Chapman and Cowling [1971], Hirschfelder, Curtiss and Bird [1964], and Ferziger and Kaper, [1972]). Pidduck [1916] solved this equation by Neumann iteration and, most recently, precise results by use of this later technique have been reported by Tompson and Loyalka [1987] for arbitrary values of the mass ratio.

The quantity of primary interest here, the condensation rate, is expressed as:

$$\tilde{u}(r) = -4\pi\, \tilde{r}^2\, m_v \int d\tilde{c}\, (\tilde{c} \cdot \mathbf{n}_r)\, f(\tilde{r}, \tilde{c}) = 4\pi\, \ell\, D\, \rho_s\, \hat{u}(r) \tag{6.56}$$

where:

$$\hat{u}(r) = -\frac{2\, r^2}{\pi^{3/2}} \int dc \exp(-c^2)\, (c \cdot \mathbf{n}_r)\, h(r,c) \tag{6.57}$$

Note that $\tilde{u}(r)$, and hence $\hat{u}(r)$, being the net current over a spherical surface of radius, r, in a source free medium, is constant. This is a consequence of the fact that mass is conserved in collisions, i.e.:

$$L(1) = 0 \tag{6.58}$$

For diffuse reflection, the boundary condition on the surface of the sphere is simply:

$$h(r,c) = 0 \quad ; \quad (c \cdot \mathbf{n}_r) > 0 \tag{6.59}$$

Because of the spherical symmetry of the problem, we consider the polar coordinates, $c=(c,\mu,)$ and $r=r$, and then note that the asymptotic solution of Eqn. (6.44), to $O(1/r)$, can be written as:

$$\lim_{r \to \infty}[h(r,c,\mu)] = \frac{n_\infty - n_s}{n_s} + \hat{u}(r)\frac{1}{r}\left(-1 + \frac{c\mu}{r}\phi_d(c)\right) \tag{6.60}$$

Aerosol Science: Theory and Practice

The boundary value problem posed here has recently been solved by Loyalka, Hamoodi, and Tompson [1989b]. One introduces the functions:

$$\psi(r,c,\mu) = \frac{1}{\hat{u}} \left\{ \frac{n_\infty - n_s}{n_s} - h(r,c,\mu) \right\}$$

$$\sigma(c) = \frac{\epsilon \, v(c)}{c}$$

$$q(r,c,\mu) = \frac{\epsilon}{c} \int dc' \exp(-c'^2) K(c,c') \psi(r,c',\mu') \tag{6.61}$$

in terms of which the problem becomes:

$$\Omega \cdot \frac{\partial}{\partial r} \psi(r,c,\mu) + \sigma(c) \psi(r,c,\mu) = q(r,c,\mu)$$

$$\psi(R,c,\mu) = \frac{1}{\hat{u}(R)} \left[\frac{n_\infty - n_s}{n_s} \right] \quad ; \quad \mu > 0$$

$$\lim_{r \to \infty} [\psi(r,c,\mu)] = \frac{1}{r} \left(1 - \frac{c\mu}{r} \phi_d(c) \right) \tag{6.62}$$

and in which:

$$\hat{u}(R) = \left(\frac{n_\infty - n_s}{n_s} \right) \left[\frac{\sqrt{\pi}}{R^2} - 4 \int_0^\infty dc\, c^3 \exp(-c^2) \int_{-1}^0 d\mu\, \mu\, \psi(R,c,\mu) \right]^{-1} \tag{6.63}$$

The boundary value problem above is actually quite difficult to solve. It involves a hyperbolic partial differential integral equation in spherical geometry with a complicated integral and discontinuous boundary condition. This is, however, a problem of the type encountered in neutron transport and in the design of nuclear explosives. In the latter case, the problem is related to calculation of eigenvalues in a multiplying system of fast neutrons, a circumstance that results in some simplification of the collision term.

The method of solution, the S_N method, is based on discretization of $[r(i),c(j),\mu(m)]$ space, where in each cell an attempt is made to carry out the integration along the direction of neutron travel. Accurate results are obtained by ensuring particle conservation, $L(1)=0$. Details of the approach are discussed in the cited references, and we only note the basic iterative formulas:

for $\mu_m = -1$:

$$\overline{\psi}_{-1j}^{i+1/2} = \frac{(A_{i+1} + A_i) \psi_{-1j}^{i+1} + V_{i+1/2} q_{-1j}^{i+1/2}}{(A_{i+1} + A_i) + V_{i+1/2} \sigma_j} \tag{6.64}$$

for $-1 < \mu < 0$:

$$\overline{\psi}_{mj}^{i+1/2} = \frac{-\mu_m (A_{i+1} + A_i) \psi_{mj}^{i+1} + \frac{1}{\omega_m} \left(a_{m+1}^{i+1/2} + a_{m-1}^{i+1/2} \right) \psi_{m-1j}^{i+1} + V_{i+1/2} q_{mj}^{i+1/2}}{-\mu_m (A_{i+1} + A_i) + \frac{1}{\omega_m} \left(a_{m+1}^{i+1/2} + a_{m-1}^{i+1/2} \right) + V_{i+1/2} \sigma_j} \tag{6.65}$$

Condensation and Evaporation

for $\mu > 0$:

$$\overline{\psi}_{mj}^{i+1/2} = \frac{\mu_m (A_{i+1} + A_i) \psi_{mj}^{i-1} + \frac{1}{\omega_m}\left(a_{m+1}^{i+1/2} + a_{m-1}^{i+1/2}\right) \psi_{m-1j}^{i+1} + V_{i+1/2}\, q_{mj}^{i+1/2}}{\mu_m (A_{i+1} + A_i) + \frac{1}{\omega_m}\left(a_{m+1}^{i+1/2} + a_{m-1}^{i+1/2}\right) + V_{i+1/2}\, \sigma_j}$$

(6.66)

with:

$$\hat{u}(R) = \left(\frac{n_\infty - n_s}{n_s}\right)\left[\frac{\sqrt{\pi}}{R^2} - 4 \sum_{j=1}^{J} c_j^3 \exp(-c_j^2)\, w_j \sum_{m=2,4,\ldots}^{N} \omega_m\, \mu_m\, \psi_{mj}^{1/2}\right]^{-1}$$

(6.67)

The quantities expressed in these formulas are defined by the following expressions:

$$\psi_{mj}^{i} = 2\, \overline{\psi}_{mj}^{i+1/2} - \psi_{mj}^{i+1}$$

$$\psi_{m+1j}^{i+1/2} = 2\, \overline{\psi}_{mj}^{i+1/2} - \psi_{m-1j}^{i+1}$$

$$\sigma_j = \sigma(\beta\, c_j)$$

$$A_i = 4\pi\, r_i^2$$

$$V_{i+1/2} = \frac{4\pi}{3}\left(r_{i+1}^3 - r_i^2\right)$$

$$a_1^{i+1/2} = a_{2N+1}^{i+1/2} = 0$$

$$a_{m+1}^{i+1/2} = a_{m-1}^{i+1/2} - \omega_m\, \mu_m\, (A_{i+1} - A_i)$$

(6.68)

For explicit calculations, one uses the expansion:

$$\int dc'\, \exp(-c'^2)\, Kh = \sum_{\ell=0}^{N} \frac{2\ell + 1}{2} P_\ell(\mu) \int dc'\, c'^2\, \exp(-c'^2)\, k_\ell(c,c') \int_{-1}^{1} d\mu'\, P_\ell(\mu')\, h(x,c',\mu')$$

(6.69)

where $P_\ell(\mu)$ are the Legendre polynomials and $k_\ell(c,c')$ are the expansion coefficients:

$$k_\ell(c,c') = \int_{-1}^{1} d\mu_0\, P_\ell(\mu_0)\, K(c,c')$$

(6.70)

μ_0 is the cosine of the angle between c and c'. Thus, the function, q, is given by:

$$q_{mj}^i = \frac{\varepsilon}{c_j} \sum_{\ell=0}^{L} \frac{2\ell + 1}{2} P_\ell(\mu_m) \sum_{j'=1}^{J} c_{j'}^2\, \exp(-c_{j'}^2)\, w_{j'}\, k_\ell(c_j,c_{j'}) \sum_{m'=2,4,\ldots}^{2N} \omega_{m'}\, \psi_{m'j'}^{i\ell}$$

(6.71)

where:

$$\psi_{mj}^{i\ell} = P_\ell(\mu_m)\tfrac{1}{2}\left(\psi_{mj}^{i-1/2} + \psi_{mj}^{\ell+1/2}\right)$$

(6.72)

and:

$$\psi_{mj}^{i} = \psi_{mj}^{i0}$$

(6.73)

Since $k_\ell(c_j,c_{j'})$ is not well behaved, use of the expression above, without modifications, leads to poor results. Thus, we actually, approximate ψ by (see Loyalka [1989]):

$$\psi(x,c,\mu) = \sum_{j=1}^{J} \psi(x,c_j,\mu) \phi_j(c)$$

(6.74)

where $\phi_j(c)$ are some basis functions. For example, one could choose these to be hat or inverted V functions. This leads to:

$$\int dc' \exp(-c'^2) K \psi \bigg|_{c=c_j} = \sum_{\ell=0}^{N} \frac{2\ell+1}{2} P_\ell(\mu) \sum_{j'=1}^{\infty} k_{\ell jj'} \int_{-1}^{1} d\mu' P_\ell(\mu') \psi(x,c_{j'},\mu')$$

(6.75)

and:

$$q_{mj}^i = \frac{\varepsilon}{c_j} \sum_{\ell=0}^{N} \frac{2\ell+1}{2} P_\ell(\mu) \sum_{j'=1}^{J} k_{\ell jj'} w_{j'} \sum_{m'=2,4,...}^{2N} \omega_{m'} \psi_{m'j'}^{i\ell}$$

(6.76)

where:

$$k_{\ell jj'} = \int_0^\infty dc' \, c'^2 \exp(-c'^2) k_\ell(c_j, c') \phi_j(c')$$

(6.77)

There remains one minor difficulty. Unless a high order quadrature is used (J=81 or so), the integrations over c′ do not turn out to be quite accurate and the condition, L(1)=0, is not satisfied. This difficulty results from the fact that v(c), which is really an integral of K, is analytically evaluated earlier and thus is more accurate. For consistency, all integrations, need to be carried out to the same degree of accuracy. This task is easily handled by computing the readjusted v(c) and $\phi_d(c)$ from the following equations:

$$v(c_j) = \sum_{j'=1}^{J} k_{0jj'}$$

$$v(c_j) c_j \phi_d(c_j) - \sum_{j'=1}^{J} k_{1jj'} c_{j'} \phi_d(c_{j'}) = c_j$$

(6.78)

We note that the density profile is expressed as:

$$\left(\frac{n(r)-n_s}{n_s}\right) = \left(\frac{n_\infty - n_s}{n_s}\right) - \hat{u} \, g(r)$$

(6.79)

in which:

$$g(r) = \frac{2}{\sqrt{\pi}} \int_0^\infty dc \, c^2 \exp(-c^2) \int_{-1}^{1} d\mu \, \psi(r,c,\mu)$$

(6.80)

and, correspondingly:

$$g(r_i) = \frac{2}{\sqrt{\pi}} \sum_{j=1}^{J} c_j^2 \exp(-c_j^2) w_j \sum_{m=2,4,...}^{2N} \omega_m \psi_{mj}^{i+1/2}$$

(6.81)

The above algorithm is easily implemented and results of a specified convergence can be obtained. It is found that the diffusion solution provides a good start for the calculations and that convergence is typically obtained within 50 to 70 iterations.

Some typical results of these calculations are shown in Tables 6.1 and 6.2. The results for the condensation rate are also compared with some available experimental data (Table 6.3).

Table 6.1: The dimensionless condensation rate, $\tilde{u}/\tilde{u}_{fm}$ (P_3 approximation). Adapted from Loyalka et al. [1989b] with permission of the American Institute of Physics.

R (Kn^{-1})	β^2			
	0.1	0.5	1.0	10.0
0.1	0.9884	0.9782	0.9770	0.9734
0.25	0.9490	0.9415	0.9401	0.9342
0.50	0.8765	0.8706	0.8688	0.8612
0.75	0.8060	0.8012	0.7990	0.7908
1.0	0.7410	0.7374	0.7353	0.7273
1.25	0.6824	0.6802	0.6783	0.6710
1.50	0.6304	0.6293	0.6279	0.6215
1.75	0.5845	0.5842	0.5832	0.5779
2.0	0.5440	0.5443	0.5436	0.5392
3.0	0.4224	0.4223	0.4226	0.4214
5.0	0.2885	0.2887	0.2891	0.2894
10.0	0.1594	0.1594	0.1595	0.1598

Table 6.2: The normalized density profile $[(n(r)-n_1)/(n_\infty-n_s)]$ for R=5.0. Adapted from Loyalka et al. [1989b] with permission of the American Institute of Physics.

r	BGK model	β^2				CCC-S model
		0.1	0.5	1.0	10.0	
5.00	0.1244	0.1189	0.1235	0.1262	0.1329	0.1348
5.05	0.1457	0.1525	0.1471	0.1479	0.1527	0.1544
5.15	0.1778	0.1895	0.1802	0.1796	0.1825	0.1840
5.25	0.1994	0.2063	0.2007	0.2000	0.2025	0.2038
5.55	0.2540	0.2535	0.2524	0.2520	0.2536	0.2546
5.75	0.2842	0.2798	0.2807	0.2807	0.2823	0.2823
6.05	0.3238	0.3156	0.3182	0.3186	0.3203	0.3211
7.05	0.4257	0.4131	0.4165	0.4177	0.4197	0.4203
8.05	0.4993	0.4862	0.4895	0.4907	0.4927	0.4933
9.05	0.5557	0.5432	0.5461	0.5473	0.5492	0.5497
10.05	0.6005	0.5888	0.5915	0.5926	0.5944	0.5949
12.05	0.6675	0.6573	0.6597	0.6606	0.6621	0.6625
15.05	0.7343	0.7260	0.7279	0.7287	0.7300	0.7303

CCC-S: Constant Collision Cross-Section model.

Table 6.3: Comparison of the theoretical dimensionless condensation rate, $\tilde{u}/\tilde{u}_{fm}$, of dioctyl phthalate (DOP) in air with the experimental results of Ray et al. [1988] ($\beta^2 = m_{N2}/m_{DOP} = 28.014/390.54 = 0.0717$) ($P_5$ approximation). Adapted from Loyalka et al. [1989b] with permission of the American Institute of Physics.

R (Kn^{-1})	Data of Ray et al. [1988]	Loyalka et al. [1989]
1.329	0.662	0.6662
0.810	0.783	0.7910
0.618	0.818	0.8447
0.515	0.896	0.8745
0.419	0.910	0.9027
0.340	0.963	0.9260
0.290	0.979	0.9409
0.233	1.000	0.9606
0.190	1.000	0.9702
0.162	1.000	0.9782
0.138	1.000	0.9848

The important point that emerges here is that the normalized mass condensation rate has only a weak dependence on the mass ratio. The normalized density profile, however, is slightly more sensitive.

During the calculations here, it was noted that as the ratio, $\beta^2 \rightarrow 0.1$, more terms in the Legendre expansion are needed. This requirement is related to the highly anisotropic scattering encountered when heavy vapor molecules collide with light background molecules. This is actually the case of a Fokker-Planck gas, where the scattering operator, L, can be written as:

$$Lh = \tfrac{1}{2} \frac{\partial^2 h}{\partial c_\alpha \partial c_\alpha} - c_\alpha \frac{\partial h}{\partial c_\alpha} \tag{6.82}$$

For spherical geometry, Eqn. (6.82) can be written as:

$$Lh = -c \frac{\partial}{\partial c} h(r,c,\mu) + \tfrac{1}{2} \left\{ \frac{1}{c^2} \frac{\partial}{\partial c}\left(c^2 \frac{\partial}{\partial c}\right) + \frac{\partial}{\partial \mu}\left((1-\mu^2)\frac{\partial}{\partial \mu}\right) \right\} h(r,c,\mu) \tag{6.83}$$

The problem now is more accurately solved by avoiding the Legendre polynomial expansion and approximating the evaluation of the operator, L, by use of splines in (c,μ). This program is being followed (Loyalka and coworkers) but it is interesting to note that the operator, L, has the spectrum:

$$L\psi_i = \lambda_i \psi_i \tag{6.84}$$

in which i is a generalized subscript :

$$\psi_{r\ell} = \left(\frac{r!\left(\ell+\tfrac{1}{2}\right)}{\pi\left(\ell+\tfrac{1}{2}+r!\right)}\right)^{1/2} c^{\ell} P_{\ell}(\mu) S^{(r)}_{\ell+1/2}(c^2) \tag{6.85}$$

$$\lambda_{r\ell} = -(2r+\ell) \tag{6.86}$$

and where the functions, S, are the Sonine polynomials. Thus, to first order, one can replace L by a model operator, L_M, such that:

$$L_M h = -h(\mathbf{r},\mathbf{c}) + \sum_{i=1}^{N}(1+\lambda_i)\psi_i(\mathbf{c})\int d\mathbf{c}'\exp(-c'^2)\psi_i(\mathbf{c}')h(\mathbf{r},\mathbf{c}') \tag{6.87}$$

which has the important property, $L_M\psi_i = \lambda_i\psi_i$, and which, for N=1 and λ=0 (r=0, ℓ=0), is precisely the BGK model. The BGK model has the same Chapman-Enskog solution as the Fokker-Planck equation and also provides results that are close to those for small β, thus confirming the utility of this model.

Note that much of the work in this section can be generalized to arbitrary molecular interaction laws and gas-surface interaction laws. The case of the arbitrary condensation coefficient, α_c, is, in fact, quite straightforward, as one finds:

$$\frac{u(a,\alpha_c)}{u(a,1)} = \frac{\left[\dfrac{u(a,1)}{u_{fm}(a,1)}\right]}{1+\dfrac{1-\alpha_c}{\alpha_c}\left[\dfrac{u(a,1)}{u_{fm}(a,1)}\right]} \tag{6.88}$$

The problem of heat transfer from a particle is solved in a manner quite analogous to that of mass transfer and, indeed, progress has been made here. Now, however, instead of the Boltzmann equation for a dilute gas, one considers the linearized Boltzmann equation (for small temperature gradients) for gas mixtures. The heat transfer problem has recently been solved for a single component gas (Loyalka, Hamoodi and Tompson [1991b]) and the work is being extended to gas mixtures. Meanwhile, within certain limits and with some adjustments, the single component results can be used for mixtures.

As noted earlier, the single component results for the heat transfer rate are actually reasonably well described by some simple interpolation formulas (Loyalka [1983]). For an arbitrary thermal accommodation coefficient, α_t, one obtains:

$$\frac{q(a,\alpha_t)}{q(a,1)} = \frac{\left[\dfrac{q(a,1)}{q_{fm}(a,1)}\right]}{1+\dfrac{1-\alpha_t}{\alpha_t}\left[\dfrac{q(a,1)}{q_{fm}(a,1)}\right]} \tag{6.89}$$

Eqn. (6.89) is quite successful in describing the experimental data of Thomas and Loyalka [1982a,b].

6.4 Nonspherical particles and shape factors

Quite often, as discussed in the earlier chapters, particles of interest are not spherical and one needs to consider arbitrary shapes. While detailed numerical calculations for such bodies remain to be carried out for arbitrary sizes, useful information can be derived in some simple ways (Loyalka, Hamoodi and Tompson [1991a]).

For illustration, let us consider the case of isothermal condensation first. We note that for a sphere, the approximation:

$$\frac{\tilde{u}}{\tilde{u}_{fm}} = \left[1 + \frac{\tilde{u}_{fm}}{\tilde{u}_{cont}}\right]^{-1} \tag{6.90}$$

provides results for the condensation rate that agree with the exact results to within about 10% for all Knudsen numbers. Thus, for an arbitrarily shaped body, we might also use this approximation provided that we are able to calculate the free molecular and continuum condensation rates. The former task is quite straightforward and continuum results for several shapes (such as oblate and prolate spheroids) are available in the literature (Williams [1986]).

Since accurate numerical results for spheres and arbitrary mass ratios are available, we can possibly improve the results for the arbitrary bodies slightly. Thus, we write:

$$\left[\frac{u}{u_{fm}}\right]_{\text{arbitrary body}} = \frac{\left[\frac{u}{u_{fm}}\right]_{\text{arbitrary body}}}{\left[\frac{u}{u_{fm}}\right]_{\text{sphere}}} \left[\frac{u}{u_{fm}}\right]_{\text{sphere}} = S \left[\frac{u}{u_{fm}}\right]_{\text{equivalent sphere}} \tag{6.91}$$

where S is defined as the shape factor:

$$S = \frac{\left[\frac{u}{u_{fm}}\right]_{\text{arbitrary body}}}{\left[\frac{u}{u_{fm}}\right]_{\text{equivalent sphere}}} \tag{6.92}$$

and the equivalent sphere is defined to be a sphere with radius, a_{eq}, such that:

$$4\pi a_{eq}^2 = \text{area of the arbitrary body} \tag{6.93}$$

or, alternatively:

$$a_{eq} = \left(\frac{A_{\text{arbitrary body}}}{4\pi}\right)^{1/2} \tag{6.94}$$

We now calculate, in S, the results for the arbitrary body and sphere to a similar accuracy by using an interpolation formula, and use precise results for the equivalent sphere in Eqn. (6.91). While the overall accuracy of this procedure remains to be determined, it might prove reasonably good.

Condensation and Evaporation

For the concept of the shape factor introduced above to be meaningful, it should not depart too far from unity. To test the concept, we consider the case of an oblate spheroid of major axis, $\tilde{a}$, and minor axis, $\tilde{b}$. Then, with $x=b/a$, the area of the spheroid is:

$$A = 2\pi \tilde{a}^2 f(x) \tag{6.95}$$

where:

$$a = \frac{\tilde{a}}{\ell_D}, \quad b = \frac{\tilde{b}}{\ell_D}$$

and:

$$f(x) = 1 + \frac{x^2}{2\sqrt{1-x^2}} \ln\left[\frac{1+\sqrt{1-x^2}}{1-\sqrt{1-x^2}}\right] \tag{6.96}$$

Further:

$$\tilde{u}_{fm} = \rho_s \frac{n_\infty - n_s}{n_s} \tfrac{1}{4} \left(\frac{8kT}{\pi m}\right)^{1/2} A \tag{6.97}$$

and:

$$\tilde{u}_{cont} = \frac{4\pi D \tilde{c}}{\cot^{-1}(\lambda_0)} \rho_s \frac{n_\infty - n_s}{n_s} \tag{6.98}$$

where:

$$\lambda_0 = \frac{x}{\sqrt{1-x^2}} \tag{6.99}$$

and:

$$\tilde{c} = \tilde{a}\sqrt{1-x^2} \tag{6.100}$$

Thus:

$$\left[\frac{\tilde{u}_{fm}}{\tilde{u}_{cont}}\right]_{\text{oblate spheroid}} = \frac{1}{2\sqrt{\pi}} \frac{a}{\sqrt{1-x^2}} \cot^{-1}(\lambda_0) f(x) \tag{6.100}$$

and:

$$a_{eq} = a\sqrt{\tfrac{1}{2} f(x)} \tag{6.102}$$

The shape factor, S, using the above equations can now be written as:

$$S_{\text{oblate spheroid}} = \frac{1 + \dfrac{a_{eq}}{\sqrt{\pi}}}{1 + \dfrac{1}{2\sqrt{\pi}}\dfrac{a}{\sqrt{1-x^2}} \cot^{-1}(\lambda_0) f(x)} = \frac{1 + \left(\dfrac{f(x)}{2\pi}\right)^{1/2} a}{1 + \left(\dfrac{f(x)}{2\pi}\right)^{1/2} a \left(\dfrac{f(x)}{2(1-x^2)}\right)^{1/2} \cot^{-1}(\lambda_0)}$$

(6.103)

Note that in the limit, x=1 (sphere), the value of S from Eqn. (6.103) also goes to 1, while in the limit, x=0 (disk), we get, with $a_{eq}=a/\sqrt{2}$:

$$S_{disk} = \frac{1+\left(\frac{1}{2\pi}\right)^{1/2} a}{1+\left(\frac{1}{2\pi}\right)^{1/2}\left(\frac{\pi}{2\sqrt{2}}\right) a} \quad (6.104)$$

This expression for S_{disk} varies from (0.9957) to (0.9025) as a varies from (0.1) to (100.0).

We can derive similar expressions for a prolate spheroid (major axis $\tilde{b}$, minor axis $\tilde{a}$) and, in particular, in the limit of needle like particles, $x=b/a \gg 1$, we find, based on the previous considerations, that:

$$f(x) \approx \frac{\pi}{2} x \quad , \quad a_{eq} \approx \frac{a}{2}\sqrt{\pi x} \quad \text{and} \quad S_{needle} = \frac{1+\frac{a}{2}\sqrt{x}}{1+\frac{a}{4}\sqrt{\pi}\ln(2x)} \quad (6.105)$$

Simple computation shows that, for values of $x=50$ and a from (0.001) to (1.0), S varies from (1.001) to (1.491) only.

The calculations of this section seem to show that the concept of the shape factor, as introduced here, is a useful one.

6.5 Nonisothermal condensation and reaction rates

We have considered the evaluation of the condensation rate in the isothermal case as well as the calculation of the conductive heat transfer rate. If the heat of vaporization is large, then the mass transfer is controlled by diffusion as well as heat conduction. Here, we can obtain an expression for the condensation rate by setting:

$$\tilde{u} L = \tilde{q} \quad (6.106)$$

where L is the heat of vaporization. Note that n_s is a function of the temperature of the particle which would, in turn, be obtained from the above equation. Thus, one has (Loyalka and Park [1988]):

$$L\left(4\pi \tilde{a} D m_v\left[n_\infty - n_s(T_a)\right] f_c(Kn_c)\right) = 4\pi \tilde{a} \lambda (T_a - T_\infty) f_h(Kn_h) \quad (6.107)$$

which gives:

$$\frac{1-\frac{n_s(T_a)}{n_\infty}}{1-\frac{T_a}{T_\infty}} = -\frac{\lambda T_\infty}{D \rho_{v,\infty} L} \frac{f_h(Kn_h)}{f_c(Kn_c)} \quad (6.108)$$

and can be solved for T_a, provided the form of $n_s(T_a)$ is known.

For example, if we take (neglecting the Kelvin effect):

Condensation and Evaporation

$$n_s(T_a) = n_s(T_\infty) + \frac{L}{kT_\infty^2}(T_a - T_\infty)n_s(T_\infty) \tag{6.109}$$

then the condensation rate is given by (the supersaturation ratio, $S = n_\infty/n_s(T_\infty)$):

$$\tilde{u} = 4\pi \tilde{a} D \rho_s (S-1) \frac{f_c(Kn_c) f_h(Kn_h)}{\frac{LD}{\lambda}\rho_s \frac{Lm_v}{kT_\infty^2} f_c(Kn_c) + f_h(Kn_h)} \tag{6.110}$$

This is actually a generalization of an expression due to Mason [1962]. It is found that the Knudsen number effects are, indeed, quite significant (orders of magnitude) and that for small particles, Mason's formula (with $f_c = f_h = 1$) should not be used.

If the condensing vapor reacts with the surface or gets absorbed/adsorbed then the condensation rate can again be calculated by employing considerations similar to those above. Suppose, for example, that k_s is the surface reaction constant of the condensing species, that N_s is its concentration on the surface, and that γ is the order of reaction, then:

$$\tilde{u} = 4\pi \tilde{a}^2 k_s N_s^\gamma \tag{6.111}$$

Now, if:

$$n_s = H(N_s) \tag{6.112}$$

some known function of N_s, then we have:

$$4\pi \tilde{a} D m_v (n_\infty - H(N_s)) f_c(Kn_c) = 4\pi \tilde{a}^2 k_s N_s^\gamma \tag{6.113}$$

which is an equation that can be solved for N_s, and hence n_s. Ultimately, $\tilde{u}$ can be obtained from Eqn. (6.110).

Clement [1984] and Barrett and Clement [1988] have considered several coupled problems of heat (conductive, convective, radiative) and mass transfer in which they use arguments of the type discussed above. It is clear that the phenomena of condensation and evaporation are quite complicated. This is particularly true when they are coupled strongly with the surrounding medium and when basic information on the relevant physico-chemical properties (*e.g.*, accommodation coefficients, adsorption isotherms, shapes) is either not available or is contradictory. We have described the early work in the field and the most recent progress. For clarity, we have confined our attention to the salient aspects of the progress, but the recent progress should be regarded as only a prelude to that which remains to be done.

References: Chapter 6

Barrett, J.C. and Clement, C.F. [1988] "Growth rates for liquid drops," *J. Aerosol Sci.* **19**, 223.

Cercignani, C. and Pagani, C.D. [1967] "Variational Approach to Rarefied Flows in Cylindrical and Spherical Geometries," in Rarefied Gas Dynamics IV, Vol I, Brundin, C.L., editor (Academic, New York, N.Y.) p. 555.

Chapman, S. and Cowling, T.G. [1971] The Mathematical Theory of Non-Uniform Gases, 3rd edition (Cambridge University Press, Cambridge, U.K.).

Cipolla, J.W., Lang, H. and Loyalka, S.K. [1974] "Kinetic theory of condensation and evaporation. II," *J. Chem. Phys.* **61**, 69.

Clement, C.F. [1984] "Aerosol Formation From Heat and Mass Transfer in Vapor-Gas Mixtures" (Atomic Energy Research Establishment, Harwell, U.K.) Report # TP-1003.

Davies, C.N., editor [1966] Aerosol Science (Academic, London, U.K.).

Ferziger, J.H. and Kaper, H.G. [1972] Mathematical Theory of Transport Processes in Gases (North Holland, Amsterdam, Netherlands).

Fuchs, N.A. [1959] Evaporation and Droplet Growth in Gaseous Media (Pergamon, New York, N.Y.).

Fuchs, N.A. and Sutugin, A.G. [1971] "High-Dispersed Aerosols," in Topics in Current Aerosol Research, Vol. 2, Hidy, G.M. and Brock, J.R., editors (Pergamon, Oxford, U.K.).

Hidy, G.M. and Brock, J.R., editors [1971-1973] Topics in Current Aerosol Research, Vols. 1-3 (Pergamon, Oxford, U.K.).

Hirschfelder, J.O., Curtiss, C.F. and Bird, R.B. [1964] Molecular Theory of Gases and Liquids (John Wiley, New York, N.Y.).

Lang, H. [1983] "Heat and mass exchange of a droplet in a polyatomic gas," *Phys. Fluids* **26**, 2109.

Loyalka, S.K. [1971] "The slip problems for a simple gas," *Z. Naturforsch.* **26a**, 964.

Loyalka, S.K. [1973] "Condensation on a spherical droplet," *J. Chem. Phys.* **58**, 354.

Loyalka, S.K. [1982a] "A model for gap conductance in nuclear fuel rods," *Nucl. Tech.* **57**, 220.

Loyalka, S.K. [1982b] "Condensation on a spherical droplet - II," *J. Colloid Interface Sci.* **87**, 216.

Loyalka, S.K. [1983] "Mechanics of aerosols in nuclear reactor safety: a review," *Prog. Nuc. Energy* **12**, 1.

Loyalka, S.K. [1986] "Rarefied Gas Dynamics Problems in Environmental Sciences," in Rarefied Gas Dynamics XV, Vol. I, Boffi, V. and Cercignani, C., eds. (B.G. Tuebner, Stuttgart).

Loyalka, S.K. [1989] "Temperature jump and thermal creep slip: rigid sphere gas," *Phys. Fluids A* **1**, 403.

Loyalka, S.K. and Park, J.W. [1988] "Aerosol growth by condensation: a generalization of Mason's formula," *J. Colloid Interface Sci.* **125**, 712.

Loyalka, S.K., Hamoodi, S.A. and Tompson, R.V. [1989a] "Isothermal condensation on a plane surface," *Phys. Fluids A* **1**, 384.

Loyalka, S.K., Hamoodi, S.A. and Tompson, R.V. [1989b] "Isothermal condensation on a spherical particle," *Phys. Fluids A* **1**, 358.

Loyalka, S.K., Hamoodi, S.A. and Tompson, R.V. [1991a] "Condensation on non-spherical particles: shape factors," *Trans. Am. Nucl. Soc.* (to appear).

Loyalka, S.K., Hamoodi, S.A. and Tompson, R.V. [1991b] "Heat transfer from a sphere: rigid sphere gas," (to be published).

Loyalka, S.K. [1990] "Slip and jump coefficients for rarefied gas flows: variational results for Lennard-Jones and n(r)-6 potentials," *Physica A* **163**, 813.

Mason, B.J. [1962] Clouds, Rain and Rainmaking (Cambridge University Press, Cambridge, U.K.).

Maxwell, J.C. [1877] "Diffusion" in Encyclopedia Brittanica, Vol. 2, p. 82. Reprinted in [1890] The Scientific Papers of James Clerk Maxwell, Vol. 2, Niven, W.D., editor (Cambridge University Press, Cambridge, U.K.) p. 625.

Monchick, L. and Blakemore, R. [1988] "A variation calculation of the rate of evaporation of small droplets," *J. Aerosol Sci.* **19**, 273.

Pazooki, N. and Loyalka, S.K. [1988] "Heat transfer from a spherical particle in a rarefied monatomic gas," *J. Thermophys. Heat Transfer* **2**, 324.

Pidduck, F.B. [1916] "The kinetic theory of the motion of ions in gases," *Proc. London Math. Soc.* **15**, 89.

Pruppacher, H.R. and Klett, J.D. [1978] <u>Microphysics of Clouds and Precipitation</u> (Reidel, New York, N.Y.).

Ray, A.K., Lee, A.J. and Tilley, H.L. [1988] "Direct measurements of evaporation rates of single droplets at large Knudsen numbers," *Langmuir* **4**, 631.

Sahni, D.C. [1966] "The effect of a black sphere on the flux distribution in an infinite moderator," *J. Nucl. Energy* **20**, 915.

Seinfeld, J.H. [1986] <u>Atmospheric Chemistry and Physics of Air Pollution</u> (Wiley, New York, N.Y.).

Sitarski, M. and Nowakowski, B. [1979] "Condensation rate of trace vapor on Knudsen aerosol from the solution of the Boltzmann equation," *J. Colloid Interface Sci.* **72**, 113.

Thomas, L.B. and Loyalka, S.K. [1982a] "Determination of thermal accommodation coefficients of helium, argon, and xenon on a surface of zircaloy-2 at about 25°C," *Nucl. Tech.* **57**, 213.

Thomas, L.B. and Loyalka, S.K. [1982b] "Determination of thermal accommodation coefficients of inert gases on a surface of vitreous UO_2 at about 35°C," *Nucl. Tech.* **59**, 63.

Thomas, L.B., Krueger, L. and Loyalka, S.K. [1988] "Heat transfer in rarefied gases: critical assessments of thermal conductance and accommodation of argon in the transition regime," *Phys. Fluids* **31**, 2854.

Tompson, R.V. and Loyalka, S.K. [1986] "Condensational growth of a spherical droplet: free molecular limit," *J. Aerosol Sci.* **17**, 723.

Tompson, R.V. and Loyalka, S.K. [1987] "Chapman-Enskog solution for diffusion: Pidduck's equation for arbitrary mass ratio," *Phys. Fluids* **30**, 2073.

Tompson, R.V. and Loyalka, S.K. [1988] "Condensation on a spherical droplet, III," *J. Aerosol Sci.* **19**, 287.

Twomey, S. [1977] <u>Atmospheric Aerosols</u> (Elsevier, New York, N.Y.).

Wagner, P.E. [1982] "Aerosol Growth by Condensation," in <u>Aerosol Microphysics II</u>, Marlow, W.H., editor (Springer-Verlag, Berlin).

Williams, M.M.R. [1975] "Condensation and evaporation of a dilute vapour on a spherical droplet," *Z. Naturforsch.* **30**, 134.

Williams, M.M.R. [1986] "Neutron diffusion in spheroidal, bispherical and toroidal systems," *Nucl. Sci. and Eng.* **94**, 251.

CHAPTER 7

Particle Deposition and Resuspension

7.1 Introduction

Aerosol particles do not remain stationary. They are always in random thermal (Brownian) motion, and they also settle under gravity. They can, in addition, move because of temperature and concentration (vapor/gas diffusion) gradients or the force exerted by electrical fields or radiation. Motion of the gas brings particles near to, or in contact with, surfaces and because of the attractive nature of the prevailing short range van der Waals forces, the particles deposit on the surfaces. Also, particles can get resuspended from surfaces because of surface heating, depressurization of a containment, or the lift forces exerted by gas flows.

Particle deposition due to gravity and diffusion is obviously important in the environment. In engineered processes, one wishes to either remove particles by deposition (filtering), or to cause them to deposit in desired manners (fabrication of optical fiber preforms, sensors, sprays, coatings). In nuclear reactor safety considerations, it is desired that particles released in an accident deposit within the reactor containment and then not get resuspended.

A substantial body of literature exists on particle deposition. The theoretical analysis is largely based on the motion of a single particle in isolation (Chapter 2) or near surfaces (Chapter 4) and neglects simultaneous coagulation or condensation.

For the purposes of our future discussions we note that, just as with particle coagulation, an accurate calculation of particle deposition will require trajectory calculations. If the inertial motion of the particle is not large compared to the motion induced by convective gas flow or diffusion, temperature and concentration gradients, and external forces, then fairly good results are also obtained by solving the continuity equation:

$$\frac{\partial n}{\partial t} + \nabla \cdot \mathbf{J} = 0 \tag{7.1}$$

where $\mathbf{J}$ is the particle current caused by the various mechanisms:

$$\mathbf{J} = \mathbf{J}_c + \mathbf{J}_d + \mathbf{J}_T + \mathbf{J}_D + \mathbf{J}_e + \cdots \tag{7.2}$$

and:

$$\mathbf{J}_c = \mathbf{v}\, n \ , \quad \mathbf{J}_d = -D\, \nabla n \ , \quad \mathbf{J}_T = \mathbf{v}_T\, n \ , \quad \mathbf{J}_D = \mathbf{v}_D\, n \ , \quad \mathbf{J}_e = \mathbf{v}_e\, n \tag{7.3}$$

are, respectively, the convective, diffusive, thermophoretic, diffusiophoretic, and external force induced currents. Here, $\mathbf{v}$ is the convective (hydrodynamic, or mean mass) velocity, D is the particle diffusion coefficient, $\mathbf{v}_T$ is the thermophoretic velocity, $\mathbf{v}_D$ is the diffusiophoretic velocity, $\mathbf{v}_e$ is the velocity of the aerosol induced by external forces (for example gravity), and $n(\mathbf{r},t)$ is the particle concentration at $(\mathbf{r},t)$. Note that the diffusive and other currents are calculated with respect to the hydrodynamic current.

Eqn. (7.2) is solved subject to suitable boundary conditions. Actually, to develop basic correlations for the deposition currents and deposition rates, one considers the steady state or time averaged equation:

$$\nabla \cdot \mathbf{J} = 0 \tag{7.4}$$

subject to appropriate boundary conditions. The complete solution would require calculation of the velocity, **v**, and temperature and molecular concentrations (for v_T and v_D) from the Navier-Stokes and/or Boltzmann equations for the gas. The latter may be coupled to the particle equations. Boundary conditions associated with Eqn. (7.4) may be specified as:

$$n(\mathbf{r}) = 0, \quad \mathbf{r} \in \partial s^+ \tag{7.5}$$

where ∂s^+ indicates an extrapolated surface of the body at which the particle concentration essentially vanishes (for example, a layer of one particle radius thickness on the surface to account for the finite size-interception-of the particle by the surface). Also, far away from the body, at the edge of the concentration boundary layer, we assume that the particle concentration has some specified value, n_∞:

$$\lim_{|\mathbf{r}| \to \infty} [n(\mathbf{r})] = n_\infty \tag{7.6}$$

Eqn. (7.4) then is solved for $n(\mathbf{r})$ and **J**. The deposition rate, j (#/time), is obtained from:

$$j = -\int \mathbf{J}(\mathbf{r}) \cdot d\mathbf{S} \tag{7.7}$$

where d**S** is the surface element. Note that the local deposition rate coefficient, $k(\mathbf{r})$ (length/time), also referred to as the deposition velocity, V_d, is defined as:

$$k(\mathbf{r}) = -\frac{1}{n_\infty}(\mathbf{J} \cdot \mathbf{e}_s) = \left| -D \frac{1}{n_\infty} \nabla n \cdot \mathbf{e}_s \right| \tag{7.8}$$

where $\mathbf{e}_s$ is a unit vector normal to the surface of the body at the point, **r**, directed into the gas. Thus:

$$j = n_\infty \int k(\mathbf{r}) \, dA \tag{7.9}$$

The case when particle inertia is significant, leads to a noticeable departure of the particle motion from the fluid streamline and requires trajectory calculations. Thus, in the absence of particle rotation, we consider here for the particle velocity, $\mathbf{u}_p$, the equations:

$$\frac{d\mathbf{u}_p}{dt} = \mathbf{F} - \mathbf{F}_D$$

and:

$$\frac{d\mathbf{r}_p}{dt} = \mathbf{u}_p \tag{7.10}$$

where **F** is the driving force on the particle, and $\mathbf{F}_D$ is the drag force on the particle due to its motion in a fluid of velocity, u_f. Again, these equations can be solved in conjunction with the Navier-Stokes (or boundary layer) equations that help determine u_f, **F**, $\mathbf{F}_D$, and the local particle deposition current.

Gravitational settling and diffusion are the more obvious mechanisms for deposition. We discuss these phenomena first, and then we discuss deposition rates due to both laminar and turbulent convective-diffusive flows. Next, convective-phoretic, convective-gravitational, and inertial depositions are considered. Finally, we discuss the resuspension of particles, give a summary of the correlations useful in engineering practice, and then describe aspects of computational fluid dynamics that provide solutions to Eqns. (7.1) and (7.10) for complex flows and geometries. Since the details of specific calculations necessarily get involved, it is useful first to review ideas that might help in keeping these details in perspective in the overall schemes of calculations.

We note that Eqn. (7.1) requires a knowledge of the hydrodynamic velocity as well as the diffusive and other velocities (currents) that might be relevant to a problem. Thus, we need information on gas/vapor concentration and temperature distributions as well as charge (ion/electron) distributions. Generally, the interest is in deposition that involves convective velocity, and thus most often we are interested in solving Eqn. (7.1) with the convective term and one or more of the other terms. These various combinations arise in a myriad of applications and make the equation rather attractive from an applied viewpoint. The literature is replete with papers that essentially solve Eqn. (7.1) with the convective and one or more of the other terms. Judging from the recent research activities in the field, it is fair to say that the interest in using Eqn. (7.1) has only increased, not diminished.

Actually, the particle continuity equation is no different from an equation we might have written for convective mass transfer or deposition of molecular species (except that instead of thermophoresis we would have thermal diffusion, *etc.*). Also, it is similar to the energy equation of the Navier-Stokes equations. Historically, it appears that much of the mass transfer work has utilized the techniques of convective heat transfer and, in terms of mathematical techniques, not much new has been necessary. This also is true of aerosol deposition calculations as far as the use of Eqn. (7.1) is concerned. Here, generally, one has followed either the mass transfer work, or some of the convective heat transfer calculations directly. It is therefore useful to keep the heat-mass transfer analogies in mind.

While it is now possible, with high speed computers, to solve the Navier-Stokes equations and the particle continuity equation in arbitrary geometries for laminar flows or with simple models of turbulence, much of the previous progress in the field has been realized through use of the boundary layer theory of Prandtl. Interesting accounts of this theory and its applications are given in most good texts on fluid dynamics (*e.g.*, Batchelor [1967], Goldstein [1965], Schlichting [1979], Bejan [1984], Arpaci and Larsen [1984]). There are several texts on mass transfer which also discuss and utilize the boundary layer theory in considerable detail (see for example Levich [1962]). Various aspects of mass transfer, approximations, dimensional analyses, and correlations are given in Cussler [1984] and Hines and Maddox [1985].

We have already used boundary layer theory in Chapter 4 with respect to coagulation calculations. We will again use it extensively in this chapter as it helps provide rich insights into the role of various physical properties and flows on deposition and also leads to the development of various correlations. Useful sources of information here are Friedlander [1977] and Davies [1966].

We note that in convective heat transfer it is the Nusselt, Stanton and Peclet numbers:

$$Nu = \frac{hL}{k} = St\, Pe_{th} \quad , \quad St = \frac{h}{\rho C_p v} \quad , \quad Pe_{th} = \frac{L v}{\alpha} \tag{7.11}$$

that are useful for correlating results. Here, h is the convective heat transfer coefficient, L is a length scale, and k is the thermal conductivity of the fluid. The dimensionless group in mass transfer corresponding to Nu is the so called 'Nusselt number for mass transfer' or the Sherwood number:

$$Sh = \frac{kL}{D} = \frac{k}{v} Pe \tag{7.12}$$

where k is the deposition rate coefficient or the deposition velocity, V_d, L is a length scale, and D is the molecular diffusion coefficient. Also, Pe=Lv/D is the Peclet number. In mass transfer the Schmidt number, Sc=ν/D, is analogous to the Prandtl number, Pr=ν/α. Also, Re=Lv/ν is the Reynolds number. The natural convection leads to use of the Rayleigh number, Ra, and/or the Grashof number, Gr. The Knudsen number, Kn, and the Mach number, Ma, arise respectively when the mean free path is comparable to flow dimensions and when the flow speed is comparable to the mean molecular speed.

All these dimensionless groups are of interest in aerosol deposition. In addition, new groups that arise because of the finite particle size (the impaction number, which is a ratio of particle size to a flow scale), the particle inertia (the Stokes number, which is a ratio of the particle stopping distance to a length scale), the particle charge, the particle motion under temperature and concentration gradients, *etc.* We should note that in the aerosol literature the Sherwood number is not used that frequently, rather, one presents results for the deposition velocity, the deposition current, or the deposition efficiency which are frequently of more direct use.

To further clarify, the heat and mass transfer analogies, we note that the energy equation is:

$$\rho C_p \frac{\partial T}{\partial t} + \nabla \cdot \mathbf{J}_q = 0 \tag{7.13}$$

where T is the temperature, ρ is the mass density, C_p is the heat capacity, and $\mathbf{J}_q$ is the energy (heat) current:

$$\mathbf{J}_q(\mathbf{r},t) = \rho C_p \mathbf{v} T(\mathbf{r},t) - k \nabla T(\mathbf{r},t) \tag{7.14}$$

The heat flux at the wall is given by:

$$J_q = \left| -k \nabla T(\mathbf{r}) \cdot \mathbf{e}_s(\mathbf{r}) \right| \quad ; \quad \mathbf{r} \in \partial S \tag{7.15}$$

with the heat transfer coefficient defined as:

$$h = \frac{J_q}{\Delta T} = \frac{J_q}{T_\infty - T(0)} \tag{7.16}$$

We consider typical flow problems for which we can assume constant properties. Introducing the definitions:

$$\hat{\mathbf{u}} = \frac{\mathbf{v}}{U_\infty} \quad , \quad \hat{\mathbf{r}} = \frac{\mathbf{r}}{L} \quad , \quad \hat{n} = \frac{n}{n_\infty} \quad , \quad \hat{T} = \frac{T - T(0)}{T_\infty - T(0)} \tag{7.17}$$

and retaining only the convective and diffusive (conductive) parts, we can write the steady state particle continuity and energy equations as:

$$\hat{\mathbf{u}} \cdot \hat{\nabla} \hat{n} = \frac{1}{Pe} \hat{\nabla}^2 \hat{n} \tag{7.18}$$

and:

$$\hat{u} \cdot \hat{\nabla}\hat{T} = \frac{1}{Pe_{th}} \hat{\nabla}^2 \hat{T} \tag{7.19}$$

which then have solutions in the form ($\hat{u}=\hat{u}(Re,\hat{r})$):

$$\hat{n} = \hat{f}(\hat{r}, Re, Pe)$$
$$\hat{T} = \hat{f}(\hat{r}, Re, Pe_{th})$$

where:

$$Sh = f(Re, Pe) \quad , \quad Nu = f(Re, Pe_{th}) \tag{7.20}$$

and we have the similarity in the solutions. Thus, not only do similar techniques of solutions apply, but also one can use experimental data on heat transfer for a knowledge of mass transfer. To give a typical example, the correlations for forced convective mass and heat transfer to a spherical body are, respectively:

$$Nu = (2.0) + (0.6) Re^{1/2} Pr^{1/3}$$
$$Sh = (2.0) + (0.6) Re^{1/2} Sc^{1/3} \tag{7.21}$$

and either could have been obtained from the other without separate calculations or experiments. In the above, we used the fact that:

$$Pe = Re\, Sc \quad \text{and} \quad Pe_{th} = Re\, Pr$$

Generally we assume that particle concentrations vanish at the surfaces of objects. Occasionally, particles will have sizes comparable with the objects on which they deposit, and then we assume that the concentration vanishes at an extrapolated surface about one particle radius distant from the actual surface of the object. To consider these cases, we introduce an interception number:

$$R = \frac{a_p}{L} \tag{7.22}$$

and then we have:

$$\hat{n} = \hat{f}(\hat{r}, Re, Pe, R)$$

and:

$$Sh = f(Re, Pe, R) \tag{7.23}$$

and, for R<<1, one can write:

$$Sh = f(Re, Sc) \tag{7.24}$$

The Schmidt number for aerosols is large (of the order 10^4 or higher). This is comparable to that for liquids and the results of mass transfer studies in liquids can be used for investigations of aerosol deposition rates. Also, Sc>>1, indicates that the diffusion boundary layer is much thinner than the viscous boundary layer and that, within the diffusion boundary layer, the velocity profile is determined by the velocity boundary layer.

7.2 Gravitational settling in a well mixed volume

Gravitational settling of an aerosol particle is calculated in terms of particle terminal velocity, V_s:

$$V_s = B F$$

where F is the gravitational force on the particle, and B is the particle's mobility. Now, for a spherical particle of radius, d, neglecting the effects of buoyancy:

$$F = m g = \frac{\pi}{6} d^3 \rho g$$

and:

$$B = \frac{C(d)}{3 \pi \mu d}$$

thus:

$$V_s = \frac{1}{18} \frac{\rho d^2 g}{\mu} C(d) \tag{7.25}$$

The dependence of V_s on the particle diameter is, however, not quite as strong as it might appear since, for small d, C increases with decreasing d. Still, it is clear that larger particles settle much more quickly than small particles which may linger in the environment for long times.

For nonspherical particles, the Cunningham correction factor, C, is not known. Some results for the mobility in the continuum regime with the slip correction are, however, known and have been discussed in Chapter 2. We note, using Eqn. (2.192):

$$B = \frac{C(d_e)}{3 \pi \mu d_e \kappa}$$

where d_e is the diameter of the volume equivalent sphere, and κ is the dynamic shape factor. This leads to the result:

$$V_s = \frac{1}{18} \frac{\rho d_e^2 g}{\mu \kappa} C(d_e) \tag{7.26}$$

which can also be expressed as:

$$V_s = \frac{1}{18} \frac{\rho_0 d_a^2 g}{\mu} C(d_a) \tag{7.27}$$

where d_a is the <u>aerodynamic</u> diameter, and ρ_0 is the unit density (see Eqns. (2.195)-(2.196)).

For gravitational settling of an aerosol in a chamber one needs to distinguish between the well mixed case and the case of stratified settling (see Section 5.4).

7.3 Diffusional deposition in a stagnant gas

The Brownian motion of particles leads to their deposition on surfaces near them. Since the surfaces act as sinks, one assumes that the concentration of particles in close proximity to the surface (about one particle radius or so) is zero, and far away from the surface some prescribed value, n_∞. Thus, in the absence of convective motion, the aerosol deposition on a surface is obtained by solving the boundary value problem (Eqns. (7.4)-(7.6)) with $J=J_d$.

For deposition on a spherical surface of radius, a, we obtain for the deposition rate (#/time):

$$j = 4\pi a D n_\infty \qquad (7.28)$$

Note that Eqn. (7.28) is similar to Eqn. (6.16) encountered in the problem of the condensation of molecules on a particle. Since the particle diffusion coefficients are much smaller than molecular diffusion coefficients, particle deposition by diffusion alone is a slow process, and it is often effective only for small particles.

For deposition on nonspherical objects, development of expressions for the deposition rates would parallel the developments of Section 6.5.

7.4 Convective-diffusive deposition

Diffusion alone is not a very effective means for removal of particles from an aerosol. It can, however, be quite effective in conjunction with gaseous flows that bring particles into the vicinity of a surface. For small particles for which diffusion is important, a diffusion boundary layer develops over the surface. A net diffusion of particles to the surface occurs, with the concentration of particles taken as zero one particle radius away from the surface (a convenient approximation) and equal to its free stream value at the edge of the boundary layer. Overall, the situation then is of a large concentration gradient near the surface thereby leading to an appreciable particle deposition.

Both laminar and turbulent flows are of interest and play a major role in the design of filters, manufacturing of optical preforms for optical fibers, and particle deposition in human lungs and in nuclear reactor vessels during accidents. The mathematical modelling here follows very closely the discussions of Chapter 4 and the results lead to interesting insights and correlations that in many instances have a good experimental basis.

For calculations of the deposition rates it is sufficient to consider Eqn. (7.4) with the convective and diffusive currents only. Generally, the particle concentration is sufficiently small that v can be obtained separately from solutions of the Navier-Stokes equation for the gaseous flow, with appropriate boundary conditions, or obtained empirically from experimental data. Eqn. (7.18) is then solved in the diffusion boundary layer.

7.4.1 Laminar flows

Flow in a cylindrical tube:
This is one of the more common and easily solved problems in the subject. Actually, the problem is analogous to the Graetz problem [1885] of heat transfer. In the aerosol field, its solution derives from the work of Gormley and Kennedy [1949]. We consider the fully developed flow of a gas in a tube of radius, a. Introducing the dimensionless variables:

$$\hat{u} = \frac{u}{U} \quad , \quad \hat{r} = \frac{r}{a} \quad , \quad \hat{x} = \frac{x}{a} \quad , \quad \hat{n} = \frac{n}{n_0} \quad , \quad Pe = \frac{aU}{D} \qquad (7.29)$$

where n_0 is the inlet aerosol concentration, and u is the radially varying axial velocity expressed as:

Particle Deposition and Resuspension

$$u = 2U_{av}(1-\hat{r}^2) = U(1-\hat{r}^2) \tag{7.30}$$

Now, Eqn. (7.18) can be written as:

$$(1-\hat{r}^2)\frac{\partial \hat{n}}{\partial \hat{x}} = \frac{1}{Pe}\frac{1}{\hat{r}}\frac{\partial}{\partial \hat{r}}\left(\hat{r}\frac{\partial \hat{n}}{\partial \hat{r}}\right) \tag{7.31}$$

and we need to solve it subject to the boundary conditions:

$$\hat{n}(\hat{r},0) = 1 \quad \text{and} \quad \hat{n}(1,\hat{x}) = 0 \tag{7.32}$$

Eqn. (7.31) is solved by separation of variables and eigenfunction expansion techniques. One obtains:

$$\hat{n}(\hat{r},\hat{x}) = \sum_{i=0}^{\infty} a_n \Psi_g(\hat{r}) \exp(-\lambda_i^2 \hat{x}/Pe) \tag{7.33}$$

where $\Psi_g(\hat{r})$ are the eigenfunctions, λ_i are the eigenvalues and a_n are the expansion coefficients. All these quantities are known. The local mass transfer coefficient is calculated from Eqn. (7.8). A quantity of particular interest here is the penetration, $P(\hat{x})$, defined as:

$$P(\hat{x}, Pe) \equiv P(\hat{x}) = \frac{n_{av}(\hat{x})}{n_0} = \frac{1}{n_0}\frac{2}{a^2 U_{av}}\int_0^a u\, n(r,\hat{x})\, r\, dr \tag{7.34}$$

Which, upon explicit evaluation, is found to be:

$$P(\hat{x}) = (0.81919)\exp[-(7.312)\hat{x}/Pe] + (0.0975)\exp[-(44.62)\hat{x}/Pe]$$
$$+ (0.0325)\exp[-(113.8)\hat{x}/Pe] + \cdots \tag{7.35}$$

Note the fractional deposition or the deposition efficiency in a length, $\hat{x}$, is:

$$\eta = 1 - P(\hat{x}) = 1 - (0.81919)\exp[-(7.312)\hat{x}/Pe] + (0.0975)\exp[-(44.62)\hat{x}/Pe]$$
$$+ (0.0325)\exp[-(113.8)\hat{x}/Pe] + \cdots \tag{7.36}$$

It is evident that the laminar convective-diffusion is not a very effective means of particle deposition. Also, for small values of $\hat{x}/Pe$, Eqn. (7.35) converges slowly, and a more useful expression is obtained by use of the boundary layer theory. For this purpose, we first reduce Eqn. (7.31) to a simpler form in the boundary layer by introducing a coordinate normal to the surface, and neglecting terms of smaller order. Thus, we set:

$$\hat{y} = 1 - \hat{r} \tag{7.37}$$

and, reduce Eqn. (7.31) to:

$$2\hat{y}\frac{\partial \hat{n}}{\partial \hat{x}} = \frac{1}{Pe}\frac{\partial^2 \hat{n}}{\partial \hat{y}^2} \tag{7.38}$$

We now reduce this equation to an ordinary differential equation by introducing the similarity transformation:

$$\hat{n}(\hat{y},\hat{x}) \Rightarrow \hat{n}(\eta) \tag{7.39}$$

with:

$$\eta = \left(\tfrac{2}{9}\right)^{1/3} Pe^{1/3} \frac{\hat{y}}{\hat{x}^{1/3}} \tag{7.40}$$

and obtain:

$$\frac{d^2\hat{n}}{d\eta^2} + \varphi(\eta)\frac{d\hat{n}}{d\eta} = 0 \tag{7.41}$$

in which:

$$\varphi(\eta) = 3\eta^2 \tag{7.42}$$

The boundary conditions associated with Eqn. (7.41) are:

$$\hat{n}(\eta) = 1 \;;\; \eta \to \infty \quad \text{and} \quad \hat{n}(\eta) = 0 \;;\; \eta = 0 \tag{7.43}$$

We will encounter equations of this type frequently. We note that the solution of Eqn. (7.41) can be written as:

$$\frac{\hat{n}(\eta) - \hat{n}(0)}{\hat{n}(\infty) - \hat{n}(0)} = \gamma \int_0^\eta \exp\left(-\int_0^\xi \varphi(\varsigma)\,d\varsigma\right) d\xi \tag{7.44}$$

where:

$$\gamma = \left[\int_0^\infty \exp\left(-\int_0^\xi \varphi(\varsigma)\,d\varsigma\right) d\xi\right]^{-1} \tag{7.45}$$

Thus, Eqn. (7.44) has the incomplete gamma function:

$$\hat{n}(\eta) = \frac{1}{\Gamma(1/3)} \int_0^{\eta^3} u^{-2/3} \exp(-u)\,du \tag{7.46}$$

as its solution. We have the local radial particle (deposition) flux as:

$$J(\hat{x}) = \left|-D\frac{\partial n}{\partial r}\right|_{r=a} \tag{7.47}$$

and thus obtain:

$$J(\hat{x}) = \frac{6^{1/3}}{\Gamma(1/3)} \frac{n_0 D}{a} \frac{Pe^{1/3}}{\hat{x}^{1/3}} = (0.678) \frac{n_0 D}{a} \frac{Pe^{1/3}}{\hat{x}^{1/3}} \tag{7.48}$$

Note that the cumulative deposition (#/sec), and the deposition efficiency in a distance, x, are defined respectively by:

$$\Pi(\hat{x}) = 2\pi a \int_0^x ds\, J(s) = 2\pi a^2 \int_0^{\hat{x}} d\hat{s}\, J(\hat{s}) \tag{7.49}$$

and:
$$\eta(\hat{x}) = \frac{\Pi(\hat{x})}{n_0 U_{av} \pi a^2} \tag{7.50}$$

Thus, we get:

$$\Pi(\hat{x}) = 2\pi a^2 \frac{6^{1/3}}{\Gamma(1/3)} \frac{n_0 D}{a} Pe^{1/3} \frac{\hat{x}^{2/3}}{2/3} = 2\pi a^2 (1.017) \frac{n_0 D}{a} Pe^{1/3} \hat{x}^{2/3} \tag{7.51}$$

$$\eta(\hat{x}) = \frac{6^{4/3}}{\Gamma(1/3)} \left(\frac{\hat{x}}{Pe}\right)^{2/3} = (4.07) \left(\frac{\hat{x}}{Pe}\right)^{2/3} \tag{7.52}$$

and:

$$P(\hat{x}) = 1 - \eta(\hat{x}) = 1 - (4.07) \left(\frac{\hat{x}}{Pe}\right)^{2/3} \tag{7.53}$$

Now, if means for measuring particle concentration were available (for example, a condensation nuclei counter), then use of Eqns. (7.35) and (7.53) enables one to find:

(a) Particle diffusion coefficient (or size) for a monodisperse aerosol.
(b) Size distribution of a polydisperse aerosol.

The first undertaking here is quite straightforward as, for given parameters and measured values:
$$\hat{P}_{theory}(\hat{x}/Pe) = \hat{P}_{measured} \tag{7.54}$$

and:
$$\hat{x}/Pe = \hat{P}_{theory}^{-1}(\hat{P}_{measured}) \tag{7.55}$$

which gives:

$$D = \frac{a^2 U}{x} \hat{P}_{theory}^{-1}(\hat{P}_{measured}) \tag{7.56}$$

and the determination is relatively quite accurate.

The second undertaking appears straightforward, but has all the pitfalls of classic inverse problems. In practice, diffusion batteries are constructed so that there are a number of parallel tubes, with parts for particle extraction (and counting, for example, by a condensation nuclei counter) at several axial locations, x. The measurements, P, are then related to the distribution by an equation of the type:

$$P(x) = \int K(v,x) n(v) dv \tag{7.57}$$

which is a Fredholm integral equation of the first kind. This equation can be inverted, but one should be aware of the ill conditioning of the problem and the resultant unreliability of the 'measured' size distribution. To clarify this, we note that if the number of particles counted at a certain stage can be expressed as:

$$b_i = \sum_{j=1}^{J} K_{ij} n_j \quad ; \quad i = 1, 2, 3, ..., I \tag{7.58}$$

where n_j is the number of particles in the jth bin (section or group), K_{ij} is an experimentally or theoretically determined response matrix, I is the number of stages, and J is the number of size specific widths. Ideally, if I=J and:

$$K_{ij} = \delta_{ij}$$

a unit matrix, then:

$$n_i = b_i$$

and the determination is unique. In general, however

$$K_{ij} \neq 0 \quad ; \quad i \neq j$$

indicating that each bin has contributions from particles of several sizes. Thus, one solves Eqn. (7.58) by writing it in the form:

$$n_j = \sum_{i=1}^{I} Z_{ji} b_i \quad ; \quad j = 1, 2, 3, ..., J$$

where Z is strictly related to K. Typical K and Z matrices, for a diffusion battery with I=11 and J=8, have been computed by Cooper and Wu [1990]. Now, let us examine the number of particles in a stage, say the fifth stage. Then:

$$n_5 = \sum_{i=1}^{I} Z_{5i} b_i$$

and if:

$$Z_{5i} < 0 \quad ; \quad b_i > 0$$

the later being the counting measurements, then we clearly have $n_5 < 0$, a nonphysical result. Various methods to limit oscillations and negative results can be attempted through smoothing and/or regularization. But, all of the methods have the nature of *ad hoc* and *ex post facto* adjustments, and are accepted / not-accepted only until the arrival of the next criticism or improvement. The best solution is to design the instruments such that the problems of this type are minimal (that is, as little ill conditioning as possible).

Flow over a flat plate:
We consider the flow of a gas with free stream velocity, U, and gas concentration, n_∞, over a flat plate. Both momentum and concentration boundary layers develop over the plate. Since the Schmidt number, Sc>>1, the momentum boundary layer is actually considerably thicker than the concentration boundary layer. The thickness of the momentum layer is given by:

$$\delta_v(x) = (1.72) \times \left(\frac{\nu}{xU}\right)^{1/2} \tag{7.59}$$

and, to the first order in η, the velocity components are given by:

$$u = \alpha U \eta \quad \text{and} \quad v = \frac{\alpha U}{4}\left(\frac{\nu}{xU}\right)^{1/2} \eta^2 \tag{7.60}$$

in which η is the dimensionless parameter:

$$\eta = \left(\frac{U}{\nu x}\right)^{1/2} y \tag{7.61}$$

and $\alpha = (0.332)$. We use these results in the particle continuity equation, Eqn. (7.18), neglect diffusion in the x-direction, and use the definition:

$$\hat{n}(x,y) = \frac{n(x,y)}{n_\infty} \tag{7.62}$$

to obtain the partial differential equation:

$$\alpha \eta \frac{\partial \hat{n}}{\partial x} + \frac{\alpha}{4}\left(\frac{\nu}{xU}\right)^{1/2} \eta^2 \frac{\partial \hat{n}}{\partial y} = \frac{D}{U}\frac{\partial^2 \hat{n}}{\partial y} \tag{7.63}$$

which is subject to the boundary conditions:

$$\hat{n}(\hat{x},0) = 0 \quad \text{and} \quad \hat{n}(\hat{x},\infty) = 1 \tag{7.64}$$

Eqn. (7.63) can be solved by use of a similarity transform:

$$\hat{n}(x,y) \Rightarrow \hat{n}(\eta) \tag{7.65}$$

which reduces the partial differential equation, Eqn. (7.63), to an ordinary differential equation:

$$\frac{d^2\hat{n}}{d\eta^2} + \varphi(\eta)\frac{d\hat{n}}{d\eta} = 0 \tag{7.66}$$

with:

$$\varphi(\eta) = \left(\frac{\alpha}{4} Sc\right)\eta^2 \tag{7.67}$$

and the conditions:

$$\hat{n}(0) = 0 \quad \text{and} \quad \hat{n}(\infty) = 1 \tag{7.68}$$

We solve this problem in the manner of Eqns. (7.41)-(7.45) and find that the local mass flux, $J(x)$, is given by:

$$J(x) = \left|-D\left(\frac{\partial n}{\partial y}\right)_{y=0}\right| = (0.339)\, n_\infty\, U \left(\frac{xU}{\nu}\right)^{1/2} \frac{Sc}{Pe_x} \tag{7.69}$$

where Pe_x is the local Peclet number:

$$Pe_x = \frac{xU}{D} \tag{7.70}$$

Also, the average flux is given by:

$$J_{av} = \frac{1}{L}\int_0^L J(x)\,dx = (0.678)\,n_\infty\,U\left(\frac{LU}{\nu}\right)^{1/2}\frac{Sc^{1/3}}{Pe} \tag{7.71}$$

The results described above are of fundamental significance to most laminar flow aerosol deposition problems.

Flow over cylindrical and other objects:
Removal of particles by filtration requires an understanding of their capture from convective flows by cylindrical (or other shaped) fibers. The problem is of immense practical significance but careful studies have been made only recently. Following Friedlander [1967] we note that for the flow over a cylinder, the particle continuity equation in the concentration boundary layer reduces to:

$$u\frac{\partial n}{\partial x} + v\frac{\partial n}{\partial y} = D\frac{\partial^2 n}{\partial y^2} \tag{7.72}$$

with the boundary conditions:

$$n(x, a_p) = 0 \quad \text{and} \quad n(x, \infty) = n_\infty \tag{7.73}$$

where x and y are orthogonal curvilinear coordinates. The x-coordinate is taken parallel to the surface of the cylinder and measured from the forward stagnation point. The y-coordinate is normal to x and is measured from the surface. u and v are the velocity components in the x and y directions, respectively.

It is convenient to define:

$$\hat{n} = \frac{n}{n_\infty} \quad , \quad \hat{y} = \frac{y}{a_p} \quad , \quad \hat{x} = \frac{x}{a_p} \quad \text{and} \quad R = \frac{a_p}{a} \tag{7.74}$$

Now, near the cylinder, u and v are approximately given by:

$$u = 4AU\hat{y}\sin(\hat{x}) \tag{7.75}$$

and:

$$v = -2AU\hat{y}^2\cos(\hat{x}) \tag{7.76}$$

where:

$$A = \frac{2}{2 - \ln(Re)} \tag{7.77}$$

Eqn. (7.72) can then be written as:

Particle Deposition and Resuspension

$$4\,\hat{y}\sin(\hat{x})\frac{\partial \hat{n}}{\partial \hat{x}} - 2\,\hat{y}^2\cos(\hat{x})\frac{\partial \hat{n}}{\partial \hat{y}} = \left(\frac{1}{R^3\,\text{Pe}\,A}\right)\frac{\partial^2 \hat{n}}{\partial \hat{y}^2} \tag{7.78}$$

with the boundary conditions:

$$\hat{n}(\hat{x},\hat{y}) = 0\;;\quad \hat{y} = 1 \quad \text{and} \quad \hat{n}(\hat{x},\hat{y}) = 1\;;\quad \hat{y} = \infty \tag{7.79}$$

The solution of Eqn. (7.78) is of the form:

$$\hat{n} = \tilde{f}(\hat{x},\hat{y},R^3\,\text{Pe}\,A) \tag{7.80}$$

The local deposition rate per unit length of the cylinder is given by:

$$J(x) = D\left.\frac{\partial n}{\partial y}\right|_{y=a_p} = \frac{D\,n}{a_p}\left.\frac{\partial \hat{n}}{\partial \hat{y}}\right|_{\hat{y}=1} \tag{7.81}$$

and the total deposition rate per unit length of the cylinder is:

$$J_{total} = 2D\left(\frac{a}{a_p}\right)n_\infty \int_0^\pi \left(\left.\frac{\partial \hat{n}}{\partial \hat{y}}\right|_{\hat{y}=1}\right) d\hat{x} \tag{7.82}$$

Thus, the removal efficiency is:

$$\eta = \frac{J_{total}}{2\,a\,U\,n_\infty} = \frac{1}{R\,\text{Pe}}\,f\!\left(R\,\text{Pe}^{1/3}\,A^{1/3}\right) \tag{7.83}$$

or:

$$\eta\,R\,\text{Pe} = f\!\left(R\,\text{Pe}^{1/3}\,A^{1/3}\right) \tag{7.84}$$

where f is a function of the dimensionless groups indicated.

Note that in the limiting case of R=0 (diffusion dominant), *i.e.* point particles or molecules, one obtains:

$$\eta = \pi\,c_1\,\text{Pe}^{-2/3}\,A^{1/3} \tag{7.85}$$

where c_1 is a constant such that:

$$\pi c_1 = (3.68)$$

In the other extreme of Pe=0, particles follow the stream line and deposit on the cylinder whenever the stream line is within a radius of the cylinder. This effect is known as direct interception. The deposition is because of the component of the velocity that is normal to the surface of the cylinder. One obtains:

$$\eta = (U\,a)^{-1}\int_0^{\pi/2} v_{y=a_p}\,dx = 2A\,R^2 \tag{7.86}$$

We note that for a given fiber and flow velocity, η first decreases with particle radius (Eqn. (7.85)) and then increases (Eqn. (7.86)). This result (a minimum in the deposition efficiency) is shown quite dramatically by the actual form of the function, f, which can be found through numerical solutions of Eqn. (7.78).

It is found (Friedlander [1977]) that experimental data are well correlated by:

$$\eta = (1.3)\, Pe^{-2/3} + (0.7)\, R^2 \tag{7.87}$$

We should note here that the result:

$$\eta = \eta(R \to 0) + \eta(Pe \to 0) \tag{7.88}$$

holds only approximately.

7.4.2 Turbulent flows

For turbulent flows, the convective mass current is expressed as:

$$\mathbf{J}_c = \mathbf{v}\, n \tag{7.89}$$

where both $\mathbf{v}$ and n are fluctuating quantities with some mean (ensemble or time averaged) values, $\bar{\mathbf{v}}$ and $\bar{n}$, respectively. The mean particle current is often assumed to be of the form:

$$\bar{\mathbf{J}}_c = -\varepsilon\, \nabla \bar{n} \tag{7.90}$$

where ε is known as the eddy diffusivity for mass transfer and is dependent on local quantities:

$$\varepsilon = \varepsilon(\bar{\mathbf{v}}(\mathbf{r}), \mathbf{r}) \tag{7.91}$$

Now let us consider flow over a surface. Just as in the laminar flow analysis, particle concentration is determined by solving the particle continuity equation in the concentration layer (assuming that $\mathbf{v}$ and hence ε are known). For this purpose, in the boundary layer, we need to solve the equation:

$$\nabla \cdot (\mathbf{J}_d + \mathbf{J}_c) = 0 \tag{7.92}$$

which can be written as:

$$\frac{\partial}{\partial y}\left\{(D+\varepsilon)\frac{\partial \bar{n}}{\partial y}\right\} = 0 \tag{7.93}$$

and is to be solved subject to the conditions:

$$\bar{n}(x,y) = 0\,;\ y=0 \quad \text{and} \quad \bar{n}(x,y) = \bar{n}_\infty\,;\ y = \infty \tag{7.94}$$

Note that Eqn. (7.93) is simpler than the continuity equation for laminar flow and that it is easily integrated. This circumstance is a direct consequence of the assumption of Eqn. (7.90), where the particle current is expressed in the form of the Brownian diffusion term. We find:

$$\bar{n}(x,y) = \left(\frac{\int_0^y \frac{dy'}{D+\varepsilon}}{\int_0^\infty \frac{dy}{D+\varepsilon}} \right) \bar{n}_\infty \tag{7.95}$$

which, for the local deposition flux, gives:

$$J(x) = \left(\int_0^\infty \frac{dy}{D+\varepsilon} \right)^{-1} \bar{n}_\infty \tag{7.96}$$

For ε, one may take (Friedlander [1977]):

$$\varepsilon = \nu \left(\frac{y^+}{14.5} \right)^3$$

where:

$$y^+ = \frac{1}{\nu} \left[y U \left(\frac{f}{2} \right)^{1/2} \right]$$

and f is the Fanning friction factor. Using the above in Eqn. (7.96), we can find the local deposition rate and, hence, the Sherwood number:

$$Sh = \frac{kd}{D} = (0.042) Re\, f^{1/2}\, Sc^{1/3} \tag{7.97}$$

where d is a characteristic dimension of the flow (e.g. pipe diameter). Results of similar nature have been obtained for some of the basic problems of deposition. We note some such results below (all for fully developed flows):

Smooth pipe (Friedlander [1977]):

$$\eta = 1 - \exp\left(-(0.0236) \frac{L}{a} Re^{-1/8} Sc^{-2/3} \right) \quad;\quad Re > 10^4 \tag{7.98}$$

Rough pipe (Yaglom and Kader [1974], Hahn et al. [1985]):

$$k(r) = (3.2)\left(K^+ \right)^{-1/4} Sc^{-1/3} u_s \tag{7.99}$$

where, $K^+ = K u_s/\nu$, with K the height of the roughness elements, and u_s the friction velocity.

7.5 Convective-phoretic deposition

Aerosol particles experience a force in thermal gradients, and their motion under these forces is known as thermophoresis. The thermophoretic velocity is expressed as:

$$v_T = -K \frac{\nu}{T} \nabla T = -H \frac{\alpha}{T} \nabla T \tag{7.100}$$

where:
$$H = Pr \, K$$

and K is a dimensionless coefficient which is a function of several parameters (particle radius, thermal conductivity, gas properties, gas-surface interaction, *etc.*). This coefficient is reasonably known only in the small and large particle limits, and then only for nonrotating spherical particles in simple gases. Approximate results are, however, available for all sizes (see Loyalka [1983, 1986, 1990]). There is a controversy over the sign of the coefficient in certain ranges of the parameters, and the problem still requires careful investigations, both experimental and theoretical. We will simply assume here that it is of order 1.

In convective flows, the existence of sizeable temperature gradients near surfaces, can lead to significant particle deposition. This feature is used in optical fiber preform manufacturing and can also be quite important in the deposition of radioactive particles in nuclear reactor piping where large bulk to surface gradients can exist. The phenomenon and its practical applications have been recognized for quite some time beginning with the work of Reynolds [1875] on thermal creep and continuing with later observations by Watson [1936] of the dust free space near hot spheres. The practical applications have led to analysis of deposition in convective flows through boundary layer analyses of the type discussed in the last section (Friedlander [1977], Goren [1977], Rosner [1986], Walker *et al.* [1979], Simpkins *et al.* [1979], Morse and Cipolla [1984], and Nazaroff and Cass [1987]). We will discuss both the case of flow in a cylindrical tube and flow over a plane vertical surface. Also in this, we will restrict the discussion to the case of small particles for which the inertial effects can or will be neglected.

Cylindrical tube:

The problem here is analogous to that of the laminar diffusive deposition. We assume that the aerosol enters the tube with a uniform particle concentration, n_∞, and temperature, T_{max}. The wall temperature of the tube also remains at T_{max} until the laminar incompressible flow is fully established. At this axial location (x=0), the wall temperature decreases to T_{min} and is kept at this level. Our purpose is to determine the particle concentration distribution, and more specifically the local as well as the global particle deposition on the tube surface.

Since the velocity profile is known:

$$u(\hat{r}) = U\left(1 - \hat{r}^2\right)$$

the temperature profile is completely determined from the energy equation. Actually, with the usual neglect of the axial conduction, what we have is again the Graetz problem (Graetz [1885]):

$$\left(1 - \hat{r}^2\right)\frac{\partial \theta}{\partial \hat{x}} = \frac{1}{Pe_{th}} \frac{1}{\hat{r}} \frac{\partial}{\partial \hat{r}}\left(\hat{r} \frac{\partial \theta}{\partial \hat{r}}\right) \quad (7.101)$$

where Pe_{th} is the thermal Peclet number:

$$Pe_{th} = \frac{U a}{\alpha} \quad (7.102)$$

α is the thermal diffusivity, and θ is a dimensionless temperature:

$$\theta(\hat{r}, \hat{x}) = \frac{T - T_{min}}{T_{max} - T_{min}} \quad (7.103)$$

Particle Deposition and Resuspension

The boundary conditions associated with Eqn. (7.101) are:

$$\theta(\hat{r},\hat{x}) = 1 \quad ; \quad \hat{x} = 0 \quad \text{and} \quad \theta(\hat{r},\hat{x}) = 0 \quad ; \quad \hat{r} = 1 \tag{7.104}$$

and we find:

$$\theta(\hat{r},\hat{x}) = \sum_{j=0}^{\infty} a_j \, \Psi_j(\hat{r}) \exp\left(-\lambda_j^2 \, \hat{x} / \text{Pe}_{th}\right) \tag{7.105}$$

We note that with the boundary layer theory we have considerations identical to those encountered earlier in solving the convection-diffusion continuity equation. Thus, with the following similarity transforms:

$$\hat{y} = 1 - \hat{r} \quad , \quad \theta(\hat{y},\hat{x}) \Rightarrow \theta(\eta) \quad , \quad \eta = \left(\tfrac{2}{9}\right)^{1/3} \text{Pe}_{th}^{1/3} \frac{\hat{y}}{\hat{x}^{1/3}} \tag{7.106}$$

the Graetz problem can be reduced to the form:

$$\frac{d^2\theta}{d\eta^2} + 3\eta^2 \frac{d\theta}{d\eta} = 0 \tag{7.107}$$

with:

$$\theta(\eta) = 1 \quad ; \quad \eta \to \infty \quad \text{and} \quad \theta(\eta) = 0 \quad ; \quad \eta = 0 \tag{7.108}$$

As before, Eqn. (7.107) yields the incomplete gamma function:

$$\theta(\eta) = \frac{1}{\Gamma(1/3)} \int_0^{\eta^3} u^{-2/3} \exp(-u) \, du \tag{7.109}$$

as its solution. For future reference, we note that the heat flux at the surface is expressed as:

$$\begin{aligned}
J_q(\hat{x}) &= \left| -k \frac{\partial T}{\partial r} \right|_{r=a} \\
&= \frac{6^{1/3}}{\Gamma(1/3)} (T_{max} - T_{min}) \frac{k}{a} \frac{\text{Pe}_{th}^{1/3}}{\hat{x}^{1/3}} \\
&= (0.678)(T_{max} - T_{min}) \frac{k}{a} \frac{\text{Pe}_{th}^{1/3}}{\hat{x}^{1/3}}
\end{aligned} \tag{7.110}$$

For particle concentration, we consider the continuity equation, Eqn. (7.4), with the convective, diffusive, and thermophoretic terms. That is:

$$\nabla \cdot (\mathbf{J}_c + \mathbf{J}_d + \mathbf{J}_T) = 0 \tag{7.111}$$

with:

$$\mathbf{J}_c = n(r,x) \, u(r) \, \mathbf{n}_z \tag{7.112}$$

$$\mathbf{J}_d = -D \, \nabla n(r,x) \tag{7.113}$$

and:

$$\mathbf{J}_T = -\left(K\frac{\nu}{T(r,x)}\nabla T(r,x)\right)n(r,x) \tag{7.114}$$

Now, introducing the dimensionless quantities:

$$\Pr = \frac{\nu}{\alpha} \quad , \quad \theta^* = \frac{T_{min}}{T_{max} - T_{min}} \quad \text{and} \quad \hat{n}(\hat{r},\hat{x}) = \frac{n(r,x)}{n_0} \tag{7.115}$$

Eqn. (7.111) can be written as:

$$(1-\hat{r}^2)\frac{\partial \hat{n}}{\partial \hat{x}} - \frac{\Pr}{\mathrm{Pe}_{th}}\left\{\frac{1}{\mathrm{Sc}}\hat{\nabla}^2\hat{n} + \hat{\nabla}\cdot\left(\hat{n}\frac{K}{\theta+\theta^*}\hat{\nabla}\theta\right)\right\} = 0 \tag{7.116}$$

which we wish to solve subject to the conditions:

$$\hat{n}(\hat{r},\hat{x}) = 1 \quad ; \quad \hat{x} = 0 \quad \text{and} \quad \hat{n}(\hat{r},\hat{x}) = 0 \quad ; \quad \hat{r} = 1 \tag{7.117}$$

The problem above can be solved by finite difference or finite element techniques, and results for $\hat{n}(\hat{r},\hat{x})$, and consequently, the local and global deposition rates, can be found. It is very instructive, however, to obtain analytical solutions of the problem by using boundary layer theory.

We note that for gases, the Prandtl number, Pr~1. Also, for aerosols the Schmidt number, Sc, is quite high (~10^4) indicating that diffusion will be important only in a very thin layer near the wall. Further:

$$\mathrm{Pe}_{th} = \mathrm{Re}\,\Pr \approx \mathrm{Re} \tag{7.118}$$

In our analysis we will assume that $\mathrm{Pe}_{th} \gg 1$ and that terms of order Pe_{th}^{-1} can be neglected. In the present problem then, we consider three regimes of particle transport for Eqn. (7.116):

(1) The core, where convection dominates, and diffusion is small. In this regime, n is essentially constant in the radial direction.

(2) The thermophoretic layer (away from the surface). Here, we neglect the diffusion term, as well as axial heat conduction and axial thermophoresis, to get:

$$(1-\hat{r}^2)\frac{\partial \hat{n}}{\partial \hat{x}} - \frac{\Pr K}{\mathrm{Pe}_{th}}\frac{1}{\hat{r}}\frac{\partial}{\partial \hat{r}}\left[\hat{r}\left(\frac{1}{\theta+\theta^*}\frac{\partial \theta}{\partial \hat{r}}\right)\hat{n}\right] = 0 \tag{7.119}$$

We obtain an approximate solution to this equation by standard boundary layer techniques. We again introduce the transformations:

$$\hat{y} = 1-\hat{r} \quad , \quad \hat{n}(\hat{y},\hat{x}) \Rightarrow \hat{n}(\eta) \quad \text{and} \quad \eta = \left(\tfrac{2}{9}\right)^{1/3}\mathrm{Pe}_{th}^{1/3}\frac{\hat{y}}{\hat{x}^{1/3}} \tag{7.120}$$

and reduce Eqn. (7.119) to the form:

344

$$\frac{d\hat{n}}{d\eta} + \varphi(\eta)\,\hat{n} = 0 \tag{7.121}$$

where:

$$\varphi(\eta) = -\frac{\left(3\eta^2 + \dfrac{1}{\theta+\theta^*}\left(\dfrac{d\theta}{d\eta}\right)\right)\dfrac{d\theta}{d\eta}}{\left(\dfrac{3\eta^2}{\Pr K} + \dfrac{1}{\theta+\theta^*}\dfrac{d\theta}{d\eta}\right)(\theta+\theta^*)} \tag{7.122}$$

with the condition:

$$\hat{n}(\eta) = 1 \quad ; \quad \eta \to \infty \tag{7.123}$$

Thus, Eqn. (7.121) has the solution:

$$\hat{n}(\eta) = g_0\,\exp\!\left(-\int_0^\eta \varphi(s)\,ds\right) \tag{7.124}$$

where:

$$g_0 = \exp\!\left(\int_0^\infty \varphi(s)\,ds\right) \tag{7.125}$$

is the apparent dimensionless concentration at the wall.

(3) The concentration (diffusion) boundary layer. A scale analysis of Eqn. (7.116) shows that there is a thin layer of thickness $\sim Sc^{-1}Pe_{th}^{-1/3}$ in which convection is negligible compared to the diffusive and thermophoretic effects. Eqn. (7.116) is thus reduced to an ordinary differential equation:

$$-K\,c_w(\hat{x})\,\frac{\partial \hat{a}}{\partial \hat{Y}} = \frac{\partial^2 \hat{a}}{\partial \hat{Y}^2} \tag{7.126}$$

with:

$$\hat{Y} = Sc\,Pe_{th}^{1/3}\,\hat{y} \tag{7.127}$$

$$c_w(\hat{x}) = \frac{6^{1/3}}{\theta^*\,\Gamma(1/3)}\,\frac{1}{\hat{x}^{1/3}} \tag{7.128}$$

and:

$$\hat{a}(0,\hat{x}) = 0 \tag{7.129}$$

where:

$$\hat{n}(\hat{y},\hat{x}) \to \hat{a}(\hat{Y},\hat{x})$$

Eqn. (7.126) has the solution:

$$\hat{a}(\hat{Y},\hat{x}) = c_1\left(1 - \exp\!\left[-K\,c_w(\hat{x})\,\hat{Y}\right]\right) \tag{7.130}$$

The constant, c_1, is determined from matching this solution at the outer edge of layer (3), with the solution at the inner edge of layer (2) discussed earlier. That is:

$$\lim_{\hat{Y}\to\infty}\left[\hat{n}(\hat{Y},\hat{x})\right]=g_0 \tag{7.131}$$

Thus:

$$\hat{n}(\hat{Y},\hat{x})=g_0\left(1-\exp\left[-K\,c_w(\hat{x})\,\hat{Y}\right]\right) \tag{7.132}$$

and, we have the local radial particle (deposition) flux as:

$$J(\hat{x})=\left|-D\frac{\partial n}{\partial r}\right|_{r=a}\approx \hat{n}(\infty,\hat{x})\,V_T \tag{7.133}$$

which, using Eqn. (7.132), gives:

$$J(\hat{x})=n_0\,\frac{Sc\,Pe_{th}^{1/3}}{a}\,D\,g_0\,K\,c_w(\hat{x})=\frac{6^{1/3}}{\Gamma(1/3)}\,\frac{n_0\,D}{a}\,\frac{Pe_{th}^{1/3}}{\hat{x}^{1/3}}\,\frac{Sc\,K}{\theta^*}\,g_0 \tag{7.134}$$

We then find that:

$$P(\hat{x})=2\pi a^2\,(1.017)\,\frac{n_0\,D}{a}\,Pe_{th}^{1/3}\,\hat{x}^{2/3}\,\frac{Sc\,K}{\theta^*}\,g_0$$

and:

$$\eta(\hat{x})=(4.07)\,\frac{Pr\,K}{\theta^*}\left(\frac{\hat{x}}{Pe_{th}}\right)^{2/3}g_0 \tag{7.135}$$

Note that g_0 is a dimensionless concentration at the wall (or rather very near the wall at the edge of the concentration layer). It can be numerically evaluated from Eqn. (7.125). Numerical solutions of Eqn. (7.116) show that:

$$\eta(\hat{x})\,\frac{\theta^*}{Pr\,K\,g_0}\sim\left(\frac{\hat{x}}{Pe_{th}}\right)^{2/3} \tag{7.136}$$

and that Eqn. (7.135) provides a fairly good description for short tubes (error less than 50% for x<<1). We should note that numerical results for the deposition rate can be obtained by calculating particle trajectories directly from the consideration that the particles move under convection (x-motion) and thermophoresis (r-motion) in the 2nd layer, and then simply diffuse to the surface through the 3rd layer. Thus, for the purpose of calculating the deposition rate, calculations in the third layer are not necessary.

Eqn. (7.135) is quite useful in providing estimates of deposition in piping. It is also interesting to compare the thermophoretic deposition efficiency with the isothermal case. For this purpose we note that, for Pr K=1, Eqn. (7.125) gives:

$$g_0=\frac{\theta^*}{1+\theta^*} \tag{7.137}$$

and thus, in this case:

$$\eta(\hat{x})=(4.07)\,\frac{1}{1+\theta^*}\left(\frac{\hat{x}}{Pe_{th}}\right)^{2/3} \tag{7.138}$$

Comparing this result with Eqn. (7.52) we get:

$$\frac{\eta_{thermophoretic}}{\eta_{isothermal}} \approx \frac{1}{1+\theta^*} Le^{2/3} \qquad (7.139)$$

where Le is the Lewis number:

$$Le = \frac{\alpha}{D} = \frac{Sc}{Pr} \approx Sc \qquad (7.140)$$

Since:

$$\frac{1}{1+\theta^*} \approx 1 \quad ; \quad Sc \geq 10^4 \qquad (7.141)$$

it is clear that the deposition efficiency is strongly affected by even small temperature gradients normal to a surface in conjunction with convective flows. Thus, it is important to accurately know the flow fields and the thermophoretic velocity.

We also note here that we could have included the Brownian diffusion term in Eqn. (7.119), and followed the above analysis to explore the synergism in deposition because of both the thermophoresis and the (Brownian) diffusion. We have not done so because the thermophoresis can be so dominant. It is clear, however, that the contributions are not additive.

It is of interest to compare the thermophoretic and isothermal deposition fluxes, as well as the forms of the thermophoretic deposition flux and the heat flux at the wall. We get, using Eqns. (7.48) and (7.134):

$$\frac{J_{thermophoretic}}{J_{isothermal}} = \frac{Sc\,K}{\theta^*} g_0 \approx \frac{Sc\,K}{1+\theta^*} \qquad (7.142)$$

and:

$$\frac{J_{thermophoretic}}{J_q} = \frac{n_0}{T_{min}} \frac{\nu}{k} K g_0 = \frac{n_0 g_0}{T_{min}} \frac{1}{\rho C_p} Pr\,K \qquad (7.143)$$

These results are rather interesting because they suggest how the thermophoretic deposition may be estimated from a knowledge of the isothermal particle flux or the gaseous heat flux to the wall. We will discuss the latter aspect in the next section.

Thermophoretic deposition has been further investigated by Morse and Cipolla [1984]. These authors have considered the effect of simultaneous radiation fields on absorbing aerosols, and concluded that laser illumination can be used to enhance deposition.

7.6 A general analysis

Thermophoretic deposition on surfaces and bodies of revolution has been considered by Batchelor and Shen [1985] who show that the particle concentration can be approximately expressed as:

$$\frac{n(0)}{n_\infty} = \frac{T(0)}{T_\infty}\left\{1+(1-H)\left(\frac{T_\infty - T(0)}{T_\infty}\right)\right\} \qquad (7.144)$$

Aerosol Science: Theory and Practice

The deposition rate can then be calculated from:

$$\int J \, dA = H \frac{n(0)}{T(0)} \frac{1}{\rho C_p} \int J_q \, dA \tag{7.145}$$

where, J_q is the heat flux:

$$T(0) = T_{min} \quad , \quad T_\infty = T_{max} \quad \text{and} \quad H = Pr \, K \tag{7.146}$$

Also, n_∞ is the free stream particle concentration and $n(0)$ is the particle concentration at the surface (or rather, very near the surface at the outer edge of the concentration layer). Thus, the concentration and deposition can be estimated from the available results for temperature profiles and heat transfer rates.

Eqn. (7.144) here is simply surmised from observation that, for H=1, the energy and the particle continuity equations (neglecting the diffusive term) have similar forms, and that the ratio of concentration and temperature in this case is a constant. For ascertaining the usefulness of these approximations, we note that there are mainly two types of problems that are of practical interest: a) Flows in tubes relating to transmission of aerosols, and; b) Flows over surfaces and bodies of revolution (*e.g.*, fibers) relating to deposition and filtration.

We consider a stream of aerosol containing particles of radius, a_p, which approach the body of deposition, of typical dimension, a, with a uniform velocity, U. The free stream particle concentration is $n(x)$. Now, if x and y are the coordinates parallel and normal to the flow (with the stagnation point being the origin of the system in flow over the body), then the energy and the particle continuity equations can be written as:

$$u\frac{\partial T}{\partial x} + v\frac{\partial T}{\partial y} - \frac{1}{\rho C_p}\frac{\partial}{\partial y}\left(k\frac{\partial T}{\partial y}\right) = 0 \tag{7.147}$$

and:

$$u\frac{\partial n}{\partial x} + v\frac{\partial n}{\partial y} - \frac{\partial}{\partial y}\left(D\frac{\partial n}{\partial y}\right) - \frac{\partial}{\partial y}\left(\frac{K \, \nu \, n}{T}\frac{\partial T}{\partial y}\right) = 0 \tag{7.148}$$

Eqn. (7.148) is simplified by using the developments of the last section. We really need consider only the second layer in which the diffusive term can be neglected. The deposition rate is simply obtained from the product of the particle concentration and the thermophoretic velocity at the inner edge of the layer. This simplification is possible because of the low values of the particle diffusivity and the small values of the flow just adjacent to the wall in the concentration layer. Using this simplification, constant properties, and dimensionless groups, we can then write Eqns (7.147) and (7.148) as:

$$u\frac{\partial T}{\partial x} + v\frac{\partial T}{\partial y} - \alpha\frac{\partial^2 T}{\partial y^2} = 0 \tag{7.149}$$

and:

$$u\frac{\partial n}{\partial x} + v\frac{\partial n}{\partial y} - H\alpha\frac{\partial}{\partial y}\left(\frac{n}{T}\frac{\partial T}{\partial y}\right) = 0 \tag{7.150}$$

Note that if we put:

$$n = CT \tag{7.151}$$

where, C is some constant, then the continuity equation would have exactly the same form as the energy equation. In particular, for H=1, they become identical. Note, however that the energy equation is of the second order, and has two boundary conditions associated with it, while the continuity equation is of the first order (assuming T is known) and has only one boundary condition associated with it (that is, the concentration far away from the surface). The heat and the particle fluxes to the wall are expressed as:

$$J_q = \left| -k \frac{\partial T}{\partial y} \right|_{y=0}$$

and:

$$J = \left| -D \frac{\partial n}{\partial y} \right|_{y=0} = n \, v_T \big|_{y=0} = n \frac{H \alpha}{kT} J_q \bigg|_{y=0} \tag{7.152}$$

which are then, clearly related through the particle concentration at the wall. If this concentration at the wall could be simply related to the temperature, then we would only need to know the results for heat transfer. This situation holds for H=1 but, in general, one can only postulate some approximate relations. To examine the adequacy of the postulates, namely Eqns. (7.144) and (7.145), we consider some typical flows.

We note that the velocities are determined from:

$$u = \frac{\partial \psi}{\partial y} \quad \text{and} \quad v = -\frac{\partial \psi}{\partial x} \tag{7.153}$$

Here, ψ are the stream functions defined as:

<u>flat plate:</u>

$$\psi = (\nu U_\infty x)^{1/2} f(\eta) \quad \text{with} \quad \eta = \tfrac{1}{2} \left(\frac{U_\infty}{\nu x} \right)^{1/2} y \tag{7.154a}$$

<u>cylinder:</u>

$$\psi = (\beta \nu)^{1/2} x \, f(\eta) \quad \text{with} \quad \eta = \left(\frac{\beta}{\nu} \right)^{1/2} y \tag{7.154b}$$

<u>body of revolution:</u>

$$\psi = (\tfrac{1}{2} \beta \nu)^{1/2} x^2 \, f(\eta) \quad \text{with} \quad \eta = \left(\frac{2\beta}{\nu} \right)^{1/2} y \tag{7.154c}$$

Here, βx is the speed of the fluid at the outer edge of the boundary layer near the stagnation point on the cylinder or body of revolution. The functions, f, are determined from the equations:

<u>flat plate:</u>
$$f''' = -f f''$$

cylinder:
$$f''' = -ff'' + f'^2 - 1$$

body of revolution:
$$f''' = -ff'' + \tfrac{1}{2}(f'^2 - 1) \tag{7.155}$$

and the boundary conditions (these are common to the three equations above):

$$f(0) = 0, \quad f'(0) = 0 \quad \text{and} \quad f'(\infty) = 1 \tag{7.156}$$

The temperature is expressed as a function of η and, in the boundary layer, it is determined from the energy equation:

$$T'' + \Pr f\, T' = 0 \tag{7.157}$$

and the boundary conditions:

$$T(0) = T_{min} \quad \text{and} \quad T(\infty) = T_{max} \tag{7.158}$$

which are the same for the three geometries. The solution of this equation is:

$$\frac{T - T_{min}}{T_{max} - T_{min}} = \gamma \int_0^\eta \exp\left(-\Pr \int_0^\xi f(\varsigma)\, d\varsigma\right) d\xi \tag{7.159}$$

where, the constant, γ, is:

$$\gamma = \left[\int_0^\infty \exp\left(-\Pr \int_0^\xi f(\varsigma)\, d\varsigma\right) d\xi\right]^{-1} \tag{7.160}$$

The heat flux is now given by:

$$J_q = -k\left.\frac{\partial T}{\partial y}\right|_{y=0} = k(T_{max} - T_{min})\gamma \left.\frac{\partial y}{\partial \eta}\right|_{y=0} \tag{7.161}$$

The heat transfer problem has been extensively studied, and results for γ are available in the literature (Goldstein [1965]). It is $(0.664)\Pr^{1/3}$ for the flat plate, $(0.570)\Pr^{(0.4)}$ for the cylinder, and $(0.540)\Pr^{(0.4)}$ for the body of revolution.

With the velocities and temperatures thus known, the task now is to use the scale analysis and similarity transforms in the particle continuity equation and reduce it to an ordinary differential equation. Here, we first consider the outer layer in which the diffusion term is small compared to the convective and thermophoretic terms which are of comparable magnitudes. We find that the concentration, n, is determined from the equation:

$$\left(\frac{T'}{T} + \frac{f}{H}\right)n' + \left(\frac{T''}{T} - \frac{T'^2}{T^2}\right)n = 0 \quad ; \quad n(\infty) = n_\infty \tag{7.162}$$

Particle Deposition and Resuspension

and one has:

$$\frac{n}{n_\infty} = \exp\left(-\int_\eta^\infty \frac{\frac{T''}{T} - \frac{T'^2}{T^2}}{\frac{T'}{T} + \frac{f}{H}} d\eta'\right) \qquad (7.163)$$

Thus, at the inner edge of this outer layer, i.e. $\eta=0$:

$$\frac{n(0)}{n_\infty} = \exp\left(-\int_0^\infty \frac{\frac{T''}{T} - \frac{T'^2}{T^2}}{\frac{T'}{T} + \frac{f}{H}} d\eta'\right) \qquad (7.164)$$

There is an inner layer adjacent to the wall where diffusion becomes significant and comparable to one or more of the terms in the continuity equation. But, for practical purposes, unless we wish to determine the concentration profile in this layer or wish to make minor corrections, the diffusion current to the wall simply equals the thermophoretic current at the inner edge of the outer layer. Thus, the deposition flux is given by $n(0)v_T$, and we find:

flat plate:

$$J(x) = \tfrac{1}{2} H \alpha \gamma \frac{T_{max} - T_{min}}{T_{min}} \left(\frac{U_\infty}{\nu x}\right)^{1/2} n(0)$$

cylinder:

$$J(x) = \tfrac{1}{2} H \alpha \gamma \frac{T_{max} - T_{min}}{T_{min}} \left(\frac{\beta}{\nu}\right)^{1/2} n(0)$$

body of revolution:

$$J(x) = \tfrac{1}{2} H \alpha \gamma \frac{T_{max} - T_{min}}{T_{min}} \left(\frac{2\beta}{\nu}\right)^{1/2} n(0) \qquad (7.165)$$

where α is a known constant for each. Comparing these results with those from Eqn. (7.161) one verifies the usefulness of Eqns. (7.144) and (7.145) mentioned at the beginning of this section.

7.7 Convective-diffusiophoretic deposition

This deposition can be very important in many industrial and nuclear related applications. The mathematical treatment is similar to that for thermophoresis. However, there are some subtleties involved. To clarify these, we first discuss some basic notions:

Diffusion of molecules due to concentration gradients in a gas mixture can cause particle motion both because of the hydrodynamic velocity of the mixture induced by the presence of a noncondensing species (the Stefan velocity), and also because of another force on the particle known as the diffusiophoretic force. The motion induced by the latter is over and above that due to the hydrodynamic velocity and it would exist even if the hydrodynamic (mixed mass mean or barycentric) velocity were zero (*e.g.*, in a mixture with all species noncondensible and no driving forces other than diffusive to cause the hydrodynamic flow).

Since in nuclear reactor accidents, several condensible and noncondensible species can be present, and also since in LWR accidents large amounts of water vapor (steam) can also be present, it is important to understand aerosol motion both in condensible vapor / noncondensible background gas environments as well as in exclusively noncondensible gas environments. Very useful reviews of the basic early work on diffusiophoresis have been given by Waldmann and Schmitt [1966], and Deryagin and Yalamov [1972]. New results since then have been reported by Loyalka [1971a,b], Williams [1972], Lang and Loyalka [1972], and Loyalka and Yuan [1985]. A review by Loyalka [1983] has summarized the available results.

Consider a gas mixture, with species concentration, n_i. Then, the hydrodynamic velocity (of the mixture) is defined as:

$$\mathbf{v} = \frac{\sum n_i m_i \bar{\mathbf{c}}_i}{\sum n_i m_i} = \frac{\sum \rho_i \bar{\mathbf{c}}_i}{\sum \rho_i} = \frac{1}{\rho} \sum \rho_i \bar{\mathbf{c}}_i \qquad (7.166)$$

where m_i is the molecular weight and $\bar{\mathbf{c}}_i$ is the mean species velocity (the subscript i refers to species i) defined as:

$$\bar{\mathbf{c}}_i = \frac{1}{n_i} \int d\mathbf{c}_i \; \mathbf{c}_i \; f_i \qquad (7.167)$$

We set:

$$\bar{\mathbf{c}}_i = (\bar{\mathbf{c}}_i - \mathbf{v}) + \mathbf{v} \qquad (7.168)$$

where, we define:

$$\mathbf{V}_i = \bar{\mathbf{c}}_i - \mathbf{v} \qquad (7.169)$$

which is known as the 'diffusion' velocity. Explicit results for this in the continuum regime, in terms of the transport coefficients and the concentration and thermal gradients, can be found in the texts by Chapman and Cowling [1970] and Ferziger and Kaper [1972] among others. Thus, we have:

$$\bar{\mathbf{c}}_i = \mathbf{V}_i + \mathbf{v} \qquad (7.170)$$

and the species current is given by:

$$\mathbf{J}_i = n_i \bar{\mathbf{c}}_i = n_i \mathbf{v} + n_i \mathbf{V}_i \qquad (7.171)$$

Note that, in general, the distribution, f_i, is determined by solving the Boltzmann equation and associated conditions at the boundaries of the gas. In the bulk of the gas the distribution assumes the Chapman-Enskog or some other asymptotic form.

Now, consider the case of a binary isothermal gaseous mixture with no external body forces. Then, we have:

$$\mathbf{V}_1 = -D_{12} \frac{n}{n_1} \frac{m_2}{\rho} \nabla n_1 \qquad (7.172)$$

and:
$$V_2 = -D_{21} \frac{ñ}{n_2} \frac{m_1}{\rho} \nabla n_2 \tag{7.173}$$

where:
$$ñ = \sum n_i$$

is the total molecular concentration. Also, for constant total pressure, p:
$$\sum \nabla n_i = 0, \tag{7.174}$$

and:
$$\nabla n_2 = -\nabla n_1 \tag{7.175}$$

Finally, note that:
$$D_{21} = D_{12} \tag{7.176}$$

We discuss implications of these notions below.

a) <u>Stefan velocity</u>
Suppose that component 1 is a vapor species that condenses on surfaces, and species 2 is a noncondensing gas. Then we have a circumstance where:
$$\bar{c}_2 = 0 \tag{7.177}$$

everywhere, that is:
$$V_2 + v = 0 \tag{7.178}$$

or:
$$v = -V_2 = D_{21} \frac{ñ}{n_2} \frac{m_1}{\rho} \nabla n_2 = -D_{12} \frac{ñ}{ñ - n_1} \frac{m_1}{\rho} \nabla n_1 \tag{7.179}$$

which is known as the Stefan velocity. We shall denote this by:
$$v_s = -D_{12} \frac{ñ}{ñ - n_1} \frac{m_1}{\rho} \nabla n_1 \tag{7.180}$$

and note that the hydrodynamic velocity in this instance is caused by the noncondensibility of species 2. In a general situation (where other forces are present), the Stefan velocity is then the component of the hydrodynamic velocity caused by the presence of a noncondensing species.

b) <u>Diffusiophoretic motion</u>
Let us consider now a particle in an isothermal gas mixture (with noncondensible gases or otherwise). The velocity, V, of the particle, due to the nonuniformity of the gas mixture, is then expressed as (besides the Brownian diffusion):
$$V = v + V_D \tag{7.181}$$

where the second term is caused by a force, known as the diffusiophoretic force, F_D. Thus:

$$V_D = B F_D$$

and is expressed as:

$$V_D = -\sigma D_{12} \frac{\nabla n_1}{n_1} \tag{7.182}$$

Note that this velocity is with respect to the hydrodynamic velocity. We can define a velocity that is composed of both the Stefan and the diffusiophoretic velocity and is expressed as:

$$V_{Ds} = v_s + V_D = -\left(\frac{n}{n - n_1} \frac{m_1}{\rho} + \sigma \frac{1}{n_1}\right) D_{12} \nabla n_1 \tag{7.183}$$

The diffusiophoretic force, like the thermophoretic force, is a subtle noncontinuum effect. Both the magnitude and direction of σ depend upon complex intermolecular and molecule-surface interactions. The effect is present even in isobaric mixtures. While σ is easily calculated in the free molecular limit, its calculation in the transition (Kn~1.0) and slip (Kn~0.1) limits, requires considerable care. Here, solutions of the model Boltzmann equation with associated conditions (kinetic model calculations) can be helpful, but have stronger limitations than for single component gases. An early discussion of this phenomenon can be found in the works of Waldmann and Schmitt [1966] and Deryagin and Yalamov [1972]. Since then, results for the slip limit have been discussed by Loyalka [1971a] and Lang and Loyalka [1972] who have compared the computed results with the experimental data available to that time. A general variational formulation has been given by Loyalka [1971b] and results in the free molecular limit have been provided by Williams [1972]. In the ensuing discussion we will assume that σ, which is a complex function of intermolecular and molecule-surface interactions, is known (although, it is not, except in the free molecular and slip limits) or can be estimated by using some interpolation formula (see Loyalka [1983]).

c) <u>Deposition of particles on a large spherical drop surrounded by a binary mixture with a condensing vapor and a noncondensing background gas</u>
The above considerations are best understood in the context of specific observed phenomena, that is, the enhanced particle deposition on a drop growing because of condensation, or the dust free space around an evaporating drop in nearly isothermal conditions. Assuming that the particles do not react with the gas or vapor, we can specify the problem mathematically and solve it in two stages. In the first stage, we solve the problem of condensation and consequently obtain the hydrodynamic (Stefan) velocity. In the second stage, we use this velocity in the particle continuity equation and solve it to obtain the particle deposition rate on the drop.

First let us consider the problem of vapor transport. For the vapor, we have the <u>continuum limit</u> equation:

$$\nabla \cdot J_1 = 0 \tag{7.184}$$

with:

$$n_1(\infty) = n_{1,\infty} \quad \text{and} \quad n_1(a) = n_{1,s} \tag{7.185}$$

Now:

$$J_1 = n_1 \left(v_s - D_{12} \frac{n}{n_1} \frac{m_2}{\rho} \nabla n_1\right) \tag{7.186}$$

with:
$$\mathbf{v}_s = -D_{12}\frac{\textit{n}}{\textit{n}-n_1}\frac{m_1}{\rho}\nabla n_1 \tag{7.187}$$

Thus:
$$\mathbf{J}_1 = -D_{12}\frac{\textit{n}}{\textit{n}-n_1}\nabla n_1 \tag{7.188}$$

Note that the Stefan velocity is quite small compared to the usual vapor diffusive term. Thus, we do not expect it to influence the vapor transport to an appreciable extent. Now, using Eqn. (7.188) in Eqn. (7.184) and noting that $\textit{n}$, the total molecular concentration (gas+vapor), is a constant, we can solve the problem for n_1 and hence, for the vapor condensation rate and the magnitude of the Stefan velocity. We find:

$$v_s \approx -\frac{D_v\,a}{r^2}\frac{m_1}{m_2}\frac{n_{v,\infty}-n_{v,s}}{\textit{n}}$$

For the condensation rate on the surface we have:

$$Q = -4\pi r^2\,J_r\big|_{r=a} = 4\pi n_{v,\infty} D_v\,a\left(1-\frac{n_{v,s}}{n_{v,\infty}}\right)f_s \tag{7.189}$$

where:

$$f_s = \frac{\textit{n}}{n_{v,\infty}}\ln\left[\frac{\textit{n}-n_{v,s}}{\textit{n}-n_{v,\infty}}\right] \tag{7.190}$$

is a 'correction' factor in the condensation rate due to the Stefan velocity. This correction is small, but as we discuss below, the Stefan velocity itself plays a role in particle deposition and cannot be ignored. Note that:

$$v_s \approx -\frac{D_v\,a}{r^2}\ln\left[\frac{\textit{n}-n_{v,s}}{\textit{n}-n_{v,\infty}}\right] \tag{7.191}$$

where the subscript, v, refers to the vapor.

Now, for stage 2, let us consider the deposition of particles on the drop. For this purpose, let us also write the Stefan velocity as:

$$v_s = -\frac{v_0\,a^2}{r^2} \tag{7.192}$$

where v_0 is a constant with dimensions of velocity:

$$v_0 = \frac{D_v}{a}\frac{m_1}{m_2}\frac{n_{v,\infty}-n_{v,s}}{\textit{n}} \tag{7.193}$$

We consider the equation:

$$\nabla\cdot\mathbf{J} = 0 \tag{7.194}$$

with:

$$n(\infty) = n_\infty \quad \text{and} \quad n(a) = 0 \tag{7.195}$$

Ignoring the diffusiophoretic term, we have:

$$\mathbf{J} = \mathbf{v}_s\, n - D\, \nabla n \tag{7.196}$$

and:

$$\mathbf{v}_s = v_s\, \mathbf{n}_r$$

where $\mathbf{n}_r$ is a unit vector in the radial direction. That is, we consider the equation:

$$\frac{1}{r^2}\frac{d}{dr}\left(r^2\left(v_s\, n - D\frac{dn}{dr}\right)\right) = 0 \tag{7.197}$$

which, upon integration, gives:

$$\frac{dn}{dr} + \mathrm{Pe}_0\left(\frac{a^2}{r^2}\, n\right) = A \tag{7.198}$$

in which A is a constant and Pe_0 is the Peclet number:

$$\mathrm{Pe}_0 = \frac{v_0\, a}{D} = \frac{D_v}{D}\frac{m_1}{m_2}\frac{n_{v,\infty} - n_{v,s}}{n} \tag{7.199}$$

Solving Eqn (7.198) together with Eqn. (7.195) we get for the particle concentration:

$$n(r) = n_\infty\, \frac{1 - \exp\!\left(-\mathrm{Pe}_0\left(1 - \dfrac{a}{r}\right)\right)}{1 - \exp(-\mathrm{Pe}_0)} \tag{7.200}$$

and for the deposition rate:

$$j = 4\pi\, n_\infty\, D\, a\, f_0 \tag{7.201}$$

where:

$$f_0 = \frac{\mathrm{Pe}_0}{1 - \exp(-\mathrm{Pe}_0)} \tag{7.202}$$

is the correction to the deposition rate due to the Stefan flow. This correction is not negligible and, in fact, can be quite important. Suppose, as a first approximation, we write:

$$f_0 \approx \mathrm{Pe}_0 \approx \frac{D_v}{D} \tag{7.203}$$

Then:

$$j \approx 4\pi\, n_\infty\, D_v\, a \tag{7.204}$$

and the particle deposition rate becomes of the same order as the vapor deposition rate and is nearly independent of the particle size. This is a rather dramatic effect on particle deposition.

Particle Deposition and Resuspension

Table 7.1: Peclet numbers and normalized deposition rates (cm²/sec) as a function of particle diameter for different supersaturations (for a spherical particle in a mixture of air and water vapor).

d_p	Pe_0	f_0	$\dfrac{j}{4\pi a n_\infty}$	Pe_0	f_0	$\dfrac{j}{4\pi a n_\infty}$
	Supersaturation=0.001%			Supersaturation=0.1%		
0.01	2.15×10^{-4}	1.000	5.31×10^{-4}	2.38×10^{-2}	1.011	5.37×10^{-4}
0.10	1.67×10^{-2}	1.008	6.89×10^{-6}	1.848	2.194	1.50×10^{-5}
1.00	4.14×10^{-1}	1.221	3.37×10^{-7}	4.58×10^{1}	4.58×10^{1}	1.26×10^{-5}
10.00	4.768	4.809	1.15×10^{-7}	5.26×10^{2}	5.26×10^{2}	1.26×10^{-5}
	Supersaturation=0.01%			Supersaturation=1.0%		
0.01	2.37×10^{-3}	1.001	5.31×10^{-4}	2.37×10^{-1}	1.123	5.96×10^{-4}
0.10	1.84×10^{-1}	1.094	7.48×10^{-6}	1.84×10^{1}	1.84×10^{1}	1.26×10^{-4}
1.00	4.561	4.609	1.27×10^{-6}	4.57×10^{2}	4.57×10^{2}	1.26×10^{-4}
10.00	5.24×10^{1}	5.24×10^{1}	1.25×10^{-6}	5.26×10^{3}	5.26×10^{3}	1.26×10^{-4}

To illustrate this, we have calculated j (Eqn (7.201)) for particles from 0.01 μm to 10.0 μm diameter in a mixture of air and water vapor. These particles are to deposit onto a droplet of 10 μm diameter. Table 7.1 shows Peclet numbers and normalized deposition rates (cm²/sec), for different degrees of supersaturation, as a function of particle diameter.

We include now the diffusiophoresis term in the expression for particle current, Eqn. (7.196). Then, we have:

$$\mathbf{J} = v_s n - D \nabla n + v_D n \tag{7.205}$$

where:

$$v_D = -\phi D_{12} \frac{\nabla n_1}{n_1} \tag{7.206}$$

is the diffusiophoretic velocity. Again, we can write:

$$v_D = -\frac{v_{D,0} \, a^2}{r^2} \tag{7.207}$$

where:

$$v_{D,0} \approx \frac{\phi D_v}{a} \frac{m_1}{m_2} \frac{n_{v,\infty} - n_{v,s}}{n} \tag{7.208}$$

Using Eqn. (7.205) in Eqn. (7.194), the boundary conditions of Eqn. (7.195), and defining:

$$Pe_D = \frac{v_{D,0} \, a}{D} = \phi \frac{D_v}{D} \frac{m_1}{m_2} \frac{n_{v,\infty} - n_{v,s}}{n} \tag{7.209}$$

we get for the particle concentration:

$$n(r) = n_\infty \frac{1-\exp\left\{-(Pe_0 + Pe_D)\left(1-\frac{a}{r}\right)\right\}}{1-\exp\left\{-(Pe_0 + Pe_D)\right\}} \qquad (7.210)$$

The particle deposition rate on the spherical drop is given by:

$$j = 4\pi n_\infty \, D a f_{0+D} \qquad (7.211)$$

where the factor f_{0+D} is now:

$$f_{0+D} = \frac{Pe_0 + Pe_D}{1-\exp\left\{-(Pe_0 + Pe_D)\right\}} \qquad (7.212)$$

Clearly, if:

$$Pe_D \ll Pe_0 \qquad (7.213)$$

then it is the Stefan flow which would play the dominant role. But this dominance is by no means certain as the two Peclet numbers above are of similar magnitudes and can have different signs. In particular, note that:

$$f_{0+D} \neq f_{0+D}(Pe_0 \to 0) + f_{0+D}(Pe_D \to 0) \qquad (7.214)$$

d) Deposition on plane vertical surfaces in presence of gas-vapor mixture

An early analysis of this problem for natural (or free) convection was given by Hales et al. [1972]. They considered the particle continuity equation:

$$u\frac{\partial n}{\partial x} + v\frac{\partial n}{\partial y} - D\frac{\partial^2 n}{\partial y^2} - K v \frac{\partial}{\partial y}\left(\frac{n}{T}\frac{\partial T}{\partial y}\right) - \phi D_v \frac{\partial}{\partial y}\left(\frac{n}{n_v}\frac{\partial n_v}{\partial y}\right) = 0 \qquad (7.215)$$

where we have included the Stefan velocity in the convective term. Hales et al. used results for velocity and temperature from Minkowycz and Sparrow [1966] who had employed boundary layer theory considerations and similarity transforms to reduce the relevant Navier-Stokes equations for the vapor-gas mixture to a set of ordinary differential equations that were solved numerically. They then reduced Eqn. (7.215) to a second order ordinary differential equation using boundary layer considerations and similarity transforms and then solved it numerically. For explicit calculations, only the free molecular expressions for thermophoresis and diffusiophoresis were used.

Actually, Eqn. (7.215) can be analyzed in the manner of the work on thermophoresis. In the second layer, the Brownian diffusion term should be neglected, and the partial differential equation can be reduced via a similarity transform to a first order linear differential equation which can be solved analytically. The particle concentration at the inner edge of this layer is then determined, and with the temperature and vapor concentrations already known, the deposition velocities and fluxes can be obtained. With the approximate expressions for thermophoresis and diffusiophoresis now available, such analyses will be very useful. One

Particle Deposition and Resuspension

could also consider using the accurate and detailed kinetic theory solutions (Cipolla *et al.* [1974]) of the condensation/evaporation problem for a multicomponent gas-vapor mixture in plane geometry. We sketch such a generalized analysis below.

e) <u>General analysis:</u>
For clarity in the analysis, we consider an isothermal case. Then, the vapor continuity equation is:

$$u \frac{\partial n_v}{\partial x} + v \frac{\partial n_v}{\partial y} - \frac{\partial}{\partial y}\left(D_v \frac{\partial n_v}{\partial y}\right) = 0 \tag{7.216}$$

and, following the analysis of Eqns. (7.156)-(7.161), we obtain:

$$\frac{n_v - n_{v,s}}{n_{v,\infty} - n_{v,s}} = \gamma \int_0^\eta \exp\left(-Sc_v \int_0^\xi f(\varsigma)\,d\varsigma\right) d\xi \tag{7.217}$$

where, the constant, γ, is:

$$\gamma = \left[\int_0^\infty \exp\left(-Sc_v \int_0^\xi f(\varsigma)\,d\varsigma\right) d\xi\right]^{-1} \tag{7.218}$$

Here, $Sc = \nu/D_v$ is the Schmidt number for the vapor. The vapor number flux, ignoring the Stefan flow for the deposition of the vapor, is now:

$$J_v = \left|-D_v \frac{\partial n_v}{\partial y}\right|_{y=0} = D_v (n_{v,\infty} - n_{v,s}) \gamma \left.\frac{\partial y}{\partial \eta}\right|_{y=0} \tag{7.219}$$

The particle continuity equation in the second layer is:

$$u \frac{\partial n}{\partial x} + v \frac{\partial n}{\partial y} - \emptyset D_v \frac{\partial}{\partial y}\left(\frac{n}{n_v} \frac{\partial n_v}{\partial y}\right) = 0 \tag{7.220}$$

which has the approximate solution:

$$\frac{n(0)}{n_\infty} = \exp\left(-\int_0^\infty \frac{\frac{n_v''}{n_v} - \frac{n_v'^2}{n_v^2}}{\frac{n_v'}{n_v} + \frac{f}{\emptyset}}\,d\eta\right) \tag{7.221}$$

The particle flux at the wall is then:

$$J = |\mathbf{J}\cdot\mathbf{e}_s| \approx |(\mathbf{v}_s + \mathbf{v}_D)\cdot\mathbf{e}_s| n(0) \tag{7.222}$$

and this can be evaluated. Comparing the vapor and particle continuity equations we note that, for $\emptyset = 1$, we have the relation:

359

$$\frac{n_v}{n} = C \tag{7.223}$$

where C is a constant. Thus, we can propose an approximate relationship of the type in Eqns. (7.144) and (7.145):

$$\frac{n(0)}{n_\infty} = \frac{n_{v,s}}{n_{v,\infty}}\left\{1+(1-\phi)\left(\frac{n_{v,\infty}-n_{v,s}}{n_{v,s}}\right)\right\} \tag{7.224}$$

and:

$$\int J\, dA = \phi\, \frac{n(0)}{n_{vs}} \int J_v\, dA + n(0) \int V_s\, dA \tag{7.225}$$

7.8 Convective-gravitational deposition

This type of deposition can be important in chambers with good mixing or in aerosol flows over surfaces such as semiconductor wafers. Convection brings particles into the vicinity of surfaces and then the natural processes of diffusion and gravitational motion cause the deposition. Within the past several years, a number of investigators (Corner and Pendlebury [1951], Pich [1972], Taulbee and Yu [1975], Taulbee [1978], Goldberg [1981], Crump and Seinfeld [1981], Okuyama et al. [1986], Pui et al. [1987], etc.) have explored the various synergistic effects and interesting understandings have been gained. All these analyses are based on the particle continuity equation with either the convective and gravitational currents or the convective, gravitational and diffusive currents. Mathematically, the details follow quite closely those of the sections above. We note below some of the final expressions only.

<u>Laminar flow-gravitational settling in a horizontal pipe</u> (Pich [1972])
Here the deposition efficiency is given by:

$$\eta = \frac{2}{\pi}\left[2S\left(1-S^{2/3}\right)^{1/2} - S^{1/3}\left(1-S^{2/3}\right)^{1/2} + \sin^{-1}\left(S^{1/3}\right)\right] \tag{7.226}$$

where:

$$S > 1 \Rightarrow \eta = 1$$

S is the settling parameter:

$$S = \tfrac{3}{8}\frac{V_s\, L}{U_{av}\, a} \tag{7.227}$$

and V_s is the settling speed.

<u>Laminar flow-diffusive gravitational settling in a channel</u> (Taulbee and Yu [1975])
Here the deposition efficiency is given by:

$$\eta \approx \sqrt{\frac{\xi}{\pi}} \exp\left(-\frac{\sigma^2\,\xi}{4}\right) - \left(\frac{1}{\sigma} + \frac{\sigma}{2}\xi\right)\mathrm{erf}\left(\frac{\sigma}{2}\sqrt{\xi}\right) \tag{7.228}$$

where:

$$\xi = \frac{2x}{h\, Pe}, \quad \sigma = \frac{h\, V_s}{D}, \quad Pe = \frac{2 U_{av}\, h}{D}$$

Particle Deposition and Resuspension

h is the channel height, V_s is the gravitational settling speed, and x is the distance along the channel. Note that the deposition efficiency is, in general, not a simple summation of the convective-diffusion and convective-gravitational contributions. That is:

$$\eta \neq \eta(D \to 0) + \eta(V_s \to 0)$$

but the error involved in using the simple summation is not very large and superposition is generally a good first approximation.

<u>Turbulent gravitational settling in a chamber</u> (Corner and Pendelbury [1951], Crump and Seinfeld [1981])
Using the eddy diffusivity model of turbulence (see Eqn. (7.90)) with a y^2 dependence, the volume averaged deposition velocity (cm/sec) is given as:

$$V_d = \frac{1}{n_\infty A} \int J \cdot dS = -\frac{1}{A} \int n(r) \cdot V_s \left[1 - \exp\left(\frac{n(r) \cdot V_s}{2\sqrt{k_e D}}\right)\right]^{-1} dA \qquad (7.229)$$

where the turbulent diffusivity is expressed as:

$$D_e = k_e y^2$$

V_s is the settling velocity, $n(r)$ is the unit normal directed into the gas, and other quantities have been discussed earlier. Note again that:

$$V_d \neq V_d(D \to 0) + V_d(V_s \to 0)$$

and hence that the simple superposition:

$$V_d = V_d(D \to 0) + V_d(V_s \to 0)$$

applies only approximately.

7.9 Convective-electrical deposition

This mechanism is central to electrostatic precipitators where particles in a flow are first charged by a corona discharge and then collected on plates. Since particle charge distributions and mobilities are dependent on particle sizes (see Chapter 3), this mechanism can also be used to classify or quantify particle distributions.

<u>Electric precipitator:</u>
We consider a typical electric precipitator where particles of initial concentration n and flow speed, U, pass between two collecting plates. The flow is generally turbulent and the particles are charged by corona discharge such that a net current of them towards the plate is established. To calculate the deposition rate, we consider the time averaged particle flux (Friedlander [1977]):

$$\bar{J} = -(D + \varepsilon)\frac{\partial \bar{n}}{\partial y} - \bar{J}_e \qquad (7.230)$$

where:
$$\bar{J}_e = v_e \bar{n} \tag{7.231}$$

the quantity, v_e, can be estimated through approximate calculations of particle charging. The continuity equation is:

$$\nabla \cdot (\bar{J} n_y) = 0 \tag{7.232}$$

that is:

$$\frac{\partial}{\partial y}\left\{-(D+\varepsilon)\frac{\partial \bar{n}}{\partial y} - v_e \bar{n}\right\} = 0 \tag{7.233}$$

with the boundary conditions:

$$\bar{n}(x,\infty) = \bar{n}_\infty \quad \text{and} \quad \bar{n}(0,y) = 0$$

Upon integration and simplification, we find:

$$|\bar{J}(x)| = \frac{v_e \bar{n}_\infty}{1 - \exp\left(-v_e \int_0^\infty \frac{dy}{D+\varepsilon}\right)} \tag{7.234}$$

Note that in actual precipitation, the exponential term is quite small and:

$$|\bar{J}(x)| \approx J_e(x) = v_e \bar{n}_\infty(x) \tag{7.235}$$

Thus, for aerosol flowing through a precipitator with a speed, U, deposition can be obtained by solving:

$$U A \frac{d}{dx} n_\infty(x) = -P^* v_e n_\infty(x) \tag{7.236}$$

where A is the cross-sectional area of the flow and P^* is the perimeter. Solving this, we get for the removal efficiency (in a length x):

$$\eta(x) = \frac{n_\infty(x)}{n_\infty(0)} = \exp\left(-\frac{v_e P^*}{U A} x\right) \tag{7.237}$$

This expression has been quite useful in the design of electric precipitators.

Electrical classifier:
Convective-electrical deposition has an important application in classification and, therefore, sizing and size distribution measurements of particles. In a classifier, aerosol is first passed through a neutralizer (Kr-85) where the particle charge distribution is brought to the Boltzmann distribution. The aerosol then flows through a cylindrical column and the number of particles that pass through a narrow slit at the bottom is dependent on their size and charge. If either the current due to collected particles is measured (by an electrometer) or if the collected particles are counted (condensation nucleus counter), then one can determine the particle size distribution by

7.10 Inertial deposition

Particles deviate from fluid streamlines whenever there is an acceleration of the stream because of surfaces and obstacles. This deviation is not significant for particles that are small or when the flow speed is small (that is low inertia) but larger particles or small particles at high speeds can deviate from the flow for appreciable periods such that their trajectories intercept a surface located in the flow. Inertial impaction is thus an important means for collecting and sizing large and small particles if the flow speed is large. Cyclones and Andersen impactors are based on this phenomenon which can also lead to the deposition of particles on substrates for sensor fabrication.

The mathematical developments here follow closely those of Chapter 4. The similitude analysis now yields, for the deposition efficiency on an obstacle of characteristic dimension, L:

$$\eta = f(Stk, Re, R) \tag{7.238}$$

where Stk is the Stokes number:

$$Stk = \frac{\rho_p d_p^2 U_\infty}{18 \mu L} \tag{7.239}$$

and R is the Impaction number:

$$R = \frac{a_p}{L} \tag{7.240}$$

Usually it is found that there exists a critical Stokes number, Stk_c, below which there is no deposition. Also, plots of η vs. Stk for different Re are S-shaped indicting sharp size cutoffs in deposition (almost no deposition below Stk_c and complete deposition above it). This characteristic is used in the design of impactors for particle sizing (see *e.g.*, Fuchs [1964], Hinds [1982], Flagan and Seinfeld [1988] and Fernandez de la Mora *et al.* [1990] for detailed discussions and recent results).

Since particle filtration and deposition in piping have strong relevance to applications, we discuss below the salient aspects of theoretical estimations for these cases.

Flow over a cylindrical fiber
We consider here Eqn. (7.10) for particle motion. Results for the fluid velocity are available in the literature while outside the velocity boundary layer these are given by inviscid flow theory. We also assume that the Stokes drag laws apply and that the particle is quite small compared to the cylinder so that it doesn't influence the flow field. The equations of motion, Eqn. (7.10), are then numerically integrated (actually the problem is much simpler than those encountered in the coagulation calculations, where the colliding particles are of comparable sizes and influence the flow field). Both the particle trajectories and the collection efficiencies can then be calculated.

Turbulent flow
The calculations of particle trajectories here pose difficulties. The particle motion is described by the equation:

$$m\frac{dv}{dt} = \mathbf{F} - \mathbf{F}_D \tag{7.241}$$

where $\mathbf{F}$ and hence v and $\mathbf{F}_D$ are fluctuating quantities. Generally, $\mathbf{F}$ includes the effects of molecular collisions (diffusion) and turbulence. One can numerically simulate the particle motion through a Monte Carlo sampling of $\mathbf{F}$, and numerous particle trajectories can be generated for a given starting state. The deposition occurs if a trajectory approaches the object of deposition within a particle radius. For each starting position, one then has an estimation of the probability of deposition. The total deposition rate, and hence the efficiency of deposition, can be simply obtained by summing over all starting positions and comparing with the ideal (total cross-sectional current) deposition rate.

While such an approach has been taken for the calculation of collisional cross-sections (Almeida [1975], Enomoto and Loyalka [1985]), it appears that it has not been pursued for deposition rate calculations. Rather, results have been obtained by using the concept of particle eddy diffusivity. Correlations based on such calculations describe the available experimental data rather well (Friedlander [1977], Liu and Agarwal [1974], Hahn et al. [1985], Fernandez de la Mora and Friedlander [1982]).

An expression of some interest here is for deposition by inertial impaction in a pipe bend:

$$\eta = \left(\frac{\theta}{2}\right) \text{Stk} \tag{7.242}$$

where θ is the angle of the bend in radians. This expression is valid when the ratio of the radius of curvature of the bend to the pipe radius is smaller than 50 (Crane and Evans [1977]).

<u>Minimum in the capture efficiency</u>
We have discussed particle deposition due to both diffusive and inertial motions in the previous sections. We note that interception (finite size of aerosols) affects both motions. The diffusive and the inertial aspects were treated separately, as calculations requiring simultaneous consideration of both would necessitate the use of stochastic equations of motion. This task can be carried out, but important information can be obtained from the separate effects calculations also. In particular, a plot of particle collection efficiency as a function of particle radius, based on the two separate calculations shows a minimum in the efficiency corresponding to particle sizes between 0.1 μm and 1.0 μm. This fact has been confirmed experimentally, and the efficiency of filters is rated with respect to their ability to remove particles of the size 0.3 μm in diameter.

It is apparent that neither diffusion nor interception is effective in removing particles around 0.3 μm. The search for effective and economic means of removing such particles remains a problem of considerable interest.

7.11 Particle resuspension

Particles deposited on surfaces can become resuspended by lift forces due to depressurization, turbulence, thermophoresis (heated surfaces), or external forces. Resuspension is the underlying cause of dust explosions. In a nuclear reactor accident, resuspensions can have considerable impact on the progression of the accident and radiation releases.

The theoretical basis for particle resuspension is somewhat less well developed than for deposition. Much of the early work has relied on force balances (lift vs. adhesion) and

Particle Deposition and Resuspension

estimation of critical values of parameters (*e.g.*, Re) below which no resuspension occurs. Above the critical values, the resuspension rate, R (#/sec), is obtained in simple deterministic manners from equations of particle motion. That is, one considers:

$$\frac{d}{dt}\mathbf{u}_p(\mathbf{r}) = \mathbf{F} - \mathbf{F}_{D+A}(\mathbf{r},\mathbf{u}_p,\mathbf{u}_f) \quad \text{and} \quad \frac{d\mathbf{r}}{dt} = \mathbf{u}_p(\mathbf{r}) \tag{7.243}$$

where $\mathbf{F}$ is the driving force, and $\mathbf{F}_{D+A}$ is a combination of the drag and adhesive forces. Then, the expression:

$$R(t) = \int_{(\mathbf{u}_p(\mathbf{r})\cdot d\mathbf{s})>0} n_w(\mathbf{r},t)\left(\mathbf{u}_p(\mathbf{r})\cdot d\mathbf{s}\right) \quad ; \quad \mathbf{r} \in \partial s \tag{7.244}$$

provides results for the resuspension rate where $n_w(\mathbf{r},t)$ is the particle concentration on the surface (#/cm^3).

The deterministic approach can provide useful results for short times (~1 sec) which may be what matters in most circumstances. However, when resuspension and deposition occur simultaneously and/or the resuspension is of a relatively long term nature, more careful attention to the particle-surface-gas interaction is required and the phenomenological considerations should be more like those of molecular desorption from surfaces. Such an approach has been taken by Reeks *et al.* [1985a,b,c]. They consider a particle distribution function $W(v(t),y)$ on the surface, which is defined such that:

$$W(v,y)\,dv\,dy = \text{Expected number of particles of a given size with velocity v in dv (normal to the surface) at y in dy} \tag{7.245}$$

where y is the direction normal to the surface. This distribution function is determined by a Fokker-Planck equation in which the drag and diffusion terms are defined by means of the Langevin equation. The lift force is recognized to be composed of a mean part and a fluctuating component and the particle is bound by an attractive potential well. It is noted that the fluctuating component can have a greater role than the mean part. Once W is known, we can write the resuspension rate, R (#/sec), as:

$$R = \lim_{y \to y_B}\left[\int_0^\infty v\, W(v,y)\, dv\right] \tag{7.246}$$

where y_B is the outer edge of the potential well. One obtains for W:

$$W(v,y_B) = \omega\,(2\pi\beta\epsilon(\infty))^{-1/2}\,(2\pi\mu(\infty))^{-1/2}\,\exp\left(-\tfrac{1}{2}\frac{v^2}{\mu(\infty)}\right)\exp\left(-\frac{Q}{m\beta\epsilon(\infty)}\right) \tag{7.247}$$

where ω is the surface potential well frequency, Q is the height of the well:

$$\beta = \frac{1}{mB} \quad , \quad \epsilon(\infty) = \frac{2\langle PE \rangle}{mB} \quad , \quad \mu(\infty) = \langle v^2 \rangle \tag{7.248}$$

and B is the particle mobility. Here, <PE> is the average potential energy of the particle in the well at equilibrium, relative to point of minimum potential, and <v^2> is the mean square velocity. Use of Eqn. (7.247) in Eqn. (7.246) gives:

$$R = \frac{\omega}{2\pi}\left(\frac{\mu(\infty)}{\beta\varepsilon(\infty)}\right)^{1/2} \exp\left(-\frac{Q}{2\langle PE\rangle}\right)$$
(7.249)

This leads to the interesting observation by Reeks et al. [1985a,b,c] that, for large t ($\geq$1 sec):

$$R \approx \frac{1}{t}$$
(7.250)

The magnitude of this result is sensitive to B (by orders of magnitude) which is only approximately known (particle near the surface).

Wen and Kasper [1989] have explored the kinetic deposition model and have also carried out extensive experimental measurements for resuspension in ultraclean gaseous flows over surfaces. Following their work, we note that if we define, F=adhesion force/removal force, and N(F,t)dF=number of particles on the surface experiencing the force ratio, F in dF, at time, t, then the resuspension rate is:

$$R = -\int_0^\infty \frac{\partial}{\partial t} N(F,t)\, dF$$
(7.251)

Now, we define, a(F)=the rate of desorption of particles on the surface experiencing a force ratio, F, and we assume the Langmuir model for desorption:

$$\frac{\partial}{\partial t} N(F,t) = -a(F)\, N(F,t)$$
(7.252)

with:

$$N(F,0) = N_0(F)$$

as a specified distribution. Solving the above problem, we get:

$$N(F,t) = N_0(F) \exp(-a(F)\, t)$$
(7.253)

and:

$$\frac{\partial}{\partial t} N(F,t) = -a(F)\, N_0(F) \exp(-a(F)\, t)$$
(7.254)

Thus, the resuspension rate is given by:

$$R(t) = \int_0^\infty a(F)\, N_0(F) \exp(-a(F)\, t)\, dF$$
(7.255)

and can be evaluated with proper assumptions regarding $N_0(F)$ and a(F). Suppose one takes a uniform distribution for $N_0(F)$:

$$N_0(F) = \frac{N_{0,total}}{F_{max} - F_{min}} \quad ; \quad F_{min} \leq F \leq F_{max}$$
$$= 0 \quad ; \quad \text{otherwise}$$
(7.256)

Then, one obtains:

Particle Deposition and Resuspension

$$R(t) = \frac{N_{0,total}}{F_{max} - F_{min}} \int_{F_{min}}^{F_{max}} a(F) \exp(-a(F) t) \, dF \tag{7.257}$$

If we further assume:

$$a(F) = A \exp(-F) \tag{7.258}$$

then:

$$R(t) = \frac{N_{0,total}}{F_{max} - F_{min}} \frac{1}{t} \left[\exp\{-At \exp(-F_{max})\} - \exp\{-At \exp(-F_{min})\} \right] \tag{7.259}$$

Defining the relaxation times:

$$\tau_1 = \frac{1}{A \exp(-F_{min})} \tag{7.260}$$

and:

$$\tau_2 = \frac{1}{A \exp(-F_{max})} \tag{7.261}$$

we write:

$$R(t) = \frac{N_{0,total}}{F_{max} - F_{min}} \frac{1}{t} \left[\exp(-t/\tau_2) - \exp(-t/\tau_1) \right] \tag{7.262}$$

which is bounded at t=0. If we assume $\tau_1 \ll \tau_2$ then for short times, $t \approx \tau_1$, we have:

$$R(t) \approx \frac{N_{0,total}}{F_{max} - F_{min}} \frac{1}{t} \left[\exp(-\tau_1/\tau_2) - \exp(-t/\tau_1) \right] \tag{7.263}$$

or:

$$R \approx \frac{N_{0,total}}{F_{max} - F_{min}} \frac{1}{t} \tag{7.264}$$

For long times:

$$R = \frac{N_{0,total}}{F_{max} - F_{min}} \frac{1}{t} \exp(-t/\tau_2) \sim \frac{1}{t} \exp(-t/\tau_2) \tag{7.265}$$

These results indicate that for short times, the resuspension rate is governed by loosely bound particles, while for long times it is governed by the tightly bound particles. The short time 1/t dependence is due to a summation of discrete exponential dependences.

Wen and Kasper [1989] have carried out extensive experiments in support of the above model. They examined resuspension in a flow system where the number density (#/cm^3) of particles in the flow due to resuspension is given by:

$$n(t) = \frac{1}{Q} R(t) \tag{7.266}$$

where Q is the volumetric flow rate (#/cm^{3}sec) over the surface. Note that the parameters, A, τ_1, and τ_2, are difficult to determine *a priori*, but they can be estimated by using the experimental data for n(t), and fitting R(t)/Q to these data. Details of such an analysis can be found in the referenced work of Kasper and Wen.

Additional analysis of resuspension is contained in the work of Braaten *et al.* [1990] who report both experimental data and the results of Monte Carlo simulations. The Monte Carlo analysis was carried out by assuming that discrete lift bursts are responsible for resuspension. The bursts occur at discrete time steps of:

$$t^* = \frac{300\nu}{u_s^2}$$

(7.267)

where ν is the kinematic viscosity and u_s is the friction velocity. If the total simulation time is t_{sim}, then the total number of bursts simulated is:

$$N_{ts} = \frac{t_{sim}}{t^*}$$

(7.268)

At each burst, the lift force is sampled from a prescribed distribution. That is, we obtain F_L from the equation:

$$r_1 = \int_{F_{L,min}}^{F_L} P_L(F_L^*) dF_L^*$$

(7.269)

where r_1 is a random number distributed uniformly between 0 and 1. We can either use the adhesive force as a fixed known quantity, or sample it similarly:

$$r_2 = \int_{F_{A,min}}^{F_A} P_A(F_A^*) dF_A^*$$

(7.270)

we define $F = F_L/F_A$ and assume that if $F > F_{min}$, then a fraction $K(F)$ of particles on the surface gets resuspended. Now, if $n_r(t)$ is the number of resuspended particles at time, t, and n_0 is the number of particles initially on the surface, then we get:

$$n_r(t_i) = n_r(t_{i-1}) + K(F)(n_0 - n_r(t_{i-1})) \quad ; \quad F_i > F$$
$$= n_r(t_{i-1}) \quad ; \quad F_i \leq F$$

(7.271)

Numerous runs of this type are made to obtain the mean values, the related statistics, and hence the resuspension rate:

$$R = -\frac{d}{dt}(n_0 - n_r(t))$$

(7.272)

With some physically meaningful forms of P(F) and K, Braaten *et al.* [1990] have obtained good simulations of resuspension and agreement with their data.

Thermophoresis is now recognized to be an important factor in resuspension and interesting results in this regard have been obtained by Talbot *et al.* [1980] and Strattman *et al.* [1988].

Table 7.2: Correlations for aerosol deposition.

Deposition Mechanism	Convective Flow and Geometry	correlation
A) Diffusion only	Laminar flow in a pipe	Eqn. (7.35)
	Laminar flow over a flat plate	Eqn. (7.69)
	Laminar flow over a cylinder	Eqn. (7.87)
	Turbulent flow in a smooth pipe	Eqn. (7.98)
	Turbulent flow in a rough pipe	Eqn. (7.99)
B) Thermophoresis	Laminar flow in a pipe	Eqn. (7.145)
	Laminar flow over a flat plate	Eqn. (7.145)
	Laminar flow over a cylinder	Eqn. (7.145)
C) Sedimentation/diffusion	Laminar flow in a pipe	Eqn. (7.226)
	Laminar flow in a channel	Eqn. (7.228)
	Turbulent flow in a chamber	Eqn. (7.229)
D) Inertial impaction only	Turbulent flow in a bend	Eqn. (7.242)

Most recently it has been reported (*Scientific American*, June 1990, p.86) that submicron particles can be removed from surfaces by first wetting the substrate with water, and then heating it with lasers. It has also been reported that heating the particles with excimer lasers causes them to jump from surfaces.

Extensive data for resuspension in nuclear reactor piping have also been reported by Rahn et al. [1988].

7.12 Correlations

Much of the material of this chapter has direct relevance to the development of correlations that can be effectively used for calculating deposition rates. We give a table of some useful correlations (see Table 7.2). There is one aspect on the use of these correlations that warrants comment. We note that for deposition in chambers, the spatially homogeneous GDE would read (in the absence of condensation):

$$\frac{\partial}{\partial t}\bar{n}(v,t) = L(\bar{n},\bar{n}) + \bar{n}\frac{A}{V}V_d \qquad (7.273)$$

where V_d is the spatially averaged deposition velocity (cm/sec) defined as:

$$V_d = \frac{1}{\bar{n}A}\int \nabla \cdot \mathbf{J}\, dV = \frac{1}{\bar{n}A}\int \mathbf{J} \cdot d\mathbf{S} \qquad (7.274)$$

and, as such, includes all the effects.

7.13 Complex flows and computer programs

Calculation of particle deposition requires first a knowledge of the flow fields and temperature distributions, and then means for solving either the particle continuity equation or the equation of motion. For the simple geometries and flows considered in this chapter, much valuable information can be obtained by simple similitude considerations and use of similarity transforms. For complicated geometries (*e.g.* lungs, bifurcations, pipe bends, odd objects, dendritic formations and nonspherical particles (fibers)) and flows one can either employ some approximations or resort to direct numerical calculations. Computational fluid dynamics programs, based on body fitted coordinates and finite element numerical techniques, are now commercially available and include good graphics capabilities. We have used two such programs, FIDAP and FLUENT, to study several problems of particle deposition. Below, we briefly discuss these programs and their capabilities.

The program FIDAP allows simulation of a wide range of incompressible flows. The Navier-Stokes equations are solved for both laminar and turbulent flows, the latter with the aid of specified models of turbulence. The program includes provisions for simultaneous treatment of 'species' equations of the form of Eqn. (7.1), although only the convective and diffusive parts are included (this is a limitation that can be easily overcome). The user at his option can study transient, steady state, developing, developed, or specified flows and can account for finite sizes of particles (still ignoring inertia) by specifying particle boundary conditions at an extrapolated surface. The program does not do trajectory calculations, but this part is carried out very nicely in the FLUENT program. Actually from the point of view of deposition calculations, the two programs complement each other very well (the version of the FLUENT program that is available to us, does not carry out the species calculations). Both programs have good documentation. In particular, we have found the FIDAP documentation to be rather outstanding and well referenced.

These programs offer the interesting possibility of solving the GDE in conjunction with complex flows. An appropriate starting point here would be to consider first either coagulation alone or condensation alone, and then to construct more general purpose programs.

References: Chapter 7

Almeida, F.C. [1975] On The Effects of Turbulent Fluid Motion in the Collisional Growth of Aerosol Particles (University of Wisconson-Madison) Ph.D. Thesis.

Arpaci, V.S. and Larsen, P.S. [1984] Convection Heat Transfer (Prentice-Hall, Englewood Cliffs, N.J.).

Batchelor, G.K. [1967] An Introduction To Fluid Dynamics (Cambridge University Press, Cambridge, U.K.).

Batchelor, G.K. and Shen, C. [1985] "Thermophoretic deposition of particles in gas flowing over cold surfaces," *J. Colloid Interface Sci.* **107**, 21.

Bejan, A. [1984] Convection Heat Transfer (Wiley, New York, N.Y.).

Braaten, D.A., Paw U, K.T. and Shaw, R.H. [1990] "Particle resuspension in turbulent boundary layer - observed and modeled," *J. Aerosol Sci.* **12**, 405.

Brock, J.R. [1980] "The Kinetics of Ultrafine Particles," in Aerosol Microphysics I, Marlow, W.H., editor (Springer-Verlag, New York, N.Y.).

Chapman, S. and Cowling, T.G. [1970] The Mathematical Theory of Non-Uniform Gases, 3rd edition (Cambridge University Press, Cambridge, U.K.).

Cipolla, J.W., Lang, H. and Loyalka, S.K. [1974] "Temperature and Partial Pressure Jumps During Evaporation and Condensation of a Multicomponent Gas Mixture," in Rarefied Gas Dynamics XI, Becker, M. and Fiebig, M., editors (AFVLR, Gottingen, Germany).

Cooper, D.W. and Wu, J.J. [1990] "The inversion matrix and error estimation in data inversion: application to diffusion battery measurements," *J. Aerosol Sci.* **21**, 217.

Corner, J. and Pendelbury, E.D. [1951] "The coagulation and deposition of a stirred aerosol," *Proc. Phys. Soc. (London)* **B64**, 645.

Crane, R.I. and Evans, R.L. [1977] "Inertial deposition of particles in a bent pipe," *J. Aerosol Sci.* **8**, 161.

Crump, J.G. and Seinfeld, J.H. [1981] "Turbulent deposition and gravitational sedimentation of an aerosol in a vessel of arbitrary shape," *J. Aerosol Sci.* **12**, 405.

Crump, J.G., Flagan, R.C. and Seinfeld, J.H. [1983] "Particle wall loss rate in vessels," *Aerosol Sci. and Tech.* **2**, 303.

Cussler, E.L. [1984] <u>Diffusion Mass Transfer in Fluid Systems</u> (Cambridge University Press, Cambridge, U.K.).

Davies, C.N. [1973] <u>Air Filtration</u> (Academic, New York, N.Y.).

Davies, C.N., editor [1966] <u>Aerosol Science</u> (Academic, London, U.K.).

Dennis, R. [1976] <u>Handbook on Aerosols</u> (U.S. Department of Energy, Washington, D.C.).

Deryaguin, B.V. and Yalamov, Yu.I. [1972] "The Theory of Thermophoresis and Diffusiophoresis of Aerosol Particles and Their Experimental Testing," in <u>Topics in Current Aerosol Research</u>, Vol. 3, part 2, Hidy, G.M. and Brock, J.R., editors (Pergamon, Oxford, U.K.).

Enomoto, T. and Loyalka, S.K. [1985] "Turbulent-Gravitational Collision Efficiency of Nuclear Aerosols," in <u>Proceedings of the CSNI Specialists Meeting on Nuclear Aerosols in Reactor Safety</u>, Report # KfK-3800, CSNI-95.

Fernandez de la Mora, J. [1986] "Inertia and interception in the deposition of particles from boundary layers," *Aerosol Sci. and Tech.* **5**, 261.

Fernandez de la Mora, J., Rao, N. and McMurry, P.H. [1990] "Inertial impaction of fine particles at moderate Reynolds numbers and in the transonic regime with a thin-plate orifice nozzle," *J. Aerosol Sci.* **21**, 889.

Fernandez de la Mora, J. and Friedlander, S.K. [1982] "Aerosol and gas deposition to fully rough surfaces: filtration model for blade-shaped elements," *Int. J. Heat Mass Transfer* **25**, 1725.

Ferziger, J.H. and Kaper, H.G. [1972] <u>Mathematical Theory of Transport Processes in Gases</u> (North Holland, Amsterdam, Netherlands).

FIDAP Rev. 5 [1990] (Fluid Dynamics, Evanston, Illinois).

Flagan, R.C. and Seinfeld, J.H. [1988] <u>Fundamentals of Air Pollution Engineering</u> (Prentice Hall, Englewood Cliffs, N.J.).

FLUENT 2.98 [1990] (Creare.X, New Hampshire).

Friedlander, S.K. [1967] "Particle diffusion in low-speed flows," *J. Colloid Interface Sci.* **23**, 157.

Friedlander, S.K. [1977] <u>Smoke, Dust and Haze</u> (Wiley, New York, N.Y.).

Friedlander, S.K. and Johnstone, H.F. [1957] "Deposition of suspended particles from turbulent gas streams," *Ind. Eng. Chem.* **49**, 1151.

Fuchs, N.A. [1959] <u>Evaporation and Droplet Growth in Gaseous Media</u> (Pergamon, New York, N.Y.).

Fuchs, N.A. [1964] <u>The Mechanics of Aerosols</u> (Pergamon, New York, N.Y.).

Gokoglu, S.A. and Rosner, D.E. [1984] "Thermophoretically augmented mass transfer rates to solid walls across laminar boundary layers," *A.I.A.A. J.* **24**, 172.

Goldberg, I.S. [1981] "Deposition of aerosol particles due to simultaneous diffusion and gravitational sedimentation within a spherical chamber - solution to the Fokker-Plank equations," *J. Aerosol Sci.* **12**, 11.

Goldstein, S., editor [1965] <u>Modern Developments in Fluid Dynamics</u> (Dover, New York, N.Y.).

Goren, S.L. [1977] "Thermophoresis of aerosol particles in the laminar boundary layer on a flat plate," *J. Colloid Interface Sci.* **61**, 77.

Gormley, P.G. and Kennedy, M. [1949] "Diffusion from a stream flowing through a cylindrical tube," *Proc. Roy. Irish Academy* **52**, 163.

Graetz, L. [1885] "Ueber die warmeleitungsfahigkeit von flussigkeiten," *Ann. Phys. Chem.* **25**, 337 (in German).

Hahn, L.A., Stukel J.J., Leong, K.H. and Hopke, P.K. [1985] "Turbulent deposition of submicron particles on rough walls," *J. Aerosol Sci.* **16**, 81.

Hales, J.M., Schwendiman, L.C. and Horst, T.W. [1972] "Aerosol transport in a naturally-convected boundary layer," *Int. J. Heat Mass Transfer* **15**, 1837.

Hinds, W.C. [1982] Aerosol Technology (Wiley, New York, N.Y.).

Hines, A.L. and Maddox, R.N. [1985] Mass Transfer: Fundamentals and Applications (Prentice Hall, Englewood Cliffs, N.J.).

Lang, H. and Loyalka, S.K. [1972] "Diffusion slip velocity: theory and experiment," *Z. Naturforsch.* **27a**, 1307.

Levich, V.G. [1962] Physicochemical Hydrodynamics (Prentice Hall, Englewood Cliffs, N.J.) (English translation).

Liu, B.Y.H. and Agarwal, J.K. [1974] "Experimental observation of aerosol deposition in turbulent flow," *J. Aerosol Sci.* **5**, 145.

Loyalka, S.K. [1971a] "The slip problems for a simple gas," *Z. Naturforsch.* **26a**, 964.

Loyalka, S.K. [1971b] "On the motion of aerosols in nonuniform gases. I," *J. Chem. Phys.* **55**, 1.

Loyalka, S.K. [1971c] "Velocity slip coefficient and the diffusion slip velocity for a multicomponent gas-mixture," *Phys. Fluids* **14**, 2599.

Loyalka, S.K. [1983] "Mechanics of aerosols in nuclear reactor safety: a review," *Prog. Nuc. Energy* **12**, 1.

Loyalka, S.K. [1986] "Rarefied Gas Dynamics Problems in Environmental Sciences," in Rarefied Gas Dynamics XV, Vol. I, Boffi, V. and Cercignani, C., editors (B.G. Tuebner, Stuttgart) p. 177.

Loyalka, S.K. [1989] "Temperature jump and thermal creep slip: rigid sphere gas," *Phys. Fluids A* **1**, 403.

Loyalka, S.K. [1990] "Slip and jump coefficients for rarefied gas flows: variational results for Lennard-Jones and n(r)-6 potentials," *Physica A* **163**, 813.

Loyalka, S.K. and Yuan, C.C. [1985] "Calculation of the Diffusion Slip Velocity for Nuclear Aerosols," in Proceedings of the CSNI Specialists Meeting on Nuclear Aerosols in Reactor Safety, KfK-3800, CSNI-95, pp. 152-156.

Minkowycz, W.J. and Sparrow, E.M. [1966] "Condensation heat transfer in the presence of noncondensibles, interfacial resistance, superheating, variable properties, and diffusion," *Int. J. Heat Mass Transfer* **9**, 1125.

Morse, T.F. and Cipolla, J.W. [1984] "Laser modification of thermophoretic deposition," *J. Colloid Interface Sci.* **97**, 137.

Nazaroff, W.W. and Cass, G.R. [1987] "Particle deposition from a natural convection flow onto a vertical isothermal plate," *J. Aerosol Sci.* **18**, 445.

Okuyama, K., Kousaka, Y., Yamamoto, S. and Hosokawa, T. [1986] "Particle loss of aerosols with particle diameters between 6 and 2000 nm in stirred tank," *J. Colloid Interface Sci.* **110**, 214.

Pich, J. [1972] "Theory of gravitational deposition of particles from laminar flows in channels," *J. Aerosol Sci.* **3**, 351.

Pui, D.Y.H., Romay-Novas, F. and Liu, B.Y.H. [1987] "Experimental study of particle deposition in bends of circular cross section," *Aerosol Sci. and Tech.* **7**, 301.

Rahn, F.J., Collen, J. and Wright, A.L. [1988] "Aerosol behavior experiments on light water reactor primary systems," *Nucl. Tech.* **81**, 158.

Reeks, M.W., Reed, J. and Hall, D. [1985a] "On the Long Term Resuspension of Small Particles by a Turbulent Flow, Part I - A Statistical Model," (Central Electricity Generating Board, U.K.) Report # TPRD/B/0638/N85, ARPWG/P(85)35.

Reeks, M.W., Reed, J. and Hall, D. [1985b] "On the Long Term Resuspension of Small Particles by a Turbulent Flow, Part II - Determination of the Resuspension Rate Constant for an Elastic Particle on a Surface Under the Influence of van der Walls Forces," (Central Electricity Generating Board, U.K.) Report # TPRD/B/0639/N85, ARPWG/P(85)36.

Reeks, M.W., Reed, J. and Hall, D. [1985c] "On the Long Term Resuspension of Small Particles by a Turbulent Flow, Part III - Resuspension by a Turbulent Flow," (Central Electricity Generating Board, U.K.) Report # TPRD/B/0640/N85, ARPWG/P(85)37.

Reist, P.C. [1984] Introduction to Aerosol Science (McMillan, New York, N.Y.).

Reynolds, O. [1875] "On certain dimensional properties of matter in the gaseous state," *Philos. Trans. Roy. Soc. London Ser. A* **170(II)**, 727.

Rosner, D.E. [1986] Transport Processes in Chemically Reacting Flow Systems (Butterworth-Heineman, Stoneham, MA.).

Rosner, D.E. and Fernandez de la Mora, J. [1982] "Correlation and Prediction of Thermophoretic and Inertial Effects of Particle Deposition from Non-Isothermal Boundary Layer Flows," in Particulate-Laden Flows in Turbomachinery, Tabakoff, W., Crowe, C.T. and Cale, D.B., editors (American Society of Mechanical Engineers, New York, N.Y.) pp. 85-94.

Schlichting, H. [1979] Boundary - Layer Theory, 7th edition (McGraw Hill, New York, N.Y.).

Shaw, D.T., editor [1978] Recent Developments in Aerosol Science (Wiley, New York, N.Y.).

Simpkins, P.G., Greenberg-Kosinski, S. and MacChesney, J.B. [1979] "Thermophoresis: the mass transfer mechanism in modified chemical vapor deposition," *J. Appl. Phys.* **50**, 5676.

Stratmann, F., Fissan, H., Papperger, A. and Friedlander, S. [1988] "Suppression of particle deposition to surfaces by the thermophoretic force," *Aerosol Sci. and Tech.* **9**, 115.

Talbot, L., Cheng, R.K., Schefer, R.W. and Willis, D.R. [1980] "Thermophoresis of particles in a heated boundary layer," *J. Fluid Mech.* **101**, 737.

Taulbee, D.B. [1978] "Simultaneous diffusion and sedimentation of aerosol particles from Poiseuille flow in a circular tube," *J. Aerosol Sci.* **9**, 17.

Taulbee, D.B. and Yu, C.P. [1975] "Simultaneous diffusion and sedimentation of aerosols in channel flows," *J. Aerosol Sci.* **6**, 433.

Waldmann, L. and Schmitt, K. [1966] "Thermophoresis and Diffusiophoresis of Aerosols," in Aerosol Science, Davies, C.N., editor (Academic, London, U.K.).

Walker, K.L., Geyling, F.T. and Nagel, S.R. [1980] "Thermophoretic deposition of small particles in the modified chemical vapor deposition (MCVD) process," *J. Am. Ceram. Soc.* **63**, 552.

Walker, K.L., Homsy, G.M. and Geyling, F.T. [1979] "Thermophoretic deposition of small particles in laminar tube flow," *J. Colloid Interface Sci.* **69**, 138.

Watson, H.H. [1936] "The dust-free space surrounding hot bodies," *Trans. Faraday Soc.* **32**, 1073.

Weinberg, M. and Subramanian, R.S. [1982] "Thermophoretic deposition in a tube with variable wall temperature," *J. Colloid Interface Sci.* **87**, 579.

Wen, H.Y. and Kasper, G. [1989] "On the kinetics of particle reentrainment from surfaces," *J. Aerosol Sci.* **20**, 483.

Williams, M.M.R. [1972] "On the motion of small spheres in gases, II: Thermo-phoresis, Diffusio-phoresis and related phenomena," *Z. Naturforsch.* **27a**, 1804.

Yaglom, A.M. and Kader, B.A. [1974] "Heat and mass transfer between a rough wall and turbulent fluid at high Reynolds and Peclet numbers," *J. Fluid Mech.* **62**, 601.

Yoshida, T., Kousaka, Y. and Okuyama, K. [1979] *Aerosol Science for Engineers* (Power Co. Ltd., Tokyo, Japan).

Zernik, W. [1957] "The dust-free space surrounding hot bodies," *Br. J. Appl. Phys.* **8**, 117.

CHAPTER 8

Nuclear Source Term

8.1 Nuclear source term and aerosols

An understanding of aerosol behavior is of primary importance in the estimation of radioactivity releases from nuclear reactor accidents. Severe core damage in light water reactors, core disruptive accidents in fast reactors and sabotage of shipping casks or waste disposal sites with explosives are potential sources of release of radioactive aerosols to the environment. It is believed that public risks, as well as risks to property and plant equipment, are dominated by accidents that involve sizeable releases of aerosols to the environment and into the reactor containment.

A materials inventory typical of a light water nuclear reactor is given in Table 8.1. Results of this type are obtained from computer programs such as ORIGEN, which can provide estimates of mass inventories as a function of fuel burn up. The computed results are known to agree with measurements within about 20 to 30 percent for most isotopes. Actually, since some 800 species can be present at any given time in the reactor, Table 8.1 lists only the more important isotopes such as the fuel materials, H_2O, Zr, Xe, I, Sr, Cs, Pu, Ag, *etc.* Table 8.2 gives the half-lives and radioactive inventories of some of these isotopes. Health hazards are largely associated with the longer lived, volatile isotopes of I, Cs, Sr, Pu, Ru, and Te.

Table 8.1: Typical material inventories for 800 MWe pressurized water reactors (PWR) (midpoint of an equilibrium fuel cycle). Adapted from Loyalka [1983].

Material	kg	Material	kg
UO_2	1.0×10^5 (fuel)	Fe	2.5×10^3 (core structure)
Zr	2.0×10^4 (clad)		2.5×10^4 (core+bottom)
Sn	3.0×10^2 (clad)	Ag	2.0×10^3 (control rods)
H_2O	2.5×10^5 (coolant)		
Fission Products:			
Xe	260	Sb	2
Kr	20	Mo	160
I	12	Ba	70
Cs	140	Sr	54
Te	23	Zr	190
Ag	3	Ru	110

The accident sequences that could result in significant vaporization of the materials in Table 8.1 in combination with simultaneous and subsequent aerosol formation include those initiated by loss of coolant accidents and severe transients (primary coolant pipe break, main steam line break in a PWR, control rod ejection, pressure vessel failure, *etc.*). Details of such sequences are given in the references and we have summarized a few of the more important sequences in Table 8.3. The notation here is that of the Reactor Safety Study (WASH-1400).

Table 8.2: Important radioactive nuclides (in a 3412 MWth PWR operated for three years; as predicted by the ORIGEN code). Adapted from NUREG-0772.

	Half Life (days)	Inventory (Ci×10⁻⁸)		Half Life (days)	Inventory (Ci×10⁻⁸)
Iodine Isotopes					
I-131	8.05	0.87	I-133	0.875	1.8
I-132	0.0958	1.3	I-135	0.280	1.7
Noble Gases					
Kr-85	3.950	0.0066	Kr-88	0.117	0.77
Kr-85m	0.183	0.32	Xe-133	5.28	1.8
Kr-87	0.0528	0.57	Xe-135	0.384	0.38
Cesium Isotopes					
Cs 134	7.5×10^2	0.13	Cs-137	1.1×10^4	0.065
Other Fission Products					
Sr 90	1.103×10^4	0.048	Ba-140	1.28×10^1	1.7
Ru-106	3.66×10^2	0.29	Ce-144	2.84×10^2	0.92
Te-132	3.25	1.3			
Actinide Isotopes					
Pu-238	3.25×10^4	0.0012	Pu-241	5.35×10^3	0.052
Pu-239	8.9×10^6	0.00026	Cm-242	1.63×10^2	0.014
Pu-240	2.4×10^6	0.00028	Cm-244	6.63×10^3	0.0084

Table 8.3: Accident sequences that could result in sizeable release of radioactive isotopes (partial list). Adapted from Loyalka [1983].

Designation	Description
1) RSS PWR Large Containment:	
a) TMLB′-$\delta,\gamma,\varepsilon$	Loss of reactor coolant system heat removal given loss of all AC power, containment failure due to H_2 overpressurization or containment melt through.
b) S_2C-δ	Failure of containment spray injection given a small LOCA containment failure due to overpressurization.
c) S_2D-ε	Failure of ECCS given a small pipe break, containment failure due to containment melt through.
2) RSS BWR:	
a) TC-γ-γ'	Failure of reactor shutdown systems given a transient event; containment failure due to overpressure, release direct to atmosphere or release through reactor building.
b) TW-γ-γ'	Failure of decay heat removal system given a transient event; containment failure due to overpressurization, release direct to atmosphere or through reactor building.

Nuclear Source Term

Under normal conditions, radioactive isotopes are contained in the fuel rods. The cladding, water coolant system, piping and pressure vessel, containment, and engineered safety features (sprays, ice condensers, suppression pools, *etc.*) are designed to limit the release of radioactive isotopes during accidents. Natural processes (physico-chemical reactions, deposition, settling, coagulation, fragmentation, aerosol growth, *etc.*) may act to reduce or enhance the release fractions.

Typically, processes contributing to releases of fission products from fuel during severe core damage accidents may be classified as:

a) <u>In-Vessel Releases:</u> involving cladding rupture, transport from the solid fuel matrix, evaporation from molten fuel in the pressure vessel, leaching of fuel following a cladding failure, and oxidation of fragmented fuel.

b) <u>Ex-Vessel Releases:</u> involving steam explosions, high pressure melt ejection, and core debris/concrete interactions.

The dynamics of these processes are strongly dependent on the type of accident which, in turn, strongly affects the development of the accident. The releases due to the processes mentioned above are dependent on the volatility of different isotopes, their physical and chemical behavior, diffusion, chemical affinities, reaction rates, and equations of state. Complete information in all these areas is not available at present and considerable difference of opinion (orders of magnitude) exists on the estimated release fractions and their chemical states. In-vessel releases have been reviewed by Allison *et al.* [1987] and a review of ex-vessel releases has been given by Brockmann [1987]. A host of processes can affect the evolution of aerosols in the Primary Coolant System (PCS) and will act in synergism. For example, condensation, evaporation, and coagulation can affect each other very substantially. Likewise coagulation processes can be affected by the motion of aerosols under thermal and concentration gradients. Thus, not only the primary processes, but also the second and third order processes, can have a strong role in the development of the accident. Release rates that create high concentrations of aerosols can arise quite differently from release rates that create small concentrations. The typical aerosol released to a PCS could range from 20–500 kg, thus leading to concentrations of 100–2500 gm/m^3 for a 200 m^3 PCS PWR, and 28–700 gm/m^3 for a 700 m^3 PCS BWR. By comparison, a very dense aqueous fog contains approximately 1 gm/m^3.

The shapes of the aerosol particles can vary from spherical to highly irregular. Humidity favors the former. As an accident progresses, the aerosols, vapors, gases, *etc.*, will be continually released from the PCS to the reactor containment. If the accident is such that the PCS is bypassed (for example, melt ejection) then the aerosols, vapors, *etc.*, can be released directly to the containment.

Releases to containment undergo further evolution under natural transport processes as well as under the influence of engineered safety features. Materials in containment, if leaked to the environment, are a potential source of public risk. If retained, they can be a hazard to the equipment inside. Typical PWR and BWR containments are shown in Figs. 8.1 and 8.2.

The amount and timing of the release of radioactive substances from a reactor plant to the environment is referred to as a <u>nuclear source term</u>. More broadly, source terms are characterized by the radionuclides that are released to the environment as well as the time dependence of the release, the size distributions of the aerosols released, the location (elevation) of the release, the time of containment failure, the warning time, and the energy and momentum released with the radioactive material. All of these quantities are needed for the calculation of ex-plant dispersion of the radioactivity. Note that in almost all calculations, the source terms are then used as input to various programs that compute ex-plant consequences.

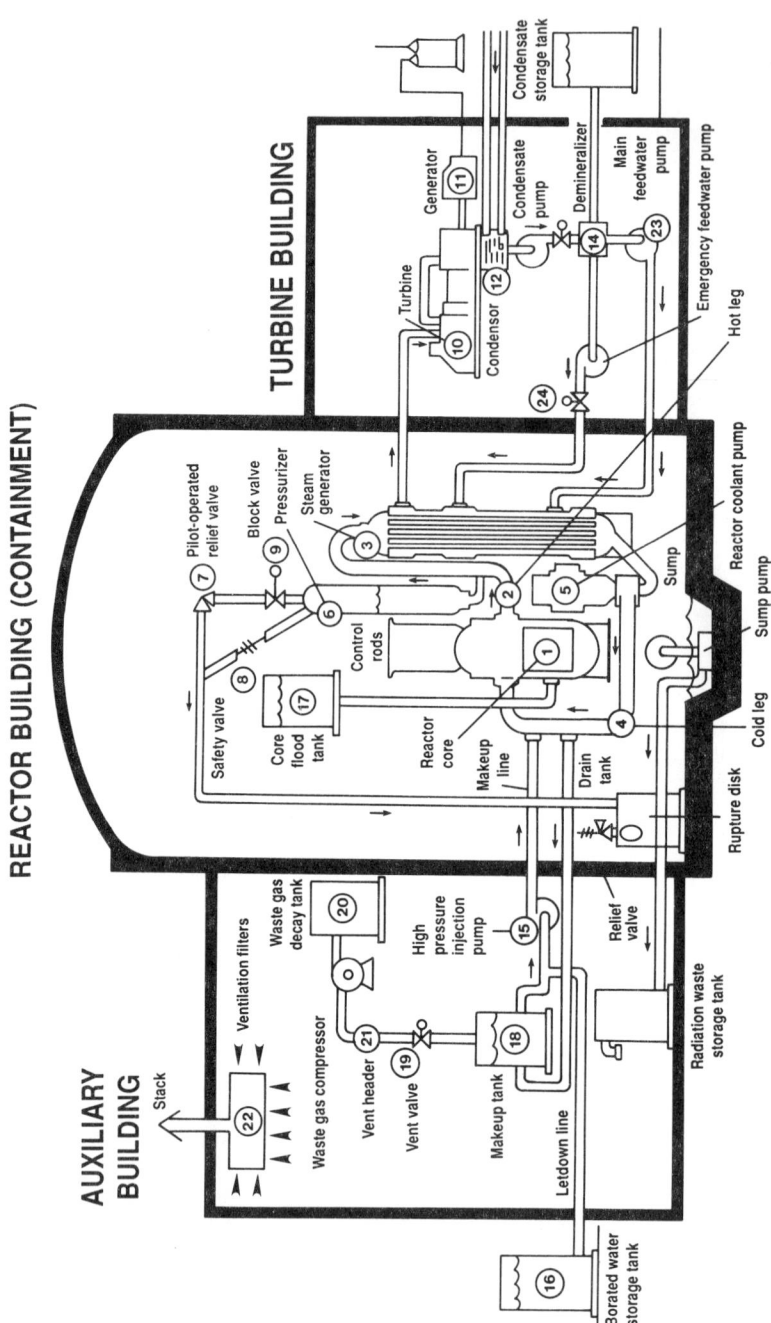

Figure 8.1: TMI-2 nuclear plant layout. TMI-2 is a typical Pressurized Water Reactor (PWR).

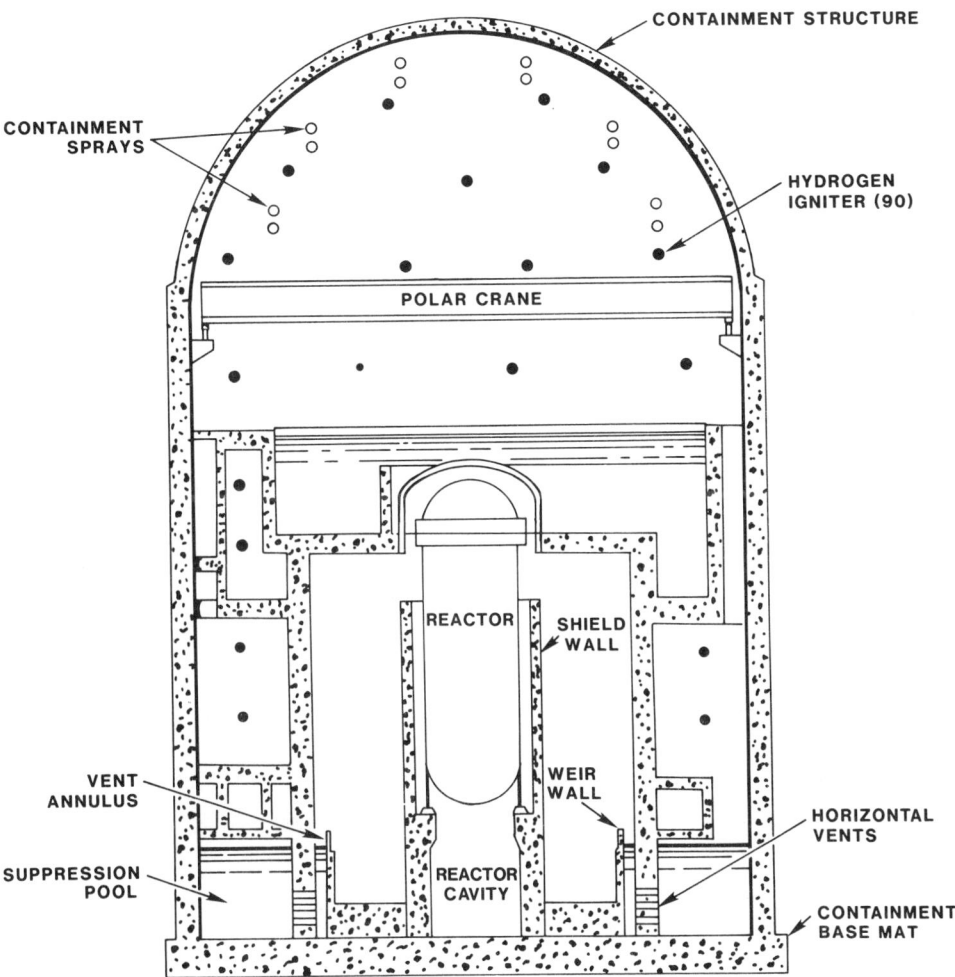

Figure 8.2: Schematic of the containment design for the Grand Gulf plant. Grand Gulf is a typical Boiling Water Reactor (BWR).

Aerosol Science: Theory and Practice

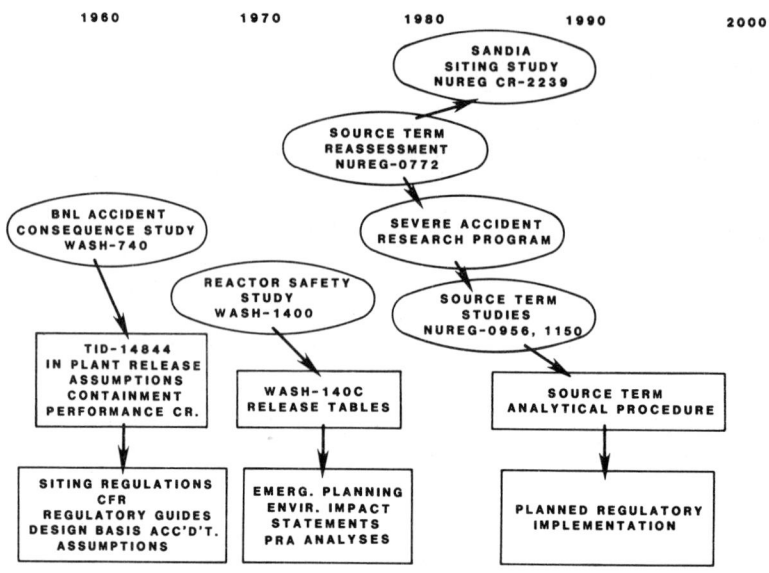

Figure 8.3: A historical perspective of U.S. computational progress.

The definition of the source term is slightly loose as different computer programs may require different inputs. Still, it is clear that source terms will be closely related to the vapors, gases and particles in suspension in the reactor containment (or building) at a given time, and the states of this suspension and the containment. If a containment does not fail (and is not bypassed) then, regardless of the complicated phenomenology that takes place inside the containment during the accident, the source term would be zero and no direct harmful effects to the public would result.

The determination of source terms within well defined bounds is not simple. First, a range of severe accident scenarios, with corresponding initiating events, must be studied. Using probabilistic methods (fault and event trees), one can then assign a probability of occurrence to the given accident scenario. Next, an integrated analysis of all that occurs in the plant needs to be carried out. Detailed physico-chemical, neutronic and thermal hydraulic models with an extensive data base (separate effects) and integrated computer programs (as verified against a range of integral experiments) are required. This task can be quite overwhelming as the number of molecular species involved is large, temperatures and pressures can be high, and the associated flows can be quite complex. High radiation fields are also present and, depending upon the specific type of accident, the situation can be very dynamic.

While efforts continue toward the development of improved mechanistic programs, considerable progress in delineating source terms has been achieved through a series of simple (perhaps simplistic) and detailed computational schemes, and limited experimental programs. These efforts have been generally supported by government agencies and more recently by private utilities, industries, and their associations. A historical perspective on the progress in the United States is shown in Fig. 8.3.

Note that in 1957, the WASH-740 reports recommended an exclusion zone of radius, r (miles), around a nuclear plant of power, P (MWth), based on the formula:

$$r = \left(\frac{P}{10}\right)^{1/2}$$

= 17 miles for a 3000 MWth plant. (8.1)

This formula is not based on realistic estimates. Rather, all material from the plant is assumed to disperse without any mitigating mechanisms. In 1957 the Windscale accident occurred, where 100% of the noble gases, 12% of the I inventory, and 10% of Cs inventory of the core were released to the environment. This accident was the basis for the TID-14844 criteria for licensing ([1962], reg. guides 1.3 and 1.4) which stipulates that release from the core to the environment will consist of 100% of noble gas, 50% of I (in gaseous form), and 10% of the nonvolatile (solids) inventory. It was also specified that the containment would retain half of the I (of that released), and all of the solids. Further retentions could occur because of particular containment designs and engineered safety features. In these guidelines, containment is assumed not to fail, but to leak.

1975 was a significant year in the history of reactor safety analysis and the source term. The WASH-1400 report provided estimates on frequency of accidents and related consequences. The analysis was specific to a PWR and a BWR. The consequence analysis used source terms that were based on computer codes that were very conservative (At least one of them, MARCH, may not have been 'conservative' at all. Among its less troublesome aspects, it violated conservation laws of mass and energy). In particular, aerosol dynamics were handled through empirical correlations in the computer program, CORRAL. Overall, consequence analysis was found to be rather weak.

The March 28, 1979 accident at the Three Mile Island-2 (TMI-2) plant near Harrisburg, Pennsylvania was rather serious. In the WASH-1400 nomenclature, the accident sequence was TMLQ (transient with loss of flow) initiated by a pressure relief valve stuck in an open position and later exacerbated by operator actions that shut down emergency cooling water. The source terms, however, were a matter of great surprise. While, all the noble gases were released (no surprise), the releases of volatiles (I, Cs) were three to four orders of magnitude smaller than those predicted by a TID-14844 type of analysis. These observations and their implications for safety analysis were noted in a series of papers. It was argued that in wet reducing environments iodine and cesium are quite reactive and do not stay in the vapor phase but either plate out or react with aerosols that settle or deposit on the walls. Thus, the chemistry of the environment and the aerosol dynamics clearly play a vital role in the estimation of source terms.

The Chernobyl accident (April 26, 1986) was characterized by a large source term (large releases). At Chernobyl, however, the situation was quite different in that the reactor building was not designed to be a containment. Also, chemical explosions rendered any containment possibilities ineffective.

Some recent studies show that typical reactor containments in the United States are well designed and that they can withstand pressures of up to 140 psia before they breach. This fact limits source terms to low values for many potential accidents of concern.

8.2 Nuclear accidents

The aim of studies of aerosol mechanics in reactor safety is to help estimate realistic (or conservative) consequences of nuclear reactor accidents. In the past, several nuclear reactors (SL-1, SNAPTRAN, Crystal River-3, Windscale-1, KIWI-TNT, HTRE-3, SPERT-1, TMI-2) have, unintentionally or otherwise, experienced core damage accidents. A summary of these accidents and the consequent releases of I and other fission products is given in Table 8.4

(adapted from Morewitz [1981]). A common observation is that large releases to the atmosphere were obtained only when 'dry' situations prevailed (Windscale-1, a graphite moderated reactor, and HTRE-3, a zirconium hybrid moderated reactor). In all light water reactor accidents releases to the atmosphere were rather small.

The Chernobyl accident is in contrast to the above accidents. The RBMK reactor #4 was destroyed, lives were lost and sizeable releases of radioactivity occurred, extending over a period of several days. For these reasons, we discuss this accident and its associated source term in greater detail below.

The Chernobyl reactor started operating in December 1983 and by April 1986 had an average fuel burnup of 10.3 MWd/kgU. The accident on April 26, 1986 was caused by operator errors that led to a prompt critical excursion exacerbated by positive void coefficients. It is estimated that within a second (1:23:44 a.m. to 1:23:45 a.m. local time) the fuel went from about 330 °C to 2000 °C and, during the next second, a substantial part of the fuel melted and vaporized (above 2760 °C). The reactor core then disassembled violently and the excursion was terminated.

In the following hours, the heat from the excursion and the radioactive decay was redistributed in the fuel and the relatively cooler graphite. Eventually, the graphite began to burn and continued to burn for several days. This exothermic reaction heated the core further. Sand and other materials (boron carbide, dolomite, Pb, *etc.*) were dropped on the core and eventually, by May 6, the fire was effectively extinguished and the release terminated.

The violent disassembly of the core and the graphite fire led to some 30 additional fires in and around the reactor. Fire fighting efforts were carried out in a radioactive environment leading to two immediate and 29 subsequent deaths of site personnel.

The radioactivity inventory of the reactor core, based on INSAG summary as prescribed by Soviet experts, is given in Table 8.5. The releases occurred over a ten day period, and can be grouped into four phases as noted in Table 8.6. The releases contained noble gases, fuel and core debris including fuel fragments and large chunks of metal and graphite, a large number of large sized particles (tens to hundreds of μm), and an even larger number of submicron aerosol particles. The latter were transported aerodynamically over large distances within the northern hemisphere. Estimates of the releases are given in Table 8.7. The proportion of volatiles to nonvolatiles varied slightly (from 7:10 to 4:6) over the time period.

We note that the releases listed in Table 8.7 are estimates obtained from near and far field measurements of radioactivity, aerosols, and debris and from atmospheric dispersion models (an inverse problem). Thus, there are considerable uncertainties and the use of different dispersion models does lead to different estimates. An entirely different way of estimating the source term would be to model the accident using thermal hydraulic and aerosol codes for the reactor and the building.

8.3 Computer programs

Computer programs that solve the General Dynamic Equation (GDE) have been written based on the methods discussed in Chapter 5. The more developed programs incorporate the thermal hydraulic equations that describe the gas-vapor mixture and couple their solutions with those of the GDE. While earlier programs such as CORRAL and more recent programs such as XSOR rely on empirical or semiempirical correlations, considerable effort has been expended on more detailed programs such as HA-3, HAARM, PARIDESKO, AEROSIM, ASTD, NAUA, MAEROS, CONTAIN, QUICK, ZONE, TRAP-MELT, *etc.* Aspects of these programs have been reviewed by Loyalka [1983], Williams *et al.* [1987], and Dunbar *et al.* [1988].

Table 8.4: Reactor Accidents. Adapted from Morewitz [1981] with permission of the American Nuclear Society.

Facility	Dry/Wet	Contained/Uncontained (C/U)	Release Iodine	Release Fission Products	Release Noble Gases	Plume Distance to Bkg. (mi)
Windscale-1	Dry	U	2.0×10^4 Ci (12%) atmosphere	1600 Ci Te 600 Ci Cs-137 80 Ci Sr-89 9 Ci Sr-90 (in atmosphere)	3.4×10^5 Ci atmosphere	200+ estimated
SL-1	Wet	C–	80 Ci (<0.5%) atmosphere	~0.1 Ci Sr-90 ~0.5 Ci Cs-137 (on the ground)	10^4 Ci atmosphere	50 ground measured
NRX	Wet	C		10^4 Ci in containment H_2O (~10^6 gal)		0.25 ground measured
TMI-2	Wet	C+	17 Ci atmosphere	none found in atmosphere	10^7 Ci atmosphere	20 air measured
WTR	Wet	C	0 Ci in atmosphere	10^4 Ci in containment H_2O (1.6×10^6)	<800 Ci atmosphere	1 ground measured
CR-3	Wet	C+	in containment 70 Ci in H_2O 2 Ci in air		1000 Ci containment air	0
PRTR	Wet	C	in containment 205 Ci in H_2O (27%); 7 Ci in air (0.9%)		~50% containment air	0
HTRE-3	Dry	U	34 Ci (~14%) atmosphere	~0.1 Ci Sr-91 ~400 Ci atmosphere		>6
ORR	Wet	C	0.15–0.2 Ci atmosphere ~300 Ci primary system	~1000 Ci primary system (~300 Ci I)		

Table 8.5: Core inventory of radionuclides at Chernobyl (based on INSAG summary as prescribed by Soviet experts). Adapted from U.S. Department of Energy [1987] (DOE/ER-0332).

Isotope	Half Life (days)	April 26 Inventory (Bq)	April 26 Inventory (MCi)	Isotope	Half Life (days)	April 26 Inventory (Bq)	April 26 Inventory (MCi)
Kr-85	3.93×10^3	3.3×10^{16}	8.9×10^{-1}	Ce-141	3.25×10^1	5.6×10^{18}	1.52×10^2
Xe-133	5.27	7.3×10^{18}	1.96×10^2	Ce-144	2.84×10^2	3.2×10^{18}	8.6×10^1
I-131	8.04	3.1×10^{18}	8.5×10^1	Sr-89	5.3×10^1	2.3×10^{18}	6.2×10^1
Te-132	3.25	3.3×10^{18}	9.0×10^1	Sr-90	1.02×10^4	2.0×10^{17}	5.4
Cs-134	7.5×10^2	1.9×10^{17}	5.0	Np-239	2.35	3.6×10^{18}	9.8×10^1
Cs-137	1.1×10^4	2.9×10^{17}	7.8	Pu-238	3.15×10^4	1.0×10^{15}	2.7×10^{-2}
Mo-99	2.8	7.3×10^{19}	1.98×10^3	Pu-239	8.9×10^6	8.5×10^{14}	2.3×10^{-2}
Zr-95	6.56×10^1	4.9×10^{18}	1.33×10^2	Pu-240	2.4×10^6	1.2×10^{15}	3.2×10^{-2}
Ru-103	3.95×10^1	5.0×10^{18}	1.35×10^2	Pu-241	4.8×10^3	1.7×10^{17}	4.6
Ru-106	3.68×10^2	2.0×10^{18}	5.4×10^1	Cm-241	1.64×10^2	2.5×10^{16}	7.0×10^{-1}
Ba-140	1.28×10^1	5.3×10^{18}	1.42×10^2				

Table 8.6: Timing and dynamics of the release at Chernobyl. Adapted from U.S. Department of Energy [1987] (DOE/ER-0332).

Phase of Release	Days of Release	Timing	Description
I	0	Apr. 26, 1:24 a.m.	Highly energetic release associated with core dispersal.
II	0 – 5	Apr. 26 – May 1	Substantial releases, declining daily because of smothering of the reactor.
III	6 – 9	May 2 – May 5	Substantial releases, increasing daily due to escape through the covering material.
IV	10 –	May 6 –	Rapid decrease in release essentially terminating the accident.

Table 8.7: Radionuclide composition of the discharge at Chernobyl. Adapted from U.S. Department of Energy [1987] (DOE/ER-0332).

Isotope	On April 26 (MCi)	(%)	By May 6 (MCi)	(%)	Isotope	On April 26 (MCi)	(%)	By May 6 (MCi)	(%)
Xe-133	5.0	--	4.5×10^{1}	100	Ce-141	4.0×10^{-1}	0.33	2.8	2.3
Kr-85	1.5×10^{-1}	--	--	100	Ce-144	4.5×10^{-1}	0.52	2.4	2.8
Kr-85m	--	--	5.0×10^{-1}	100	Sr-89	2.5×10^{-1}	0.45	2.2	4.0
I-131	4.5	5.3	7.3	20	Sr-90	1.5×10^{-2}	0.27	2.2×10^{-1}	4.0
Te-132	4.0	4.4	1.3	15	Np-239	2.7	0.72	1.2	3.2
Cs-134	1.5×10^{-1}	3.0	5.0×10^{-1}	10	Pu-238	1.0×10^{-4}	0.38	8.0×10^{-4}	3.0
Cs-137	3.0×10^{-1}	3.9	1.0	13	Pu-239	1.0×10^{-4}	0.43	7.0×10^{-4}	3.0
Mo-99	4.5×10^{-1}	0.35	3.0	2.3	Pu-240	2.0×10^{-4}	0.60	1.0×10^{-3}	3.0
Zr-95	4.5×10^{-1}	0.38	3.8	3.2	Pu-241	2.0×10^{-2}	0.43	1.4×10^{-1}	3.0
Ru-103	6.0×10^{-1}	0.54	3.2	2.9	Pu-242	3.0×10^{-7}	0.45	2.0×10^{-6}	3.0
Ru-106	2.0×10^{-1}	0.36	1.6	2.9	Cm-242	3.0×10^{-3}	0.43	2.1×10^{-2}	3.0
Ba-140	5.0×10^{-1}	0.65	4.3	5.6					

None of the above programs are capable of solving the GDE accurately and efficiently in the presence of both strong coagulation and condensation/evaporation. Also, radiation/charge aspects have not been adequately considered. Shape effects are considered in a limited fashion, but the data base is neither extensive (see Loyalka [1983] for the data base used in the programs) nor current. The programs have several other limitations when modelling reactions, coagulation of mixed species, and thermodynamic equilibrium (relating to absorption /adsorption) as well as in their integration with thermal hydraulics. Because of the complexity of the phenomenology that the programs seek to model, verification against the limited available data is the principle means for establishing confidence in their usage. Systematic and extensive theoretical, computational, and experimental verification has not been possible.

For simulation of experiments or accidents, the aerosol computer programs (codes) need to be used in an integrated (coupled) manner with thermal hydraulic codes that simultaneously simulate the fluid dynamics of the environment. The computer programs CONTAIN (see Williams et al. [1987]) and SCDAP (Severe Core Damage Analysis Package; see Allison et al. [1988]) address integrated analysis respectively for the containment and the primary coolant system progression of the accident. Programs still in the preliminary stages of development, would integrate both the PCS (in-vessel) and containment (ex-vessel) thermal hydraulics with aerosol evolution in a severe accident. A schematic diagram of the CONTAIN code is shown in Fig. 8.4.

Meanwhile, progress has been made with a sequential coupling of separate effect codes. The analyses that have received the most attention are those carried out with the BMI-2104 suite of codes and the STCP (Source Term Code Package). A schematic diagram of the BMI-2104 suite of codes is shown in Fig. 8.5. Note that the computer programs CORSOR, CORCON, TRAP-MELT, VANESA, NAUA, SPARC, and ICEDF involve aerosol transport and evolution in substantial ways.

Aerosol Science: Theory and Practice

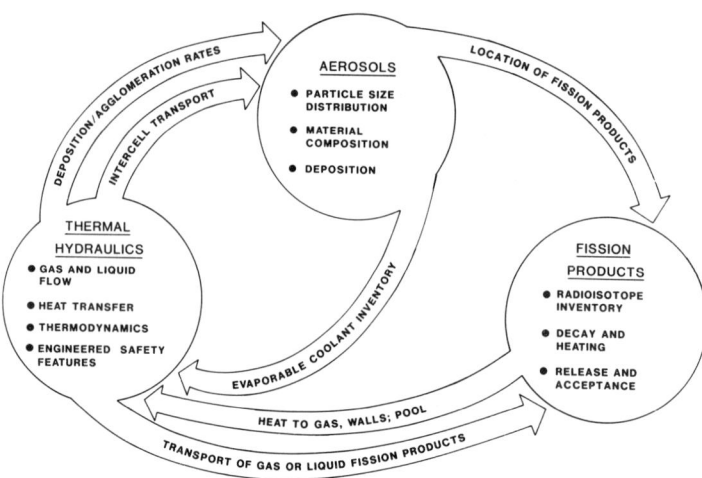

Figure 8.4: A schematic diagram of the CONTAIN code emphasizing the three basic phenomenological areas and their intercoupling.

The BMI-2104 calculations involve neither good physics nor good chemistry. While individual component codes may be quite good, overall reliability of the total calculations cannot be good except under very fortunate circumstances. One can hardly have a feel for the manner in which complex phenomena develop when there is little empirical (experimental) experience and when the situation is highly dynamic.

Actually, for regulatory purposes, it appears that even this sequential analysis is too cumbersome and too mechanistic (see NUREG-1150). A simple code XSOR has been constructed which is based on a combination of conjecture and correlations derived from BMI-2104 numerical simulations. Here, for example, transmission in piping is expressed in terms of certain parameters, the distributions of which are recommended by groups of "expert" panelists. For a particular accident, the source term is then calculated as a distribution with a mean and associated spread. The accuracy of such an approach is difficult to assess, but the circumstance reflects the complexity of the problem.

8.4 Experimental data base

Table 8.8 provides a summary of the characteristics of the various experimental facilities. These range in size from 0.53 to 850 m³. By comparison, typical containment volumes are: ~10^5 m³ (PWR) and ~10^4 m³ (BWR). The Containment System Test Facility (CSTF) at the Hanford Engineering Development Laboratory (HEDL) and the National Safety Pilot Plant (NSPP) at the Oak Ridge National Laboratory (ORNL) have provided most of the data. Recently, data have also been acquired at both the MARVIKEN and DEMONA test facilities.

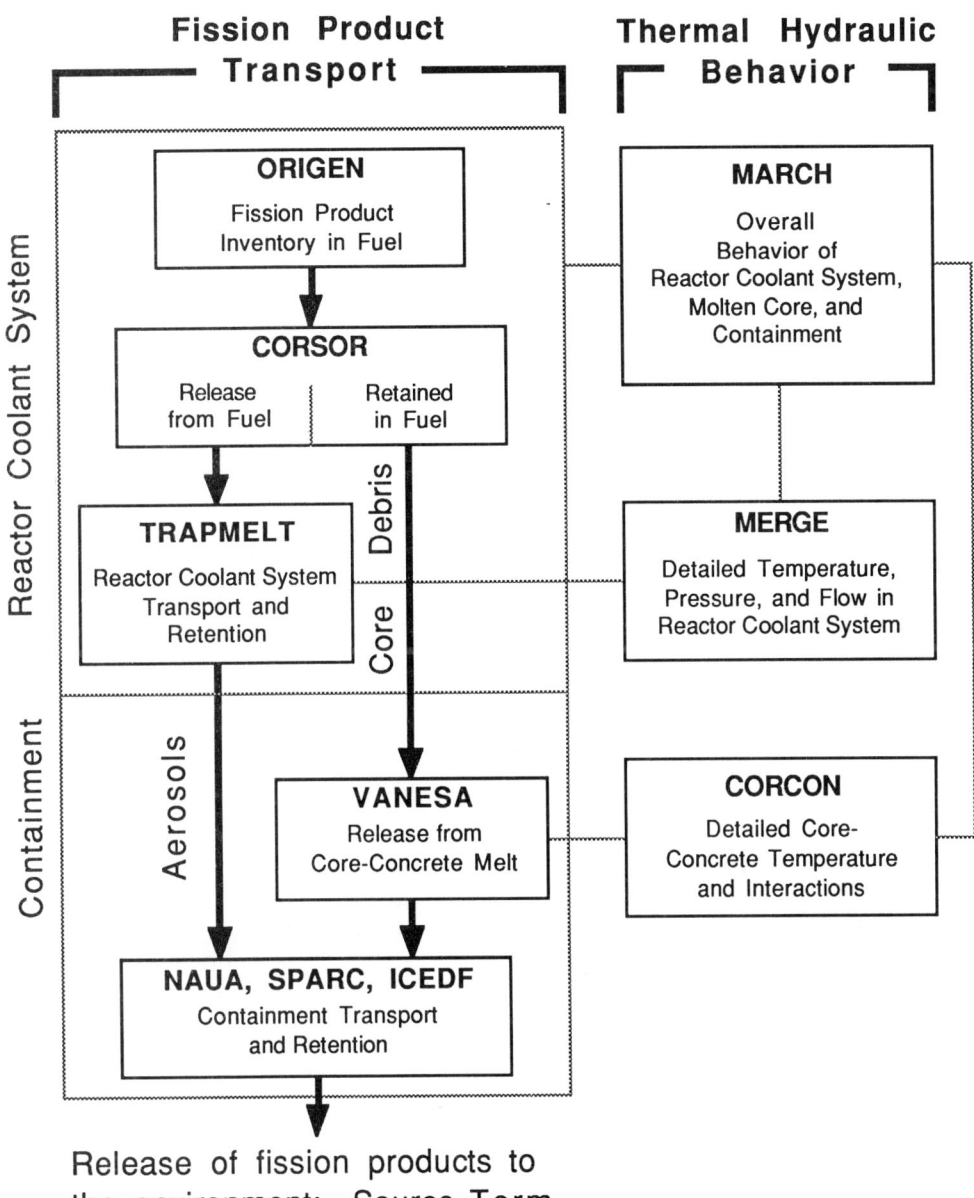

Figure 8.5: The BMI-2104 suite of codes.

Table 8.8: Experimental characteristics of LWR aerosol tests performed, in progress, or planned. Adapted from OECD [1985].

Test Name	Facility/ (Country)	Vessel Volume (m³)	Total Height (m)	Vessel Diameter (m)	Wall Material	Aerosol Source/Type
CSTF	HEDL (US)	850	20.3	7.6	steel	$UO_2/CsOH/RuO_4$
NSPP	ORNL (US)	38.3	5.2	3.1	steel	U_3O_8–Fe_2O_3
CRI-II	ORNL (US)	4.5	2	1.7	steel	UO_2+FP
DEMONA	KfK (Ger.)	640	--	--	concrete	metal oxide mix
NAUA	KfK (Ger.)	3.4	2	1.5	steel	UO_2–NO_3NaPtO_2
GRACE	Netherlands	0.3	1	1.2	glass	Au
PERVEX	Netherlands	0.15	0.6	box		Cu
SAUNA	Netherlands	0.15	0.6	box	steel	metal oxide
PITEAS	CEA (Fra.)	3	2.8	1.2	steel	$CsI/CsOH/Co_3Cs_2$
MARVIKEN V	STUDSVIK (Swed.)	primary circuit	--	--	steel	Fissium Corium

In these facilities, aerosols generated by arc vaporizers, plasma torches, exploding wires, Na fires, *etc.*, are used to simulate aerosol concentrations typical of nuclear reactor accidents. The effects of natural processes and engineered safeguards have been studied. Aerosol size distributions are measured by instruments such as impactors, electrical aerosol analyzers, centrifuges, and electro-optical devices.

Experiments relating to the PCS have also been performed. Most were done under the LACE program at the CSTF, at the Idaho National Engineering Laboratory, and at the Oak Ridge National Laboratory. We discuss below the the CSTF and the NSPP facilities.

i) <u>CSTF Experiments:</u> Fig. 8.6 shows a schematic view of the facility. From 1964 to 1971, Hilliard, Postma, and their associates conducted experiments to simulate PWR and BWR configurations. Provisions were made to inject fission products (I vapor, Cs_2O particles, I particles, Te, Ru, Ba, Xe, and particles) via a fission product (FP) simulator. Steam was injected before the tests to establish steady state thermal and convective conditions. Typical steam-air atmospheres were maintained at 250 °F (122 °C) and 50 psia (0.345 MPa). Generally, the injection of FPs in the water-steam environment created fogs which dissipated in a few hours due to particle growth, agglomeration, and settling. Use of sprays usually led to the dissipation of fogs in a few minutes. Time dependent concentrations of FPs in the vessel, on the containment walls, and in the containment environment were measured.

The CSTF facility has been used to conduct tests on Na fire aerosols in order to develop air cleaning systems for removal of such aerosols from the vessel atmospheres. Either spray fires or pool fires are used. Quantities as large as 1250 kg can be released in times as short as 100 seconds in pool fires or as long as 40 hr in small spray fires. Both the containment vessel and the air cleaning systems are equipped with multiple sampling stations for aerosol concentration, aerosol size, atmospheric composition, and temperature. The chemical forms of the suspended and deposited aerosol are analyzed by chemical and spectrometric methods.

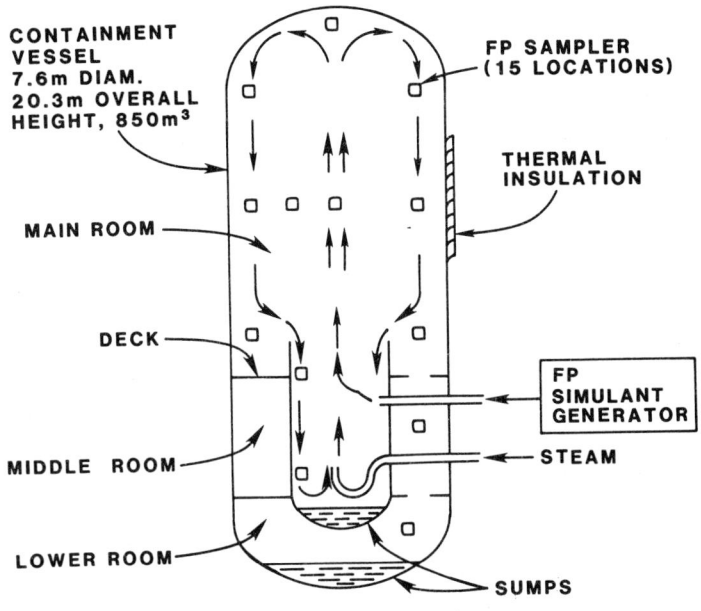

Figure 8.6: A schematic view of the CSTF facility.

Most recently the CSTF facility has been used to conduct tests for fast (AB series) as well as light water reactors (LACE Series). In the latter tests, mixed aerosols were studied in dry as well as wet environments. These tests encompassed hygroscopic aerosols as well as transport in piping.

ii) <u>NSPP Experiments:</u> The NSPP facility (Fig. 8.7) was built in 1963 and was originally used for fission product transport and deposition experiments. During the period 1967–1970, the facility was used extensively to conduct experiments on the removal of elemental I, organic iodides, and assorted particulates by means of sprays. An excellent summary of this work has been given by Roberts *et al.* [1969]. The aim of this work, like the work at the CSTF, was largely to develop correlations for use in the development of regulatory guidelines for the licensing of light water reactors. These studies noted the effectiveness of sprays containing borax, boric acid, and borax plus sulfate in the removal of I and of borax thiosulfate in the removal of methyl iodide. Particle removal experiments showed the effectiveness of condensing steam environments in removing particles from containment.

In more recent years, the NSPP facility has been used for both LMFBR and LWR aerosol tests. This work has been carried out by Kress, Adams, Tobias, and their associates and has been particularly valuable in providing verification tests for the various computer programs. Single species and mixed aerosols in both dry and wet environments have been considered and sophisticated particle analysis instrumentation has been employed. Concentration, particle size, fallout and plateout rates, vessel atmospheric temperature, pressure and moisture content, and steam condensation rate at the vessel wall have all been measured.

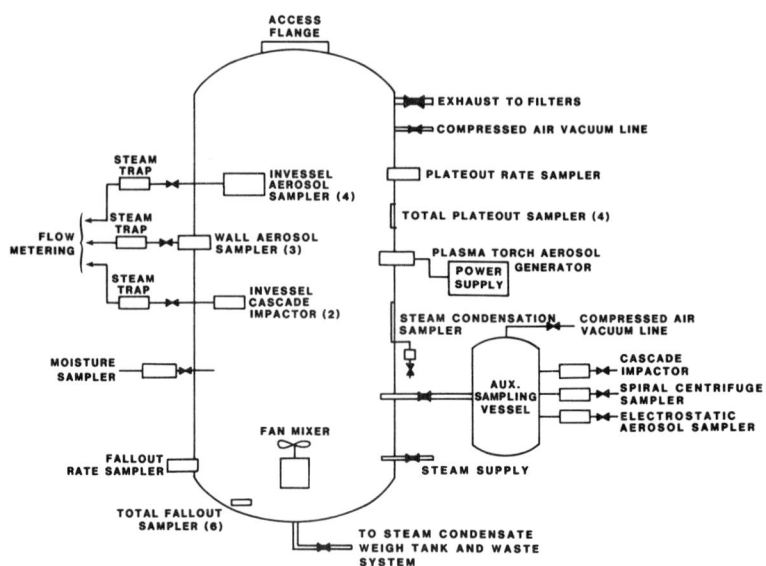

Figure 8.7: A schematic view of the NSPP facility.

iii) <u>Other Experiments:</u> Experiments have also been conducted at other facilities listed in Table 8.7. The information thus gathered has been essentially similar to that obtained from the CSTF and NSPP experiments. In particular, NAUA uses a small vessel (3.7 m³) and complements both the CSTF and NSPP experiments. A schematic of the DEMONA facility is shown in Fig. 8.8.

8.5 Simulation of experiments

<u>Early Experiments:</u> The early experiments at both the CSTF and the NSPP facilities were on I and fission product removal by natural processes and engineered safeguards. In Fig. 8.9 some typical results from these experiments are reported. This figure shows that sprays are initially very effective in reducing airborne I concentrations (one order of magnitude) but that after about 3 hr their effectiveness is marginal. It appears that particle size distributions were not measured in the test and that the experiment was not analyzed by mechanistic approaches. Instead, simple correlations were developed which appeared to describe the attenuation rates reasonably well. For example, the airborne I (i) and particulate (p) concentrations are described, respectively, by the equations:

$$c_i(t) = c_i(0) \exp\left(-\frac{k_c A_a}{V_c} t\right) \qquad (8.2a)$$

and:

$$c_p(t) = c_p(0) \exp\left(-\frac{V_p A_s}{V_c} t\right) \qquad (8.2b)$$

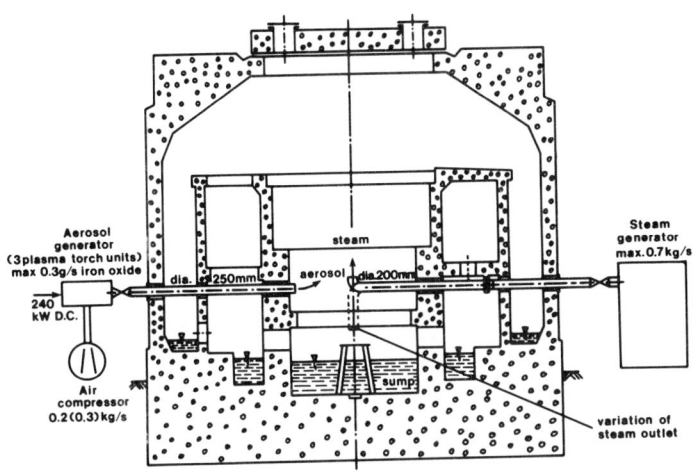

Figure 8.8: A schematic view of the DEMONA facility.

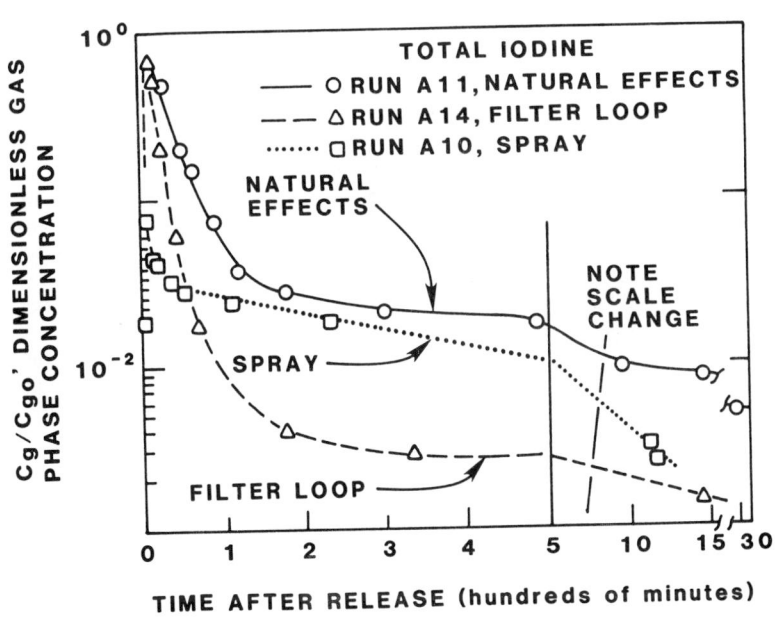

Figure 8.9: Iodine removal experiments in the CSTF.

where $c_i(t)$ and $c_p(t)$ are the airborne concentrations at time t and $c_i(0)$ and $c_p(0)$ are the initial airborne concentrations. A_a is the total area of absorption (surface of containment, droplets, and particulates), A_s is the surface area of the base of the containment, V_c is the volume of the containment, V_p is the Stokesian particle settling speed discussed earlier:

$$V_p = \tfrac{2}{9} \frac{\rho_p a^2 g}{\mu} \tag{8.3}$$

and k_c is a mass transfer coefficient:

$$k_c = (0.13)(Gr\ Sc)^{1/3} \frac{D}{L} \tag{8.4}$$

Here Gr is the Grashof number and Sc the Schmidt number. D is the diffusion coefficient of I in the background gas mixture (air-steam) and L is the thickness of the boundary layer. The above correlations for I should be used with some caution. Eventually, thermodynamic equilibrium between the I in the air and water borne phases may be established and then Eqn. (8.2a) would no longer be of use.

Note that Eqns. (8.2a) and (8.2b) correspond to the solutions of the simple rate and dynamical equations with coagulation and particle growth (except for particles acting as absorbers) neglected. These CSTF correlations formed the basis of the CORRAL program that was used in the Reactor Safety Study (RSS).

In Fig. 8.10, typical CSTF experimental data (Table 8.9) are shown and compared to the corresponding HAA-3B predictions.

The computer program appears to describe the suspended mass quite well provided the parameters are chosen to fit the experimental data. A more severe test, however, would be to compare the measured and calculated particle size distributions. Postma and Owen [1980] found that HAA-3B provided fairly good estimates of the mean aerodynamic diameter of the settled particles after nearly 200 sec in the test. Results for earlier times, however, were not that good as the measured diameters were much higher (an order of magnitude) than the predicted values. This indicates that HAA-3B may contain some underestimation of the coagulative terms or that the input parameters (initial conditions) may not have been well prescribed.

A summary of some experimental results on the behavior of Na_2O, UO_2, and mixed Na_2O–UO_2 aerosols in NSPP experiments and a comparison of these results with HAARM-3 calculations have been given by Adams et al. [1980] for 14 single component and six multicomponent experiments. The results of three of these tests are shown in Figs. 8.11-8.13 and are compared with the HAARM-3 predictions. Test 303 (Fig. 8.11) corresponds to nonspherical aerosols. Here, if the collision parameters are 'properly' chosen, then the results of the experiments can be described quite well by the HAARM-3 program (the aerosols are highly nonspherical in this case) at least as far as the total suspended mass is concerned. Figs. 8.12 and 8.13 describe the mixed oxide aerosol experiments. Clearly HAARM-3 has significant limitations in describing these experiments. The difficulties are numerous. First principles information on mixed aerosols is practically nonexistent.

<u>Recent Experiments (1980–)</u>: Tests in condensing steam environments have been conducted at both the NSPP and the CSTF facilities. We discuss these below. Some of the test results from the NSPP vessel are described in Fig. 8.14 (test 401). Here, steam was injected

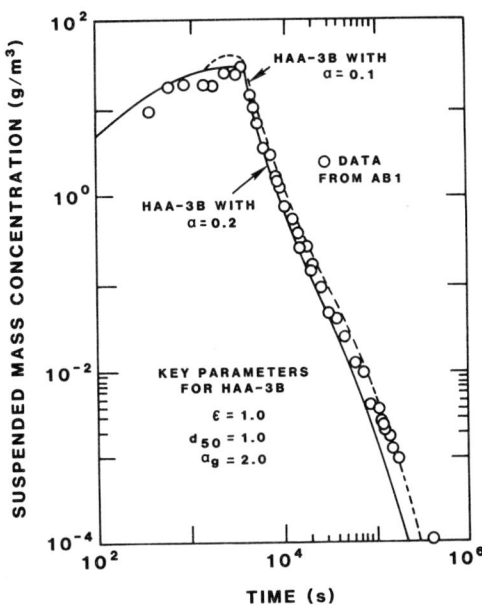

Figure 8.10: CSTF experiments AB1 and AB2 with the corresponding HAA-3B predictions.

Table 8.9: Test conditions for CSTF tests AB1 and AB2. Adapted from Postma and Owen [1980] (NUREG/CR-1724, ORNL/NUREG/TM-404, CSNI-45).

	AB1	AB2		AB1	AB2
Steam Addition			**Sodium Spill**		
Flow started (sec after t_0)	–	960	Mass Na spilled (kg)	410	472
Flow started (sec after t_0)	–	4560	Na burn pan surface (m^3)	4.4	4.4
Flow rate (kg/sec)	0	0.019	Initial Na temperature (°C)	600	600
Initial Containment Atmosphere			Na fire duration (sec)	3600	3600
O_2 (Vol%)	19.8	20.9	Total Na oxidized (kg)	157	175
Dew Point (°C)	10.0	7.6			
Temperature (°C)	26.5	20.5			
Pressure (MPa, absolute)	0.125	1.128			

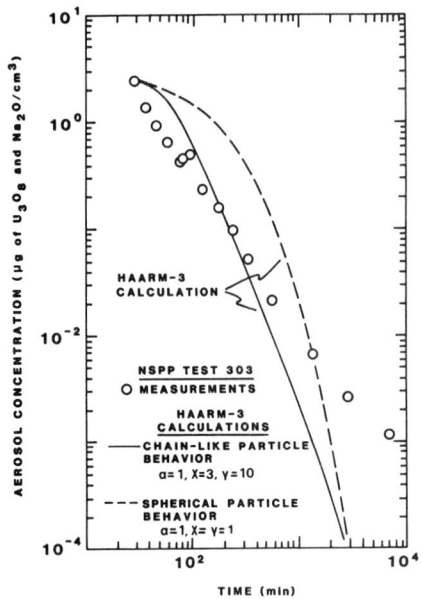

Figure 8.11: NSPP Test 303 with the corresponding HAARM-3 predictions.

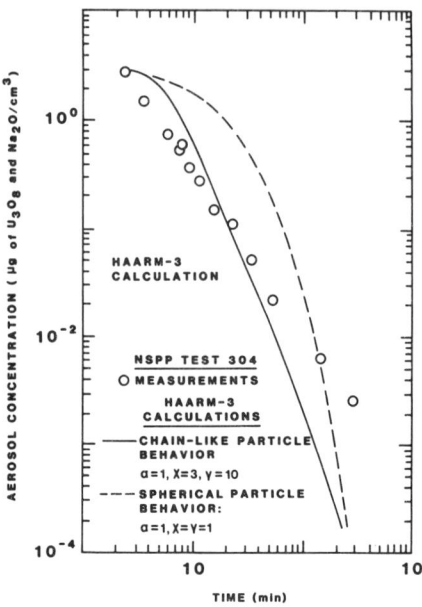

Figure 8.12: NSPP Test 304 with the corresponding HAARM-3 predictions.

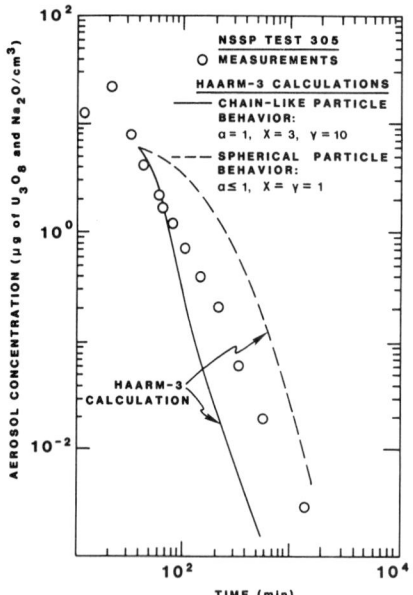

Figure 8.13: NSPP Test 305 with the corresponding HAARM-3 predictions.

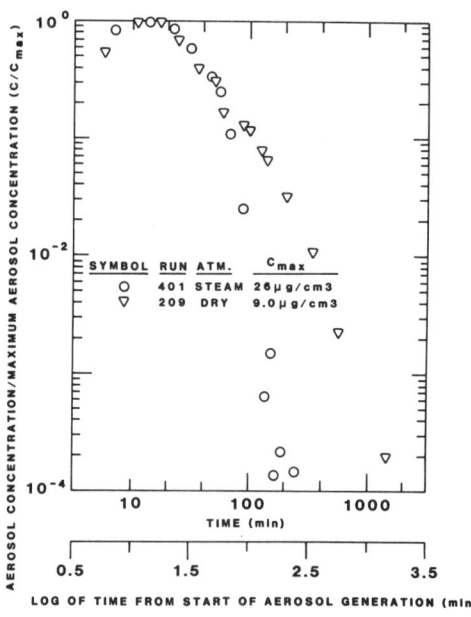

Figure 8.14: NSPP Tests 401 and 209.

Table 8.10: LACE tests and objectives.

Test	Objective
LA1 (Containment bypass)	Determine the retention and fate of aerosol materials flowing through a vented downstream auxiliary building under severe accident conditions.
LA2 (failure to isolate containment)	Provide aerosol measurements in a containment with prescribed leak paths.
LA3 (Containment bypass)	Determine retention and fate of aerosols flowing through simulated LWR piping. Three separate runs with different flow rates and soluble/insoluble aerosol mass.
LA4 (Late containment failure with overlapping injection periods)	Determine the effect of releasing two aerosol species at different times.
LA5 (Rapid depressurization with a spiked pool)	Study aerosol reentrainment from flashing pools caused by rapid depressurization.
LA6 (Rapid depressurization with aerosol injection)	Study pool flashing aerosol reentrainment and resuspension from dry surfaces during depressurization.

into the NSPP vessel and the environment was brought to a temperature of approximately 364 K and gage pressure of 0.093 MPa. After steady state conditions were achieved, the steam injection was reduced and the water condensate was removed from the vessel. The aerosol was generated next and was introduced into the vessel over a period of 10 min. Over the next 110 min, steam was slowly added to compensate for losses due to surface condensation within the vessel while the temperature and the gage pressure within the vessel were increased to 376 K and 0.143 MPa. The vessel and its contents were then allowed to cool for about 22 hr after termination of the steam injection. During the test, measurements were made of aerosol mass. Clearly, after steam injection, aerosol mass is removed rapidly from suspension. This was most likely due to the Stefan velocity (cool vessel walls) carrying aerosol to the walls.

In the DEMONA experiments, fog formation, high aerosol concentrations, and inhomogeneous aerosol distributions were identified as enhancing fog removal (see Figs. 8.15 and 8.16). However, good agreement between the data and the code results was obtained only for simple conditions (homogeneous atmosphere, steady state operation). Complex situations led to large differences between the data and the code results.

The LACE series consisted of the six tests listed in Table 8.10 all of which dealt with MnO/CsOH aerosols in condensing environments. Their purpose was to obtain data for some dynamic thermal hydraulic conditions and compromised reactor coolant boundaries (a leak or bypass path from containment to the auxiliary building or environment). The tests LA1 and LA3 deal with containment bypass, LA2 and LA4 relate to aerosol behavior in a leaking containment building, and LA5 and LA6 model delayed containment failure to provide data on aerosol resuspension. The tests represent circumstances that can lead to source terms larger than in many other accident sequences. Hence, understandings of aerosol behavior here can, perhaps, lead to some empirical bounds on the source term.

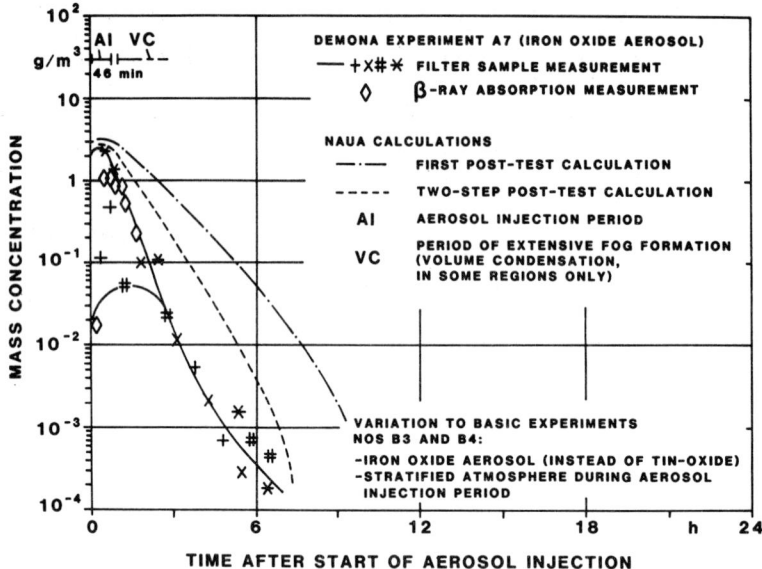

Figure 8.15: DEMONA experiment A7 (iron oxide aerosol).

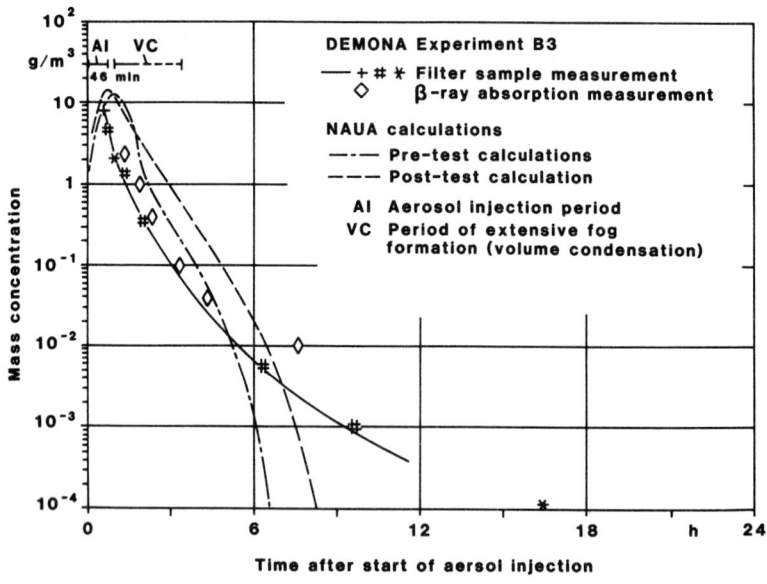

Figure 8.16: DEMONA experiment B3.

The experiments were all conducted in two CSTF vessels with modifications and additions appropriate to each test. The aerosol mixture consisted of CsOH aerosol to simulate volatile, hygroscopic, fission product species and MnO to simulate semi-volatile, nonhygroscopic, fission products and structural compounds. The CsOH aerosol was generated through a reaction of steam with Cs vapor (obtained by heating elemental Cs in a furnace). The MnO aerosol was generated through vaporization of metallic Mn by a plasma torch and the reaction of the resulting Mn vapor with steam. The two aerosols were generated separately and mixed in a 5.2 m^3 settling vessel before injection into the test piping or the CSTF vessel. In all of these tests, thermal hydraulic conditions and aerosol behavior were characterized (measured) and the predictive capabilities of the available computer codes were assessed. Tables 8.11 and 8.12 list the participants and the computer codes that were used.

First, let us discuss the containment bypass tests, LA1 and LA3. A schematic of LA1 is shown in Fig. 8.17, while the modifications carried out in LA3 are shown in Fig. 8.18. These tests used a test pipe 63 mm in diameter and 27 m in length with five 90° bends, four horizontal sections, and two vertical sections.

The conditions for the LA1 test are given in Table 8.13. The conditions for the LA3 test were similar (Fig. 8.18) except that the auxiliary building (the CSTF vessel) was not used and the gas inlet velocity and aerosol generation rates were lowered. Each test was performed in three stages including a warm up stage to establish the initial thermal hydraulic conditions, a 60 min period of aerosol injection, and a cooling down period.

In the LA1 test, it was found that almost 98% of the aerosol material was retained in the 63 mm test pipe. This result is largely due to the nature of the aerosol. Its surface tension, viscosity, and melting point led to a viscous film that tightly adhered to the pipe surface (note that in the same setup, earlier scoping tests with NaOH and $Al_2(OH)_3$ aerosols led to little retention).

The LA3 test results are shown in Figs. 8.19 and 8.20. Note again that aerosol retention is quite large (60 to 80%) with the bends playing a significant role. Also shown in these figures are the predictions of several computer programs. Clearly, the codes underpredicted the deposition, indicating that the deposition process needs to be modeled better (inter-code comparisons indicate that some codes did not properly account for the enhanced deposition at the bends).

The objective of the LA2 test was to study aerosol evolution in a simulated containment with the presence of leaks (failure to isolate). Size and composition distributions of both the suspended and leaked aerosols were measured and the effects of the leak location on the distributions were also assessed. The in-vessel sampling was not extensive (~4 locations) and inferences regarding well-mixing (spatial homogeneity) of the aerosols were simply drawn from measurements at the two leak locations (see Fig. 8.21).

Some data of the LA2 test are shown in Figs. 8.22-8.24. These figures compare the measurements of total suspended aerosol concentration with the predictions of several codes. Clearly, the codes did a poor job of describing the data.

Since the subject aerosols were hygroscopic, they absorbed/adsorbed water vapor quite rapidly even though the environment was not saturated (the saturation density near the surface is lower for hygroscopic aerosols than for nonhygroscopic; see Eqn. (6.1)). This growth caused rapid sedimentation which, when coupled with gravitational agglomeration, led to significant aerosol removal from suspension. This situation is simple to model but most of the codes had neglected it. Not surprisingly, the failure of the codes to model mixed aerosols appropriately is even more noticeable.

Table 8.11: Participants and aerosol behavior computer codes used. Adapted from Rahn [1988] with permission of the Electric Power Research Institute.

Participant	Computer Code
Finland	RETAIN-2C
	NAUA-5
Italy	TRAP-MELT 2
	NAUA-5
Japan	REMOVAL
	TRAP-MELT 2
Sweden	AUX 2.9
	RETAIN-S
United Kingdom	AEROSIM-M
	CONTAIN
	TRAP-MELT 2
United States	
New York Power Authority	MCT-2
NRC/ORNL	CONTAIN
NRC/Battelle, Columbus Division	QUICK-M
	NAUA-4
	TRAP-MELT 2
EPRI	NAUA-4
EPRI/AI	HAA-4
EPRI/Stone & Webster	NAUA-4

Table 8.12: Participants and thermal-hydraulic computer codes used. Adapted from Rahn [1988] with permission of the Electric Power Research Institute.

Participant	Computer Code
Commission of European Communities	CONTAIN
France	JERICHO
Germany	FIPLOC
Italy	CONTEMPT-LT
	ARIANNA
Japan	CONTAIN
United Kingdom	APRITE
	CONTAIN
United States	
New York Power Authority	MARCH
EPRI/Intermountain Technologies	CONTEMPT-LT
EPRI/MIT	LIMIT
NRC/Battelle, Columbus Division	QUICK-M

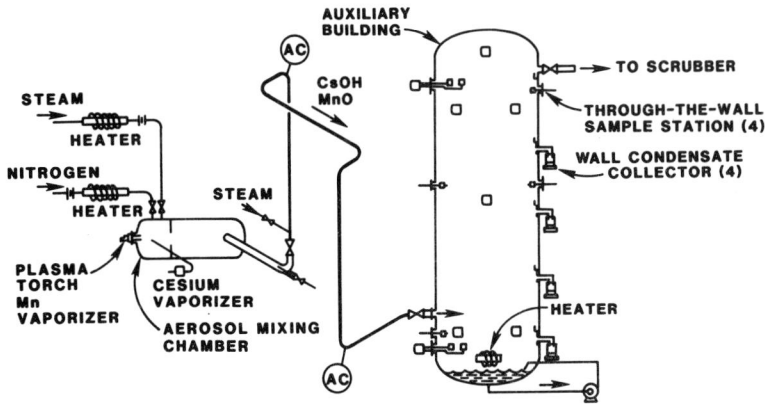

Figure 8.17: Instrument locations for the LA1 test. Adapted from Rahn [1988] with permission of the Electric Power Research Institute.

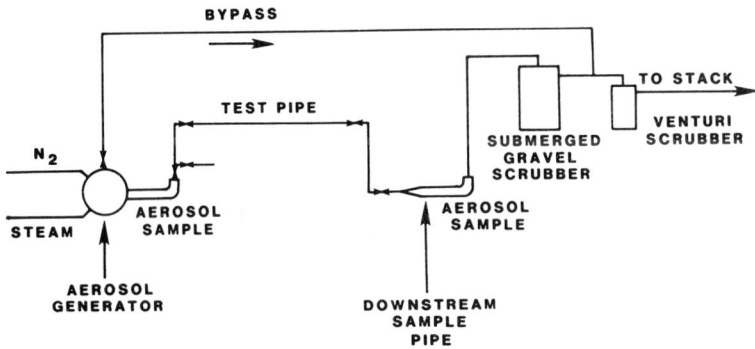

Figure 8.18: Layout of the experimental apparatus for the LA3 tests. Adapted from Rahn [1988] with permission of the Electric Power Research Institute.

The LA4 was similar to the LA2 test except for some additional instrumentation (three turbine anemometers, a photometer, two calorimeters, and a steam fraction station) and the absence of the lower leak path. The focus was again on the evolution of mixed aerosols in a condensing environment but now coupled to a late containment failure (see Fig. 8.25). Significant differences from the LA2 test included overlapping injections of CsOH and MnO aerosols (CsOH alone for 30 min and in combination with MnO for 20 min) and late venting. During the injection period the pressure in containment was increased to about 3 atm which was maintained for about 200 min. After this, the containment was vented. Some of the data on the total suspended aerosol concentration and the suspended mass ratio are shown in Figs. 8.26 and 8.27. Again, comments similar to those on LA2 apply.

In the tests LA5 and LA6 a flashing pool tank (3.0 m pool depth, 3.1 m in freeboard, and 1.5 m in diameter) was prefabricated and moved inside the CSTF vessel. The tank was the largest tank that could be moved inside the vessel. The test configuration is shown in

Table 8.13: LA1 containment bypass test conditions. Adapted from Rahn [1988] with permission of the Electric Power Research Institute.

Aerosol Source at Test Pipe Inlet

CsOH rate	0.48 g/sec
MnO rate	0.65 g/sec
Source size, AMMD	1.64 µm
Source geometric standard deviation	1.91
Suspended concentration:	
CsOH	1.8 g/m^3
MnO	2.3 g/m^3
Duration	60 min

Test Pipe Conditions

ID	63 mm
Length	29 m
# of 90° bends	6
Ball valves	4
Composition	56% steam / 44% nitrogen
Inlet temperature	247 °C
Superheat	141 °C
Velocity at inlet	97 m/sec
Velocity at outlet	193 m/sec
Pressure drop	125 kPa (average)

Auxiliary Building Conditions at Time Zero

Volume	852 m^3
Heat transfer	Approximately steady state
Vent	Open
Atmospheric composition	56% steam / 44% air
Temperature	118 °C
Superheat	31 °C
Pressure	1.1 atm

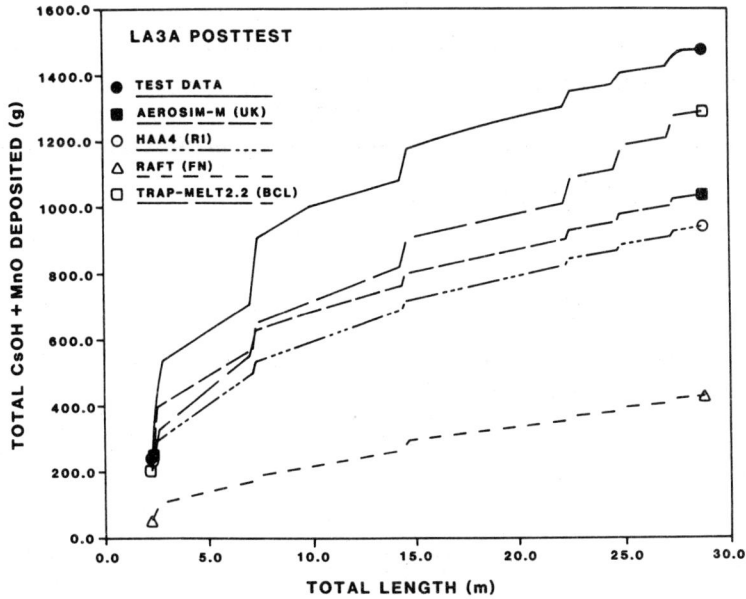

Figure 8.19: LA3 data and posttest code predictions. From A.L. Wright, private communication.

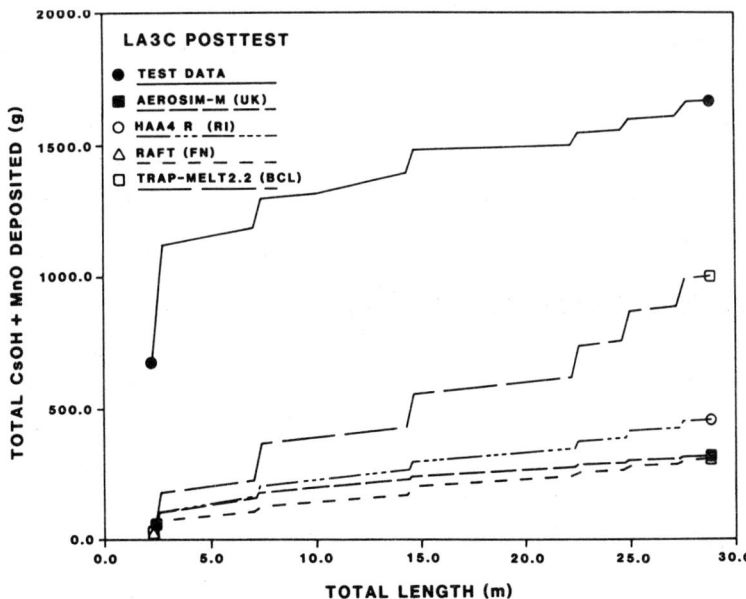

Figure 8.20: LA3c data and posttest code predictions. From A.L. Wright, private communication.

Aerosol Science: Theory and Practice

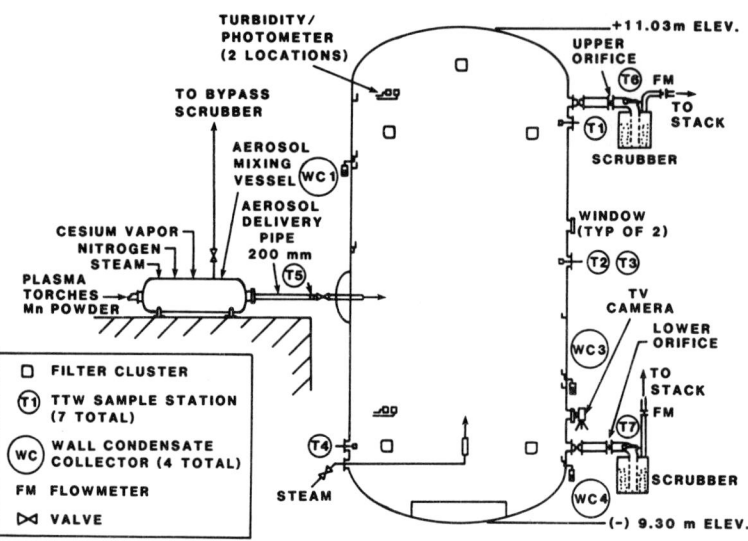

Figure 8.21: Experimental arrangement for the LA2 test. Adapted from Rahn [1988] with permission of the Electric Power Research Institute.

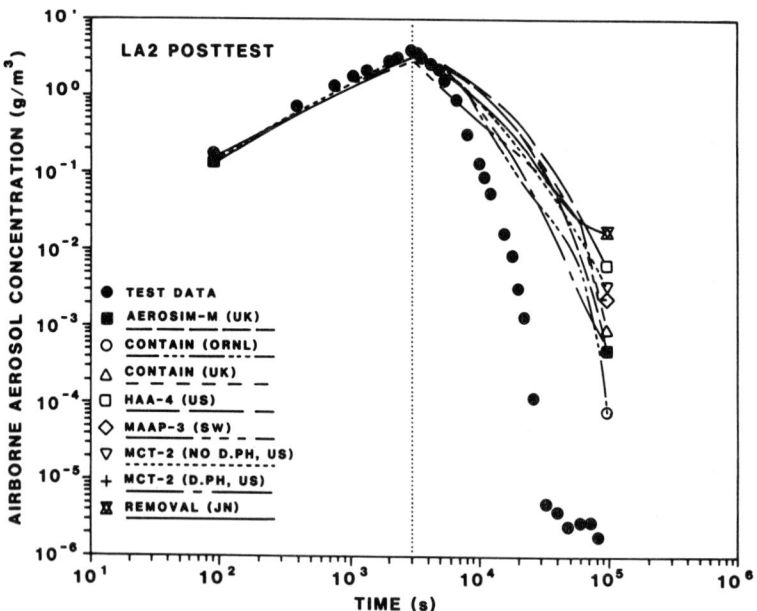

Figure 8.22: LA2 data and posttest code predictions. From A.L. Wright, private communication.

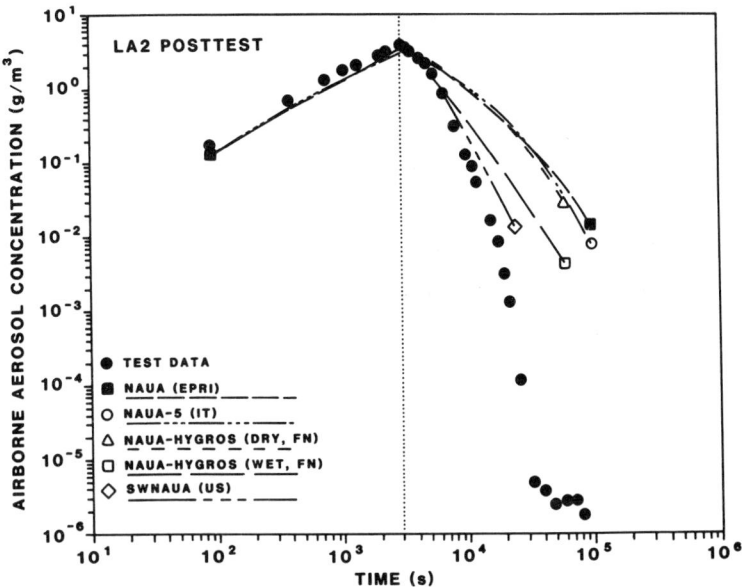

Figure 8.23: LA2 data and posttest code predictions. From A.L. Wright, private communication.

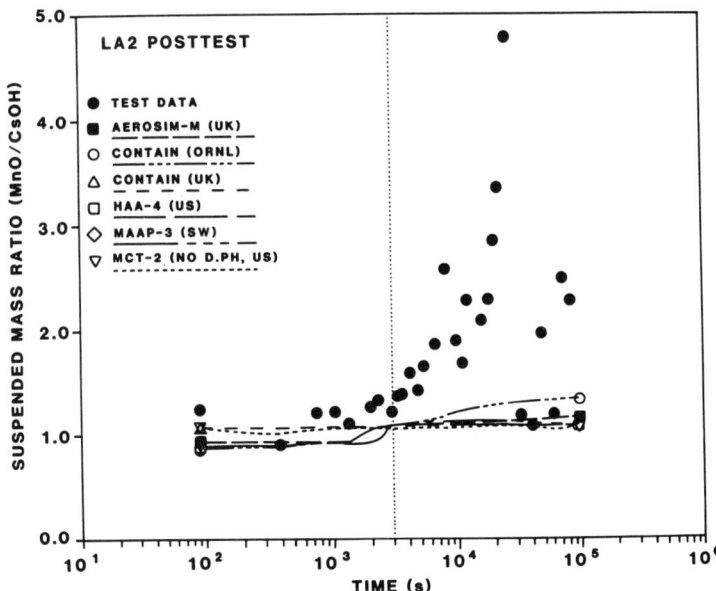

Figure 8.24: LA2 data on suspended mass ratio and posttest code predictions. From A.L. Wright, private communication.

Fig. 8.28. In both tests the pool was spiked with Li_2SO_4 and ZnO particles. In LA6 CsOH and MnO aerosols were pre-injected into the containment. The data for LA6 are shown in Figs. 8.29 and 8.30 and are compared with the code predictions. Again, the codes are not quite successful, especially over long periods of time.

8.6. Discussions and conclusions

We believe that nuclear power plants are quite safe compared to the risks posed by most other competitive sources of energy. A high fraction of the public risk posed by nuclear plants, however, is estimated to arise from accident sequences that lead to severe core damage and possible aerosol releases to the primary coolant system and the reactor containment. If there are leaks in the containment, or if the containment is breached, then these aerosols are a potential source of public risk. Accidents, with sizeable releases of aerosol to the environment, although rare, have occurred. Some of these have had very unfortunate consequences and it is important, first to prevent the core damage accidents and, next to limit the source terms that may be associated with such accidents. Aerosol science plays a vital and central role in all source term estimations, and understandings here will play an important role in the future of nuclear power.

As evident in the text, aerosol science is a multidisciplinary field, and the subject has elicited contributions from both scientists and engineers. Since the subject spans studies of interactions and interaction physics, chemistry, and mathematics from the molecular to the system level, it is necessary to employ many approximations, especially at the interfaces between the different length and time scales that are involved. Formulation of consistent (realistic or conservative) approximations is a challenging task that often strains scientific and technical imagination and creativity as well as institutional resources. The task becomes virtually impossible when results of a certain nature are sought within a time span not compatible with the scientific state-of-the-art. There is no doubt, however, that such demands do lead to rapid developments.

Aerosol science has benefitted greatly from the compelling interests of the environmental, industrial, and nuclear communities. Within the past several years, more generalized equations for physico-chemical descriptions have been developed and new analytical and numerical methods of solution have become available. Progress has been made at all levels of interaction, from molecule-particle to particle-system. New experimental data that provide enhanced understandings, as well as challenges to theoretical efforts, have been obtained. Still, it must be realized that the state-of-the-art is not such that one can be confident, within reasonable error bounds, of predictions of a phenomenon as complex as the source term. Put simply, the robustness of the various estimations remains quite questionable.

While it is possible that the question will never be resolved to everyone's satisfaction, it is imperative that considerable attention be paid to the many aspects of aerosol science at all levels. For example, there is a strong need for experiments at the bench scale level to study single and two particle physics and chemistry in different environments. One does not need a CSTF vessel to study the effects of humidity on a particle. The task is much better carried out in a small cell the size of a tea cup and, indeed, such experiments are in progress. It is also clear that particles are but one constituent of the aerosol. One cannot ignore the state and the dynamics of the host gas. Thus, future computer programs must not decouple the two or make the coupling so weak that it does not remain very meaningful. Rapid progress in computational aerosol mechanics is now occurring, and correspondingly, there will be a greater need for system scale experiments that can provide the benchmark tests of the phenomenology and the computational schemes. Overall, one can be optimistic that this progress will lead to the elimination of *ad hoc* assumptions and to the development of tools that will help in providing robust estimates of the source term.

Nuclear Source Term

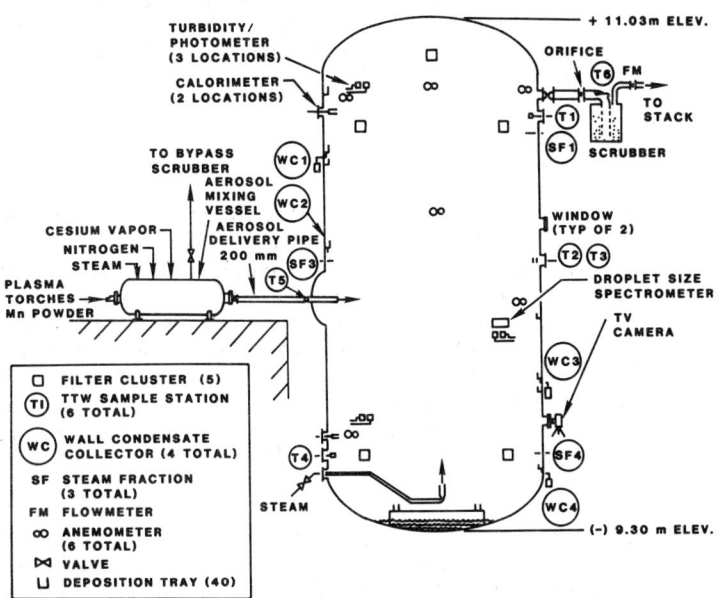

Figure 8.25: A schematic diagram of the 800 m³ containment vessel arrangement for the LA4 test. Adapted from Rahn [1988] with permission of the Electric Power Research Institute.

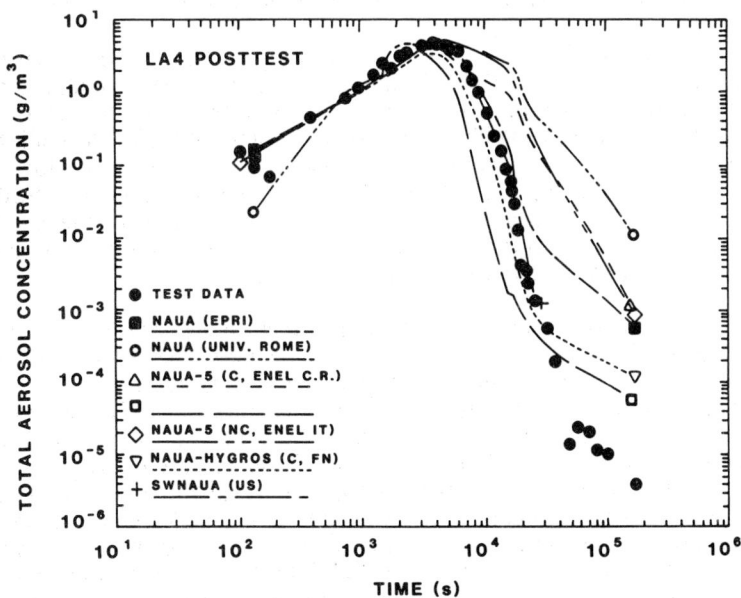

Figure 8.26: LA4 data and posttest code predictions. From A.L. Wright, private communication.

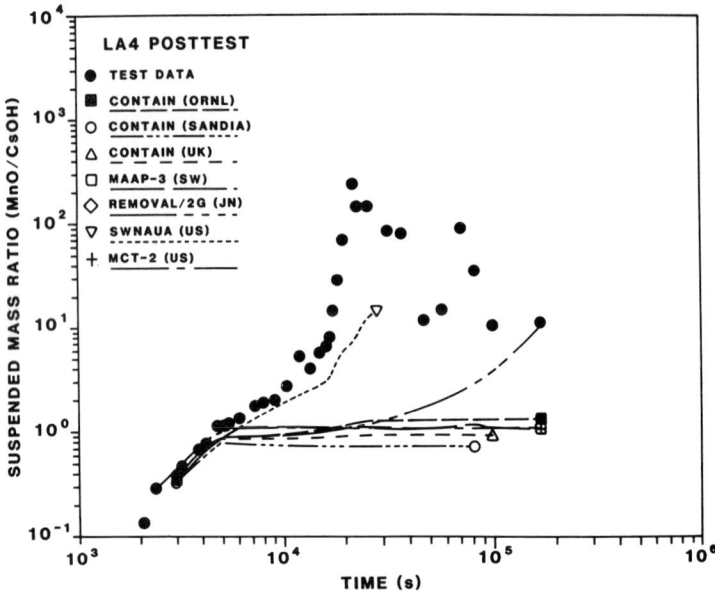

Figure 8.27: LA4 data and posttest code predictions. From A.L. Wright, private communication.

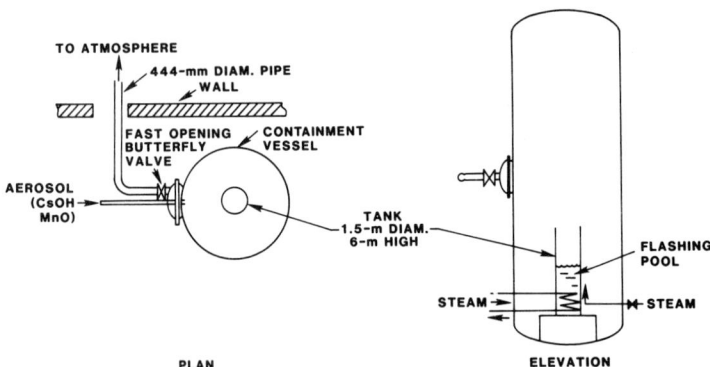

Figure 8.28: Test arrangements for LA5 and LA6. Adapted from Rahn [1988] with permission of the Electric Power Research Institute.

Nuclear Source Term

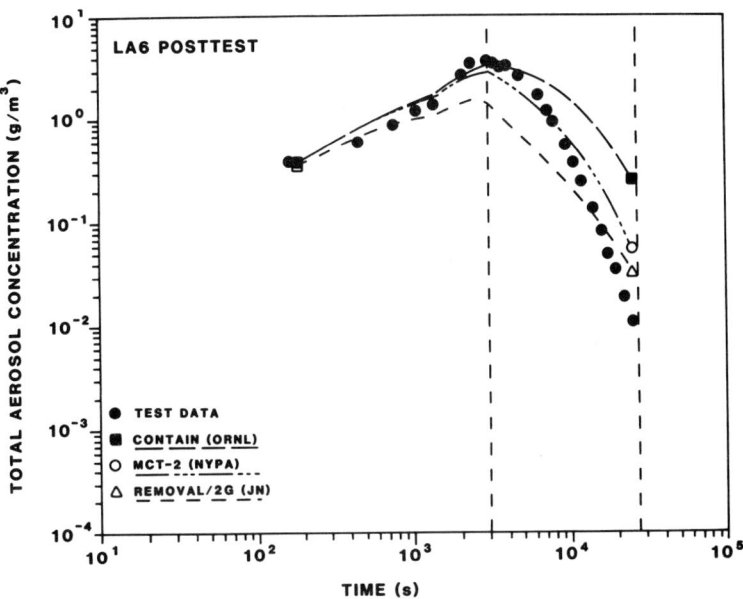

Figure 8.29: LA6 data and posttest code predictions. From A.L. Wright, private communication.

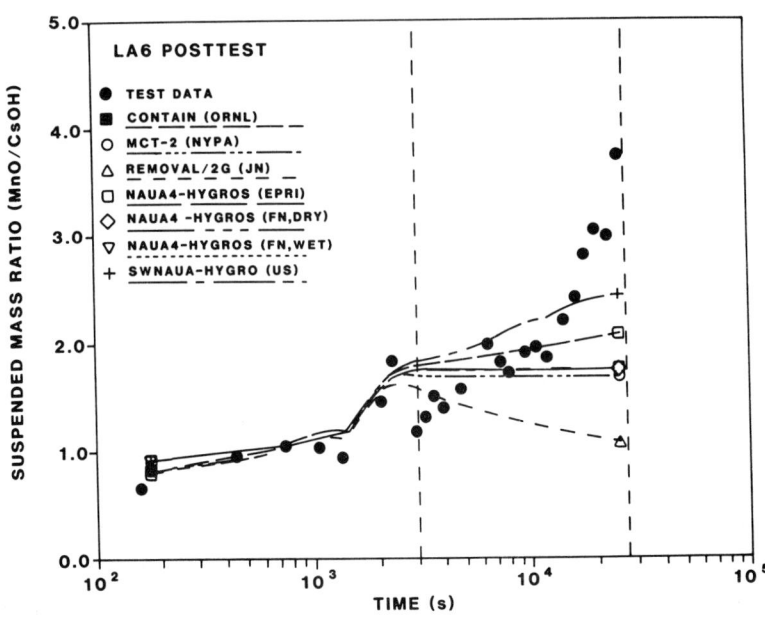

Figure 8.30: LA6 data on suspended mass ratio and posttest code predictions. From A.L. Wright, private communication.

Aerosol Science: Theory and Practice

References: Chapter 8

Adams, R.E., Kress, T.S., Han, J.T. and Silberberg, M. [1980] "Behavior of Sodium Oxide, Uranium Oxide and Mixed Sodium Oxide-Uranium Oxide Aerosols in a Large Vessel," in Proceedings of the CSNI Specialists Meeting on Nuclear Aerosols in Reactor Safety (Oak Ridge National Laboratory, Oak Ridge, Tennessee) Report # NUREG/CR-1724, ORNL/NUREG/TM-404, CSNI-45, p. 497.

Allison, C., Rest, J., Lorenz, R., Hagrman, D., Carlson, E. and Broughton, J. [1987] "Severe core damage and associated in-vessel fission product release," *Prog. Nuc. Energy* **20**, 89.

Blond, R., Taylor, M., Margulies, T., Cunningham, M., Baranowsky, P., Denning, R. and Cybulskis P. [1982] "The Development of Severe Accident Source Terms: 1957-1981" (U.S. Nuclear Regulatory Commission) Report # NUREG-0773 R1.

Bowsher, B.R. [1987] "Fission-product chemistry and aerosol behaviour in the primary circuit of a pressurized water reactor under severe accident conditions," *Prog. Nuc. Energy* **20**, 199.

Brockmann, J.E. [1987] "Ex-vessel releases: aerosol source terms in reactor accidents," *Prog. Nuc. Energy* **19**, 7.

Buhl, A. *et al.* [1984] "Nuclear Power Plant Response to Severe Accidents" (Edison Electric Institute, Washington, D.C.) Final technical summary, Industry Degraded Core Rulemaking Hearing (IDCOR program).

Bunz, H., Koyro, M. and Schoeck, W. [1983] "NAUA Mod4" (Kernforschungzentrum Karlsruhe (KfK), Germany) Report # KfK-3554.

Dunbar, I.H., Fermandjian, J. and Gauvain, J. [1988] "The intercomparison of aerosol codes," *Nucl. Tech.* **82**, 36.

Emami, A. [1980] "Effect of Condensation on Aerosol Behavior in Post-hcda LMFBR Containment" (University of Missouri-Columbia) Ph.D. Thesis.

Emami, A. and Loyalka, S.K. [1981] "Role of condensation in aerosol source term for liquid-metal fast breeder reactor containment," *Nucl. Tech.* **52**, 162.

Gelbard, F. [1982] "MAEROS User Manual" (Sandia National Laboratories) Report # SAND80-0822, NUREG/CR-1391.

Gieseke, J.A., Cybulskis, P., Denning, R.S., Kuhlman, M.R., Lee, K.W. and Chen, H. [1984] "Radionuclide Release Under Specific LWR Accident Conditions," Vols. I-VI (Battelle Columbus Laboratories) BMI-2104.

Gieseke, J.A., Lee, K.W. and Reed, L.D. [1978] "HAARM-3 User's Manual" (Battelle Memorial Institute) Report # BMI-NUREG-1991.

Hilliard, R.K., Postma, A.K., McCormack, J.D. and Coleman, L.F. [1971] "Removal of iodine and particles by sprays in the containment systems experiment," *Nucl. Tech.* **10**, 499.

Hubner, R.S., Vaughn, E.V. and Banume, L. [1973] "HAA-3 User's Report," (Atomics International) Report # AI-AEC-13038.

Kress, T.S. and Tobias, M.L. [1981] "LMFBR Aerosol Release and Transport Program, Quarterly Report, January-March 1981," Report # NUREG/CR-2299. Vol. 1, ORNL/TM-7946 and subsequent quarterly reports by Kress, T.S. and Tobias, M.L. and by Adams, R.E. and Tobias, M.L.

Levenson, M. and Rahn, F. [1981] "Realistic estimates of the consequences of nuclear accidents," *Nucl. Tech.* **53**, 99.

Loyalka, S.K. [1983] "Mechanics of aerosols in nuclear reactor safety: a review," *Prog. Nuc. Energy* **12**, 1.

Loyalka, S.K. [1987] "Recent progress on source-term containment analysis," *Prog. Nuc. Energy* **19**, 1.

Morewitz, H.A. [1981] "Fission product and aerosol behavior following degraded core accidents," *Nucl. Tech.* **53**, 120.

Murata, K.K., Carroll, D.E., Washington, K.E., Gelbard, F., Valdez, G.D., Williams, D.C. and Bergeron, K.D. [1989] "User's Manual for CONTAIN 1.1: A Computer Code for Severe Nuclear Reactor Accident Containment Analysis" (Sandia National Laboratories) Report # NUREG/CR-5026, SAND87-2309/R4.

Niemczyk, S.J. and McDowell-Boyer, L.M. [1982] "Technical Considerations Related to Interim Source Term Assumptions for Emergency Planning and Equipment Qualification" (Oak Ridge National Laboratory, Oak Ridge, Tennessee) Report # ORNL/TM-8275.

OECD [1985] "Nuclear Aerosols in Reactor Safety" (Nuclear Energy Agency, Paris, France). Available from Director of Information, OECD, 2 rue Audre Pascal, 75775 Paris Cedex 16, Fr.

Postma, A.K. and Johnson, B.M. [1971] "Containment Systems Experiment Final Program Summary" (Battelle Pacific Northwest Laboratories) Report # BNWL-1592.

Postma, A.K. and Owen, R.K. [1980] "Comparison of Aerosol Behavior During Sodium Fires in CSTF with the HAA-3B Code," in Proceedings of the CSNI Specialists Meeting on Nuclear Aerosols in Reactor Safety (Oak Ridge National Laboratory, Oak Ridge, Tennessee) Report # NUREG/CR-1724, ORNL/NUREG/TM-404, CSNI-45, p. 517.

Powers, D.A., Brockmann, J.E. and Shiver, A.W. [1985] "VANESA; A Mechanistic Model of Radionuclide Release and Aerosol Generation During Core Debris Interactions with Concrete" (Sandia National Laboratories) Report # NUREG/CR-4308, SAND85-1370 (to be published).

Rahn, F. [1988] "The LWR Aerosol Containment Experiments (LACE) Project Summary Report" (Electric Power Research Institute, Palo Alto, California) Report # NP-6094-D.

Robert, B.F., Freid, S.H., Parker, G.W., Pasley, L.F. and Row, T.H. [1969] "Evaluation of Various Methods of Fission Product Aerosol Simulation" (Oak Ridge National Laboratory, Oak Ridge, Tennessee) Report # ORNL-TM-2628.

Silberberg, M., Mitchell, J.A., Meyer, R.O. and Ryder, C.P. [1986] "Reassessment of the Technical Bases for Estimating Source Terms; Final Report" (U.S. Nuclear Regulatory Commission) Report # NUREG-0956.

U.S. Atomic Energy Commission [1957] "Theoretical Possibilities and Consequences of Major Accidents on Large Nuclear Power Plants," Report # WASH-740.

U.S. Atomic Energy Commission [1962] "Calculation of Distance Factors for Power and Test Reactor Sites," Report # TID-14844.

U.S. Department of Energy [1987] "Health and Environmental Consequences of the Chernobyl Nuclear Power Plant Accident," Report # DOE/ER-0332, UC-41 & 48.

U.S. Nuclear Regulatory Commission [1975] "Reactor Safety Study: An Assessment of Accident Risks in U. S. Power Plants," Report # WASH-1400, NUREG-75/014.

U.S. Nuclear Regulatory Commission [1981] "Technical Basis for Estimating Fission Product Behavior During LWR Accidents," Report # NUREG-0772.

U.S. Nuclear Regulatory Commission [1990] "Severe Accident Risks: An Assessment For Five U.S. Nuclear Power Plants" (Final Summary Report) NUREG-1150, Vols. 1-2.

Warman, E.A. [1986] "Realistic Assessment of Postulated Accidents at Light Water Reactor Nuclear Power Plants," in Advances in Nuclear Science and Technology, Lewins, J. and Becker, M., editors (Plenum, New York, N.Y.).

Williams, D.C., Bergeron, K.D., Rexroth, P.E. and Tills, J.L. [1987] "Integrated phenomenological analysis of containment response to severe core damage accidents," *Prog. Nuc. Energy* 19, 69.

Williams, M.M.R. [1990] "Nuclear aerosol behaviour during reactor accidents," *Prog. Nuc. Energy* 23, 101.

Wilson, R., Araj., K., Allen, A.O., Auer, P., Boulware, D.G., Finlayson, F., Goren, S., Ice, C., Lidofsky, L., Seesoms, A.L., Shoaf, M.L., Spiewak, I. and Tombrello, T. [1985] "Report to the American Physical Society of the study group on radionuclide release from severe accidents at nuclear power plants," *Rev. Mod. Phys.* 57(3), Part II, pp.S1-S154.

Wooton, R.O., Cybulskis, P. and Quayle, S.F. [1984] "MARCH 2 Code Description and User's Manual" (Battelle Columbus Laboratories) Report # BMI-2115, NUREG/CR-3988.

APPENDIX A

Common Dimensionless Groups

Table A.1: Common dimensionless groups.

GROUP	SYMBOL	DEFINITION	SIGNIFICANCE
Courant #	C_0	$= \dfrac{c \Delta t}{\Delta v}$	signal travel distance / discretization distance
Grashof #	Gr	$= \dfrac{L^3 g \beta \Delta T}{\nu^2}$	bouyancy force / viscous force
Gravity #	Gr	$= \dfrac{V_s (a_p + b_p)}{2D}$	gravitational mass transfer / diffusive mass transfer
Interception # or Impaction #	R	$= \dfrac{a_p}{L}$	characteristic particle dimension / characteristic flow dimension
Knudsen #	Kn	$= \dfrac{\lambda_g}{a_p}$	mean free path of molecules / characteristic dimension (*e.g.* radius of the particle)
Lewis #	Le	$= \dfrac{\alpha}{D}$	thermal diffusivity / mass diffusivity
Mach #	Ma	$= \dfrac{U}{V_a}$	flow speed / sonic speed
Nusselt #	Nu	$= \dfrac{hL}{k}$	total heat transfer / conductive heat transfer
Peclet #	Pe	$= \dfrac{LU}{D}$	bulk mass transfer / diffusive mass transfer
Peclet # (local)	Pe_x	$= \dfrac{xU}{D}$	bulk mass transfer / diffusive mass transfer
Peclet # (thermal)	Pe_{th}	$= \dfrac{LU}{\alpha}$	bulk heat transfer / conductive heat transfer
Prandtl #	Pr	$= \dfrac{\nu}{\alpha}$	momentum diffusivity / thermal diffusivity
Rayleigh #	Ra	$= Gr\, Pr$	bouyancy force / thermal force
Reynolds #	Re	$= \dfrac{a_p U}{\nu}$	inertia force / viscous force
Reynolds # (space)	Re_s	$= \dfrac{a_p U}{\nu}$	inertia force / viscous force

Common Dimensionless Groups

Name	Symbol	Formula	Meaning
Reynolds # (time)	Re_T	$= \dfrac{a_p^2 \omega}{\nu}$	inertia force / viscous force
Schmidt #	Sc	$= \dfrac{\nu}{D}$	momentum diffusivity / mass diffusivity
Sherwood #	Sh	$= \dfrac{k^* L}{D}$	total mass transfer / diffusive mass transfer
Stanton #	St	$= \dfrac{h}{\rho C_p U}$	heat transferred / heat content of the fluid
Stokes #	Stk	$= \dfrac{\tau U}{L}$	stopping distance / characteristic flow dimension

a_p particle radius (also 'a')   (L)
B particle mobility   (T M^{-1})
C_p specific heat at constant pressure   (L^2 T^{-2} K^{-1})
D mass diffusivity (diffusion coefficient)   (L^2 T^{-1})
d_p particle diameter   (L)
g acceleration due to gravity   (L T^{-2})
h convective heat transfer coefficient   (M T^{-3} K^{-1})
k fluid thermal conductivity   (M L T^{-3} K^{-1})
k^* deposition velocity (also 'k')   (L T^{-1})
L characteristic dimension of a body or flow (also 'a')   (L)
m particle mass   (M)
U characteristic flow speed (also 'v', 'U_∞', and 'v_0')   (L T^{-1})
V_a sonic speed   (L T^{-1})
V_s sedimentation velocity   (L T^{-1})
x some arbitrary distance   (L)
$\alpha = \dfrac{\rho C_p}{k}$ thermal diffusivity   (L^2 T^{-1})
β coefficient of thermal expansion   (K^{-1})
ΔT temperature differential   (K)
λ_g molecular mean free path (also 'ℓ', 'ℓ_t', and 'ℓ_D')   (L)
μ fluid dynamic viscosity   (M L^{-1} T^{-1})
$\nu = \dfrac{\mu}{\rho}$ kinematic viscosity   (L^2 T^{-1})
ρ fluid mass density   (M L^{-3})
$\tau = m B$ relaxation time (τU is the stopping distance)   (T)
ω characteristic frequency for a time dependent disturbance   (T^{-1})
Units   K–temperature,   L–length,   M–mass,   T– time

APPENDIX B

Commonly Encountered Constants

Boltzmann's constant	k_B	1.3807×10^{-23}	J K^{-1}
		1.3807×10^{-16}	erg K^{-1}
Elementary charge	e	1.6022×10^{-19}	C
		4.8032×10^{-10}	statcoulomb
Gravitational constant	G	6.6720×10^{-11}	m^3 sec^{-2} kg^{-1}
		6.6720×10^{-8}	dyne cm^2 g^{-2}
Planck's constant	h	6.6262×10^{-34}	J sec
		6.6262×10^{-27}	erg sec
	$\hbar = h/2\pi$	1.0546×10^{-34}	J sec
		1.0546×10^{-27}	erg sec
Speed of light in vacuum	c	2.9979×10^{8}	m sec^{-1}
Permittivity of free space	ε_0	8.8542×10^{-12}	F m^{-1}
Permeability of free space	μ_0	$4\pi \times 10^{-7}$	H m^{-1}
Stefan-Boltzmann constant	σ	5.6703×10^{-8}	W m^{-2} K^{-4}
		5.6703×10^{-5}	erg cm^{-1} sec^{-1} K^{-4}
Avogadro's number	A or N_A	6.0220×10^{23}	mol^{-1}
Gas constant	$R = N_A k_B$	8.3144	J K^{-1} mol^{-1}
		8.3144×10^{7}	erg K^{-1} mol^{-1}
Loschmidt's number	n_0	2.6868×10^{25}	m^{-3}
Standard temperature	T_0	273.16	K
Atmospheric pressure	$p_0 = n_0 k_B T_0$	1.0133×10^{5}	Pa
		1.0133×10^{6}	dyne cm^{-2}
Gravitational acceleration	g	9.8067	m sec^{-2}

$$\pi = 3.1416$$
$$1/\pi = 0.3183$$
$$\pi^2 = 9.8696$$
$$\ln(\pi) = 1.1447$$
$$\pi^{1/2} = 1.7725$$
$$e = 2.7183$$
$$1/e = 0.3679$$
$$e^2 = 7.3891$$
$$e^{1/2} = 1.6487$$
$$\pi^e = 22.4592$$
$$e^\pi = 23.1407$$
$$e^{-\pi} = 0.0432$$
$$\sqrt{2} = 1.4142$$
$$\ln(2) = 0.6931$$
$$\log_{10}(2) = 0.3010$$
$$\sqrt{3} = 1.7321$$

$$\ln(3) = 1.0986$$
$$\log_{10}(3) = 0.4771$$
$$\gamma = 0.5772$$
$$\ln(\gamma) = -0.5495$$
$$\Gamma(1/6) = 5.5663$$
$$\Gamma(1/5) = 4.5908$$
$$\Gamma(1/4) = 3.6256$$
$$\Gamma(1/3) = 2.6789$$
$$\Gamma(2/5) = 2.2182$$
$$\Gamma(1/2) = 1.7725$$
$$\Gamma(3/5) = 1.4892$$
$$\Gamma(2/3) = 1.3541$$
$$\Gamma(3/4) = 1.2254$$
$$\Gamma(4/5) = 1.1642$$
$$\Gamma(5/6) = 1.1288$$
$$\Gamma(1) = 1.0$$

APPENDIX C

Typical Aerosol Properties

Table C.1: Aerosol properties as a function of size of a unit density sphere in air at Standard Temperature and Pressure (STP).

Particle Diameter (μm)	Sedimentation Velocity* (cm/sec)	Reynold's Number* (Re)	Diffusion Coefficient* (cm²/sec)	Particle Mobility* (sec/gm)	Relaxation Time* (sec)
d_p	$V_s = \dfrac{m\,g\,C_c}{3\pi \mu d_p}$	$Re = \dfrac{V_s\,d_p}{\nu}$	$D = B\,k_B\,T$	$B = \dfrac{V_s}{m\,g}$	$\tau = m\,B$
0.001	6.5530E–07	4.3494E–13	5.1084E–02	1.2719E+12	6.6595E–10
0.01	6.6901E–06	4.4538E–11	5.2312E–04	1.3025E+10	6.8197E–09
0.1	8.6316E–05	5.7466E–09	6.7494E–06	1.6804E+08	8.7988E–08
1.0	3.5054E–03	2.3338E–06	2.7410E–07	6.8245E+06	3.5733E–06
10.0	3.0605E–01	2.0375E–03	2.3931E–08	5.9583E+05	3.1198E–04
100.0	2.4844E+01	1.6538E+00	2.3583E–09	5.8717E+04	3.0744E–02

*Corrected values based on the Cunningham factor, C_c.

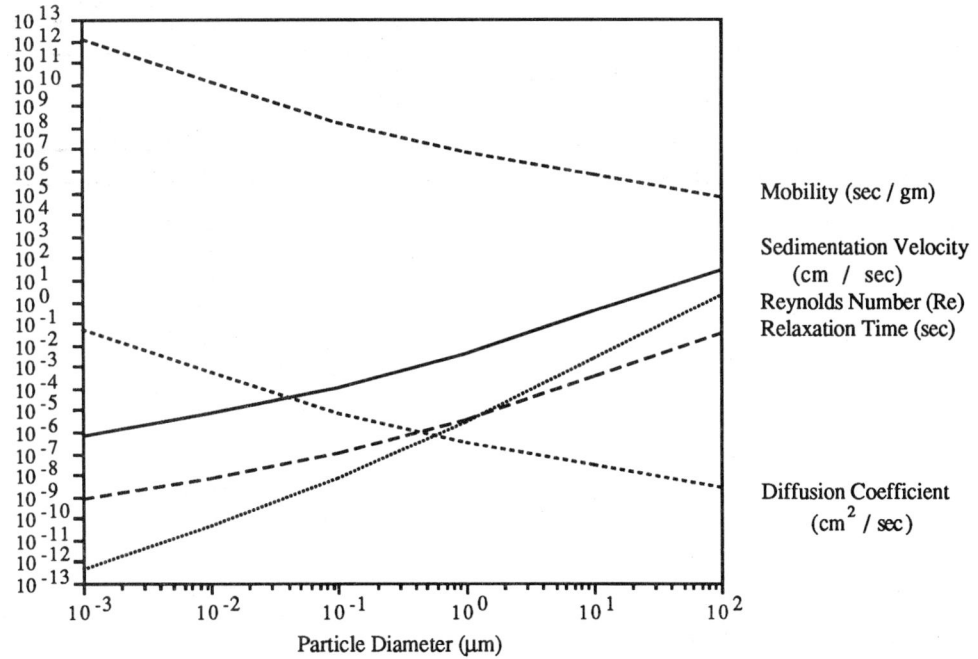

Figure C.1: Aerosol properties as a function of size of a unit density sphere in air at Standard Temperature and Pressure (STP).

APPENDIX D
Differential Operators

f is a scalar, **A** is a vector, **T** is a tensor

Rectangular Coordinates – (x,y,z)

Divergence of a vector:

$$\nabla \cdot \mathbf{A} = \frac{\partial A_x}{\partial x} + \frac{\partial A_y}{\partial y} + \frac{\partial A_z}{\partial z}$$

Gradient:

$$\nabla f = \frac{\partial f}{\partial x}\mathbf{i} + \frac{\partial f}{\partial y}\mathbf{j} + \frac{\partial f}{\partial z}\mathbf{k}$$

Curl:

$$\nabla \times \mathbf{A} = \left(\frac{\partial A_z}{\partial y} - \frac{\partial A_y}{\partial z}\right)\mathbf{i} + \left(\frac{\partial A_x}{\partial z} - \frac{\partial A_z}{\partial x}\right)\mathbf{j} + \left(\frac{\partial A_y}{\partial x} - \frac{\partial A_x}{\partial y}\right)\mathbf{k}$$

Laplacian of a scalar:

$$\nabla^2 f = \frac{\partial^2 f}{\partial x^2} + \frac{\partial^2 f}{\partial y^2} + \frac{\partial^2 f}{\partial z^2}$$

Laplacian of a vector:

$$\nabla^2 \mathbf{A} = \nabla^2 A_x \, \mathbf{i} + \nabla^2 A_y \, \mathbf{j} + \nabla^2 A_z \, \mathbf{k}$$

Cylindrical Coordinates – (r,ϕ,z)

Divergence of a vector:

$$\nabla \cdot \mathbf{A} = \frac{1}{r}\frac{\partial}{\partial r}(r A_r) + \frac{1}{r}\frac{\partial A_\phi}{\partial \phi} + \frac{\partial A_z}{\partial z}$$

Gradient:

$$(\nabla f)_r = \frac{\partial f}{\partial r}, \quad (\nabla f)_\phi = \frac{1}{r}\frac{\partial f}{\partial \phi}, \quad (\nabla f)_z = \frac{\partial f}{\partial z}$$

Curl:

$$(\nabla \times \mathbf{A})_r = \frac{1}{r}\frac{\partial A_z}{\partial \phi} - \frac{\partial A_\phi}{\partial z}$$

$$(\nabla \times \mathbf{A})_\phi = \frac{\partial A_r}{\partial z} - \frac{\partial A_z}{\partial r}$$

$$(\nabla \times \mathbf{A})_z = \frac{1}{r}\frac{\partial}{\partial r}(r A_\phi) - \frac{1}{r}\frac{\partial A_r}{\partial \phi}$$

Differential Operators

Laplacian of a scalar:
$$\nabla^2 f = \frac{1}{r}\frac{\partial}{\partial r}\left(r\frac{\partial f}{\partial r}\right) + \frac{1}{r^2}\frac{\partial^2 f}{\partial \phi^2} + \frac{\partial^2 f}{\partial z^2}$$

Laplacian of a vector:
$$\left(\nabla^2 A\right)_r = \nabla^2 A_r - \frac{2}{r^2}\frac{\partial A_\phi}{\partial \phi} - \frac{A_r}{r^2}$$

$$\left(\nabla^2 A\right)_\phi = \nabla^2 A_\phi + \frac{2}{r^2}\frac{\partial A_r}{\partial \phi} - \frac{A_\phi}{r^2}$$

$$\left(\nabla^2 A\right)_z = \nabla^2 A_z$$

Divergence of a tensor:
$$(\nabla \cdot T)_r = \frac{1}{r}\frac{\partial}{\partial r}(r\, T_{rr}) + \frac{1}{r}\frac{\partial T_{\phi r}}{\partial \phi} + \frac{\partial T_{zr}}{\partial z} - \frac{T_{\phi\phi}}{r}$$

$$(\nabla \cdot T)_\phi = \frac{1}{r}\frac{\partial}{\partial r}(r\, T_{r\phi}) + \frac{1}{r}\frac{\partial T_{\phi\phi}}{\partial \phi} + \frac{\partial T_{z\phi}}{\partial z} + \frac{T_{\phi r}}{r}$$

$$(\nabla \cdot T)_z = \frac{1}{r}\frac{\partial}{\partial r}(r\, T_{rz}) + \frac{1}{r}\frac{\partial T_{\phi z}}{\partial \phi} + \frac{\partial T_{zz}}{\partial z}$$

Spherical Coordinates – (r,θ,ϕ)

Divergence of a vector:
$$\nabla \cdot A = \frac{1}{r^2}\frac{\partial}{\partial r}(r^2 A_r) + \frac{1}{r\sin(\theta)}\frac{\partial}{\partial \theta}[A_\theta \sin(\theta)] + \frac{1}{r\sin(\theta)}\frac{\partial A_\phi}{\partial \phi}$$

Gradient:
$$(\nabla f)_r = \frac{\partial f}{\partial r}, \quad (\nabla f)_\theta = \frac{1}{r}\frac{\partial f}{\partial \theta}, \quad (\nabla f)_\phi = \frac{1}{r\sin(\theta)}\frac{\partial f}{\partial \phi}$$

Curl:
$$(\nabla \times A)_r = \frac{1}{r\sin(\theta)}\frac{\partial}{\partial \theta}[A_\phi \sin(\theta)] - \frac{1}{r\sin(\theta)}\frac{\partial A_\theta}{\partial \phi}$$

$$(\nabla \times A)_\theta = \frac{1}{r\sin(\theta)}\frac{\partial A_r}{\partial \phi} - \frac{1}{r}\frac{\partial}{\partial r}(r A_\phi)$$

$$(\nabla \times A)_\phi = \frac{1}{r}\frac{\partial}{\partial r}(r A_\theta) - \frac{1}{r}\frac{\partial A_r}{\partial \theta}$$

Laplacian of a scalar:
$$\nabla^2 f = \frac{1}{r^2}\frac{\partial}{\partial r}\left(r^2 \frac{\partial f}{\partial r}\right) + \frac{1}{r^2 \sin(\theta)}\frac{\partial}{\partial \theta}\left(\sin(\theta)\frac{\partial f}{\partial \theta}\right) + \frac{1}{r^2 \sin^2(\theta)}\frac{\partial^2 f}{\partial \phi^2}$$

Laplacian of a vector:

$$(\nabla^2 \mathbf{A})_r = \nabla^2 A_r - \frac{2 A_r}{r^2} - \frac{2}{r^2}\frac{\partial A_\theta}{\partial \theta} - \frac{2\cot(\theta) A_\theta}{r^2} - \frac{2}{r^2 \sin(\theta)}\frac{\partial A_\phi}{\partial \phi}$$

$$(\nabla^2 \mathbf{A})_\theta = \nabla^2 A_\theta + \frac{2}{r^2}\frac{\partial A_r}{\partial \theta} - \frac{A_\theta}{r^2 \sin^2(\theta)} - \frac{2\cos(\theta)}{r^2 \sin^2(\theta)}\frac{\partial A_\phi}{\partial \phi}$$

$$(\nabla^2 \mathbf{A})_\phi = \nabla^2 A_\phi - \frac{A_\phi}{r^2 \sin^2(\theta)} + \frac{2}{r^2 \sin(\theta)}\frac{\partial A_r}{\partial \phi} + \frac{2\cos(\theta)}{r^2 \sin^2(\theta)}\frac{\partial A_\theta}{\partial \phi}$$

Divergence of a tensor:

$$(\nabla \cdot \mathbf{T})_r = \frac{1}{r^2}\frac{\partial}{\partial r}(r^2 T_{rr}) + \frac{1}{r \sin(\theta)}\frac{\partial}{\partial \theta}[T_{\theta r} \sin(\theta)] + \frac{1}{r \sin(\theta)}\frac{\partial T_{\phi r}}{\partial \phi} - \frac{T_{\theta\theta} + T_{\phi\phi}}{r}$$

$$(\nabla \cdot \mathbf{T})_\theta = \frac{1}{r^2}\frac{\partial}{\partial r}(r^2 T_{r\theta}) + \frac{1}{r \sin(\theta)}\frac{\partial}{\partial \theta}[T_{\theta\theta} \sin(\theta)] + \frac{1}{r \sin(\theta)}\frac{\partial T_{\phi\theta}}{\partial \phi} + \frac{T_{\theta r}}{r} - \frac{T_{\phi\phi}\cot(\theta)}{r}$$

$$(\nabla \cdot \mathbf{T})_\phi = \frac{1}{r^2}\frac{\partial}{\partial r}(r^2 T_{r\phi}) + \frac{1}{r \sin(\theta)}\frac{\partial}{\partial \theta}[T_{\theta\phi} \sin(\theta)] + \frac{1}{r \sin(\theta)}\frac{\partial T_{\phi\phi}}{\partial \phi} + \frac{T_{\phi r}}{r} + \frac{T_{\phi\theta}\cot(\theta)}{r}$$

APPENDIX E
Acronyms and Abbreviations

Selected Acronyms and Abbreviations

AB	Aerosol Behavior
AC / DC	Alternating Current / Direct Current
AI	Atomics International
ASTD	Aerosol Source Term Development
BGK	Bhatnagar-Gross-Krook
BMI	Battelle Memorial Institute
C	Contained
CEA	Comissariat a l'Energie Atomique
CPU, C.P.U.	Central Processing Unit
CR-3	Crystal River unit-3
CSNI	Committee on the Safety of Nuclear Installations
CSTF	Containment System Test Facility
DEMONA	Demonstration Experiment for Modeling Of Nuclear Aerosols
DOE	Department of Energy
DOP	DiOctyl Phthalate
ECCS	Emergency Core Cooling System
EPRI	Electrical Power Research Institute
FP	Fission Product
GDE	General Dynamic Equation
HEDL	Hanford Engineering Development Laboratory
HTRE	Heat Transfer Reactor Experiment
INSAG	International Nuclear Safety Advisory Group
KfK	Kernforschungszentrum Karlsruhe
L	Length
LA	Light water reactor Aerosol
LACE	Light water reactor Aerosol Containment Experiments
LHS	Left Hand Side
LMFBR	Liquid Metal Fast Breeder Reactor
LOCA	Loss Of Coolant Accident
LWR	Light Water Reactor
MIT	Massachusetts Institute of Technology
MWd	MegaWatt days
MWe	MegaWatts electric
MWth	MegaWatts thermal
NRC	Nuclear Regulatory Commision
NRX	NRX natural uranium heavy water reactor, Chalk River, Ontario, Canada
NSPP	National Safety Pilot Plant
NUREG	NUclear REGulatory
ODE	Ordinary Differential Equation
OECD	Organization for Economic Cooperation and Development
ORNL	Oak Ridge National Laboratory
ORR	Oak Ridge Research Reactor
PCS	Primary Coolant System
PDE	Partial Differential Equation
PE	Potential Energy
PRTR	Plutonium Recycle Test Reactor, Hanford, Washington
PWR	Pressurized Water Reactor
RH	Relative Humidity
RHS	Right Hand Side
RSS	Reactor Safety Study
SNAPTRAN	Systems for Nuclear Auxilliary Power TRANsient reactor test series

SPERT	Special Power Excursion Reactor Test
STP	Standard Temperature and Pressure
T	Time
TMI-2	Three Mile Island, unit 2
U	Uncontained
WASH-1400	WASHington-1400
WTR	Westinghouse Testing Reactor, Waltz Mills, Pennsylvania

Selected Computer Codes

AEROMECH	AEROsol MECHanics
AEROSIM	AEROsol SIMulation
AEROSIM-M	AEROsol SIMulation-Multicomponent
AGRO	Aerosol GROwth
COAGUL	COAGULation
CONFEMM	CONdensation Finite EleMent Modified
CONTAIN	(not an acronym)
CORCON	CORe CONcrete
CORRAL	(not an acronym)
CORSOR	CORe SOuRce
FIDAP®	(a trademark of Fluid Dynamics International, Inc.)
FLUENT®	(a trademark of Creare.x, Inc.)
GCEFF	Gravitational Collision EFFiciency
HA-3	Heterogeneous Aerosol-3
HAA-3B	Heterogeneous Aerosol Agglomeration-3B
HAA-4	Heterogeneous Aerosol Agglomeration-4
HAARM	(derived from HAA codes)
HAARM-3	(derived from HAA codes)
ICEDF	ICE Decontamination Factor
JERICHO	(not an acronym)
LIMIT	(not an acronym)
MAEROS	Multicomponent AEROSols
MARCH	Meltdown Accident Response CHaracteristics
MERGE	(not an acronym)
NAUA	NAchUnfall Atmosphare (post accident atmosphere)
NAUA-4	NAchUnfall Atmosphare-4
NAUA-5	NAchUnfall Atmosphare-5
ORIGEN	Oak Ridge Isotope GENeration
QUICK	(not an acronym)
QUICK-M	(not an acronym)
REMOVAL	(not an acronym)
RETAIN-2C	(not an acronym)
RETAIN-S	(not an acronym)
SCDAP	Severe Core Damage Analysis Package
STCP	Source Term Code Package
TRAP-MELT	(not an acronym)
TRAP-MELT 2	(not an acronym)
VANESA	(not an acronym)
XSOR	X SOuRce
ZONE	(not an acronym)

Foreign Expressions

e.g.	exempli gratia (LATIN, for example)
i.e.	id est (LATIN, that is)
et al.	et alii (LATIN, and others)
etc.	et cetera (LATIN, and so forth)
ad hoc	(LATIN, for this; for a particular end or case without consideration of wider application)
a priori	(LATIN, from the former; presupposed by experience)
ex post facto	(LATIN, from a thing done afterward; after the fact)

APPENDIX F
Journal Abbreviations

Abh. Theoret. Phys.	Abhandlungen der Theoretische Physik
Aerosol Sci. and Tech.	Aerosol Science and Technology
A.I.A.A. J.	American Institute of Aeronautics and Astronautics Journal
A.I.Ch.E. J.	American Institute of Chemical Engineering Journal
Aircr. Eng.	Aircraft Engineering
Ann. Nucl. Energy	Annals of Nuclear Energy
Ann. Phys. Chem.	Annalen der Physik und Chemie
Ark. Matemat.	Arkiv foer Matematik
Astrophys. Space Sci.	Astrophysics and Space Science
Br. J. Appl. Phys.	British Journal of Applied Physics
Camb. Philos. Soc. Trans.	Cambridge Philosophical Society Transactions
Can. J. Chem.	Canadian Journal of Chemistry
Chem. Eng. Prog.	Chemical Engineering Progress
Chem. Eng. Prog. Symp. Ser. 1	Chemical Engineering Progress Symposium Series 1
Chem. Eng. Sci.	Chemical Engineering Science
Chem.-Ing.-Tech.	Chemie-Ingenieur-Technik
Colloid and Polym. Sci.	Colloid and Polymer Science
Comput. Methods Appl. Mech. and Eng.	Computer Methods in Applied Mechanics and Engineering
Faraday Society Discussions	Faraday Society Discussions
I.J.M.E.E.	International Journal of Mechanical Engineering Education
Ind. Eng. Chem.	Industrial and Engineering Chemistry
Int. J. Heat Mass Transfer	International Journal of Heat and Mass Transfer
Int. J. Numer. Methods Fluids	International Journal for Numerical Methods in Fluids
J. Aerosol Sci.	Journal of Aerosol Science
J. Am. Ceram. Soc.	Journal of the American Ceramic Society
J. Appl. Phys.	Journal of Applied Physics
J. Atmos. Sci.	Journal of the Atmospheric Sciences
J. Chem. Phys.	Journal of Chemical Physics
J. Colloid Interface Sci.	Journal of Colloid and Interface Science
J. Colloid Sci.	Journal of Colloid Science
J. Comput. Phys.	Journal of Computational Physics
J. Eng. Math.	Journal of Engineering Mathematics
J. Fluid Mech.	Journal of Fluid Mechanics
J. Geophys. Res.	Journal of Geophysical Research
J. Inst. Nuc. Engrs.	Journal of the Institution of Nuclear Engineers
J. Meteor.	Journal of Meteorology
J. Nucl. Energy	Journal of Nuclear Energy
J. Phys. A Math. Nucl. Gen.	Journal of Physics A: Mathematical, Nuclear and General
J. Phys. D Appl. Phys.	Journal of Physics D: Applied Physics
J. Phys. Soc. Japan	Journal of the Physical Society of Japan
J. Phys. Radium	Journal de Physique et le Radium [France]
J. Thermophys. Heat Transfer	Journal of Thermophysics and Heat Transfer
Kolloid Z.	Kolloid Zeitschrift [West Germany]
Kolloidchem. Beih.	Kolloidchemische Beihefte

Langmuir	Langmuir
Mathematika	Mathematika
Math. Ann.	Mathematische Annalen
Monthly Weather Rev.	Monthly Weather Review
Nature Mag.	Nature Magazine
Nucl. Tech.	Nuclear Technology
Nucl. Sci. and Eng.	Nuclear Science and Engineering
Opt. Acta	Optica Acta
Philos. Mag	Philosophical Magazine
Philos. Trans. Roy. Soc. London Ser. A	Philosophical Transactions of the Royal Society of London Series A: Mathematical and Physical Sciences
Phys. Fluids	Physics of Fluids
Phys. Fluids A	Physics of Fluids A
Phys. Rev.	Physical Review
Phys. Rev. Lett.	Physical Review Letters
Phys. Z.	Physikalische Zeitschrift
Physica	Physica
Phys. Rep.	Physics Reports. Review Section of Physics Letters Section C [Netherlands]
Proc. Camb. Philos. Soc.	Proceedings of the Cambridge Philosophical Society
Proc. Inst. Mech. Engrs.	Proceedings of the Institution of Mechanical Engineers
Proc. London Math. Soc.	Proceedings of the London Mathematical Society
Proc. Roy. Soc. (London)	Proceedings of the Royal Society of London
Proc. Roy. Irish Academy	Proceedings of the Royal Irish Academy
Prog. Biophys. and Biophys. Chem.	Progress in Biophysics and Biophysical Chemistry
Prog. Nuc. Energy	Progress in Nuclear Energy [England]
Pure and Appl. Geophys.	Pure and Applied Geophysics
Q. J. Math.	Quarterly Journal of Mathematics
Q. J. Mech. and Appl. Math.	Quarterly Journal of Mechanics and Applied Mathematics
Q. J. Roy. Meteor. Soc.	Quarterly Journal of the Royal Meteorological Society
Reine Angew. Math.	Reine und Angewandte Mathematik
Rev. Mod. Phys.	Reviews of Modern Physics
Sci. Prog. (Oxford)	Science Progress (Oxford)
Sitzungsber. Math.-Phys. Kl. Bayer. Akad. Wiss. München	Sitzungsberichte Mathematisch-Physicalischen Klasse Bayerischen Akademie der Wissenschaften zu München
Staub J.	Staub Journal
Trans. Am. Nucl. Soc.	Transactions of the American Nuclear Society
Trans. Faraday Soc.	Transactions of the Faraday Society
Z. f. Phys. Chemie	Zeitschrift fuer Physikalische Chemie
Z. Naturforsch.	Zeitschrift fuer Naturforschung
Z. Physik	Zeitschrift fuer Physik
Z.A.M.P.	Zeitschrift fuer Angewandte Mathematik und Physik

Author Index

Bold page numbers refer to the references at the end of each chapter. *Italic* page numbers refer to occurences within the text.

A

Abramowitz, M, *247*, **302**
Acrivos, A.- *102*, **106**
Adams, R.E.- **392**, **408**
Adler, P.M.- *172*, *208*, **236**
Agarwal, J.K.- *364*, **372**
Aguirre, J.L.- *181*, **238**
Aitchison, J.- *271*, **302**
Alam, M.K.- *184*, *185*, *187*, *188*, **236**
Allen, A.O.- **409**
Allen, M.D.- *161*, **236**
Allison, C.- *377*, *385*, **408**
Almeida, F.C.- *364*, **370**
Andrade, E.N da C.- *212*, **236**
Araj.- K.- **409**
Arp, P.A.- *172*, **236**
Arpaci, V.S.- *328*, **370**
Atkinson, W.R.- *218*, **236**
Auer, P.- **409**

B

Banume, L.- *245*, **302**, **408**
Baranowsky, P.- **408**
Barnes, H.M.- **303**
Barrett, J.C.- *323*, **323**
Bart, E.- *86*, **104**
Bassanini, P.- **105**, **237**
Batchelor, G.K.- *99*, *102*, **104**, *181*, *190*, *196*, *203*, *204*, *206*, *208*, *219*, **236**, *328*, *347*, **370**
Baum, H.R.- **303**
Beard, K.V.- *202*, *218*, **236**, **237**
Becker, M.- **370**, **409**
Bejan, A.- *328*, **370**
Bergeron, K.D.- **408**, **409**
Berry, E.X.- *288*, **302**
Berry, M.V.- *65*, **104**
Bird, R.B.- *313*, **324**
Bjerknes, C.A.- *217*, **236**
Blakemore, R.- *310*, **324**
Blond, R.- **408**
Boffi, V.- **324**, **372**
Booth, F.- *154*, *231*, *233*, **236**
Boulware, D.G.- **409**
Bowsher, B.R.- **408**
Braaten, D.A.- *368*, **370**

Brandt, O.- *212*, *216*, **236**
Brenner, H.- **10**, *19*, *24*, *26*, *28*, *34*, *36*, *49*, *84*, *88*, *91*, **104**, **105**
Bricard, J.- *142*, **151**
Brock, J.R.- **10**, *159*, **237**, *252*, *253*, *288*, *296*, *299*, *302*, *303*, *304*, *305*, *324*, *370*, **371**
Brockmann, J.E.- *377*, **408**, **409**
Broughton, J.- **408**
Brown, J.A.C.- *271*, **302**
Brundin, C.L.- **323**
Buckley, R.L.- *300*, **302**
Buhl, A.- **408**
Bunz, H.- **408**

C

Cale, D.B.- **373**
Carlson, E.- **408**
Carrier, G.F.- *223*, **236**
Carroll, D.E.- **408**
Cass, G.R.- *342*, **372**
Cassell, J.S.- **239**
Cawood, W.- *212*, **238**
Cercignani, C.- *14*, *22*, *23*, *24*, **105**, *155*, *161*, **236**, **237**, *310*, **323**, **324**, **372**
Chandrasekhar, S.- *156*, **237**
Chapman, S.- *13*, *16*, **105**, *313*, **323**, *352*, **370**
Chen, H.- *300*, **303**, **408**
Cheng, R.K.- **373**
Chien, C.W.- **303**
Cipolla, J.W.- **323**, *342*, *347*, *359*, **370**, **372**
Clement, C.F.- **323**, **323**, **324**
Coleman, L.F.- **408**
Collen, J.- **372**
Collins, W.D.- *55*, **105**
Cooley, M.D.A.- *42*, *55*, *80*, *81*, *84*, *102*, **105**
Cooper, D.W.- *133*, *134*, **151**, *336*, **371**
Corner, J.- *360*, **371**
Coull, J.- **106**
Cowling, T.G.- *13*, *16*, **105**, *313*, **323**, *352*, **370**
Cox, R.G.- *84*, *88*, **105**

Crane, R.I.- *364*, **371**
Crowe, C.T.- **373**
Crump, J.G.- *360*, *361*, **371**
Cunningham, E.- *160*, **237**
Cunningham, M.- **408**
Curtis, A.S.G.- *172*, *203*, **237**
Curtiss, C.F.- *313*, **324**
Cussler, E.L.- *328*, **371**
Cybulskis, P.- **408**, **409**

D

Dahneke, B.E.- *54*, **105**
Davies, C.N.- **10**, *136*, *142*, **151**, *306*, *324*, *328*, **371**, **373**
Davis, M.H.- *83*, *85*, *86*, *87*, *91*, *94*, *96*, *97*, **105**, *218*, *223*, *224*, **237**
Dean, W.R.- *91*, **105**
Deirmendjian, D.- *271*, **302**
Denning, R.S.- **408**
Dennis, R.- *4*, **10**, **371**
Deryaguin, B.V.- *352*, *354*, **371**
Deutch, J.M.- *181*, **237**
Dorrepaal, J.M.- *41*, *55*, *70*, **105**
Drake, R.L.- *268*, **302**
Dunbar, I.H.- *165*, **237**, *382*, **408**

E

East, T.W.R.- *172*, **237**
Einstein, A.- *156*, **237**
El-Hakeem, A.S.- *125*, **151**
Emami, A.- *288*, *299*, **302**, **408**
Enomoto, T.- *364*, **371**
Ernst, M.H.- *249*, **302**
Evans, R.L.- *364*, **371**
Eveson G.F.- *95*, **105**

F

Fedoseev, V.A.- **10**
Fermandjian, J.- **408**
Fernandez de la Mora, J.- *363*, *364*, **371**, **373**
Ferziger, J.H.- **302**, *313*, **324**, *352*, **371**
Fiebig, M.- **370**
Finlayson, F.- **409**

421

Finn, W.D.L.- *294*, **304**
Fissan, H.- **373**
Flachsbart, H.- **107**
Flagan, R.C.- *363*, **371**
Flanagan, V.P.V.- *133*, **152**
Fogler, H.S.- *209*, *211*, **238**
Freid, S.H.- **409**
Frick, G.M.- *136*, **152**
Friedlander, S.K.- **10**, *22*, **105**, *113*, *114*, *132*, **152**, *263*, *264*, *267*, *269*, *270*, *279*, *281*, *302*, *303*, *304*, *328*, *338*, *340*, *341*, *342*, *361*, *364*, **371**, **373**
Fuchs, N.A.- **10**, *162*, *167*, *236*, **237**, *249*, *250*, *302*, *324*, *363*, **371**

G

Gans, R.- *55*, **105**
Gauvain, J.- **408**
Gelbard, F.M.- *132*, **152**, *249*, *250*, *252*, *258*, *260*, *290*, *298*, *299*, *300*, *302*, **408**
Geyling, F.T.- **373**
Gieseke, J.A.- **152**, *302*, *303*, **408**
Gillespie, T.- *140*, *141*, **152**
Glass, J.S.- *125*, **152**
Gluckman, M.J.- *49*, *50*, *51*, *52*, *53*, *54*, **105**
Gokoglu, S.A.- **371**
Goldberg, I.S.- *360*, **371**
Goldman, D.E.- *111*, **152**
Goldstein, S.- *328*, *350*, **371**
Goren, S.L.- *342*, *372*, **409**
Gormley, P.G.- *332*, **372**
Graetz, L.- *332*, *342*, **372**
Green, J.T.- *203*, *204*, *206*, *208*, **236**
Greenberg-Kosinski, S.- **373**
Grover, S.N.- *202*, *218*, *236*, **237**, *239*, **240**
Gunn, R.- *133*, **152**

H

Hagrman, D.- **408**
Hahn, L.A.- *341*, *364*, **372**
Halbritter, J.- *68*, *70*, *71*, *73*, **105**
Hales, J.M.- *358*, **372**
Hall, D.- **373**
Hall, E.W.- **105**
Hamielec, A.E.- **239**
Hamoodi, S.A.- *307*, *311*, *314*, *319*, *320*, **324**
Han, J.T.- **408**
Hansford, R.E.- *84*, *85*, **105**

Happel, J.- **10**, *19*, *24*, *26*, *28*, *34*, *36*, *91*, **105**
Hausdorff, F.- *63*, **105**
Heiss, J.F.- **106**
Hidy, G.M.- **10**, *159*, *237*, *270*, *302*, *303*, *305*, *324*, **371**
Hiedemann, E.- *212*, *216*, **236**
Hill, G.W.- *218*, **238**
Hill, R.- *44*, *49*, **106**
Hilliard, R.K.- **408**
Hinds, W.C.- **10**, *363*, **372**
Hines, A.L.- *328*, **372**
Hirschfelder, J.O.- *313*, **324**
Hochrainer, D.- **107**
Hocking, L.M.- *84*, *85*, *87*, *88*, *91*, **106**, *172*, *187*, *203*, *218*, *220*, *221*, *222*, *223*, *224*, *225*, **237**
Homsy, G.M.- **373**
Honig, E.P.- *187*, **237**
Hopke, P.K.- **372**
Hoppel, W.A.- *136*, **152**
Horlock, J.H.- *125*, **152**
Horst, T.W.- **372**
Hosokawa, T.- **372**
Huang, L.K.- *296*, *298*, **304**
Hubner, R.S.- *302*, **408**
Hudishweskyj, A.B.- **304**

I

Ice, C.- **409**

J

Jacobi, W.- *137*, **152**
Jayaweera, K.O.L.F.- *95*, **106**, *224*, **237**
Jeffery, G.B.- *51*, *55*, *66*, *69*, *70*, *71*, *73*, *89*, *90*, *93*, **106**
Jeffrey, D.J.- *99*, **106**
Johnson, B.M.- **409**
Johnson, D.L.- *56*, *57*, *58*, **106**
Johnstone, H.F.- **371**
Jonas, P.R.- *218*, *220*, *221*, *222*, *225*, **237**
Jordan, H.- *117*, **152**, *299*, **302**
Junge, C.- *137*, **152**

K

Kader, B.A.- *341*, **373**
Kanwal, R.P.- *67*, **106**
Kaper, H.G.- *313*, *324*, *352*, **371**
Kasper, G.- *366*, *367*, **373**
Keefe, D.- *132*, *137*, **152**
Kennard, E.H.- *22*, **106**, *160*, *162*, **237**
Kennedy, M.- *332*, **372**

Kestin, J.- **152**
Kim, Y.P.- *299*, **303**
Kirby, C.R.- *165*, **237**
Klett, J.D.- **10**, *104*, **107**, *167*, *218*, *223*, *224*, *225*, **237**, **238**, *306*, **325**
Konig, W.- *217*, **237**
Kops, J.A.M.M.- *54*, *57*, *59*, *62*, **106**, *150*, **152**
Kousaka, Y.- **11**, *153*, *372*, **374**
Koyro, M.- **408**
Kress, T.S.- **408**
Krueger, L.- **325**
Kuhlman, M.R.- **408**
Kunkel, W.B.- *52*, *54*, **106**, *142*, **152**

L

Lai, F.S.- *270*, *279*, *281*, *282*, **303**
Lang, H.- *323*, *324*, *352*, *354*, *370*, **372**
Langmuir, I.- *218*, **237**
Larsen, P.S.- *328*, **370**
Lassen, L.- *203*, **237**
Law, W.S.- *24*, **106**, *155*, *161*, **238**
Lea, K.C.- *24*, **106**, *155*, *161*, **238**
Lee, A.J.- **325**
Lee, K.J.- **238**
Lee, K.W.- **152**, *275*, *302*, *303*, **408**
Lee, P.S.- *235*, **238**
Lee, S.C.- *218*, **238**
Leith, D.- *55*, *57*, *58*, **106**
Leong, K.H.- **372**
Levenson, M.- **408**
Levich, V.G.- *176*, **238**, *328*, **372**
Lewins, J.- **409**
Lidofsky, L.- **409**
Lin, C.J.- *203*, *208*, **238**
Lin, C.L.- *218*, **238**
Liu, B.Y.H.- *136*, *142*, **153**, *364*, **372**
Lorentz, H.A.- *99*, **106**
Lorenz, R.- **408**
Loyalka, S.K.- **10**, *24*, **106**, *155*, *161*, *164*, *218*, *225*, *235*, **238**, *239*, *252*, *288*, *291*, *299*, *300*, *302*, *303*, *307*, *310*, *311*, *313*, *314*, *315*, *317*, *318*, *319*, *320*, *322*, **323**, *324*, **325**, *342*, *352*, *354*, *364*, *370*, **371**, **372**, *375*, *376*, *382*, *385*, **408**

Author Index

Lushnikov, A.A.- *132*, **152**, *255*, **303**

M

MacChesney, J.B.- **373**
Mackay, G.D.M.- *88*, **106**
Maddox, R.N.- **328**, **372**
Majumdar, S.R.- *42*, *55*, *91*, *94*, *95*, *96*, *102*, **105**, **106**, **107**
Mandelbrot, B.B.- *63*, *64*, **106**
Margulies, T.- **408**
Marlow, W.H.- **10**, **325**, **370**
Marshall, J.S.- *172*, **237**
Mason, B.J.- **10**, **106**, **237**, **306**, **323**, **324**
Mason, S.G.- **106**, *172*, *203*, *206*, *208*, **236**, **239**
Maude, A.D.- *80*, *84*, **106**
Maxwell, J.C.- **324**
McCormack, J.D.- **408**
McDowell-Boyer, L.M.- **409**
McLeod, J.B.- *243*, **303**
McMurry, P.H.- **371**
Mednikov, E.P.- *215*, *216*, **238**
Melik, D.H.- *209*, *211*, **238**
Mercer, T.T.- *61*, **106**
Meyer, R.O.- **409**
Middleton, P.- *288*, *299*, **303**
Millikan, R.A.- *161*, **238**
Minkowycz, W.J.- *358*, **372**
Mitchell, J.A.- **409**
Monchick, L.- *310*, **324**
Morewitz, H.A.- *382*, *383*, **408**
Morgan, K.Z.- *135*, **152**
Morlock, *165*, **238**
Morrow, P.E.- **106**
Morse, T.F.- *342*, *347*, **372**
Mountain, R.D.- *65*, **106**, *283*, **303**
Mulholland, G.W.- *65*, **106**, *245*, *283*, **303**
Muller, H.- *111*, **152**
Murata, K.K.- **408**
Murphy, T.J.- *181*, **238**

N

Nagamoto, C.T.- **153**
Nagel, S.R.- **373**
Nazaroff, W.W.- *342*, **372**
Neiburger, M.- *218*, **239**, *300*, **303**
Neumann, E.P.- *216*, **238**
Niemczyk, S.J.- **409**
Nir, A.- *102*, **106**
Niven, W.D.- **324**
Nolan, P.J.- **152**
Norton, J.L.- *216*, **238**
Nowakowski, B.- **325**

O

O'Brien, V.- *45*, *47*, *48*, *54*, **106**
O'Connor, T.C.- *133*, **152**
O'Neill, M.E.- *42*, *55*, *80*, *81*, *84*, *91*, *94*, *95*, *96*, *102*, **105**, **106**, **107**
Oberbeck, A.- *38*, **107**
Okuyama, K.- **11**, *151*, **153**, *360*, **372**, **374**
Onishi, Y.- *99*, **106**
Oppenheim, I.- *181*, **237**
Oseen, C.W.- *19*, **107**
Owen, R.K.- *392*, *393*, **409**

P

Pagani, C.D.- *22*, **105**, **237**, *310*, **323**
Paluch, I.- *218*, **236**
Papperger, A.- **373**
Park, J.W.- *252*, *291*, *299*, **303**, *322*, **324**
Parker, G.W.- **409**
Parker, R.C.- *212*, *212*, **238**
Pasley, L.F.- **409**
Patterson, H.S.- *139*, **153**, *212*, **238**
Paw U, K.T.- **370**
Payakakes, A.C.- **153**
Payne, L.E.- *29*, *31*, *37*, *55*, *67*, **107**
Pazooki, N.- *310*, **324**
Pearcey, T.- *218*, **238**
Pearson, J.R.A.- *19*, **107**
Pell, W.H.- *29*, *31*, *37*, *55*, *67*, **107**
Pendelbury, E.D.- *360*, *361*, **371**
Percival, I.C.- *65*, **104**
Perrin, F.- *234*, **238**
Pertmer, G.A.- *218*, *225*, **238**
Peterson, T.W.- **303**
Pfeffer, R.- **105**
Phillips, W.F.- *161*, **238**
Pich, J.- *250*, *267*, **303**, *360*, **372**
Pidduck, F.B.- *313*, **325**
Pitter, R.L.- *218*, *235*, **238**
Plomp, A.- **108**
Postma, A.K.- *392*, *393*, **408**, **409**
Power, G.- *44*, *49*, **106**
Powers, D.A.- **409**
Pradel, J.- *142*, **151**
Proudman, I.- *19*, **107**
Pruppacher, H.R.- **10**, *104*, **107**, *167*, *218*, *225*, *235*, **238**, **239**, **240**, *306*, **325**
Pui, D.Y.H.- *360*, **372**

Q

Quayle, S.F.- **409**

R

Raabe, O.G.- *161*, **236**
Rahn, F.J.- *369*, **372**, *398*, *399*, *400*, *402*, *405*, *406*, *408*, **409**
Rajendran, N.- *216*, **239**
Ramabhadran, T.E.- *252*, *253*, *255*, **303**
Ramarao, B.V.- *52*, **107**
Ramsdale, S.A.- **303**
Ranger, K.B.- **105**
Rao, A.- *291*, *296*, *299*, *300*, **304**
Rao, N.- **371**
Ray, A.K.- *318*, **325**
Reed, J.- **373**
Reed, L.D.- *302*, **303**, **408**
Reeks, M.W.- *365*, *366*, **373**
Reist, P.C.- **11**, *133*, *134*, **151**, **373**
Rest, J.- **408**
Rexroth, P.E.- **409**
Reynolds, O.- *342*, **373**
Rich, T.A.- **152**
Richardson, L.F.- *63*, **107**
Robert, B.F.- *389*, **409**
Roebersen, G.J.- **237**
Romay-Novas, F.- **372**
Rosinski, J.- *142*, **153**
Rosner, D.E.- *342*, **371**, **373**
Row, T.H.- **409**
Rozenberg, L.D.- **239**
Russell, A.- *137*, **153**
Ryder, C.P.- **409**
Ryley, D.J.- *125*, **151**

S

Sadron, A.- *234*, **239**
Saffman, P.G.- *172*, *195*, *196*, *198*, *200*, **239**
Sahni, D.C.- **325**
Sakurai, A.- *125*, **153**
Sampson, R.A.- *46*, **107**
Sartor, J.D.- *218*, *224*, **237**
Sather, N.F.- *219*, **238**, **240**
Savic, P.- *45*, *46*, **107**
Schefer, R.W.- **373**
Schlamp, R.J.- *218*, **239**
Schlichting, H.- *328*, **373**
Schmidt, E.- *125*, **153**
Schmitt, K.- *352*, *354*, **373**
Schoeck, W.- **408**
Schowalter, W.R.- *203*, **240**
Schumaker, P.M.- **152**, **302**

Schumann, T.E.W.- *111*, **153**, *243*, *264*, **303**
Schwendiman, L.C.- **372**
Scott, W.T.- *244*, **303**
Sedunov, Yu.S.- **11**
Seesoms, A.L.- **409**
Segal, A.- *32*, *49*, **107**
Seigneur, C.- *299*, **304**
Seinfeld, J.H.- **11**, *132*, **152**, **239**, *249*, *250*, *252*, *258*, *260*, *290*, *298*, *299*, *302*, *303*, *304*, *305*, *325*, *360*, *361*, *363*, **371**
Shafrir, U.- *218*, **239**
Shahub, A.M.- *202*, **239**
Shapiro, A.H.- *123*, **153**
Shaw, D.T.- **11**, *216*, *235*, *238*, **239**, **373**
Shaw, R.H.- **370**
Shen, C.- *347*, **370**
Shirokova, N.L.- *216*, **239**
Shiver, A.W.- **409**
Shoaf, M.L.- **409**
Silberberg, M.- **408**, **409**
Simons, S.- *65*, **107**, *127*, *128*, *130*, *145*, **153**, *194*, *215*, **239**, *254*, *260*, *263*, *264*, **304**
Simpkins, P.G.- *342*, **373**
Simpson, D.R.- *121*, **153**, *299*, **304**
Sitarski, M.- **239**, **325**
Slack, G.W.- **106**, **237**
Smit, H.C.D.- **108**
Smith, W.E.- **303**
Smolarkiewicz, P.K.- *294*, **304**
Smoluchowski, M. von- *109*, **153**, *170*, **239**, *241*, **304**
Sneddon, I.N.- *75*, **107**, *247*, *248*, **304**
Sparrow, E.M.- *358*, **372**
Spielman, L.A.- *80*, **107**, *187*, **239**
Spiewak, I.- **409**
St. Clair, H.W.- *212*, **239**
Stöber, W.- *61*, **106**, **107**
Stegun, I.A.- **302**
Stimson, M.- *51*, *55*, *73*, **107**
Stokes, G.G.- *21*, **107**
Stratmann, F.- *368*, **373**
Stukel J.J.- **372**
Subramanian, R.S.- **373**
Suck, S.H.- *288*, **304**

Sutugin, A.G.- **10**, *249*, *302*, **324**
Suzuki, M.- **106**
Swift, D.L.- *264*, **304**

T

Tabakoff, W.- **373**
Takayama, F.- *125*, **153**
Talbot, L.- *368*, **373**
Taulbee, D.B.- *360*, **373**
Taylor, G.I.- *174*, *200*, **239**
Taylor, M.- **408**
Thomas, L.B.- *319*, **325**
Tien Chi, *52*, **107**
Tilley, H.L.- **325**
Tills, J.L.- **409**
Tobias, M.L. **408**
Tombrello, T.- **409**
Tompson, R.V.- *307*, *310*, *311*, *313*, *314*, *319*, *320*, *324*, **325**
Tsang, T.H.- *291*, *296*, *298*, *299*, *300*, **304**
Tu, K.W.- *216*, **239**
Turner, J.E.- *135*, **152**
Turner, J.S.- *172*, *195*, *196*, *198*, *200*, **239**
Tuttle, R.F.- *235*, **239**
Twomey, S.A.- **11**, *115*, **153**, *305*, *306*, **325**

U

V

Valdez, G.D.- **408**
van de Vate, J.F.- *33*, **108**, *115*, **153**
van de Ven, T.G.M.- *172*, *203*, *206*, *208*, **239**
van Kampen, N.G.- **240**
van Leeuwan, W.F.- **108**
Varoḡlu, E.- *296*, **304**
Vaughn, E.V.- *302*, **408**
Voloschuk, V.M.- **11**

W

Wacholder, E.- *219*, **240**
Wagner, P.E.- **325**
Wakiya, S.- *91*, **108**
Waldmann, L.- *352*, *354*, **373**

Walker, K.L.- *342*, **373**
Wang, C.S.- *263*, *267*, *269*, *270*, *279*, *281*, *302*, **304**
Wang, P.K.- *218*, **240**
Ward, S.G.- **105**
Warman, E.A.- **409**
Washington, K.E.- **408**
Watson, H.H.- *342*, **373**
Weickemann, H.- **303**
Weinbaum, S.- **105**
Weinberg, M.- **373**
Wen, H.Y.- *366*, *367*, **373**
Werle, D.- **153**
Whitby, E.R.- **303**
Whitby, K.T.- *136*, *142*, **153**, **303**
Whytlaw-Gray, R.- *139*, **153**, **238**
Wiersema, P.H.- **237**
Williams, D.C.- *382*, *385*, **408**, **409**
Williams, M.M.R.- *14*, *15*, *20*, *22*, *42*, *44*, *54*, *70*, *88*, **108**, **153**, *159*, *198*, *202*, *215*, **239**, *240*, *244*, *252*, *256*, *258*, *275*, *276*, *279*, *280*, *281*, *282*, **304**, *310*, *320*, *325*, *352*, *354*, **373**, **409**
Willis, D.R.- **373**
Wilson, R.- **409**
Woods, W.A.- *125*, **152**
Wooton, R.O.- **409**
Wright, A.L.- **372**
Wu, J.J.- *336*, **371**

X

Y

Yaglom, A.M.- *341*, **373**
Yalamov, Yu.I.- *352*, *354*, **371**
Yamamoto, S.- **372**
Yoshida, T.- **11**, **374**
Yu, C.P.- *352*, *360*, **373**
Yuan, C.C.- **372**

Z

Zebel, G. von, *111*, *117*, *142*, *144*, *145*, *146*, *147*, **153**
Zeichner, G.R.- *203*, **240**
Zernik, W.- **374**

Subject Index

A

Absorption, 2, 305, 385
  total area of, 392
Acceleration, 12
  due to gravity, 201, 218
  effective, 201
  of a pocket of air, 196
Accident (al),
  core disruption, 1
  nuclear reactor, 1
  progression, 377
  release, 132
  scenarios, severe, 380
  sequences, 375-376, 395, 404
  simulation of, 385
Accommodating spheres, 161
Accommodation, perfect, 72
  coefficient, 68, 323
Acid rain, 305
Acoustic, forces, 154
  coagulation, 212, 215, 216
    kernel, 216
  theory, 212
  turbulence, 216
Actinide isotopes, 376
Additive approximation, 212
Additivity assumption, 193
Adhesion force, 366, 368
Adiabatic analysis, 122
Adsorption, 385
  isotherms, 323
Advection in fluid flows, 294
Aerodynamic diameter, 54, 55, 57, 59, 331, 392
AEROSIM, 278, 382, 398
Aerosol
  age, 265, 283, 287
  analyzers, electrical, 388
  applications, 311
  atmospheric, 1, 202, 244
  balance equation, 111
    for pure coagulation, 283
  behavior 397
    computer codes, 398
    understandings of, 395
  calculation, 225
  charge, 136
  cloud, 154, 176
  coagulation, general theory, 179
  computations, 294
  computer programs, 382, 385

concentration, 117, 332, 388
  total suspended, 397
CsOH, 395-399, 404
currents, 299
density, 214, 251
deposition, 329, 332
  calculations, 328
  rates, 330
distribution, 109, 110, 283, 289, 294
  function, 125, 271
  with electric charge, 142
dynamics, 65, 211, 291, 381
evolution, 1, 146-147, 287, 292, 377, 397
  in a severe accident, 385
flow, 122
  over surfaces, 360
formation, 375
generation rates, 397
growth, 294, 377
high concentrations of, 377
history of, 145
hygroscopic nature of, 306
mass, 395
material, 395-397
measurements, 395
mechanics, 305, 381
mixture, 397
MnO, 395-399, 404
particles, 52, 57, 89, 117, 132, 136, 145, 154, 170, 202, 212, 231, 260, 305, 326, 341
  distribution function, 155
  volume distribution, 128
population, 147
problems, 176, 208
production, 144
radioactive, 1, 130, 260
rate processes, 287
reactor, 2
resuspension, 395
retention, 397
science, 404
shape, 2
size, 2, 288-289, 388
  distributions, 388
source, 388
  at test pipe Inlet, 400
strongly charged, 147
transmission of, 348

transport, 122, 385
type, 150
velocity distribution function, 159
weakly charged, 146
Agglomerates, 235
  chain-like, 2
Agglomeration, 1, 134, 388
  kernel, 154
  of fuel constitutes, 1
Aggregates, 61, 65
Air
  bubbles, 89
  cleaning systems, 388
  density, 135
  molecules, 160
  polluted, 1
Airborne radioactivity, 130-131, 261
$Al_2(OH)_3$ aerosols, 397
Algebraic equations, linear, 94
Algorithm, 289, 316
Ambient, flow, 204
  gas mixture, 305-306
American Nuclear Society, 383
Amplitude
  difference, 215
  of vibration of a particle, 213
Analogies, heat and mass transfer, 328-329
Analysis
  theoretical, 326
  trajectory, 217
Analytical solution (result), 209, 241, 256-258, 261, 270, 278, 292, 301, 344
Analyzer, electrical, 5, 388
Andersen impactors, 363
Anemometers, turbine, 399
Angle
  dependent relative velocities, 168
  of a bend, 364
Angular
  dependence, 235
  speed, 89-90, 94
  velocity, 66, 71-73, 95
Anisotropic scattering, 318
Ansatz, for the volume
  distribution function, 275
Antidiffusion velocity, 295

Antiparallel motion, 87, 96-97, 224
Apparent mean free path, 163
Approximate (Approximation)
  Brownian kernel, 138
  equivalent sphere, 49
  results (solutions), 287, 342
  techniques, 91
APRITE, 398
Aqueous particles, 305
Arbitrary (Arbitrarily),
  condensation coefficient, 319
  constants, 288
  geometries, 328
  Knudsen number, 311
  mass ratios, 320
  molecular interaction laws, 319
  orientation of spheres, 96-98
  polar direction, 195
  shaped bodies, 72, 320
  size ratio, 96
  thermal accommodation coefficient, 319
Arc vaporizers, 388
Area, effective, 164
ARIANNA, 398
Arrays of spheres, linear, 49
Artificial constraints, 267
Associated Legendre polynomials, 93
ASTD, 382
Asymmetry, 19, 52
Asymptotic
  analysis, 194, 211
  behavior, 46
  expansions, 206
  fall velocity, 188
  fluid velocity components, 204
  formula, 84
  results, 84
  shape, 264, 279
  solution, 185, 313
Atmospheric
  aerosols, 1, 202, 244
  composition, 388, 400
  dispersion models, 382
  ions, 132
  sciences, 225, 230
Attenuation rates, 390
Attractive, force, 137, 187
  hydrodynamic interactions, 216
AUX 2.9, 398
Auxiliary
  building, 395-397
  conditions at time zero, 400
  factors, 311
Average
  charge on a particle, 138

density, 118-119, 129
diffusion coefficients, 138
  for spheroids, 233
flow velocity, 20
mass, 129
volume, 113, 129, 265
Axi-symmetric body, 26, 34, 49
Axial, conduction, 342-344
  symmetry, 24-26, 66
  thermophoresis, 344
Axially symmetric, body, 35, 45, 66, 73
  problem, 73
Axis
  of rotation, 91
  of symmetry, 70
Azimuthal
  angle, 209
  symmetry, 195
  body, 35
  problem, 191

B

Ba, 388
Background
  gas, 132, 306, 311
  radiation, 132
Backward difference, 294
Bacteria, 1
Balance
  equation, 110, 113, 116, 137, 150, 184
  relationship, 13
Basic, equations of mass and heat transfer, 306
  functions, 289, 316
Bath of ions, 145
Benchmark
  results, 299
  solutions, 252
  tests, 404
Bernoulli
  forces, 217
  law, 217
Bessel functions, 42
BGK model, 24, 161, 311, 317-319
Bifurcations, 370
Bimodal distribution, 287
Binary
  aerosol, 298
  isothermal gas mixture, 352
  mixture, 354
Bioaerosols, 2
Bipolar
  charging, 136, 139
  ions, 133
  transformations, 74

Bispherical units, 55
Bjerknes, theory of, 217
Black sphere, 311
BMI-2104
  calculations, 386
  numerical simulations, 386
  suite of codes, 385-387
Body
  fitted coordinates, 370
  irregularly shaped, 32, 55
  of arbitrary shape, 72
  of deposition, 348
  of revolution, 347-350, 351
Boiling water reactor, 379-381
  configurations, 388
  containment, 377
Boltzmann
  collision operator,
    dimensionless, linearized, 312
  constant, 109
  distribution, 180, 362
  law, 202
  equation, 15, 19-21, 162-164, 180, 305, 308, 319, 327, 352
    numerical solution of, 310
  equilibrium, 148
  statistics, 132
Borax thiosulfate, 389
Boric acid, 389
Boron carbide, dolomite, 382
Boundary conditions, 13, 18-21, 24-27, 31-37, 41-46, 50-51, 66-73, 76-77, 81, 85-90, 93-94, 134, 157-158, 184, 189-192, 197, 206, 210, 231, 255, 287, 307, 313-314, 326-327, 332-334, 337-339, 343, 349-350, 357, 362, 370
  no-slip, 69
  slip, 24, 31-32, 35, 68, 69, 85, 88
Boundary layer, 332, 340
  analyses, 342
  theory, 328, 333-334, 358
  velocity, 330, 363
  viscous, 330
Boundary value problem, 314, 332
Bronchial tract of the lung, 63
Brownian
  agitation, 156
  coagulation, 65, 113, 127, 145, 155, 162, 179, 190, 193-194, 203, 209, 258, 263, 270, 277-279, 287

Subject Index

continuum limit, 266
kernel, 144, 178, 184, 187
Knudsen limit, 266
of fibrous-like particles, 235
of spheroidal particles, 231
diffusion, 115, 137, 177-181, 197-198, 209, 340, 347, 353, 358
coefficient, 113, 176, 179, 184, 236
coefficients of randomly oriented rods, 236
forces, 197
interaction, 148
kernel, 137, 267
motion, 40, 109, 137, 142, 154-155, 176-181, 190, 203, 206, 231-232, 331
theory of, 163
particle, rotating, 72
trajectory of, 155
plus gravitational kernels, 212
rotation, 61, 235
self-similarity, 269
theory, 181
to gravitational settling, changes from 193
Buoyancy, effects of, 331

C

Calculation of ex-plant dispersion of radioactivity, 377
Canopies, fractal, 64
Capacitance, electric, 236
Capture of particles, rate of, 137
Carbon whiskers, 1
Carrier-modified Oseen
approximation, 223
forces, 225, 228
drag, 230
Cartesian coordinates, 167, 205, 218, 219, 221
CEA, 388
Central differencing, 293-294
Centrifuges, 388
Cesium isotopes, 376
Change in drag with Knudsen number, 161
Chapman-Enskog, function, 312
asymptotic form, 352
solution, 308, 319
Characteristic
coagulation, 117
dimension of a flow, 341
mixing time, 117
time, 158
constant, 134
Charge (Charged), 385

aerosol
moderately, 146
particles, 132, 135-137, 143, 202
unipolar, 138, 145
correction, 139
distribution, 133, 140, 328
effect on coagulation, 136
electrostatic, 142
neutralization, 133
state, 202
total positive and negative, 149
Charging rate, 202
Chemical (Chemistry)
affinities, 377
and spectrometric methods, 388
composition, 1, 260
conditions, 287
forms, 388
of the environment, 381
reaction, 2, 115
states, 377
Chernobyl accident, 381-382
isotopes composition, 385
Chimney effect, 117
Circumscribed spheres, 49
Cladding failure (rupture), 377
Classic inverse problems, 335
Classifier, 5
electrical, 362
Clausius-Clapeyron equation, 306
Climatic consequences, 1
Close separation, 221
Closed
container, 175
packed structure, 63
streamlines, 208
Cloud droplets, 132-133, 222
Clusters, 65
formation, 305
of vapor molecules, 305
Coagulation, 110, 113, 127, 131, 136-140, 143, 154, 241, 248, 252-254, 259, 263, 274, 289-290, 298-299, 377
and condensation phenomena, 252
and deposition history of an aerosol, 271
and particle growth, 392
and scavenging of radioactive aerosols, 142
basic properties of, 288
between dendritic shapes, 236
between spheroids and spheres, 235

Brownian, 65
and gravitational, 190, 193, 194, 209
and turbulent, 179
by diffusion, 216
calculations, 328, 363
characteristic, 117
constant, 140, 141,142, 175, 286
degree of, 136
effect of charge on, 136
enhanced, 136
kernel, 113, 127-129, 137, 145, 148-151, 154, 159-164, 168-170, 177-178, 190-191, 197-202, 206-216, 231-232, 236, 241-243, 258, 266, 276, 298
for Brownian motion, 184
for turbulent diffusion, 176
generalized, 155
hydrodynamic, 280-281
net, 287
neutral, 140
two region problem, 177
mechanisms, 110, 140, 150, 154, 266, 275, 279, 287
ways of interaction, 180
of large and small spheroids, 235
of large spheres and small spheroids, 231-233
of mixed species, 385
of nonspherical particles, 231
of small and large spheroids, 231
of small spheroids with large spheres, 231
particles, 109
problems, 110
process, 94, 110, 113, 134, 142, 145, 154, 198, 254, 255, 377
influence of nonsphericity, 231
rate, 111, 139-140, 165, 186, 193, 198, 212-214, 274
of change due to, 144
strong, 385
terms, 288, 392
in the dynamic equation, 283
times, 117, 134
unified theory of, 179
volume, 216
with deposition, condensation, and a source, 273
Coalescence, 109, 131, 144
Coarse aerosols, 142

427

Coefficient
  accommodation, 68
  Brownian diffusion, 113
  force, 224
  friction, 156, 160
  local mass transfer, 333
  metric, 25
  of diffusion, 194, 199, 233
    translational diffusion, 234
  of thermal conductivity, 16
  resistance, 83, 97
  supersaturation, 267
  transport, 352
Collection, efficiency, 2, 363
  points, 46
Colliding particles, 363
Collision, 131, 148-150
  behavior, 85
  between two rods, 235
  cross-section, 159, 311-312, 364
  diameter, effective, 235
  dynamics, 217
  efficiency, 154, 158-159, 165, 169-173, 208, 215-218, 220-226, 235
    spherical gravitational, 235
  free, 71
  frequency, 132
  of supercooled water droplets with ice plates, 235
  of two particles, 73
  parameters, 392
  probability, 217
  rate, 172, 184, 203, 214-215
    between aerosol particles, 170
    between two aerosol clouds, 159
    for turbulent diffusion, 172
    instantaneous, 214
    of uniformly distributed particles, 173
  shape factor, 235
  term, 314
  time, 88
Colloid (al)
  aggregation, 249
  science, 203
Combustion, process, 1
Commission of European Communities, 398
Compactness, of aggregate, 225
Comparative study of the moments method, 278
Complex
  flows, 370
    and computer programs, 369
  and geometries, 328
  shape, 21, 154
Composite particles, 263
  radioactivity, 263
Composition
  chemical, 1
  distribution, 397
  radionuclide, 385
Computational
  aerosol mechanics, 404
  costs, 297
  fluid dynamics, 300, 328
  programs, 370
  grid, 297
  schemes, 380
  step, 292
  time, 290
Computer programs, 288-291, 375, 380-382, 385, 389, 397, 404
  AEROSIM-M, 385
  and benchmarking, 299
  APRITE, 385
  ARIANNA, 385
  AUX 2.9, 385
  CONTAIN, 385
  CONTEMPT-LT, 385
  FIPLOC, 385
  HAA-4, 385
  JERICHO, 385
  LIMIT, 385
  LSODE, 289
  MARCH, 385
  MCT-2, 385
  NAUA-4, 385
  NAUA-5, 385
  QUICK-M, 385
  REMOVAL, 385
  RETAIN-2C, 385
  RETAIN-S, 385
  TRAP-MELT2, 385
Computing time, number and mass concentrations, 300
Concentration, 126, 306, 349-350, 389
  aerosol, 117, 332, 388
  and deposition, 348
  boundary layer, 327, 336-338, 345
  distribution, 182
  gradient, 181, 195, 326, 332, 352, 377
    water vapor, 142
  in a gas mixture, 351
  high aerosol, 395
  layer, 340, 348
    edge of, 346
  of FPs, time dependent, 388
  of particles, 173, 180, 331-332
  of species, 306
  on a surface, 323
  profile, 351
Condensation, 57, 113-115, 127, 241, 249, 255, 264, 267, 289, 305, 354, 369, 377
  /evaporation, 290-292, 298, 323, 359, 385
  and coagulation, 260
  nuclei counter, 335, 362
  of molecules on a particle, 332
  of water vapor, 306
  on an aerosol, 249
  rate, 127, 250, 255, 308, 313, 316, 322-323, 355
  strong and weak, 300
Condensible and noncondensible species, 352
Condensing, environment, 389, 392, 395, 399
  species, 323
  vapor, 323, 352-354
Conducting spheres
  approximation, 202
Conduction (Conductivity)
  electrical, 133
  coefficient of, 16
  heat current, 309
  heat transfer in a background gas, 308
  heat transfer rate, 322
  thermal (heat), 16, 305-306, 250, 328, 342
Confocal system, 52
Conformal transformation, 34
Consequences
  analysis, 381
  of nuclear reactor accidents, 381
  ex-plant, 377
Consequent releases, 381
Conservation
  condition, 283
  equations, 290
  form, 294
  laws of mass and energy, 381
  of effective volume, 284
  of energy, 15
  of mass, 15
    in a collision, 127
    in coagulation, 288
  of momentum, 15
  of solid volume, 284
  of volume, 112, 150, 291
    in a collision, 127
    in coagulation, 288
  relationships, 150

Subject Index

requirement, 295
Constant, coagulation kernel, 149, 264, 274
Collision Cross-Section model, 317
CONTAIN, 382, 385-386, 398
Containment
  thermal hydraulics, 385
  air, 383
  atmosphere, 117
  building, 395
  bypass, 395
    tests, 397
  design for the Grand Gulf plant, 379
  designs and engineered safety features, 381
  environment, 388
  failure, 376-377, 395
  melt through, 376
  sub-volumes, 176
  System Test Facility (CSTF), 386
  transport and reaction, 387
  vessel, 117, 132, 388
    arrangement, 405
  volume producing turbulence, 176
  walls, 388
CONTEMPT-LT, 398
Continuity
  condition, 177
  equation, 15-19, 91-93, 122, 173, 184, 189, 326, 343, 349-351, 354, 362
    for laminar flow, 340
    particle, 340
    vapor and particle, 359
Continuous, function, 111, 297
  quantity, 145
  source, 129, 250
  variable, 243
Continuum
  approximation, 253
  condensation rate, 320
  flow, 162
  limit, 307
    equation, 354
    near, 308
  regime, 24, 53, 159, 188, 352
  results, 320
  theory, 20, 309
Control
  rod ejection, 375
  volume, 122-123
Convection, 175, 344-346
  -diffusion continuity equation, 343

Convective
  and diffusive currents, 332, 370
  -diffusion and convective-gravitational contributions, 361
  -diffusiophoretic deposition, 351
  -diffusive deposition, 332
  -diffusive flows, Laminar and turbulent, 327
  -electrical deposition, 361-362
  -gravitational deposition, 327, 360
  -phoretic deposition, 327, 341
  and gravitational currents, 360
  and thermophoretic terms, 350
  current, 326
  flow, 118, 338, 342, 347
    and Geometry, 369
  gas flow, 326
  gravitational and diffusive currents, 360
  heat and mass transfer, 305-306
    transfer coefficient, 328
  loops, 117
  mass current, 340
  mass transfer, 328
  motion, 113, 188, 332
  term, 118, 198
  velocity, 326-328
Conventional distribution function, 126
Convergence, 49, 316
  for certain coagulation processes, 278
  numerical, 84
  rate of, 51
Convolution transform, 243
Cooling down period, 397
Coordinate (s)
  curvilinear, 27
  cylindrical, 24
  intrinsic, 26
  natural, 49
  of two particle motion in a flow, 166
  orthogonal curvilinear, 25
  spherical, 67
  system for laminar shear flow, 171
  transformation, 81
CORCON, 385, 387
Core
  damage accidents, 1, 375-377, 381, 404
  debris, 382, 387
    -concrete interactions, 377
  disruptive accidents, 1

  in fast reactors, 375
  inventory of radionuclides at Chernobyl, 384
  -concrete temperature and interactions, 387
Corona discharge, 361
CORRAL, 381-382, 392
Correction
  factor, 87, 178, 202, 355
    Cunningham, 22, 54, 87, 161-164, 190, 213, 231
    drag, 53
    slip, 21, 54, 70, 91
Corrective step, 288, 295-296
Correlations for aerosol deposition, 369
CORSOR, 385
Cosmic rays, 132
Couette flow, 170
Coulomb's law, 137, 202
Coupled
  aerosol equations, 117
  calculations, 306
  chambers, 122
  first order, nonlinear differential equations, 277
  problems, 323
  zones, 118, 122
Coupling coefficients, 122
Courant number, 294
Cr-3, 383
Creeping motion, 18
CRI-II, 388
Criteria for licensing, 381
Critical
  impact parameter, 166
  pressure ratio, 123
  Stokes number, 363
  streamline, grazing collision, in the equatorial plane, 166
  trajectory, 165, 216
  volume, 216
Cross section (al)
  current, total, 364
  for collision, 159
  differential scattering, 12
  effective, 154
  of air molecules, 160
  transfer, 290
Crystal River-3, 381
Cs
  inventory, 381
  vapor, 397
Cs_2O particles, 388
CsOH aerosol, 397-399, 404
CSTF, 388-404
Cubic spline, fit, 225, 288-289

Cumulative deposition, 334
Cunningham correction factor, 22, 54, 87, 161-164, 190, 213, 331
Current
  incident on hemisphere, total, 168
  of particles, 162, 168, 236
  of vapor molecules, 306
  onto a test sphere, 177
  over a whole sphere, total, 170
  through a surface, 162
Curvilinear, coordinate system, 27
Cutoffs, mass ratio, 225
Cyclones, 363
Cylindrical
  fibers, 338
  tube, 342
  coordinate system, 24, 92

D

Data base, 385
Decay heat removal system, failure of, 376
Delayed containment failure, 395
Delta, function, 112, 127-130, 150-151, 283
  source at short times, 257
DEMONA, 386-396
Dendritic formations, 370
Density
  equilibrium, 305
  gradients, 154
  of a liquid, 250
  of a particle, 188
  profile, 316
  unperturbed, 137
  variation, 286
Deposited aerosol, 388
Deposition, 110, 113-114, 127-129, 147, 241, 248, 252, 263, 365, 377, 290
  and resuspension, particle, 326
  basic problems of, 341
  body of, 348
  by inertial impaction in a pipe bend, 364
  calculations, 370
  constant, 250
  convective-diffusiophoretic, 351
  convective-diffusive, 332
  convective-electrical, 361-362
  convective-gravitational, 327, 360
  convective-phoretic, 327, 341
  current, 326, 329

due to both diffusive and inertial motions, 364
efficiency, 329, 333-334, 340, 347, 360-364
experiments, 389
flux, 351, 358
fractional, 333
in convective flows, 342
in human lungs, 332
in piping, 346, 363
laminar diffusive, 342
mechanism, 251, 369
of molecular species, 328
of particles, 354-355, 363
of radioactive particles, 342
of small particles onto a large particle, 194
of a vapor, 359
on a spherical surface, 332
on nonspherical objects, 332
on plane vertical surfaces, 358
on surfaces, 331
process, 397
rate, 142, 326-327, 332, 346-348, 356, 364, 369
  calculations, 364
  coefficient, 329
  due to Stefan flow, 356
  global, 344
  local, 327, 339, 344
  normalized, 357
  total, 339, 364
  term, 145
  velocity, 327-329, 358
  volume averaged, 361
Depressurization, 364
  of a containment, 326
Derivation of diffusion equation, 180
Descriptors, 151
Determination of source terms, 380
Deviations
  from hydrodynamic theory, 203
  from Stokes flow, 218
Dielectric constant, 208
Difference equation, 293
Differential equation, first order linear, 358
Diffuse reflection, 306-313
Diffusing species, 115
Diffusion, 113-114, 134, 294-296, 305, 322, 326-327, 344, 351, 364, 377
  -advection equation, 217
  -convection equation, 194-195, 201
  and gravitational motion, 360

battery, 335-336
boundary layer, 330-332
coefficient, 109, 156-160, 176, 181, 184-185, 231-233, 249, 253, 312, 392
  average, 138
  effective, 145, 194, 199
  modified turbulent, 72
  of a vapor, 250, 307
  Stokes-Einstein value for, 158
controlled growth, 249
current, 351
equation, 156, 172, 176, 184-190, 194, 198, 201, 231-233
  derivation of, 180
flow, 118
of a very light species in a heavy background gas, 311
of molecules, 351
process, 109, 181
  with a superimposed drift velocity, 180
  isotropic in space, 157
solution, 316
term, 344, 350
theory, 155, 162, 178, 201
velocity, 295, 352
Diffusional deposition, 194
  in a stagnant gas, 331
Diffusiophoretic, 358
  current, 326
  force, 351-356
  motion, 353
  velocity, 326, 357
Diffusive
  current, 326
  effect, 198, 345
  flow, 194
  flux, 182, 194
  force, 181, 193
  mechanisms, 127
  motion, 155, 188, 365
  term, 118, 343
  velocity, 328
Diffusivity
  of vapor molecules in a background gas, 249
  relative, 183
Digamma function, 277
Dilute gas mixture, 306
Dimensional molecular position and velocity vectors, 311
Dimensionless
  coefficient, 342
  concentration, 345-346
  condensation rate, 317-318

Subject Index

constants, 265
groups, 329, 348
jump
  coefficient, 308
  distances, 311
  linearized Boltzmann collision operator, 312
  parameter, 337
  particle separation, 185
  relative diffusion coefficient, 185
  temperature, 342
  time, 258
Dioctyl phthalate, 318
Direct
  interception, 339
  numerical calculations, 370
  numerical solution of the Boltzmann equation, 310
  transformation, 116
Direction of diffusion, 231, 234
Discontinuity, 251
  of the molecular velocity distribution, 308
Discontinuous boundary condition, 314
Discrete
  distribution function, 111
  equations of aerosol coagulation, 270
  lift bursts, 368
  representation, 253
  time steps, 368
  volumes, 241
Discretization, 314
Dispersion, 294
  and deposition, 144
  and diffusion, 290
  electrostatic, 138-139, 143-147
  models, 382
  of radioactivity, ex-plant, 377
Dispersivity of a distribution, 274
Dissipation
  of fogs, 388
  rate, energy, 174-175
Distortion
  in particle shape, 294
  of fluid, 172
  of stream lines, 173
Distribution, 133, 287, 296
  change in, 271
  for outgoing and incoming molecules, 306
  log-normal, 271-273, 287
  absolute velocity, 174
  of charge, 140, 202
  of particle sizes, 271

Distribution function, 110, 114, 155, 128, 154, 174, 202, 260, 298
  extended, 131, 142, 258
  generalized, 126
  volume, 110, 116, 250, 254, 271
Divergent integral, 187
Drag
  and adhesive forces, 365
  coefficients, 224
  force, 150, 218, 225, 327
  on a sphere, 156
Drift velocity, 180
Driving force, 327, 365
Dry and wet environments, 389
Dust
  explosions, 364
  free space, 342, 354
Dyadic notation, 184
Dynamic (al) Equation, 109, 126, 132, 145, 150-151, 241-243, 248-249, 256, 271-272, 287-290, 297-299, 392
  for charged aerosols, 140-142
  for nonspherical particles, 150
  for pure coagulation, 241
  generalized, 150
  methods of solving the, 241
Dynamic
  shape factor, 150, 331
  thermal hydraulic conditions, 395

E

Eccentric oblate, 46
Eccentricity, 44-46, 72
ECCS, failure of, 376
Eddy diffusivity, for mass transfer, 340
  model of turbulence, 361
Eddy viscosity, 114
Edge of the concentration layer, 346
Effective
  acceleration, 201
  area, 56, 164
  radius, 40-42, 232, 283
  size, 65
  volume, 283
  weighted cross section of air molecules, 160
Effects
  of humidity on a particle, 404
  of molecular collisions, 364
  of natural processes, 388
Efficiency

of deposition, 364
of filters, 364
Eigenfunctions, 333
  Eigenvalues, 314, 333
  expansion, 333
Einstein
  formula, 160
  relation, 40
Electric (al)
  aerosol analyzers, 5, 388
  capacitance, 236
  charge, 132, 140, 143, 202
  classifier, 362
  flow of, 236
  field, 114, 143, 203, 208, 326
  forces, 218
  Power Research Institute, 398-402, 405-406
  precipitator, 361-362
Electro-optical devices, 388
Electromagnetic scattering in polydispersions, 271
Electrostatic
  barrier, 211
  charge on radioactive aerosols, 142
  dispersion, 138-139, 143-147
  energy, 132
  forces, 201-202
  interactions for a two body system, 183
  precipitators, 212, 361
Ellipse, 44
Ellipsoid, 44-45
  in free molecular flow, 71
Emergency cooling water, 381
Empirical correlations, 381-382
Energy
  balance, 122
  current, 329
  conservation of, 15
  dissipation, 44, 135
  rate, 175
  electrostatic, 132
  equation, 122-124, 328-329, 342, 348-350
  of a particle, 132
Engineered
  processes, 326
  safeguards, 388-390
  safety features, 377
Enhanced
  coagulation, 136
  particle deposition, 354
Enhancement factor, 139-140, 187-188
Enthalpy, 122
Entrained particles, 212

Entrainment
  by turbulent eddies, 172
  degree of, 213
Environment, 1, 117, 287, 331
EPRI, 398
Equation
  of motion, 196, 217-219, 225, 363, 369
  Boltzmann, 15, 19-21, 162-164, 180, 305, 308, 319, 327, 352
  continuity, 15-19, 91-92, 122
  continuum, 24
  energy, 15
  hydrodynamic, 16, 19-21
  Langevin, 65
  momentum, 15-16
  Navier-Stokes, 17, 87
  of state, 377
  Slow viscous flow, 18, 24
  transport, 14, 24
Equations, linear algebraic, 94
  of particle motion, 365
Equatorial radius, 45
Equilibrium, 133, 136, 145, 249, 305, 365
  charge distribution, 132-133
  conditions, 142
  density, 305
  fuel cycle, 375
  ion density, 134
  system, 180
  theory, 140
Equivalent
  diameter, 56
  sphere, 2, 53-54, 60, 65, 116, 129, 320
    approximation, 32, 40, 44, 49
  volume, 56
Error function, 312
Euler's forward difference method, 293
Evaporating drop, 354
Evaporation, 113, 127, 290, 305, 377
  of an aerosol, 249
Evolution of aerosols, 260, 377, 399
Ex-plant
  consequences, 377
  dispersion of radioactivity, 377
Ex-vessel releases, 377
Exact
  formula, 84
  kernel, 193, 198, 275
  result, 42, 44, 46,

solution, 69, 88, 98, 142, 145, 241, 250, 255, 268, 287-289, 293-294
Exclusion zone, 380
Exothermic
  reaction, 382
  coefficients, 315, 333
Expansion, 124
  inner and outer, 84, 211
  Taylor series, 295
Experimental
  characteristics of LWR aerosol tests, 388
  data, 354, 392
  base, 386
  results, 142, 392
  simulations, 385, 390
Exploding wires, 388
Explosions, 381
Explosives, 375
Exponential integral, 198
Extended
  distribution function, 131, 142, 258
  spheroid, 231
External forces, 181, 188, 326, 352, 364
Extrapolated surface, 327, 330, 370
Extremum principles, 44

F

Factor
  correction, 87
  drag correction, 53
  form, 48
Failure
  of decay heat removal system, 376
  of ECCS, 376
  of reactor shutdown systems, 376
Fallout rate, 389
Fanning friction factor, 341
Fast reactors, 1
  safety, 225
  core disruptive accidents in, 375
Fick's law, 307
FIDAP, 370
Filtering, 326
Filters, 332
  efficiency of, 364
Filtration, 2, 348
Finite
  dielectric constant, 208
  difference, 241, 293, 299, 344

element, 49, 241, 289, 296-299, 344, 370,
  method based on moving grid characteristic, 296
FIPLOC, 398
Fire fighting efforts, 382
Fission Products, 375-377, 381-383, 388, 397
  inventory in fuel, 387
  removal, 390
  simulator, 388
  species, 397
  transport, 387-389
Flame zone, 117
Flashing pool tank, 399
Flat plate, 349-351
  flow over a, 336
Flow
  field, 170-172, 235, 347, 363, 369
  about a sphere, 165
  in a cylindrical tube, 332, 342
  nonuniform, 19
  of electric charge, 236
  over a cylindrical fiber, 363
  over a flat plate, 336
  over a plane vertical surface, 342
  over cylindrical and other objects, 338
  over surfaces and bodies of revolution, 348
  past two spheres, 52
  rate, volumetric, 368
  regime, 214
  slip, 20
  speed, 329, 361-363
  Stokes, 18, 24
  transient, 370
  velocity, 340
    average, 216
Fluctuating
  quantities, 364
  velocity, gradient of, 174
FLUENT, 370
Fluid
  continuum, 156
  density, 206
  dynamics, 73, 91, 296
    ideal, 217
    of the environment, 385
  field, 174, 218, 225
  flow
    inertial effects neglected, 165
    three dimensional, 24
  forces, 172, 208, 215-217
  inertia, 223
  streamline, 327

velocity, 167, 223, 363
Flux
  of particles, 195, 310
  of spheres onto a spheroid, 232
  heat, 16
  turbulent, 79-82, 288
Fog
  formation, 395
  removal, 395
Fokker-Planck
  equation, 319, 365
  gas, 318
Force (s) (d)
  balances, 364
  coefficients, 224
  convective mass and heat
    transfer to a spherical body, 330
  drag, 96
  electrical, 218
  electrostatic, 201, 202
  external, 181, 188. 326, 352, 364
  fluid, 172, 208, 215, 217
  frictional, 19
  law, 13, 16
  of attraction, 136
  of gravity, 96, 188
  random, 65
  ratio, 366
  repulsive, 86
Form factor, 48
Fourier components, 294
Fractal
  index, 283, 286-287
  canopies, 64
  dimensionality, 63
  geometries, 65
  particles, 65
  structure, 286
    particles with, 61
  umbrella trees, 64
Fractional deposition, 333
Fragmentation, 1, 377
Fredholm integral equation of the
    first kind, 335
Free molecular
  limit, 309-311, 354
  coagulation, 279
    Brownian, 275
    kernel, 281-282
  condensation rate, 65, 68-71, 307, 320
  expressions, 358
  flow, 159, 162
  form of the Brownian kernel, 279
  regime, 68, 89

results, 164
Free stream
  particle concentration, 348
  value, 332
  velocity, 336
Frequency of accidents, 381
Friction (al)
  coefficient, 65, 156, 160
  force, 19
    retarding, 155
  velocity, 341, 368
Fuchs
  formula, 167
  -Sutugin formula, 311
  jump distance, 162
  method, 162, 203
  technique, 162
Fuel
  burn up, 375, 382
  cycle, equilibrium, 375
  fragments, 382
  materials, 375
  rods, 377
Fully developed flow, 332, 341
Function (al)
  Bessel, 42
  equation, 247
  form, 306
  of log-normal, 271
  shape, 265

G

Gage pressure, 395
Gamma
  distribution, 145, 244, 271-272, 279, 287
  initial condition, 268
  modified, 271
  with self-similarity, 279
  function, 272, 278
Gap
  distance, 96, 187, 221-222
  separation, 187
Gas
  concentration, 336
  constant, 250
  density, 212
  dynamics, 117, 121
  inlet velocity, 397
  mixture, 311, 319, 351, 352
  phase concentration, 249
  pressure, 122
  properties, 342
  temperature, 122
  -surface interaction, 310, 342
  laws, 319
  properties, 305
  -vapor mixture, 358, 382

/vapor concentration, 328
Gaseous
  flow, 332
  /vaporous environment of
    particles, 287
  heat flux, 347
  mixture, binary isothermal, 352
  system, 2
Gauss' theorem, 114, 118
Gaussian
  distribution, 201
  form, 199
Gear's method, 225, 289
Gegenbauer polynomials, 46, 50
Gelation, 249
General dynamic (al) equation, 382
Generalized
  coagulation kernel, 155
  distribution function, 126
  dynamic equation, 150
  equations, 404
  theory of aerosol coagulation, 179
  turbulent coagulation, 194
Geometric
  cross-section, 159
  mean, diameter, 60
  Stokes radius, 151
  standard deviation, 60
Geometry of a two particle
    collision, 165
Global
  climatic changes, 1
  deposition rate, 344
  particle deposition on a tube
    surface, 342
Gold aerosols, 142
Government agencies, 380
GRACE, 388
Graetz problem, 342-343
  heat transfer, 332
Graphite
  fire, 382
  moderated reactor, 382
Grashof number, 329, 392
Gravitational
  agglomeration, 397
  coagulation, 164, 172, 198, 216, 266, 287
  kernel, 202
  collision efficiency, 225-229
  effect, 194
  field, 216
  force, 201, 331
  motion, 360
  sedimentation, 215

settling, 19, 126-131, 154, 188-190, 194, 198, 209, 216, 256, 327, 331
   in a well mixed volume, 331
   speed, 361
   velocity, 216
Gravity
   number, 209, 310
   diffusion and interparticle forces, 189
   force of, 96, 188
Grazing
   collision, 165, 216-218
   impact parameter, 217
Grid intervals, 300
Group
   method, 290, 298
   of particles, 195
   in equilibrium, 181
Growth
   and deposition, 249
   diffusion controlled, 249
   factor, 254
   law, 127, 249, 255, 258
   parameter, 252
   of clusters through vapor condensation, 305
   of rain drops, 305
   rate, 113-115, 258, 294
Gunn's theory, 133

H

HA-3, 382
HAA-3B, 392-393
HAA-4, 398
HAARM, 382, 392-394
Hail, 271
Half life, 375-376, 384
Hamaker constant, 187
Hanford Engineering Development Laboratory (HEDL), 386
Hard spheres
   collisions, 159
   interaction, 160, 186
   two particles overlap, 182
Hat functions, 297, 316
Hazard
   to equipment, 377
   to health, 1, 375
Haze, 271
Health hazards, 1, 375
Heat
   and mass transfer, 305
   analogies, 328-329
   transition regime, 311
   and particle fluxes, 349
   capacity, 329
   conduction, 16, 322

conductivity of a gas, 309
flow, 16
flux, 16, 329, 347, 350
   vector, 15
of vaporization, 322
sources, 175
transfer, 311, 323, 330, 349
   coefficient, 329
   expression, 311
   problem, 319, 350
   radiative, 305
   rate, 308-309, 319, 348
Heated surfaces, 364
Heavy
   contaminants, 311
   vapor molecules, 318
HEDL, 388
Hemisphere, 195
Heterogeneous aerosols, 111
High (ly)
   aerosol concentrations, 377, 395
   distorted plate-like particles, 235
   energetic release, 384
   ionization, 132
   irregular, 377
   order quadrature, 316
   pressure melt ejection, 377
   speed computers, 328
   Historical perspective of U.S. computational progress, 380
History of
   aerosol, 145
   reactor safety analysis, 381
Homogeneity, 268
   condition, 267
Homogeneous
   assumption, 266
   atmosphere, 395
   kernels, 266
   model, 117
   nucleation, 1
   turbulence, theory of, 196
Homogenization, 256
   procedure, 113
Host gas molecules, 305
HTRE-3, 381-383
Human respiration, 1
Humidity, 377
   effects on a particle, 404
Hydrodynamic
   velocity, 354
   coagulation kernel, 280-281
   current, 326
   equations, 16, 19-21
      continuum, 24

flow, 352
force term, 187
forces, 190, 201, 206
interaction, 181, 184
limit, 15
resistance of rigid pair, 181
theory, 19, 156, 222
   deviations from, 203
velocity, 328, 351-354
   of a mixture, 351
Hydrosols, 109
Hygroscopic aerosols, 306, 389, 397
Hyperbolic partial differential integral equation in a spherical geometry, 314
Hypothetical sphere, 216

I

Ice condensers, 377
ICEDF, 385-387
Idaho National Engineering Laboratory, 388
Ideal fluid dynamics, 217
Ill conditioning, 335, 336
Image force, 137, 202
Imaginary number, 293
Impact parameter, 166, 217
Impaction number, 329, 363
Impactors, 363, 388
Imposed source term 273
Improved mechanistic programs, 380
In-vessel
   releases, 377
   sampling, 397
Incident distribution, 307
Incoming molecular current, 305
Incomplete gamma function, 334, 343
Incompressible
   flow, 117, 370
   fluid, 17
Independent source term, 113
Indoor air
   pollutants, 1
   quality, 1
Inertia (al)
   of a particle, 189, 206-208
   deposition, 327, 363
   effect, 172, 199-200, 225, 342
   impaction, 363, 369
   impactors, 5
   motion of a particle, 326
   turbulent
      coagulation, 199
      forces, 197
Inertialess problem, 206

Subject Index

Inhibited (ing)
  coagulation, 202
  effect of a fluid, 191
  relative Brownian motion, 181
Inhomogeneous aerosol
  distributions, 395
Initial
  airborne concentrations, 392
  burst, 129-131
  charge, 140
  concentration, 361
  conditions, 109-110, 140, 144, 208, 243-244, 247, 250, 253, 256, 259-260, 268, 279, 282, 289, 392
  containment atmosphere, 393
  distribution, 110, 252, 259, 292
    parameter, 279
  particles, 263
    radioactive, 263
  radioactivity, 263
  separation, 85
  size
    distribution, 244, 301
    of particles, 246
  thermal hydraulic conditions, 397
  transient, 184, 213
  value, 145
    problem, 250-252
  volume, 252
    of particles, 259
Injection
  of FPs in a water-steam environment, 388
  of material, 125
  period, 395, 399
Inlet aerosol concentration, 332
Inner and outer expansions, 84, 211
INSAG summary, 384
Inscribed
  body, 44
  spheres, 49
Instantaneous
  collision rate, 214
  velocity, 163, 196
Integrated analysis, 380, 385
Integration
  along a constant characteristic, 292
  along the direction of neutron travel, 314
  with thermal hydraulics, 385
Integro-differential equation, 111, 130, 265
  nonlinear, 241

Inter-code comparisons, 397
Inter-particle fluid forces, 218
Interaction (s) (tive), 2, 90, 202
  between positive and positive particles, 148
  electrostatic, 183
  vapor-vapor, 305
  effect, 218
  force between particles, 181
  gas-atom, 20
  gas-surface, 12, 21
  of particles in a flow field, 170
  probability, 154
  vapor-solid, 127
  wall-gas, 22
  with gas molecules, 305
Interception number, 330
Intermolecular interaction, 354
Internal heat generation, 306
Interparticle
  distance, 181, 211, 219, 310
    average, 214
  forces, 190, 198, 206, 209-212, 218, 233, 235
  law, 13, 218
  van der Waals, 181
  orientation, 204
  potential, 137, 184-186, 204-206, 211, 310
  separation, 84, 204
Interpolation
  formula, 87, 289, 310, 319-320, 354
  linear, 222
Intrinsic
  activity, 142
  coordinates, 26
Inventory, 376, 384
Inverse
  power law, 211, 137
  particle density, 286
Inviscid flow theory, 363
Inward heat current., 309
Iodine
  isotopes, 376
  inventory, 381
  particles, 388
  removal experiments in the CSTF, 391
  vapor, 388
Ion
  concentration, 134, 148
    average, 134
  density, 134-135
    equilibrium, 134
  pair, 134-136
Ionic concentration, 133-134
Ionization, 135, 142

Ionizing radiation, 134
Ions
  positive and negative, 133-134
Iron oxide aerosol, 396
Irregular
  bodies, 32, 55, 63
  objects, 63
Irregularly shaped aerosols, 63
Isentropic flow, 123
Isobaric mixtures, 354
Isochoric specific heat, 122
Isothermal
  conditions, 354
  deposition flux, 347
  gas mixture, 353
  mass transport of a vapor, 306
  particle flux, 347
Isotropic
  eddies, 174, 200
  turbulence, 176
Isotopes, 385
Iterative formulas, 314

J

J-space transform, 288, 297-299
Jacobian of transformation, 297
JERICHO, 398
Jump
  coefficient, 308
  limit, 309

K

Kelvin effect, 249, 305, 322
kernel
  approximation, 241
  differential scattering, 16
  exact, 193, 198
KFK, 388
kinematic viscosity, 174, 368
Kinetic
  deposition model, 366
  model calculations, 354
  theory, 22, 32, 69, 85-86, 253
    calculation, 155, 187
    of gases, 162, 195, 199
    slip, 190
    solutions, 359
KIWI-TNT, 381
Knudsen
  -Weber factor, 22
  form of the Brownian kernel, 270
  number, 14, 21-24, 32, 54, 155, 160-161, 164, 306, 310, 320, 329
    effects, 323
    value, 249
Kundt's tube experiment, 212

L

LA1-LA6, 395-407
LACE, 389, 395
Lagrange's expansion, 247
Laminar
 and turbulent convective-
  diffusive flows, 327
 and turbulent flows, 332, 370
 convective-diffusion, 333
 diffusive deposition, 342
 flow, 328, 332
  -diffusive gravitational
   settling in a channel, 360
  -gravitational settling in a
   horizontal pipe, 360
 aerosol deposition problems,
  338
 in a channel, 369
 in a pipe, 369
 over a cylinder, 369
 over a flat plate, 369
 incompressible flow, 342
 shear, 170, 188, 203, 267
  kernel, 175, 267
Langevin, equation, 65, 159, 365
 instantaneous equation of
  motion, 155
Langmuir model for desorption,
 366
Laplace
 equation, 236
 inverse, 247
 inversion, 263
 transform, 243-244, 247-248,
  254, 259-262, 268
Laser illumination, 347
Late containment failure, 399
Latent heat, 250
Leak
 location, 397
 paths, 395
 probability, 117
Least squares fit, 56
Legendre
 expansion, 318
 functions, 69
 polynomials, 90, 94, 193, 315
  associated, 93
  expansion, 318
Length scale of turbulence, 174
Lewis number, 347
Li_2SO_4 and ZnO particles, 404
Licensing of light water reactors,
 389
Lift force, 364-365, 368
Light
 background molecules, 318
 species, 311
 water reactors, 375, 389
  accidents, 306, 382
  regulatory guidelines for the
   licensing of, 389
  severe core damage in, 375
LIMIT, 398
Limit, hydrodynamic, 15
Linear
 absorption coefficient, 135
 aggregates, 61
 Boltzmann equation, 306, 311
  for gas mixtures, 319
 chains, 61
 combination, 267
 dimensions, 283
 flow field, 204
 interpolation, 222
 transport
  equations, 244
  theory, 13
Liquid
 droplets, 154
 metal cooled fast reactor, 130
 particles, 2
LMFBR aerosol test, 389
Local
 deposition, flux, 341
  rate, 327, 339, 344
 gas velocity, 212
 mass
  flux, 337
  transfer coefficient, 333
 Peclet number, 338
 radial particle flux, 334, 346
Localized aerosol sources, 299
Log-normal
 distribution, 271-273, 287
 form, 284
 gamma, 278
 shape, 271
London-van der Waals, 183, 211
Loss (es)
 mechanism, 113
 of all AC power, 376
 of coolant accidents, 375
 of reactor coolant system, 376
 due to settling and deposition,
  126
Low inertia, 363
Lower leak path, 399
LSODE, 289
Lubrication theory, 87
Lungs, bronchial tract of, 63
LWR
 accidents, 352
 aerosol test, 388-389

M

Mach number, 329
Macroscopic Brownian motion,
 172
MAEROS, 382
Main steam line break in a PWR,
 375
Major
 and minor radii, 36
 axis, 321-322
MARCH, 381, 387, 398
MARVIKEN V, 386-388
Mason's formula, 323
Mass
 and heat transfer
  basic equations of, 305-306
 concentration of a species, 298
 condensation rate, 306
  normalized, 318
 conservation in the case of
  coagulation, 290
 density, 329
 flow rate, 122-125
 fraction, 126
 inventories, 375
 ratio, 95, 311-313, 318
  cutoffs, 225
  of background to vapor gas
   molecules, 308
 transfer, 305, 322-323, 330
  eddy diffusivity for, 340
  jump coefficient, 311
  studies in liquids, 330
  transfer coefficient, 392
  local, 333
Material inventories, 375
Mathematical modelling, 332
Matrix coefficients, 25
Maxwell-Boltzmann distribution,
 13, 71, 159, 306-307, 311
MCT-2, 398
Mean
 free path, 20-22, 65, 71, 155,
  159-160, 163, 195, 203,
  222, 310-312, 329
 of atoms in a background gas,
  249
 vapor molecular, 306
 free time between collisions,
  163
 gas velocity, 114
  molecular speed, 329
 particle, current, 340
 particle, flux, 180
 primary particle diameter, 61
 species velocity, 352
 speed of gas molecules, 68

Subject Index

square distance of travel, 156, 158
square velocity, 365
velocity gradient, 174, 199
Mechanistic
　approaches, 390
　programs, 380
Melt ejection, 377
MERGE, 387
Method (s)
　of characteristics, 247, 250, 255-256, 260
　of Charpit, 248
　of moments 273
　of residues, 268
　of solving the dynamic equation, 241
Millikan's
　formula, 23
　data, 161
Minimum
　gap distance
　impact parameter, 168
　in the capture efficiency, 364
　particle size, 251
Minor axis, 321-322
Mixed
　aerosol condensation/ evaporation problem, 299
　aerosols, 389, 392, 397
　mass mean velocity, 352
　model, 117
　oxide aerosol experiments, 392
Mixing length theory, 176, 195
Mixture of air and water vapor, 357
Mn vapor, 397
MnO aerosol, 397-399, 404
MnO/CsOH aerosols, 395
Mobility, 40, 132-134, 138, 143, 331, 361
　of a particle 157
　tensor, 181-182, 189
Model
　Boltzmann equation, 354
　equations, 164
　operator, 319
Modelling reactions, 385
Moderately charged aerosol, 146
Modified
　Bessel function, 193, 247, 259
　coagulation kernel, 161
　Coulomb law, 137, 202
　diffusion coefficient, 161
　gamma distribution, 271-272, 275-279
　Stokes' law, 160-161
　stream function, 168

turbulent diffusion coefficient, 172
Molecular
　concentration, 327
　　total, 353-355
　desorption, 365
　diffusion coefficient, 329, 332
　distribution, vapor, 306, 311
　impacts, 155
　mean free path, vapor, 306
　properties, 305
　regime, 65
　species, 380
　transport in a gas, 305
　velocity, 306
　volume, 249
　weight, 250, 352
Molecule-surface interaction, 354
Molten
　core and containment, 387
　fuel, 377
Moments method, 271, 279, 283-284
Momentum
　boundary layer, 336
　equation, 16
　layer, thickness of the, 336
　rate of change of, 71
Monodisperse
　aerosol, 216, 235
　distribution, 252
　initial condition, 251
　particles, 262
　polystyrene spheres, 61
Monte Carlo
　calculations, 283
　sampling, 364
　simulations, 368
Most probable charge, 134
Motion
　anti-parallel, 96-97
　of a single particle, 326
　of aerosols, 377
　parallel, 86, 96-97
　relative, 91
　rotational, 66
　translational, 95
Moving
　grid characteristic based finite element method, 296, 300
　sectional method, 299
　sections, 292
Multicomponent
　experiments, 392
　aerosol, 258, 297
　codes, 300
　equation, 117
　gas-vapor mixture, 359

Multiple
　arrays, 52
　energy group equations, 290
　sampling stations, 388
　sphere system, 50
Multiplying system of fast neutrons, 314
Multipole, 50-51
Multispecies aerosol, 125, 132, 144, 260, 288
Mutual electrostatic dispersion, 143

N

Na fire aerosols, 388
Na_2O, UO_2, and mixed Na_2O-UO_2 aerosols, 392
Na_2O-UO_2 aerosols, 392
NaOH and $Al_2(OH)_3$ aerosols, 397
Natural
　convection, 329, 358
　processes, 377, 390
　effects of, 388
　transport processes, 377
NAUA, 382-388, 398
Navier-Stokes equations, 17, 87, 218, 223, 327-328, 332, 370
Needle
　axis, 40
　like particles, 322
Negative
　charge, 137
　ions, 132-134
　results, 336
Net
　current, 158, 184
　　over a spherical surface, 313
　relative velocity, 189
Neumann iteration, 313
Neutral atoms, 132
Neutron
　density, 290
　transport, 314
Neutronic models, 380
New York Power Authority, 398
Newtonian viscous fluids, 44
No slip, 86, 89, 221
　boundary conditions, 69
　condition, 32, 51
Noble gases, 376, 381-383
Noise source term, 155
Nonconcentric spheres, 71
Noncondensibility, 353
Noncondensible background gas environments, 352-354
Noncondensing

gas, 353
species, 351
Noncontinuum effect, 354
Nonequilibrium situation, 305
Nonhygroscopic, 397
Noninteracting
  cloud, 180
  particles, 216
Nonisothermal condensation and reaction rates, 322
Nonlinear
  coagulation term, 115
  differential equations, 272
  equations, first order, 274
  integro-differential equation, 143, 241
Nonradioactive gold aerosols, 142
Nonrotating spherical particles, 342
Nonspherical
  aerosols, 392
  objects, 56
  particles, 2, 150, 305, 320, 331, 370
  rotating bodies, 69
  shape, 41
  symmetry, 231
Nonuniform
  disturbance 154
  flow, 19, 188
Nonvolatiles, 382
  inventory, 381
Normal
  conditions, 377
  gamma distribution, 281-282
  stress, 28
  velocity, 26, 35
Normalized
  density profile, 317-318
  deposition rates, 357
  mass condensation rate, 318
NRC
NRX, 383
NSPP, 388-395
Nuclear
  accidents, 381
  aerosols, 202
  explosives, 314
  industry, 287
  plant, 380
  power plants, 404
  reactor, 132, 381
    physics, 290
    piping, 342, 369
    safety, 1, 326
    vessels during accidents, 332
  reactor accident, 145, 305, 352, 364, 375, 388
  consequences of, 381
  source term, 375-377
  winter, 1
Nucleation, 1
Number
  Courant, 294
  density, 199, 249, 306
    of particles, 367
  distribution function, 301
  Grashof, 329, 392
  gravity, 209, 310
  imaginary, 293
  impaction, 329, 363
  interception, 330
  Knudsen, 14, 21-24, 32, 54, 155, 160-161, 164, 306, 310, 320, 329
  Lewis, 347
  Mach, 329
  Nusselt, 328
  of aerosol particles, 260
  of ions, 134
  of particles, 335-336, 366
    in a stage, 336
  of size specific widths, 336
  of stages, 336
  Peclet, 206, 328-329, 356-358
    local, 338
    thermal, 342
  Prandtl, 329, 344
  Raleigh, 329
  Reynolds, 18-19, 199, 206, 214, 217, 224-226, 329
  Schmidt, 329-330, 336, 344, 359, 392
  Sherwood, 328
  Stanton, 328
  Stokes, 329, 363
Numerical
  approximation, 46
  aspects, 292
  calculations, 320
  convergence, 84
  differentiation, 289
  diffusion, 291
  method, 241, 287-288, 293
  quadrature, 289
  results for deposition rate, 346
  solutions, 216, 256, 290, 293, 346
    for condensation/evaporation without coagulation, 292
    of the GDE, 292
NUREG, 386
Nusselt number for mass transfer, 328

O

Oak Ridge National Laboratory, 386-388
Object of deposition, 364
Objects
  less regular, 57
  spherical, 65
Oblate
  eccentric, 46
  spheroid, 34, 37-44, 49, 68-70, 73, 233-235, 320-321
    coordinate system for, 38
    drag on, 37
Odd objects, 370
Open trajectories, 208
Operator errors, 382
Opposite polarity, 136
Optical fiber, 1, 326, 332, 342
Optimum phase, 215
Order of reaction, 323
Ordinary differential equation, 334, 337, 345, 350, 358
Organic iodides, 389
Orientation (s)
  of a spheroid to the direction of diffusion, 234
  arbitrary, 96, 98
  of particles, 231
  of velocity, 195
  probability, 235
Orifice, 122
ORIGEN, 375, 387
ORNL, 388
ORR, 383
Orthogonal (ity)
  curvilinear coordinates, 25, 338
  properties, 50
Orthokinetic interaction, 215, 216
Oscillations, 336
Oscillatory motion, 212
  relative, 215
Oseen ('s)
  approximation, 223
  calculation, 224
  drag forces, 225
  equation, 223
  flow, 226
  theory, 223-224
Outer
  containment, 117
  expansion, 84, 211
Outward current
  molecular, 305
  of vapor molecules emitted from a particle, 306
Overpressure, 376

Oxidation of fragmented fuel, 377
Ozone depletion, 1

P

Parakinetic, 216
Parallel motion, 86, 97
Parallelepipeds, 63
Parametric studies, 287
PARIDESKO, 382
Partial
  differential equation, 208, 260, 288, 299, 337, 358
  pressures, 305
Particle
  analysis instrumentation, 389
  -surface-gas interaction, 365
  beams
    distribution, 5
    settling speed, 5
    shape, 2
    size, 2
  boundary conditions, 370
  charge, 143, 329
    distribution, 361-362
  charging, 362
  coagulation, 326
  collection efficiency, 364
  composite, 63
  composition, 298
  concentration, 109, 326-335, 340, 343, 347-348, 356-358, 365
    at a surface, 348
    at a wall, 349
    distribution, 342
    uniform, 342
  configuration, 181
  conservation, 314
  constituents, 305
  continuity equation, 328, 337-340, 348-350, 354, 358-360, 369
  current, 326, 340, 357
  density, 206, 225, 258, 285-287
    variance with time, 278
    total, 259
  deposition, 326, 332-333, 342, 354-356, 369-370
    and resuspension, 326
    by diffusion, 332
    due to gravity and diffusion, 326
    on a tube surface, 342
    rate, 354-358
  diameter, 331, 357
    effective, 53
  diffusion coefficient, 326, 332, 335
  diffusivity, 348
  distribution, 361
    function, 110, 114-116, 365
  dynamics, 286
  eddy diffusivity, 364
  equations, 327
  evolution, 287
  extraction, 335
  filtration, 363
  flux, 359
    diffusion coefficient, 180
    time averaged, 361
  growth, 251, 388
    kinetics, 305
  highly irregular, 54
  inertia, 215, 327-329
  interactions with gas constituents, 305
  mobility, 331, 365
    tensor, second rank tensor, 180
  motion, 351, 363-365
    in a curved path, 165
    under temperature and concentration gradients, 329
  orientation, 49
  population, 139
  primary, 63-65
  radius, 306, 342, 364
  removal experiments, 389
  resuspension, 364
  rotation, 327
  separation, dimensionless, 185
  size, 241-243, 263, 305, 329, 361, 389
    average, 193
    distribution, 59, 362, 390-392
    range, 1
    variations, 288
  sizing, 363
  stopping distance, 329
  surface, 305
  terminal velocity, 331
  trajectory, 169, 216, 346, 363-364
  transport, 344
  velocity, 213-214, 327
  volume, 159, 258, 263
Particles
  deposited on surfaces, 364
  depletion of, 310
  distance between 176
  entrained, 212
  finite sizes of, 370
  flux of, 195
  fraction of, 368
  gaseous environment of, 287
  groups of, 195
  initial radioactive, 263
  loosely bound, 367
  positive and negative, 147-148
  primary, 63-65
  rotation of, 95
  tightly bound, 367
  total number of, 261
  uniformly distributed, 173
  vaporous environment of, 287
Particular solution, 266
Particulate contamination, 2
PCS, 388
  thermal hydraulics, 385
Peclet number, 206, 328-329, 356-358
  local, 338
  thermal, 342
Perfect gas equation, 123
Perfectly accommodating spheres, 161
Period of aerosol injection, 397
Permanent doublet, 208
Perturbation
  analysis, 211
  of fluid by a particle, 170
  parameter, 206, 310
  theory, singular, 209-211
PERVEX, 388
Phase shift, 213
Photometer, 399
Physical
  and chemical behavior, 377
  interpretation, 263
  processes, 154
Physico-chemical
  descriptions, 404
  models, 380
  properties, 323
  reactions, 377
Piecewise polynomial, 297
Pipe
  bends, 370
  break, 376
Planar geometry, 134-135, 359
Plasma torches, 388, 397
Plate like objects, 52
Plateout rate, 389
Plume distance, 383
Point charge approximation, 202
Poisson's equation, 138, 143
Polar
  angle, 233
  coordinates, 157, 167, 190, 313

Pollen, 1
Pollution control, 212
Polydisperse, 274
Polydispersions, electromagnetic scattering in, 271
Polynomials
 associated Legendre, 93
 Gegenbauer, 46, 50
 Legendre, 90, 94
Pool
 fires, 388
 flashing, 395
Population of spherical aerosol particles, 150
Porous aggregate, 283
Position vector, 306
Positive
 charges, 137-138
 ions, 132-134
 particles, 147-148
 void coefficients, 382
Posttest code predictions, 401-407
Potential
 accidents, 381
 energy, average, 365
 source
  of public risk, 377, 404
  of radioactive release, 375
 well, 365
Practical applications, 288
Prandtl number, 329, 344
Precipitation, 362
Precipitators, electric, 212, 361
Predictive step, 295
Pressure
 and moisture content, 389
 drop, 400
 of a vapor, 250
 relief valve, 381
 tensor, 17-19
 vessel, 377
  failure, 375
Pressurized water reactor, 375, 378
Primary
 circuit, 176, 388
 coolant system, 377, 385
  pipe break, 375
 particles, 63-64
  diameter, 61
 processes, 377
 system, 383
Private utilities, 380
Probabilistic methods, 380
Probability
 density function, 180-181
 distribution of velocity, 195

Prolate spheroid, 36-40, 49, 68-69, 231-232, 236, 320-322
 drag on, 37
 surface of, 37
Prompt critical excursion, 382
Properties, dynamical, 5
Proximity to a surface, 331
PRTR, 383
Pseudo-self-similar form, 272, 275
Public risk, 375, 404
Pure coagulation, 113, 264, 283
 Brownian, 274
 and homogeneous kernels, 268
PWR, 381
 containments, 377
 configurations, 388

Q

Quadrature, 251
 formulas, 288
Quasi-continuous function, 111
Quasi-static flow, 44
QUICK, 382, 398
Quiescent fluid, 172, 188

R

Radial
 current, 306-307
 direction, 344
 potential, 235
 velocity, effective, 209
Radiation, 148, 326
 field, 145, 380
 releases, 364
 /charge aspects, 385
Radiative
 absorption, 306
 heat transfer, 305
 particles, 263
Radioactive
 decay, 382
 aerosols, 1, 130, 260, 375
  electrostatic charge on, 142
 characteristics, specific, 260
 contamination, 132
 content of an aerosol, 260
 environment, 382
 fuel
  aerosol, 263
  particles, 130
 gold aerosols, 142
 inventories, 375
 isotopes, 377
 material, 130-132, 135
  energy and momentum released with, 377
  particles, 306

substances, 377
 systems, 142
Radioactivity, 1, 131, 260-263
 amount of, 262
 calculation of ex-plant dispersion of, 377
 calculations, 264
 initial, 263
 inventory of a reactor core, 382
 label, 260
 releases, 375
 total airborne, 261
Radionuclides, 377
 composition of the discharge at Chernobyl, 385
Radius
 distribution function, 249
 of curvature, 115
 of gyration, 65, 151
 ratio, 185
 vector, 195
Rain, 271
Raleigh number, 329
Random
 direction, 163
 force, 65
  due to molecular bombardment, 155
  impact of molecules, 155
 motion, 172, 180
  thermal, 326
 walk, 155
Range-energy relationships, 135
Rapid
 Brownian motion, 131
 decrease in release, 384
 depressurization, 395
 sedimentation, 397
 vaporization, 305
Rate
 expressions, 305
 of accretion, 253
 of capture of particles, 137
 of change due to coagulation, 144
 of change of momentum, 71
 of coagulation, 140, 193, 214
 of coalescence of raindrops, 218
 of collision, 199, 203
 of desorption of particles, 366
 of growth, 113-115, 258
 of production, 134
 of strain tensor elements, 204
 processes, 287, 299, 305
Ratio
 mass, 95
 of concentration, 348

Subject Index

RBMK reactor #4, 382
Reactant species concentration, 305
Reaction rates, 377
Reactor
  accident, 136, 383
  building, 376, 381
  containment, 113, 175, 326, 375-377, 380-381, 404
  coolant
    boundaries, 395
    system, 387
  core, 382
    radioactivity inventory of, 382
  environments, 116
  plant, 377
  Safety Study, 375, 392
    analysis, history of, 381
  shutdown systems, failure of, 376
Reciprocity, 13
Recombination coefficient, 134, 136
Reduced
  distribution, 126, 130, 277
  interaction, 136
Reference temperature, 123
Refill factor, 216
Reflection
  diffuse, 13, 22
  specular, 13
Regime, slip flow, 160
Regularization, 336
Regulatory guidelines, 389
Relative
  diffusion coefficient, dimensionless, 185
  diffusivity of two spheres, 183
  motion, 172-173
    between two particles, 173
    fluid, 216
  position vectors, 181
  velocity, 154, 164, 167, 170-172, 189, 195-199, 203-204, 212, 216, 221
    radial component of, 173
Relaxation time, 181, 196, 367
Release
  fractions, 377
  ex-vessel, 377
  from core-concrete melt, 387
  of fission products to the environment, 387
  of radioactive isotopes during accidents, 377
  rates, 377
  to containment, 377

to the atmosphere, 382
Removal
  efficiency, 339
  force, 366
  of aerosols from vessel atmospheres, 388
  of iodine, 389
  of methyl iodide, 389
  of particles
    by filtration, 338
    from an aerosol, 332
  rate, 150
  term, 267
REMOVAL, 398
Repulsion, 138
Repulsive forces, 86
Residence time, 2
Residual charge, 132
Resistance
  coefficient, 83, 97
  matrix elements, 99
  tensors, 98
Response matrix, 336
Restrictive conditions, 271
Results
  approximate, 42
  asymptotic, 84
  exact, 42, 51, 84
Resuspension, 326-328, 364-368
  from dry surfaces, 395
  rate, 365-369
  particle, 364, 368
RETAIN, 398
Return loop, 117
Reynolds number, 18-19, 199, 206, 214, 217, 224-226, 329
Rigid
  body, 90
  spheres, 181, 312
Risks to property and plant equipment, 375
Rod-like structures, 235
Root mean square velocity, 195
  turbulent, 175
Rotating
  Brownian particle, 72
  sphere, 67, 89
Rotation, 19, 66, 91, 96
  Brownian, 61, 233
  friction coefficient, 234
  of particles, 95, 327
Rough pipe, 341
Roughness elements, 341
Ru, 388
Runge-Kutta method, 278, 289

S

Sabotage
  of shipping casks, 375
  of waste disposal sites, 375
Safety analysis, 381
Saturation
  density, 306, 312, 397
  vapor pressure, 250
SAUNA, 388
Scattered light, 2
Scattering, 2
  gas-atom, 21-22
  kernel, 13
  operator, 318
  particle counter, 5
Scavenging of particulates, 1
SCDAP, 385
Schmidt number, 329-330, 336, 344, 359, 392
Scrubbers, 212
Second order
  central difference, 293
  ordinary differential equation, 358
Sectional method, 299-301
Sedimentation, 257, 369
  flux, 194
  velocity differential, 140
Self-preserving
  form, 268, 271
  particle size distribution, 265, 270
  solution, 264-266
Self-similarity, 65, 275, 279
  form, 275
  solution, 268
Semi-empirical
  method for extending continuum results, 160
  result, 56-57
Semi-major axis, 236
Semivolatile, 397
Semianalytical
  results, 54
  study, 45
Semiconductor, 360
Semiempirical
  correlations, 382
  description, 53
Sensor
  fabrication, 363
  fiberoptic, 1
Separation, 96
  distance, 85
  infinite, 80, 85
  interparticle, 80-84
  of variables, 333
  particle, 185

small, 88
wide, 100
Set of
  descriptors, 150
  differential equations, 291
Settled particles, 392
Settling, 110, 113, 144, 377, 388
  parameter, 360
  rate, 55
  speed, 2
  times, 134
  velocity, 114, 164
  vessel, 397
  gravitational, 19
Severe accident
  conditions, 395
  core damage, 1, 375-377, 404
  scenarios, 380
Severe transients, 375
Shape
  arbitrary, 49, 72
  complex, 21
  dependence, 283
  effects, 52, 385
  factor, 56, 231, 320-322
    dynamic, 54
  highly irregular, 32
  irregular, 52, 63
  nonspherical, 53
  of an aerosol, 150, 377
  regular, 54
Shear
  and Brownian motion, 203
  coagulation, 140
  effect, 203
  flow, 206
  kernels, 275
    Brownian and laminar, 206
  motion, 208
  rate, 267
  velocity, unperturbed, 172
Sherwood number, 328
Shipping casks, 375
Short
  range van der Waals forces, 326
  tubes, 346
Silicon, 1
Similarity
  function, 281-282
  transform, 334, 337, 343, 350, 358, 370
Simplified kernels, 241
Simulated (simulation)
  accidents, 385
  containment, 397
  experiments, 385, 390
  LWR piping, 395

Simultaneous
  coagulation and condensation, 253
  radiation fields on absorbing aerosols, 347
Single
  and two particle physics and chemistry, 404
  component
    aerosol, 298-299
    gas, 319
    results, 319
  particle, 155, 182, 326, aerosol, 125
  species, 125, 154, 288, 389
    aerosol, 131, 259, 297-298
  zone, 118
Singly charged positive ions, 145
Singular perturbation theory, 209-211
Singularity, 222
Sink of particles, 310
Size
  and composition distributions, 397
  cutoffs in deposition, 363
  distribution, 255, 264, 377
    effects, 140
    of a polydisperse aerosol, 335
  of aerosol particles, 264
  of aerosols, 249
  range of aerosol particles, 160
  ratio, 187, 222, 225
  spectra, 279, 290
Sizeable
  release of radioisotopes, 376, 382
  release of aerosols, 375
Skin drag, 55
SL-1, 381-383
Slender body theory, 52
Slip, 18, 44-46
  at a surface, 66
  boundary conditions, 24, 31-32, 35, 68-69, 85, 88
  corrected values, 221
  correction, 32, 53-54, 69-70, 198, 222, 331
  effect, 21, 89, 224
  flow, 20-21, 88
    regime, 160
  limit, 354
  type, 27
Slow viscous flow, 32, 66, 182
  theory, 180, 214
  equation, 18, 24
Small

pockets of fluid, 195
settling speeds, 164
Smolarkiewicz method, 294-296
Smooth
  nozzle, 124
  pipe, 341
Smoothing, 336
S_N method, 314
SNAPTRAN, 381
Snowflakes, 63
Sodium
  oxide, 130, 263
  spheres, 225
  spill, 393
Solar-radiative, 1
Solid fuel matrix, 377
Soluble /insoluble aerosol mass, 395
Solution of the steady state diffusion equation, 162
Sonic
  coagulation, 212, 216-217
  field, 216-217
Sonine polynomials, 319
  expansion, 313
Sound
  field, 212, 216
  intensity, 213, 216
  pressure, 212
  wave, 154
Source
  free medium, 313
  reinforcement, 287
  term, 129, 144, 150, 377, 380-381, 387, 395
    code package, 385
    estimations, 381, 404
    imposed, 273
    independent, 113
Space dependent balance equation, 113
SPARC, 385, 387
Spatial (ly)
  -temporal elements, 297
  averaged deposition velocity, 369
  constant aerosol density, 113
  constant number density, 192
  dependence, 110
  derivatives, 114
  distribution of particles, 257
  domain, 299
  effects, 256
  homogeneity, 127, 288, 299
  homogeneous, 369
  homogenization, 113
  inhomogeneous, 288
Species

Subject Index

calculations, 370
concentration, 352
current, 352
equations, 370
Specific
  heats, 213
  ionization, 135
Specular reflection, 13
Speed
  of sound, 212
  of a fluid, 349
SPERT-1, 381
Sphere (s)
  arbitrary orientation of, 96
  equal size, 82, 90, 101
  equivalent, 49, 54, 65
  identical, 51
  interacting, 73
  isolated, 51
  monodisperse polystyrene, 61
  motion of, 91
  nonconcentric, 71
  of equal radius, 206
  perfectly accommodating, 161
  reference, 55
  rotating, 67
  rotating, 95, 104
  touching, 51-52
  translating, 95
  widely separated, 51
Spherical (ly)
  body, 330
  caps, hollow, 55
  coordinates, 67
    polar, 184, 191
  drop, 311, 354, 358
  gravitational collision
    efficiencies, 235
  object, 65
  particle, 127, 231, 235-236,
    305-306, 331, 357
  symmetric diffusion, 184
  symmetry, 313
Sphericity, 61
Spheroid (al), 32, 36, 39-40, 49,
  55-57, 154, 233-235
  arrays of, 52
  coordinates, 231
  drag force on, 34
  extended, 231
  oblate, 34, 40, 44, 49, 52, 70-72
  prolate, 40, 49, 52, 72
  surface of, 36, 231
  surfaces, 235
Splines, 318
Spray fires, 388
Sprays, 377, 388

effectiveness of, 389
Stability, 293-296
  limit, 294
  maximum, 49
Stagnation point, 338, 348-349
Standard
  boundary layer techniques, 344
  conditions, 305
  orifice, 124
Stanton number, 328
Starting position, 364
Static gas pressure, 213
Stationary spherical particle, 133
Statistical
  approach, 195
  arguments used by Saffman and
    Turner, 178
  asymmetry, 140
  average
    of the convective term, 201
    over many realizations of the
      turbulence, 173
  averaging, 201
  mechanical averaging, 159
  properties of the velocity, 181
Statistically
  averaged fluid velocities 172
  averaged velocity, 199
Steady
  flow around an isolated sphere,
    208
  source problem, 251
  state, 117, 134, 326
    conditions, 395
    diffusion equation, 184
    equations, 305
    flow, 189, 370
    operation, 395
    particle continuity equations,
      329
    particle distribution, 310
    particle energy equations,
      329
    solution, 211, 251
    thermal and convective
      conditions, 388
    thermodynamic force, 181
    velocity of a particle, 180
Steam
  explosions, 377
  -air atmospheres, 388
  condensation rate, 389
  fraction station, 399
  injection, 395
Stefan
  and diffusiophoretic velocity,
    354
  flow, 358-359

  deposition rate due to, 356
  velocity, 351-355, 358-359
Stochastic
  differential equation, 195
  forcing term, 201
  equations of motion, 364
Stokes (s') (ian)
  approximation, 217, 225
  -Cunningham factor, 22
  -Einstein value for the diffusion
    coefficient, 158
  diameter, 54-55
  drag, 225, 230
  forces, 227
  laws, 363
  dynamics, 225
  equation, 84, 358
  flow, 18, 24, 165, 168, 188,
    216-219, 224-226
    around sphere, 101
  forces, 225
  pure, 87
  formula, 22, 55
  hydrodynamics, 224
  law, 24, 53, 100, 110, 156,
    160, 181, 213
  no-slip theory, 88
  number, 329, 363
  particle settling speed, 392
  radius, 150-151
  regime, 225
  resistive force, 196
  results, 225
  stream function, 24, 31-34, 73,
    77, 81, 165
  stream lines and particle
    trajectories, 169
  theory, 223-224
  value, 31, 51, 226
  velocity, 114, 129, 206, 256
Stratified settling, 331
Stream
  function, 24-25, 29, 32-34, 88,
    166-168, 225, 349
    Stokes, 73, 77
    superposition of, 49
  line (s), 168, 339
  motion, 167
    distortion of, 173
    of relative motion between
      two particles, 173
    of aerosol, 348
Stress
  tensor, 21
  dyadic, 28
  normal, 28, 31, 55
  tangential, 28, 31, 55
Strong

coagulation, 385
turbulence, 196, 299
Strongly charged aerosol, 147
Structural compounds, 397
Structure
  closed packed, 63
  fractal, 63
STUDSVIK, 388
Submicron particles, 369
Subsaturated environments, 306
Substantial releases, 384
Subvolumes, 299
Sulphur and nitrogen oxides, 305
Superposition, 39
  method, 101, 225, 229, 235
  of stream functions, 49
  technique, 52
Supersaturation, 250, 357
  coefficient, 267
Suppression pools, 377
Surface
  area of interaction, 2
  condensation, 395
  element., 327
  exposed to diffusion., 115
  heating, 326
  loss constant, 141
  of containment, 392
  of a spheroid, 231
  potential well frequency, 365
  reaction constant, 323
  source, 136
  temperature, 306
  tension, 249
Suspended
  aerosol, 388, 397
  concentration, 400
  mass ratio, 399, 403, 407
  mass, total, 392
Suspension, 136
  of particles, 1
Symmetry, 91-93, 155, 269, 273
  axial, 24
  axially, 43
  axis of, 70
  bipolar, 139
Systems of equations, 289
  partial differential, 288

T

Tangential
  stress, 21, 26, 35
  velocity, 28
Taylor series expansions, 295
Te, 388
Temperature, 327, 358
  distributions, 369

gradient, 154, 319, 326, 342, 347
jump distance, 310
of a particle, 306, 322
profiles, 348
reference, 123
Tensor
  pressure, 17-19
  resistance, 98
  stress, 21
Terminal velocity, 55
Test
  particle, 189
  pipe, 397
    conditions, 400
  sphere, 195
Tests LA1-LA6, 395-407
Theoretical analysis, 326
Theory
  of Bjerknes, 217
  continuum, 20
  hydrodynamic, 19
  kinetic, 32
  lubrication, 87
  of Brownian motion, 163
  of homogeneous turbulence, 196
Thermal
  accommodation coefficient, 319
  conductivity, 342
    of a drop, 250
    of a fluid, 328
  creep, 342
  diffusion, 328
  diffusivity, 342
  gradients, 306, 341, 352, 377
  hydraulic
    codes, 382, 385, 398
    behavior, 387
    conditions, 287, 397
    equations, 382
    models, 380
    PCS, 385
  motion, 155
  Peclet number, 342
  velocity, average, 162
Thermodynamic
  forces, 181-182
  equilibrium, 159, 180, 385, 392
Thermophoresis, 328, 346-347, 351, 358, 364, 368-369
Thermophoretic
  current, 326, 351
  deposition, 347
    efficiency, 346
    flux, 347

effect, 345
flux, 347
force, 354
layer, 344
term, 343
velocity, 326, 341, 347-348
Thickness of the
  boundary layer, 392
  momentum layer, 336
Three Mile Island-2, 378, 381-383
TID-14844 criteria for licensing, 381
Tightly
  adhered, 397
  bound particles, 367
Time averaged
  equation, 326
  particle flux, 361
Time
  constants, 136
  dependent concentrations of FPs, 388
  transients, 189
Timing and dynamics of the release at Chernobyl, 384
TMI-2, 378, 381-383
Topological dimensions, 61
Toroids, closed, 41-44, 55
Torque, 20, 65-72, 89, 93-95
  on a body, 19
  on a sphere, 91
  on an ellipsoid, 71
  on rotating spheres, 91
Total
  aerosol density, 251
  airborne radioactivity, 261
  current on a hemisphere, 168
  pressure, 353
  radioactivity, 131, 261
  simulation time, 368
Touching spheres, limiting case of, 81
Trajectory (ies), 85, 363
  analysis, 217
  calculation, 85, 326-327, 370
  closed, 208
  of a Brownian particle, 155
  open, 208
  particle, 169, 215-216, 346, 363-364
Transcendental equation, 145
Transfer cross sections, 290
Transformations, 68-70, 159, 192, 260, 269, 288, 344
  bipolar, 74
  conformal, 34
  coordinate, 81

Subject Index

Transients, 268
  event, 376
  flow, 370
  with loss of flow, 381
Transition
  limit, 354
  from Brownian to turbulent diffusion, 178
  regime, 164, 310
    between continuum and free molecular flow, 162
    heat and mass transfer, 311
Translating spheres, 96
Translational motion, 95
Transmission of aerosols, 348
Transport
  equations, 290
  coefficients, 352
  in piping, 389
  of dust and smoke, 1
  processes, 1
TRAP-MELT, 382, 385-387, 398
Trapezoidal elements, 297
Trimodal distribution, 287
Truncation errors, excessive, 294
Turbine anemometers, 399
Turbulence, 154, 174, 195, 364
  eddy diffusivity model of, 361
  isotropic, 174-176
  shear, 188
  simple models of, 328
  specified models of, 370
  statistical average, 173
  strong, 196, 299
  theory of homogeneous, 196
Turbulent
  action, 172
  coagulation, 172
    by the Saffman and Turner method, 198
    generalized, 194
    inertial, 199
  component, 114
  diffusion, 172, 177-178, 194, 197, 267
    coagulation kernel, 175
    coagulation rate, 199
    coefficient, 172, 176, 179
    coefficient of touching spheres, 179
    coefficient, modified, 172
  diffusive forces, 197
  diffusivity, 361
  eddies, 172
  energy dissipation rate, 174
  field, 196
  flow, 340, 363
    in a bend, 369
    in a chamber, 369
    in a rough pipe, 369
    in a smooth pipe, 369
  fluctuations, 199
  fluid, 114
  gravitational settling in a chamber, 361
  inertial, 266-267
    effect, 172
    forces, 197
    motion, 194, 197
  information, 194
  motion, 178, 194, 199
  realizations, 195
  velocity
    average, 195
    root mean square, 175
Two
  body system, 183
  component aerosol, 129
  particle
    coagulation, 157
    interaction, 188
    motion in laminar shear flow, 171
  species equations, 128
  sphere
    interactions, 218
    problem, 224
Type of accident, 377
Typical energy dissipation rates, 176

U

U.S. Department of Energy, 384
Uncoagulated source particles, 263
Unequal spheres, 187
Uniform distribution, 157, 366-367
Unipolar
  aerosols, 147
  charged aerosol, 145
  charged particles, 138, 143
  charging, 136, 139
Unit
  matrix, 336
  normal, 306
  tensor, 181
  vector, 72, 114-115, 191, 203
    in the radial direction, 356
    normal to surface, 168, 327
Unperturbed
  density, 137
  shear velocity, 172
  uniform distribution, 157
UO_2, and mixed Na_2O-UO_2 aerosols, 392
Upstream unperturbed velocity, 165
Upwind
  difference approximation, 294
  differencing and sectional method, 301
  method, 294-296
  schemes, 294
    criterion, 296

V

van der Waals, 181
  -London forces, 85
  forces, 140, 188
    short range, 326
  potential, 187
VANESA, 385-387
Vapor
  -gas mixture, 306, 358
  -solid interaction, 127
  -vapor interactions, 305
  concentration, 358
  condensation, 113
    rate, 355
    growth of clusters through, 305
  continuity equation, 359
  density, 307
  deposition rate, 356
  diffusive term, 355
  evaporation, 113
  metallic, 57
  molecular distribution, 311
  molecular mean free path, 306
  molecules, 115, 145
  number flux, 359
  phase, 381
  pressures, 115
  species, 127, 353
  transport, 354-355
Vaporization
  of metallic Mn, 397
  of materials, 375
  heat of, 322
Vaporous environment of particles, 287
Variance of a distribution, 285
Variational
  formulation, 354
  method, 22, 42-44
  principle, 88
  results, 23-24
Vector
  heat flux, 15
  unit, 72
  vorticity, 25

Velocity (ies)
  at the edge of an orifice, 122
  angular, 66, 71-73, 95
  average flow, 20
  boundary layer, 330, 363
  components, 336-338
    obtained from a stream
      function, 166
  differential, 154, 164, 215
  distribution, 71
  due to gravitational settling
    167
  equal, 101
  gradient
    constant, 170
    effective, 175-176
  in the particle continuity
    equation, 354
  net relative, 189
  normal, 26, 35
  of colliding particles, 172
  of an aerosol, 326
  of a gas, 213
  of surrounding air, undisturbed,
    196
  opposite, 101
  profile, 330, 342
  radial, effective, 209
  settling, 55
  slip, 308
  tangential, 21, 26, 35
  terminal, 55
  uniform, 348
Verification tests, 389
Very dense aqueous fog, 377
Vessel
  atmospheric temperature, 389
  diameter, 388
  volume, 388
  wall, 389

Viscosity, 16, 77, 163
  of a fluid 156
Viscous
  boundary layer, 330
  flow
    equations, 66, 91
    slow, 66
  fluid, 44
  forces, 187-188
Volatiles, 381, 382, 397
  isotopes, longer lived, 375
Volatility of isotopes, 377
Volume
  averaged deposition velocity,
    361
  distribution function, 110,
    116, 250, 254, 271
  equivalent sphere, 331
  fraction, 264
  of solid particulate, 285
  of a containment, 392
  radius, 150
  ratio, 124
  reactions, 115
  sources, 136
Volumetric flow, 117
  rate, 368
von Neumann
  stability analysis, 293-294
Vorticity vector, 25

W

Waiting time, 268
Wall
  -particle scattering kernel, 13
  deposition, 145
    by diffusion, 115
  material, 388
  temperature, 342

Warm up stage, 397
Warning time, 377
Wash-740, 380
Wash-1400, 381
Waste disposal sites, 375
Water
  coolant system, 377
  droplets in air, 206, 225
  vapor concentration gradient,
    142
Weak condensation, 299
Weakly charged aerosol, 146
Well
  -mixed hypothesis, 115-117,
    120, 126, 256-257
  -mixing of aerosols, 397
  stirred atmosphere, 113
Wet reducing environments, 381
Wiebull or modified gamma
  distribution, 271
Windscale-1, 381-383
  accident, 381
WTR, 383

X

Xe, 388
XSOR, 382, 386

Y

Z

Zero
  deposition, 149
  -Reynolds number limit, 226
Zirconium hybrid moderated
  reactor, 382
Zonal methods, 299
ZONE, 382